NEW AND FUTURE DEVELOPMENTS IN MICROBIAL BIOTECHNOLOGY AND BIOENGINEERING

NEW AND FUTURE DEVELOPMENTS IN MICROBIAL BIOTECHNOLOGY AND BIOENGINEERING

From Cellulose to Cellulase: Strategies to Improve Biofuel Production

Edited by

Neha Srivastava

Manish Srivastava

P.K. Mishra

P.W. Ramteke

Ram Lakhan Singh

ELSEVIER

Elsevier
Radarweg 29, PO Box 211, 1000 AE Amsterdam, Netherlands
The Boulevard, Langford Lane, Kidlington, Oxford OX5 1GB, United Kingdom
50 Hampshire Street, 5th Floor, Cambridge, MA 02139, United States

Library of Congress Cataloging-in-Publication Data
A catalog record for this book is available from the Library of Congress

British Library Cataloguing-in-Publication Data
A catalogue record for this book is available from the British Library

ISBN: 978-0-444-64223-3

Publisher: Susan Dennis
Acquisition Editor: Kostas Marinakis
Editorial Project Manager: Michael Lutz
Production Project Manager: Prem Kumar Kaliamoorthi
Cover Designer: Greg Harris

Typeset by SPi Global, India

Contents

14. Genetic Engineering Applications to Improve Cellulase Production and Efficiency: Part II

PRAGATI MISRA, PRADEEP KUMAR SHUKLA, K. PRASADA RAO, P.W. RAMTEKE

15. Biofuel Cells With Enzymes as a Catalyst

KELVII WEI GUO

16. Algal Cellulases

PRIYANKA ARORA, VIPIN KUMAR SHUKLA, ARCHANA TIWARI

Contributors

Priyanka Arora School of Sciences, Noida International University, Greater Noida, India

Kulsoom Bano Protein Research Laboratory, Department of Bioengineering, Integral University, Lucknow, India

S.M. Bhatt Biotechnology Department, SBBS University, Jalandhar, India

Subhojit Chakraborty Department of Microbiology, University of Delhi South Campus, New Delhi, India

Jairam Choudhary Indian Institute of Farming Systems Research, Modipuram, India

Misbah Ghazanfar Department of Biotechnology, University of Sargodha, Sargodha, Pakistan

Kelvii Wei Guo Department of Mechanical and Biomedical Engineering, City University of Hong Kong, Kowloon Tong, Hong Kong

Vijai Kumar Gupta Department of Chemistry and Biotechnology, ERA Chair of Green Chemistry, School of Sciences, Tallinn University of Technology, Tallinn, Estonia

Hemansi Department of Microbiology, School of Interdisciplinary & Applied Life Sciences, Central University of Haryana, Mahendergarh, India

Muhammad Irfan Department of Biotechnology, University of Sargodha, Sargodha, Pakistan

Subburamu Karthikeyan Tamil Nadu Agricultural University, Coimbatore, India

Farha Khan Protein Research Laboratory, Department of Bioengineering, Integral University, Lucknow, India

Rahul Kumar Kharwar Department of Economics, Banaras Hindu University, Varanasi, India

Mohammed Kuddus Department of Biochemistry, University of Hail, Hail, Saudi Arabia

Ramesh Chander Kuhad Department of Microbiology, University of Delhi South Campus, New Delhi; Central University of Haryana, Mahendergarh, India

Ajay Kumar Division of Microbiology, ICAR-Indian Agricultural Research Institute, New Delhi, India

Lalthafala Molecular Microbiology and Systematic Laboratory, Department of Biotechnology, Mizoram University, Aizawl, India

Vincent Vineeth Leo Molecular Microbiology and Systematic Laboratory, Department of Biotechnology, Mizoram University, Aizawl; Department of Biotechnology, J.J College for Arts and Science, Pudukkottai, India

Navodita Maurice Laboratory of Immunology, Institute of Genetics, Biological Research Centre, Hungarian Academy of Sciences, Szeged, Hungary

Ramchander Merugu Department of Biochemistry, Mahatma Gandhi University, Nalgonda, India

P.K. Mishra Department of Chemical Engineering and Technology, Indian Institute of Technology (Banaras Hindu University), Varanasi, India

Kajal Mishra Department of Chemical Engineering and Technology, Indian Institute of Technology (Banaras Hindu University), Varanasi, India

Pragati Misra Department of Molecular and Cellular Engineering, Jacob Institute of Biotechnology and Bioengineering, Sam Higginbottom University of Agriculture, Technology and Sciences, Allahabad, India

Iniya Kumar Muniraj Kumaraguru Institute of Agriculture, Erode, India

Muhammad Nadeem Food & Biotechnology Research Center, PCSIR Labs Complex, Lahore, Pakistan

Lata Nain Division of Microbiology, ICAR-Indian Agricultural Research Institute, New Delhi, India

Sadaf Parveen Protein Research Laboratory, Department of Bioengineering, Integral University, Lucknow, India

Enosh Phillips Department of Biotechnology, St. Aloysius College (Autonomous), Jabalpur, India

K. Prasada Rao Department of Biological Sciences, Sam Higginbottom University of Agriculture Technology & Sciences (Formerly Allahabad Agricultural Institute), Allahabad, India

N. Ramesh Department of Biotechnology, J.J College for Arts and Science, Pudukkottai; PG and Research Department of Botany, Govt. Arts College for Men, Krishnagiri, India

Desikan Ramesh Tamil Nadu Agricultural University, Coimbatore, India

P.W. Ramteke Department of Biological Sciences, Sam Higginbottom University of Agriculture Technology & Sciences (Formerly Allahabad Agricultural Institute), Allahabad, India

Roohi Protein Research Laboratory, Department of Bioengineering, Integral University, Lucknow, India

Darshan M. Rudakiya Bioconversion Technology Division, Sardar Patel Renewable Energy Research Institute; Department of Microbiology, N. V. Patel College of Pure and Applied Sciences, Anand, Gujarat, India

Jitendra Kumar Saini Department of Microbiology, School of Interdisciplinary & Applied Life Sciences, Central University of Haryana, Mahendergarh, India

Sreedevi Sarsan Department of Microbiology, St. Pious X Degree & P.G. College, Hyderabad, India

Abha Sharma Division of Microbiology, ICAR-Indian Agricultural Research Institute, New Delhi, India

Shilpa Department of Biotechnology, CGC Landron, Chandigarh, India

Pradeep Kumar Shukla Department of Biological Sciences, Sam Higginbottom University of Agriculture Technology & Sciences (Formerly Allahabad Agricultural Institute), Allahabad, India

Vipin Kumar Shukla Department of Biotechnology, C.C.S University, Meerut, India

Surender Singh Department of Microbiology, Central University of Haryana, Mahendergarh, India

Balkar Singh Department of Botany, Arya PG College, Panipat, India

Bhim Pratap Singh Molecular Microbiology and Systematic Laboratory, Department of Biotechnology, Mizoram University, Aizawl, India

Kumar Rohit Srivastava Department of Chemical Engineering and Technology, Indian Institute of Technology (Banaras Hindu University), Varanasi, India

Neha Srivastava Department of Chemical Engineering and Technology, Indian Institute of Technology (Banaras Hindu University), Varanasi, India

Manish Srivastava Department of Physics and Astrophysics, University of Delhi, Delhi, India

Quratulain Syed Food & Biotechnology Research Center, PCSIR Labs Complex, Lahore, Pakistan

Kiruthika Thangavelu Tamil Nadu Agricultural University, Coimbatore, India

Archana Tiwari Amity Institute of Biotechnology, Amity University, Noida, India

Garima Yadav Department of Microbiology, School of Interdisciplinary & Applied Life Sciences, Central University of Haryana, Mahendergarh, India

Mohammed Rehan Zaheer Department of Chemistry, Gagan College of Management and Technology, Aligarh, India

Zothanpuia Molecular Microbiology and Systematic Laboratory, Department of Biotechnology, Mizoram University, Aizawl, India

Foreword

The ever-increasing population growth and related development is heavily dependent on energy resources. The need to tap renewable energy has been realized over the past few decades but the technologies associated with its production have always been expensive until recently. However, newer and more efficient technologies have become available that make use of renewable energy as an alternative source at an affordable price.

The production of biofuels from cellulosic biomass is a renewable, low-cost, and environmentally friendly process. The process of generating biofuel from cellulosic biomass has received global attention. The major bottleneck to making this technology more affordable has been the cost of cellulases and associated enzymes. Newer technologies, especially biotechnology, have made this process economically viable. In this bioprocess there are multiple variables that play key roles in both efficiency and cost of the technology. To develop this process sustainably there is a need to focus more on individual factors and bring them together. Concerted global initiatives can help to address this issue from different directions.

From Cellulose to Cellulase: Strategies to Improve Industrial Production is a great effort in this direction. I write this foreword with great satisfaction as a researcher interested in biofuels from biomass. This book essentially presents 16 chapters focusing on various low-cost strategies to produce economic cellulolytic enzymes vital for sustainable biofuel production. The book is targeted at various factors of cellulase production process improvement using cellulosic biomass. It also covers existing loopholes in the currently used strategies with proposed sustainable approaches, which may overcome the issue of the cost of biofuel production. I anticipate that the book will be considered as a unique collection of practical information for scientists, researchers, teachers, students, and industries who are interested in biofuel production.

I appreciate the efforts of Dr. Neha Srivastava (IIT (BHU), Varanasi), Dr. Manish Srivastava (DU, Delhi), Prof. (Dr.) P.K. Mishra (IIT (BHU), Varanasi), Prof. (Dr.) Pramod W. Ramteke (SHUATS, Allahabad), and Prof. (Dr.) Ram Lakhan Singh (RMLAU, Faizabad) for their help with this book. The efforts made by the editors will certainly fill a gap and meet the demands of industries, scientists, teachers, researchers, and students. This book will be a useful resource for start-ups in cellulosic biomass conversion through enzymatic processes. I appreciate the meticulous work of the editors in bringing the final shape to this book. I am sure readers will find it highly useful reading such an informative book, which will also be useful as reference material.

P.K. Jain
IIT (BHU) Varanasi, Varanasi, India

CHAPTER

1

Cost Economy Analysis of Biomass-Based Biofuel Production

Neha Srivastava, Rahul Kumar Kharwar†, P.K. Mishra**

*Department of Chemical Engineering and Technology, Indian Institute of Technology (Banaras Hindu University), Varanasi, India †Department of Economics, Banaras Hindu University, Varanasi, India

1.1 INTRODUCTION

Energy plays a vital role in the world economy by contributing ~90% of commercially made energy from nonrenewable fossil fuels used for the transporter sector (Bio diesel, http://bioethanol-np.blogspot.com/). Though the demand for fossil fuels is global, there are several drawbacks related to fossil fuels such as limited lifespan and tremendous air pollution associated with them. The huge supply and demand of energy and limited existence of fossil fuels have encouraged the search for similar or more efficient energy sources for a continuous and balanced supply of energy under sustainable environmental control (Ritchie and Roser, 2018). To reinforce energy security, several countries are emphasizing production and use of renewable energy sources corresponding to biofuels, which is increasing as an industry within the current economic surroundings. Biofuels are fuels that are derived from biomass conversion, similar to solid biomass, liquid fuel, and various biogases (U.S. Energy Information Administrator, 2018).

Biofuels are the potential green alternative to replace fossil fuels because they are eco-friendly, simply available, have low carbon content, completely combustible, and nontoxic (Wyman and Hinman, 1990). In addition, the increasing demand for energy has led to the search for substitutes in the form of biofuels for replacement of fossil fuels due to their restricted availability. Biofuel can be produced from carbon-rich sources such as plant biomass by fermentation and the photosynthesis process. Since it can be produced from plant waste biomass, biofuel production is economical too. When biofuels are burnt, they emit low amounts of carbon and simultaneously control the production of poisonous substances, contributing to low air pollution levels (Wyman and Hinman, 1990).

Today, biofuels can be efficiently used in the transportation market. They are cleaner, which suggests that they produce fewer emissions on burning. Biofuel properties are suitable for all existing transportation engines under most conditions. In addition, engines that use biofuels need less maintenance, which means low cost, no pollution, and hence a longer life for vehicles as well as the environment (Limayema and Ricke, 2012). From an economic point of view, cellulose-rich agriculture waste from local areas with local labor will also contribute to reducing the cost of biofuel production because economic biofuel production depends significantly on raw materials.

Biofuel can be found in three forms: liquid, solid, and gas such as wood charcoal, bioethanol, biobutanol, biodiesel, biohydrogen, and biogas. Bioethanol is one of the simplest types and the most commonly used is bioalcohol (http://biofuel.org.uk/types-of-biofuels.html). Ethanol is obtained by the fermentation method through a biochemical process in which starch is converted to sugar followed by ethanol by fermentative microorganisms (Balan et al., 2012). The United States was the first country to use ethanol as a biofuel. In different ratios, ethanol is also used as a mixture agent with gasoline. This mixture is also used to increase the octane content, and it reduces toxic fumes

From Cellulose to Cellulase: Strategies to Improve Biofuel Production
https://doi.org/10.1016/B978-0-444-64223-3.00001-1

in excessive amounts, whereas butanol can be used directly in petrol engines as fuel. At present a gasoline/ethanol mixture containing 15% ethanol is used to run a gasoline engine without any technical change (Coyle, 2007).

Another future liquid fuel is biodiesel, which is a nontoxic fuel created by the chemical process between alcohol and fatty acid base-forming esters from vegetable oils, animal fats, or plant extracts (Mata et al., 2013). Unwanted products like alcohol, alkyl group and organic compounds are needed to be effectively removed for transestrification. To confirm the correct performance, fuel-grade biodiesel should be created for strict business specifications (ASTM D6751 within the United States and EN 41214 in Europe). Biodiesel is more consistent than fermentation alcohol and may be used as a fuel directly in any unmodified diesel engine; it may be mixed with mineral diesel at any level. It should also be mentioned that cleaner and greener biodiesel plays a commendable role in reducing harmful emissions (Ahindra, 2008). The attraction of biofuels in the field of transportation is not only because they are a reusable resource, but also because they decrease the emission of harmful substances and increase the performance of the engine by increasing the octane level (http://biofuel.org.uk/second-generation-biofuels.html).

Among gaseous fuels, biogas is produced by the anabolic process of organic matter such as agricultural waste, animal waste, weeds and other organic sources, which are available at no cost or at a lower cost. Biogas is basically a composition of methane, carbon dioxide, and hydrogen. To complete this process, a digester and other devices are all that is needed to develop a cheaper and healthier source of energy. Biogas is landfill gases that are mainly produced by the anabolic process, largely due to the high growth of biogas usage in developing countries.

Apart from physical status, biofuels can be categorized into three different classifications known as generations. First-generation biofuels were obtained from different primary edible sources such as vegetable oil, animal fats, rice, sugar, or starch; this is a completely nonviable technology due to the use of human edible primary crops. Therefore, focus was shifted toward starch or carbohydrate sources, which are from nonedible crops generally considered as agricultural waste. This possibility opens the door for "second-generation biofuels," which represent nonedible food crops such as cellulosic biofuels. Cellulosic biofuels belong to a wide range of carbon-rich waste substrates such as cellulosic and lignocellulosic plant biomass, which can be efficiently converted into sugar by the bioconversion process and finally into biofuel through microbial fermentation (Fulton et al., 2004). Additionally, these waste carbon raw materials, which are used for biofuel production, are available at very low cost and are known as cellulosic ethanol, cellulosic hydrogen, and biodiesel. One of the advantages of second-generation biofuels is that there is no need to develop new crops every year for biofuel production and low waste management contributes to the development of a sustainable environment by reducing harmful emission of fossil fuels (Wyman and Hinman, 1990).

The third generation of biofuels is derived from algae; these biofuels are also called "oil-gas." The algae can be harvested in only 5–6 days and can be converted into biohydrogen and biobutanol using the fermentation method. One healthy aspect of biofuel produced from algae is that they protect lakes and rivers, and therefore marine vegetation and other organisms, because these algae use more nitrogen and phosphorus (Biofuel, 2018). Therefore, the objective of this chapter is to evaluate and analyze the cost of bioethanol production technology to reduce the cost of ethanol and make it available in a sustainable way so it can be utilize as biofuel instead of using fossil fuels. Fig. 1.1 shows an overview of the bioethanol process (Kumar and Sharma, 2017).

1.2 PRETREATMENT PROCESS OF CELLULOSIC BIOMASS

Though lignocellulosic biomass-based biofuel is considered as a potential alternative fuel, its complex structure significantly increases major technical and economic challenges for the sustainable production process of biofuel (Claassen et al., 1999). The objective of biomass is to reduce cellulose crystallization, remove lignin and hemicellulose, and increase the void fraction of cellulose content. An effective pretreatment process ensures the following advantages: (1) improving the sugar content due to high availability of cellulose, (2) avoiding degradation or loss of carbohydrate, (3) preventing the production of subproducts, basically interceptors, and thereafter avoiding fermentation and hydrolysis, and (4) making it economically feasible (Uni Assignment, 2018). Protective methods of pretreatment can be categorized on the basis of primary characteristics such as physical (milling and grinding) and physical/chemical (steam protective) strategies, or possibly classified as physical (milling and grinding), physical/chemical (steam protective/autoreaction, hydrothermolysis, and wet oxidation), chemical (alkali, dilute acid, reaction agent, and organic solvents), and biological implant strategies (hydrolysis, hydrothermolysis, and wet oxidation). Obstetrics in many ways has made the transplantation of lignocellulosic biomass into chemicals

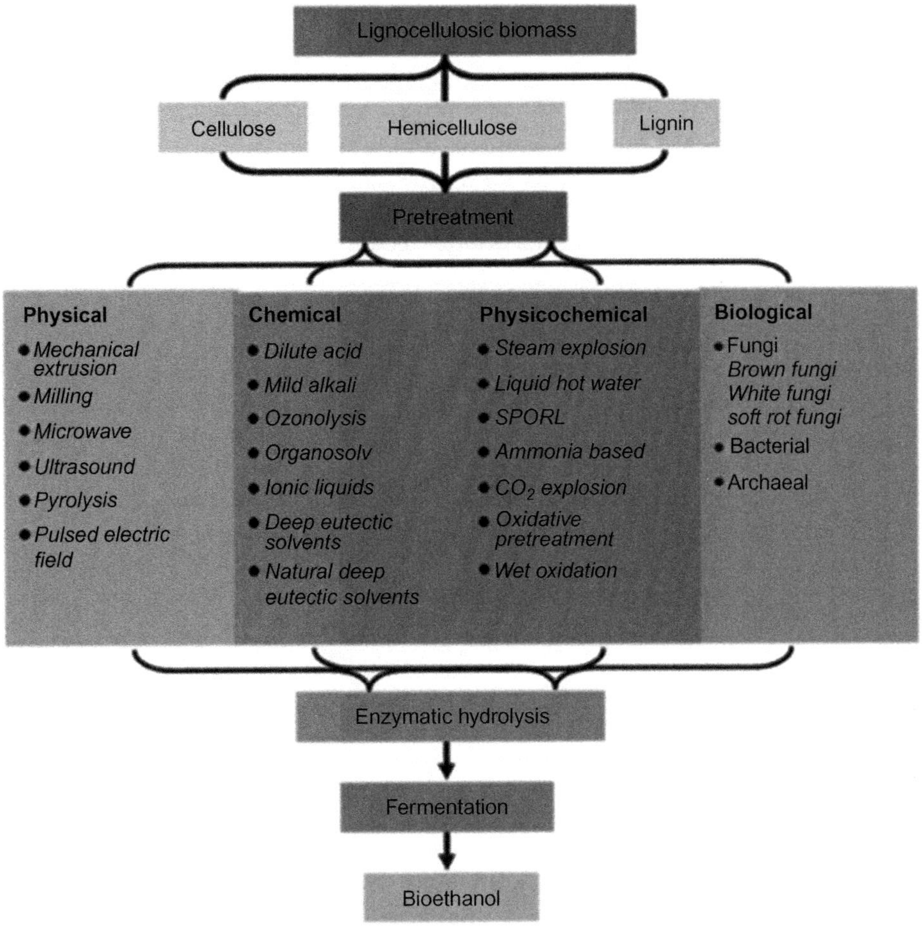

FIG. 1.1 The complete bioethanol process (Kumar and Sharma, 2017).

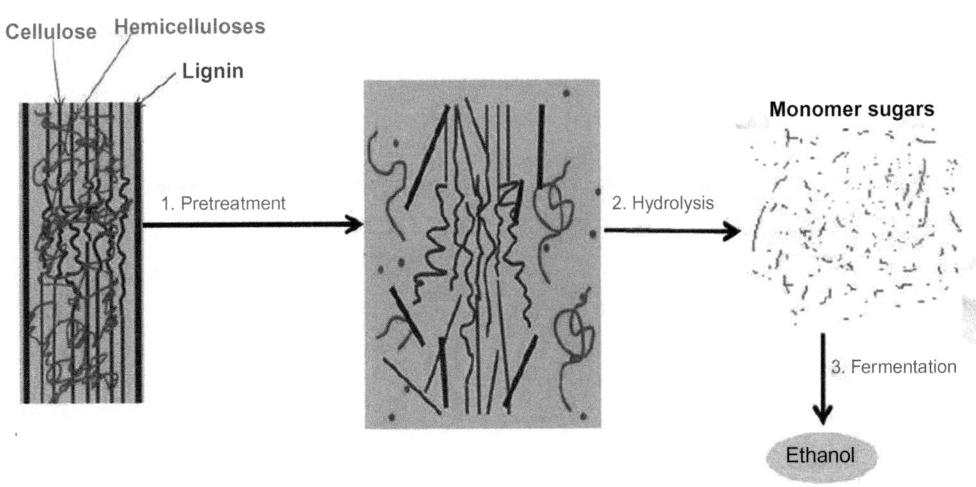

FIG. 1.2 The pretreatment process (Madadi et al., 2017).

and fuels economically feasible and achievable, in which the conversion process of biomass into fuels is biological (Che Kamarludin et al., 2014). Fig. 1.2 explains the pretreatment process of agricultural biomass (Madadi et al., 2017).

The pretreatment process of lignocellulosic biomass can be divided into the following types: physical and mechanical pretreatment of lignocellulosic biomass and physical pretreatment, which explores and measures the dry matter of biomass at the first level to renovate the total strength of the downstream process. The physical method of pretreatment includes commination (mechanical reduction in size of biomass particulates).

1.3 PRETREATMENT

The method of common size is used for compression milling, vibratory ball milling, wet milling, and dry milling. The consumption of energy for grinding the LB (lignocellulosics biomass) depends on various machine variables: humidity, initial particle size, material properties content, and machine variability. The classification of grinders depends on its mechanism; the most efficient way to reduce the size of the particle is splitting or cutting with knives but this also damages the geometry of the particle due to impact or compression. Table 1.1 shows a literature report on how the size of a particle is reduced using various grinder techniques (Binod et al., 2012).

1.3.1 Microwave Treatment of Biomass

The optimal choice for conventional heating in various domains is microwave radiation because it is more efficient to use less energy and provide faster heating (Xiong et al., 2002). As explained by Azuma et al. (1984), the whole ultrastructure of cellulose is changed by breaking the layer of hemicellulose and lignin by using a process of microwave heating, which also enhances enzymatic hydrolysis (Zhu et al., 2006). Microwave heating in addition to chemical transplantation is efficient and the reaction rate is increased (Hu and Wen, 2008; Zhu et al., 2015). Microwave radiation of lignocellulosic feedstock pretreatment is widely used because: (1) it is less technical, i.e., easy to use; (2) it requires low energy input; (3) it is efficient in terms of time and production (thermal); (4) the generation of blocks is minimal; and (5) the organizational structure of the fraction is reduced. In addition, a more effective breakdown of light alkali reagents is preferred. On a microwave-based base transplant of switch grass, a study produced 70%–90% of sugars (Zhu et al., 2015). A broad study of the effect of microwave radiation on chemical pretreatment was done by Boonmanumsin et al. (2012). In comparison to the standard heating of H_2SO_4 and NaOH, sugar output was increased by 12 times. The use of orthogonal design in the microwave pretreatment of wheat straw optimized the ethanol yield from 2.678% to 14.8% in the study by Chang et al. (1997).

1.3.2 Breaking or Milling

Mechanical grinding (milling) is used to reduce cellulose crystallization. It mostly contains adhesive, grinding, and/or milling techniques (Kumar and Sharma; 2017). By adhesion the biomass size can be reduced to only 10–30 mm, while by grinding and milling the particle size can be reduced to 0.2 mm. However, studies have shown that the reduction of biomass particles below 0.4 mm and lack of hydrolysis (Taherzadeh and Karimi, 2008) have no significant effect on the rate and yield. Due to the reduction of particle size and crystalline cellulose, effectively due to the shear forces generated during milling, the heat and mass transfer limitations decrease due to the adhesive. The type and duration of milling increase the specific surface area of biomass, the final degree of polymerization, and the determination of pure reduction in crystalline cellulose. Different milling methods such

TABLE 1.1 The Cost of Bioethanol by Using Yeast (*Saccharomyces cerevisiae*)

Ethanol, L/dry ton sugarcane	**45 L**		
	US$/L	**100 L/US$**	**Indian National Rupee (INR)/L**
Sugarcane bagasse	0.000	0	0
Enzymes	0.341	34.1	23.4
Acid	0.080	8	5.48
Base	0.020	2	1.37
Water consumption	0.045	4.5	3.08
Other raw materials	0.000	0	0
Labor, maintenance, insurance	0.148	14.8	10.14
Electricity consumption	x	x	x
Yeast (*Saccharomyces cerevisiae*)	0.190	19	13.02
Capital cost	0.560	56	38.37
Minimum ethanol selling price	1.384	138.4	94.8

leaves is just Rs. 52/L, which is much less than the current market prices of oil and petroleum. By achieving cost optimization the low cost of bioethanol is achievable as a long-term environmentally sustainable fuel.

Acknowledgments

Authors N.S. and P.K.M. acknowledge the Department of Chemical Engineering and Technology, IIT (BHU), Varanasi, India for providing research facilities. N.S. also acknowledges IIT (BHU), Varanasi for providing Institute PDF fellowship.

References

Ahindra, N., 2008. Biofuels Refining and Performance. McGraw Hill, USA.

Azuma, J., Tanaka, F., Koshijima, T., 1984. Enhancement of enzymatic Susceptibility of lignocellulosic wastes by Microwave irradiation. J. Ferment. Technol. 62, 377–384.

Bai, F.W., Chen, L.J., Zhang, Z., Anderson, W.A., Moo-Young, M., 2004. Continuous ethanol production and evaluation of yeast cell lysis and viability loss under very high gravity medium conditions. J. Biotechnol. 110 (3), 287–293.

Balan, V., Kumar, S., Bals, B., Chundawat, S.P.S., Jin, M., Dale, B.E., 2012. Biochemical and thermochemical conversion of switchgrass to biofuels. In: Monti, A. (Ed.), Switchgrass: A Valuable Biomass Crop for Energy. Springer, London, UK, pp. 153–186 (Chapter 7).

Baptista, C.M.S.G., Cóias, J.M.A., Oliveira, A.C.M., et al., 2006. Natural immobilisation of microorganisms for continuous ethanol production. Enzym. Microb. Technol. 40 (1), 127–131.

Bayrock, D.P., Michael Ingledew, W., 2001. Application of multistage continuous fermentation for production of fuel alcohol by very-high-gravity fermentation technology. J. Ind. Microbiol. Biotechnol. 27 (2), 87–93.

Binod, P., Satyangalakshmi, K., Sindhu, R., Janu, K.U., Sukumarani, R., Panday, A., 2012. Short duration microwave assisted pretreatment enhance the enzymatic saccharification and fermentatble sugar yield from sugarcane bagasse. Renew. Energy 37, 109–116. https://doi.org/10.1016/j.renene.2011.06.007.

Biofuel, 2018. Biofuel.org.uk. Retrieved from: http://biofuel.org.uksecond-generation-biofuels.html

Boonmanumsin, P., Treeboobpha, S., Jeamjumnunja, K., Luengnaruemitchai, A., Chaisuwan, T., Wongkasemjit, S., 2012. Release of monomeric sugars from *Miscanthus sinensis* by microwave-assisted ammonia and phosphoric acid treatments. Bioresour. Technol. 103 (1), 425–431.

Cazetta, M.L., Celligoi, M.A.P.C., Buzato, J.B., Scarmino, I.S., 2007. Fermentation of molasses by *Zymomonas mobilis*: effects of temperature and sugar concentration on ethanol production. Bioresour. Technol. 98 (15), 2824–2828.

Chang, V.S., Burr, B., Holtzapple, M.T., 1997. Lime pretreatment of switchgrass. Appl. Biochem. Biotechnol. 63–65, 3–19. https://doi.org/10.1007/BF02920408.

Che Kamarludin, S.N., et al., 2014. Mechanical pretreatment of lignocellulosic biomass for biofuel production. Appl. Mech. Mater. 625, 838–841.

Claassen, P.A.M., Contreras, A.M.L., Sijtsma, L., Weusthuis, R.A., Van Lier, J.B., Van Niel, E.W.J., Stams, A.J.M., de Vries, S.S., Weusthuis, R.A., 1999. Utilisation of biomass for the supply of energy carriers. Appl. Microbiol. Biotechnol. 52, 741–755. https://doi.org/10.1007/s002530051586.

Coyle, W., 2007. The future of biofuels: a global perspective. Amber Waves 5 (5). Available online at www.ers.usda.gov amber waves November 07 Features Biofuels.htm.

Deesuth, O., Laopaiboon, P., Jaisil, P., Laopaiboon, L., 2012. Optimization of nitrogen and metal ions supplementation for very high gravity bioethanol fermentation from sweet sorghum juice using an orthogonal array design. Energies 5 (9), 3178–3197.

Dhaliwal, S.S., Oberoi, H.S., Sandhu, S.K., Nanda, D., Kumar, D., Uppal, S.K., 2011. Enhanced ethanol production from sugarcane juice by galactose adaptation of a newly isolated thermotolerant strain of *Pichia kudriavzevii*. Bioresour. Technol. 102 (10), 5968–5975.

Fulton, L., Howes, T., Hardy, J., 2004. Biofuels for Transport: An International Perspective. International Energy Agency, Paris, France. Available online at http://www.iea.org/textbase/nppd/free/2004/biofuels2004.pdf.

Gnansounou, E., Dauriat, A., 2005. Ethanol fuel from biomass: a review. J. Sci. Ind. Res. 64 (11), 809–821.

Hu, Z.H., Wen, Z.Y., 2008. Enhancing enzymatic digestibility of switchgrass by microwave-assisted alkali pretreatment. Biochem. Eng. J. 38, 369–378. https://doi.org/10.1016/j.bej.2007.08.001.

Ingram, L., Gomez, P., Lai, X., et al., 1998. Metabolic engineering of bacteria for ethanol production. Biotechnol. Bioeng. 58 (2–3), 204–214.

Kosaric, N., Velikonja, J., 1995. Liquid and gaseous fuels from biotechnology: challenge and opportunities. FEMS Microbiol. Rev. 16 (2–3), 111–142.

Kourkoutas, Y., Komaitis, M., Koutinas, A.A., Kanellaki, M., 2001. Wine production using yeast immobilized on apple pieces at low and room temperatures. J. Agric. Food Chem. 49 (3), 1417–1425.

Kumar, P., Barrett, D.M., Delwiche, M.J., Stroeve, P., 2009. Methods for pretreatment of lignocellulosic biomass for efficient hydrolysis and biofuel production. Ind. Eng. Chem. Res. 48, 3713–3729. https://doi.org/10.1021/ie801542g.

Kumar, R., Wyman, C.E., 2009. Effects of cellulase and xylanase enzymes on the deconstruction of solids from pretreatment of poplar by leading technologies. Biotechnol. Prog. 25 (2), 302 14.

Kumar, Sharma, 2017. Recent updates on different methods of pretreatment of lignocellulosic feedstocks: a review. Bioprocess 4, 7.

Liang, L., Zhang, Y.-P., Zhang, L., Zhu, M.-J., Liang, S.-Z., Huang, Y.-N., 2008. Study of sugarcane pieces as yeast supports for ethanol production from sugarcane juice and molasses. J. Ind. Microbiol. Biotechnol. 35 (12), 1605–1613.

Limayema, A., Ricke, S.C., 2012. Lignocellulosic biomass for bioethanol production: current perspectives, potential issues and future prospects. Prog. Energy Combust. Sci. 38, 449–467. https://doi.org/10.1016/j.pecs.2012.03.002.

Lin, Z., Huang, H., Zhang, H., Zhang, L., Yan, L., Chen, J., 2010. Ball milling pretreatment of corn stover for enhancing the efficiency of enzymatic hydrolysis. Appl. Biochem. Biotechnol. 162, 1872–1880. https://doi.org/10.1007/s12010-010-8965-5.

Madadi, M., Tu, Y., Abbas, A., 2017. Recent status on enzymatic saccharification of lignocellulosic biomass for bioethanol production. Electron. J. Biol. 13, 2.

Maiorella, B.L., Blanch, H.W., Wilke, C.R., 1984. Economic evaluation of alternative ethanol fermentation processes. Biotechnol. Bioeng. 26 (9), 1003–1025.

Makebiofuel, 2018. Makebiofuel.co.uk. bioethanol production.

Macrelli, S., Mogensen, J., Zacchi, G., 2012. Techno-economic evaluation of 2nd generation bioethanol production from sugar cane bagasse and leaves integrated with the sugar-based ethanol process. Biotechnol. Biofuels 5, 22.

Mata, T.M., Martins, A.A., Caetano, N.S., 2013. Valorization of waste frying oils and animal fats for biodiesel production. In: Advanced Biofuels and Bioproducts. Springer, Berlin, Germany, pp. 671–693. (Chapter 4).

McIntosh, S., Vancov, T., 2010. Enhanced enzyme saccharification of *Sorghum bicolor* straw using dilute alkali pretreament. Bioresour. Technol. 101 (17), 6718–6727.

Najafpour, G., Younesi, H., Ismail, K.S.K., 2004. Ethanol fermentation in an immobilized cell reactor using *Saccharomyces cerevisiae*. Bioresour. Technol. 92 (3), 251–260.

Olsson, L., Hahn-Hägerdal, B., 1993. Fermentative performance of bacteria and yeasts in lignocellulose hydrolysates. Process Biochem. 28 (4), 249–257.

Öztop, H.N., Öztop, A.Y., Karadağ, E., Işikver, Y., Saraydin, D., 2003. Immobilization of *Saccharomyces cerevisiae* on to acrylamide-sodium acrylate hydrogels for production of ethyl alcohol. Enzym. Microb. Technol. 32 (1), 114–119.

Prasad, B., Mishra, I.M., 1995. On the kinetics and effectiveness of immobilized whole-cell batch cultures. Bioresour. Technol. 53 (3), 269–275.

Ritchie, H., Roser, M., 2018. Fossil Fuels. Published online at OurWorldInData.org. Retrieved from https://ourworldindata.org fossil fuels Online Resource.

Rodríguez, E., Callieri, D.A.S., 1986. High yield conversion of sucrose into ethanol by a flocculent *Zymomonas* sp isolated from sugarcane juice. Biotechnol. Lett. 8 (10), 745–748.

Sassner, P., Mårtensson, C.G., Galbe, M., Zacchi, G., 2008. Steam pretreatment of H_2SO_4 impregnated Salix for the production of bioethanol. Bioresour. Technol. 99 (1), 137–145.

Sills, D.L., Gossett, J.M., 2011. Assessment of commercial hemicellulases for saccharification of alkaline pretreated perennial biomass. Bioresour. Technol. 102 (2), 1389–1398.

Taherzadeh, M.J., Karimi, K., 2008. Pretreatment of lignocellulosic wastes to improve ethanol and biogas production: a review. Int. J. Mol. Sci. 9, 1621–1651. https://doi.org/10.3390/ijms9091621.

Tzeng, J.-W., Fan, L.-S., Gan, Y.-R., Hu, T.-T., 1991. Ethanol fermentation using immobilized cells in a multistage fluidized bed bioreactor. Biotechnol. Bioeng. 38 (10), 1253–1258.

U.S. Energy Information Administrator, 2018. Retrieved from https://www.eia.gov/energyexplained/index.php?page=about_home.

Uni Assignment, 2018. Lignin is Amorphous and Highly Complex Biology Essay.

Vats, S., Maurya, D.P., Shaimoon, M., Negi, S., 2013. Development of a microbial consortium for the production of blend enzymes for the hydrolysis of agricultural waste into sugars. J. Sci. Ind. Res. 72, 585–590.

Vitolo, M., 1996. Production of ethanol and invertase by *S. cerevisiae* grown in blackstrap molasses. In: Proceedings of the 7th Biomass for Energy and the Environment. Pergamon Press, Copenhagen, Denmark, pp. 1477–1481.

Wang, W., Yuan, T., Wang, K., Cui, B., Dai, Y., 2012. Combination of biological pretreatment with liquid hot water pretreatment to enhance enzymatic hydrolysis of *Populus tomentosa*. Bioresour. Technol. 107, 282–286.

Wyman, C.E., 2004. Ethanol fuels. In: Encyclopedia of Energy. Elsevier, New York, NY, pp. 541–555.

Wyman, C.E., Hinman, N.D., 1990. Ethanol: fundamentals of production from renewable feedstocks and use as transportation fuel. Appl. Biochem. Biotechnol. 24, 735–775. https://doi.org/10.1007/BF02920291.

Xiong, C., He, X., Zhang, Z., 2002. Microwave assisted extraction or saponitication combined with microwave assisted decomposition applied in pretreatment of soil or mussel sompus for the determination of polychlorinated biphenryus. Anal. Chem. Acta 413, 49–56. https://doi.org/10.1016/S0003-2070-(00)00754-6.

Zabed, H., Faruq, G., Sahu, J.N., Azirun, M.S., Hashim, R., Boyce, A.N., 2014. Bioethanol production from fermentable sugar juice. Sci. World J. 2014. http://dx.doi.org/10.1155/2014/957102.

Zhu, S.D., Wu, Y., Zhang, X., Wang, C., Yu, F., Jin, S., 2006. Production of ethanol from microwave assisted alkali pretreatment wheat straw. Process Biochem. 41, 869–873. https://doi.org/10.1016/j.procbio.2005.10.024.

Zhu, Z., Macquarrie, D.J., Simister, R., Gomez, L.D., McQueen-Mason, S.J., 2015. Microwave assisted chemical pretreatment of Miscanthus under different temperature regimes. Sustain. Chem. Process. 3, 15–27. https://doi.org/10.1186/s40508-015-0041-6.

CHAPTER

2

Cellulose as Potential Feedstock for Cellulase Enzyme Production: Versatility and Properties of Various Cellulosic Biomasses

*Abha Sharma**, *Jairam Choudhary*[†], *Surender Singh*[‡], *Balkar Singh*[§], *Ramesh Chander Kuhad*[¶,‖], *Ajay Kumar**, *Lata Nain**

***Division of Microbiology, ICAR-Indian Agricultural Research Institute, New Delhi, India [†]Indian Institute of Farming Systems Research, Modipuram, India [‡]Department of Microbiology, Central University of Haryana, Mahendergarh, India [§]Department of Botany, Arya PG College, Panipat, India [¶]Department of Microbiology, University of Delhi South Campus, New Delhi, India [‖]Central University of Haryana, Mahendergarh, India**

2.1 INTRODUCTION

Rising concerns over the exhaustion of nonrenewable energy resources as well as emission of greenhouse gases by their burning have encouraged the production of alternative renewable energy resources. In this regard, production of bioethanol provides a feasible, sustainable, and eco-friendly alternative to fossil fuels (Kim, 2018). First-generation bioethanol is principally produced by fermenting sugary/starchy substrates like wheat, corn, and cane sugar. However, these crops are food crops and thus cannot be used for fuel generation, especially in developing countries to ensure food security. Alternatively, lignocellulosic substrates can be effectively used for fuel production, which is called second-generation bioethanol (Choudhary et al., 2016). Utilization of lignocellulosics for the production of biofuel provides many advantages such as being abundant, inexpensive, easily available, having a higher productivity per hectare, and more significantly do not threaten food supplies and biodiversity (Takano and Hoshino, 2018). Lignocellulose comprises carbohydrates (cellulose, hemicelluloses) and lignin (aromatic polymer). For bioethanol production from lignocelluloses, the substrate is pretreated (alkali/acid/steam) to remove lignin and hemicelluloses, thereby making cellulose accessible for hydrolysis by cellulase enzymes into glucose units (the saccharification step). The resulting monomeric glucose units are subsequently fermented to ethanol with the aid of fermentative microbes. The economic viability of this bioprocess is highly dependent on the cellulase enzymes because a huge amount of this enzyme cocktail is used in the saccharification step that contributes to 25%–35% of the total process cost (Liguori et al., 2016). However, enzyme cost can be reduced by utilization of inexpensive feedstock, use of novel microorganisms producing high titers of cellulolytic enzymes, use of holocellulase-conserving pretreatment processes, and in-house enzyme production that causes maximum release of reducing sugars with the generation of minimum inhibitory compounds (Johnson, 2016).

Cellulases hydrolyze β-1,4 linkages in cellulose chains to release glucose units. Cellulases are synthesized mainly by bacteria, fungi, protozoans, plants, and animals (Kuhad et al., 2011). Industrial cellulase production is carried out from a variety of fungi and bacteria. These include members of the genera *Clostridium*, *Cellulomonas*, *Thermomonospora*, *Trichoderma*, and *Aspergillus* (Kuhad et al., 2016). Although cellulases have been produced by the use of avicel and other crystalline celluloses as carbon sources, inexpensive carbon and nitrogen sources are required to reduce the cost

From Cellulose to Cellulase: Strategies to Improve Biofuel Production
https://doi.org/10.1016/B978-0-444-64223-3.00002-3

of enzyme production. In this regard, lignocellulosic material is the suitable alternative to pure substrates used for the production of cellulases due to their plentiful availability and lesser cost (Sukumaran et al., 2009). Lignocellulosic biomass is broadly categorized into three classes: (1) virgin biomass, which includes trees, bushes, and grasses, (2) waste biomass, which includes by-products of various industries such as corn stover, sugarcane bagasse, wheat bran, etc., and (3) high-productivity energy crops (HPEC), which includes switch grass and elephant grass. Among these three categories of cellulosic biomass, employing agro-industrial waste biomass is an attractive approach for the reduction of cost associated with culture medium formulation. The production of both virgin biomass and HPEC will require conventional forestry on forest lands. Forest lands have low productivity and will also lead to loss of wildlife habitats, therefore they are not suitable for the development of a sustainable bioprocess.

This chapter discusses the structure, function, and diversity of cellulase enzymes as well as microorganisms, which produce and secrete this complex enzyme system extracellularly. The chapter specifically deals with various substrates that are used for microbial cellulase production, both pure and lignocellulosic, as well as the production methods employed. Various transcriptome and quantitative proteomic approaches to study the secretomes of cellulolytic microorganisms in the presence of specific carbon sources are also discussed. Finally, commercial formulations of cellulase available in the market are also discussed.

2.2 CELLULASE: STRUCTURE, FUNCTION, AND DIVERSITY

Cellulase is a set of three enzymes that hydrolyze cellulose, and are categorized based on mode of action: exoglucanases (EC 3.2.1.91 and 3.2.1.74), endoglucanases (EC 3.2.1.4), and β-glucosidases (BGs) (EC 3.2.1.21). Exoglucanase or cellobiohydrolase (CBH) acts on unprotected ends of the chain and chops off cellobiose. On the other hand, carboxymethyl cellulase (CMCase) or endoglucanase cuts β-1,4-bonds of cellulose chains at random and thus produces new ends. Cellobiose, being the smallest representative of cellulose microfibrils, is consequently hydrolyzed by BGs to glucose. An appreciable level of association between endoglucanase and CBH results in an efficient hydrolysis of cellulose (Gupta and Lee, 2009). However, the products (cellodextrin and cellobiose) of endoglucanases and CBHs, respectively, inhibit activity of the enzyme. Nevertheless, these products can be very well cleaved by BGs to release glucose. Therefore synergistic action of all three enzymes causes complete and proficient hydrolysis of cellulosic biomass.

The structure of cellulases depicts two separate domains: a catalytic domain and a cellulose-binding module (CBM), joined by a flexible linker region. There are around 35 amino acids in CBM, while high amounts of serine and threonine are present in the linker. A cleft/groove-shaped active site is present in the catalytic module of most of the endoglucanases, allowing the enzyme to bind as well to release glucose, cellodextrins, and insoluble fragments by cleaving the chain (Mejia-Castillo et al., 2008). On the other hand, two surface loops form a tunnel structure in the catalytic module of CBHs (exoglucanase) (Muñoz et al., 2001). Interestingly, BGs do not contain a CBM and thus activity on insoluble cellulose is insignificant. However, BGs release glucose by hydrolyzing cellobiose and soluble cellodextrins. The importance of this degradation lies in the fact that cellobiose is inhibitory to CBH as well as endoglucanase. A variety of BGs with different catalytic modules are synthesized by a number of bacteria, archaea, fungi, plants, and animals.

Microbes that produce cellulases employ two mechanisms for utilizing cellulose: (1) distinct uncomplexed cellulases characteristically produced by aerobic bacteria as well as fungi, and (2) cellulosomes, which are complexed cellulases being expressed characteristically on the surface of anaerobic bacteria and fungi. The mechanism of decomposition of cellulose is analogous in aerobic fungi and bacteria, while a different system is present in anaerobic microbes (Kuhad et al., 2016). Generally, a set of individual cellulases, each containing a CBM joined by a linker to the catalytic module, is secreted by most of the aerobic cellulolytic microbes (Wilson, 2008). On the contrary, most of the anaerobic microorganisms secrete cellulosomes, which are very large multienzyme complexes (>1 million molecular mass) (Bayer et al., 2004). The enzymes of cellulosomes do not contain a CBM but are attached to scaffolding protein containing a CBM (Schwarz, 2001). Interestingly, the cellulosomal enzymes also comprise of a dockerin domain or cellulose-binding domain (CBD) (Bayer et al., 2004).

Preliminary research by Bayer and colleagues (Bayer et al., 1998) demonstrated that the CBD has a planar configuration involving amino acids, tyrosine, tryptophan, arginine, histidine, and aspartic acid, which interacts with cellulose causing binding of the scaffolding protein to cellulose. After substrate binding, a supramolecular reorganization occurs in the cellulosome, allowing its subunits to redistribute and thus interact with different target substrates.

Catalytically, cellulases use acid-base catalysis to cause cleavage of glycosidic bonds. This catalysis is carried out by the enzyme's two catalytic residues: acid and base (Davies and Henrissat, 1995). Based on the spatial arrangement of the residues, either inversion or retention of the anomeric configuration carries out the hydrolysis reaction. For the "retaining ones," the anomeric "C" carrying the target glycosidic bond keeps a similar arrangement after a double-displacement hydrolysis with the two main glycosylation/deglycosylation steps, while in case of "inverting cellulases," configuration of anomeric "C" is inverted after a single nucleophilic displacement hydrolysis (Vocadlo and Davies, 2008). Saharay and coworkers have elaborated and reviewed very well the catalytic mechanism of cellulases (Saharay et al., 2010).

Glycoside hydrolases are classified in 153 families on the basis of similarities in the sequence of amino acids as well as crystalline structures. A number of cellulase genes have been cloned and characterized. The Carbohydrate-Active Enzymes database (CAZy; http://www.cazy.org) furnishes constantly updated information of the glycoside hydrolase (EC 3.2.1.–) families (Cantarel et al., 2009).

2.3 CELLULASE-PRODUCING MICROORGANISMS

Fungi: Among microorganisms, fungi are the most prolific degraders of cellulose, possibly decomposing 80% of cellulose on earth (Sajith et al., 2014). The major cellulose decomposing fungi belong to ascomycetes, basidiomycetes, deuteromycetes, and some chytrids residing in animals' rumen. Some proficient cellulolytic fungi belong to the genera *Aspergillus, Trichoderma, Penicillium, Fusarium, Chaetomium, Cladosporium, Stachybotrys, Acremonium, Alternaria, Humicola, Myrothecium*, and *Ceratocystis* (Wood, 1985). For industrial applications, aerobic fungal cellulases are usually chosen because they are produced extracellularly in large amounts. In contrast, most of the anaerobic fungal as well as bacterial cellulases are present as tight multienzyme complexes, mostly bound to membrane as cellulosome and thus difficult to extract, thereby making them economically less important (Mathew et al., 2008).

For commercial production of cellulolytic enzymes, a number of aerobic cellulolytic fungi, belonging to the genera *Aspergillus, Fusarium, Trichoderma, Penicillium*, and *Sclerotium* are exploited. Among these, the most comprehensively studied fungus is *Trichoderma reesei*, which can also efficiently hydrolyze native cellulose. In addition to soft rots, white rot and brown rots are also major contributors to cellulose degradation in nature, although the mechanism of their degradation is different to that of soft rot fungi (Kuhad et al., 1994). Interestingly, brown rot fungi are known to quickly degrade cellulose at the time of early wood decay and the explanation for this is considered to be lack of exoglucanase in brown rot fungi (Kuhad et al., 1997). Nevertheless, a few reports of brown rot fungi secreting exoglucanase are also present (Deswal et al., 2011). Some well-known cellulolytic brown rot fungi include *Fomitopsis* sp. (Deswal et al., 2011), *Poria placenta* (Ryu et al., 2011), *Lenzites trabea* (Hulme and Stranks, 1974), *Coniophora puteana* (Riley et al., 2014), and *Tyromyces palustris* (Hishida et al., 1999). The ability of white rot fungi to produce cellulases is heterogeneous because these microorganisms are mainly lignin degraders. Some common white rots that produce cellulases are *Phanerochaete chrysosporium, Sporotrichum thermophile, Sporotrichum pulverulentum, Bjerkandera adusta, Pleurotus ostreatus*, and *Trametes versicolor* (Salinas et al., 2011; Tirado-Gonzalez et al., 2016).

Furthermore, thermophilic fungi have also been studied extensively for the production of cellulases because of the swelling of cellulose fibers at elevated temperatures making them easily accessible for hydrolytic enzymes (Li et al., 2011). A typical thermophilic fungus, *Talaromyces emersonii*, produces thermostable cellulase (active at 70°C), which efficiently decomposes intact cellulose (Murray et al., 2001). *Penicillium* sp. isolated from decaying plant material was found to produce thermostable cellulases (Santa-Rosa et al., 2018).

Besides, anaerobic fungi also carry out a significant part in biodegradation of plant cellulose and produce a range of cellulolytic enzymes. The most extensively studied anaerobic fungi are *Piromyces communis* (Ali et al., 1995), *Orpinomyces* sp. (Hodrová et al., 1998), *Neocallimastix frontalis* (Srinivasan et al., 2001), etc.

Bacteria: Higher growth rate compared to fungi make bacteria prospective microorganisms to be employed for the production of cellulases. However, cellulolytic bacterial cultures are rarely used for cellulase production because the enzyme is produced in low titers by them compared to those in fermentation with cellulolytic fungal cultures. On the contrary, isolation and characterization of novel cellulases from bacteria are also being widely exploited. This is because bacteria inhabit a broad range of environmental niches and thus are resistant to various environmental stresses. In addition to their survival in harsh environments, these strains secrete enzymes that are stable under extreme conditions, which are also persistent in industrial bioprocesses and thus increase process efficiency (Maki et al., 2009). Cellulolytic enzymes from bacteria have been reported from *Bacillus* (Reddy et al., 2018), *Paenibacillus* (Ko et al., 2007), *Acinetobacter* (Lo et al., 2010), *Cellulomonas* (Poulsen et al., 2016), *Pseudomonas* (Sethi et al., 2013), *Serratia* (Sethi et al., 2013), *Streptomyces* (Singh et al., 2014), and *Clostridium* (Leis et al., 2017). Furthermore, rumen bacteria also show potential to produce cellulolytic enzymes (Kuhad et al., 1994). In this regard, Nyonyo et al. (2014) cultured cellulolytic bacteria from bovine rumen using azo-CMC as carbon source and isolated

129 strains (Nyonyo et al., 2014). Out of these, 51 and 117 showed FPase and CMCase activities, respectively, and belonged to the genera *Ruminococcus*, *Fibrobacter*, *Pseudobutyrivibrio*, *Streptococcus*, *Enterococcus*, *Saccharofermentas*, and *Ruminococcus*. Asem et al. (2017) isolated *Serratia rubidaea* and *Aneurinibacillus aneurinilyticus* as potent cellulase and xylanase producers from rumen of swine and goat (Asem et al., 2017). Interestingly, a cellulolytic strain identified as *Escherichia coli* was isolated from bovine rumen. The bacterium secreted 9.13 U of exoglucanase, 5.13 U of endoglucanase, 7.27 U of BG, and degraded 14.3% of cellulose in rice straw (Pang et al., 2017). The results paved the way for further investigation into the evolutionary mechanisms of rumen microorganisms, thereby providing immense potential of engineered *E. coli* in biorefineries.

2.4 SUBSTRATES FOR CELLULASE PRODUCTION

Cellulases are biomass-hydrolyzing enzymes that are produced by microorganisms when a carbon source in the form of complex cellulosic substrate is provided (Li et al., 2018). Cellulase production by microbes is inducible because these enzymes are produced extracellularly in the presence of cellulosic substrate. These substrates can be either purified cellulose powder or lignocellulosic biomass. The desirable properties of a substrate for cellulase production are listed as follows:

- It should be cheap and easily available throughout the year in plentiful amount.
- It should have high cellulose and low lignin content.
- It should not be toxic for the microorganism producing cellulase.
- It should be ecologically safe.
- It should not create any problem during the separation of enzyme during downstream processing.
- It should not encourage production of toxic by-products.
- It should facilitate the lesser effluents' release.
- It should facilitate sufficient nutrient supply for microbial growth.

2.4.1 Availability of Cellulosic Feedstocks for Cellulase Production

Agricultural activities generate lots of lignocellulosic biomass in the form of crops' postharvest residues like straw and processing residues like bran, hull, husk, and perishable waste from the food industries. A significant portion of these materials find their way into animal feed, mulching material, energy generation, and compost making, while some are simply disposed of in the field by burning (Singh et al., 2014). The approximate quantity of this material generated annually worldwide has been reviewed by various researchers (Saini et al., 2015b; Pandiyan et al., 2019) and is depicted in Fig. 2.1. Enzyme production would require only a small fraction of the total residues generated as the lignocellulosic substrates are generally added at 1%–2% w/v of medium in submerged fermentation (SmF). Similarly for solid substrate fermentation (SSF), the amount of substrate required is very low due to high productivity under SSF.

2.4.2 Pure Substrates for Cellulase Production

Researchers have used several pure substrates, namely carboxymethyl cellulose, filter paper, avicel, ball-milled cellulose, sigmacell, phosphoric acid, swollen cellulose, salicin, etc. for cellulase production from microorganisms (Table 2.1). The production of different cellulases ranged between 0.023 and 566 IU/mL depending on substrate and microorganisms used.

2.4.3 Lignocellulosic Substrates for Cellulase Production

Enzymes produced by utilizing the pure substrate may not be very efficient for a bioprocess because there is limited activity of either cellulolytic or hemicellulolytic enzymes. On the other hand, lignocellulosic biomass contains both cellulose and hemicellulose and thus promotes production of a complete set of enzymes for biomass degradation. Nevertheless, use of purified substrates also significantly adds to the total cost of the production process. As a result, use of agricultural and household wastes instead of expensive synthetic fermentation media is one of the most sought-after approaches being employed for reducing enzyme production cost. Lignocellulosic biomass presents the cheapest and an attractive source for enzyme production due to its abundance as well as diversity. Moreover,

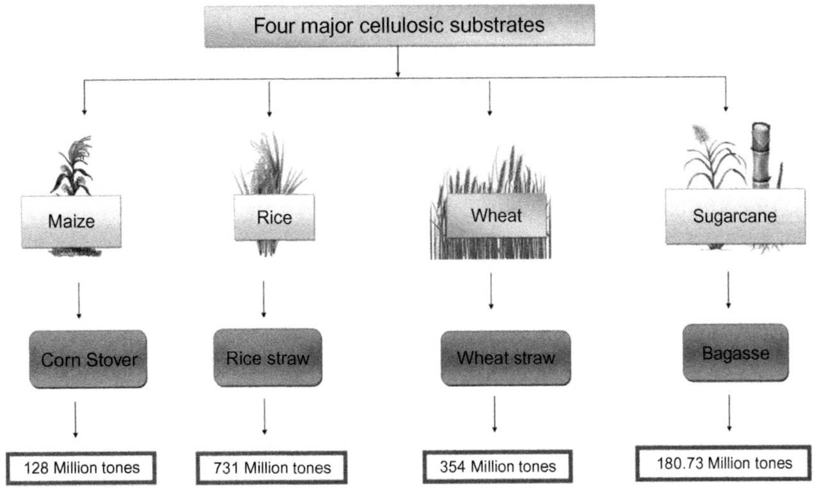

FIG. 2.1 Annual production of crop residues from four major agricultural crops (Saini et al., 2015a,b; Pandiyan et al., 2019).

TABLE 2.1 Pure Substrates for Cellulolytic Enzyme Production

Substrate	Microorganism	Incubation time	Enzyme activity	References
Cellulose 123 powder	*Penicillium janthinellum* NCIM 1171	10 days	FPase—2.3 IU/h/L, CMCase—111 IU/h/L, β-glucosidase—2.4 IU/h/L	(Adsul et al., 2004)
Birchwood xylan	*Trichoderma harzianum*	5 days	CMCase—0.58 IU/mL, FPase—0.64 IU/mL, β-glucosidase—0.54 IU/mL	(Ahmed et al., 2009)
Glucose	*T. harzianum*	5 days	CMCase—0.023 IU/mL, FPase—0.03 IU/mL, β-glucosidase—0.05 IU/mL	(Ahmed et al., 2009)
CMC	*T. harzianum*	5 days	CMCase—0.79 IU/mL, FPase—7.8 IU/mL, β-glucosidase—0.92 IU/mL	(Ahmed et al., 2009)
CMC	*Trichoderma viride*	6 days	CMCase—173 IU/mL	(Neethu et al., 2012)
CMC	*T. harzianum*	6 days	CMCase—150 IU/mL	(Rubeena et al., 2013)
α-Cellulose (1%)	*Phoma exigua* ITCC 2049	8 days	CMCase—37.00 IU/mL, β-glucosidase—3.5 IU/mL	(Tiwari et al., 2013)
CMC	*Aspergillus flavus*	5 days	CMCase—2793 IU/mL	(Sajith et al., 2014)
CMC	*Aspergillus terreus* MS105	150 h	CMCase—59 IU/mL	(Sohail et al., 2016)
Ball-milled cellulose	*A. terreus* MS105	150 h	FPase—0.184 IU/mL, CMCase—0.32 IU/mL, β-glucosidase—0.15 IU/mL	(Sohail et al., 2016)
Phosphoric acid swollen cellulose	*A. terreus* MS105	150 h	FPase—0.08 IU/mL, CMCase—0.12 IU/mL, β-glucosidase—0.07 IU/mL	(Sohail et al., 2016)
Filter paper	*A. terreus* MS105	150 h	FPase—0.15 IU/mL, CMCase—0.28 IU/mL, β-glucosidase—0.18 IU/mL	(Sohail et al., 2016)
Cellulose acetate	*A. terreus* MS105	150 h	FPase—0.10 IU/mL, CMCase—0.18 IU/mL, β-glucosidase—0.08 IU/mL	(Sohail et al., 2016)
Sigmacell	*A. terreus* MS105	150 h	FPase—0.11 IU/mL, CMCase—0.26 IU/mL, β-glucosidase—0.16 IU/mL	(Sohail et al., 2016)
Salicin	*A. terreus* MS105	150 h	FPase—0.22 IU/mL, CMCase—0.38 IU/mL, β-glucosidase—0.56 IU/mL	(Sohail et al., 2016)
Crystalline cellulose	*Clostridium thermocellum* ATCC 31924	120 h	FPase—0.33 U/mg protein, CMCase—7.12 U/mg protein	(Singh et al., 2018)
CMC	*Bacillus subtilis* MUS1	24 h	CMCase—566.66 U/mL	(Sreena and Sebastian, 2018)

CMCase, Carboxymethyl cellulase.

utilization of these substrates for enzyme production also helps in alleviating the problem of pollution caused by their disposal. Researchers have utilized a large number of lignocellulosic substrates for producing cellulases such as crop residues (corn stover, rice straw, wheat straw, rice hull, soybean hull, grasses, corn cob, etc.), agro-processing by-products (wheat bran, maize bran, sugarcane bagasse, olive pomace, oil cakes, corn steep liquor, distilled dried products), and food industry waste (fruit and vegetable peels). The amount of cellulases (exoglucanase, endoglucanase, and BG) or hemicellulases (xylanases) produced by fermenting strains on lignocellulosic substrate is determined by the proportion of cellulose and hemicellulose in raw substrate (Ghanem et al., 2000). The higher cellulose contents in grasses (32%–43%), sugarcane bagasse (35%–45%), and corn cob (38%–69%) make them potential substrates for cellulase production (Siqueira et al., 2013; Ververis et al., 2004). Furthermore, a lower lignin-to-cellulose ratio in lignocellulosic materials is also preferred for cellulase production because it makes cellulose in the substrate accessible and thereby induces cellulase production (Li et al., 2018). The cellulose, hemicellulose, and lignin content of various lignocellulosic materials is summarized in Table 2.2. The crystallinity of cellulose as well as amount of

TABLE 2.2 Composition of Different Crop Residues

Substrate	Cellulose (%)	Hemicellulose (%)	Lignin (%)	References
AGRICULTURAL BIOMASS				
Rye straw	33–35	27–30	16–19	(Rowell, 1992)
Oat straw	31–37	27–38	16–19	-do-
Barley straw	31–34	24–29	14–15	-do-
Bamboo	26–43	15–26	21–31	-do-
Elephant grass	22	24	23.9	-do-
Paper	85–99	0	0–15	(Howard et al., 2003)
Cotton seed hairs	80–95	5–20	0	-do-
Leaves	15–20	80–85	0	-do-
Corn stover	38	26	19	-do-
Nut shells	25–30	25–30	30–40	-do-
Bermuda grass	26–43	15–26	21–31	-do-
Rice straw	32.1	24	18	-do-
Corn cobs	45	35	15	-do-
Banana waste	13.2	14.8	14	-do-
Duckweed	30.4	23.6	1.5	(Li et al., 2018)
AGRO-PROCESSING BY PRODUCTS				
Coffee pulp	35	46.3	18.8	(Sánchez, 2009)
Rice bran	27	37	5	(Dhingra et al., 2012)
Wheat bran	6.8 g/kg	20.0 g/kg	5.6	(Sandberg et al., 1982)
Sugarcane bagasse	30.2	56.7	13.4	(El-Tayeb et al., 2012)
Sawdust	45.1	28.1	24.2	(El-Tayeb et al., 2012)
Sugar beet waste	26.3	18.5	2.5	(El-Tayeb et al., 2012)
Cotton stalks	58.5	14.4	21.5	(Nigam et al., 2009)
FOOD INDUSTRY WASTE				
Orange peel	9.2	10.5	0.84	(Rivas et al., 2008)
Pineapple peel	18.11	–	1.37	(Paepatung et al., 2009)
Potato peel waste	2.2	–	–	(Weshahy and Rao, 2012)
Coffee skin	23.77	16.68	28.58	(Lina et al., 2014)

lignin in these substrates varies resulting in differential expression of cellulase/hemicellulase enzymes as discussed in the next section.

Various agricultural residues have been used extensively by researchers for cellulase production from fungi. For instance, wheat straw (Jecu, 2000; Singh et al., 2009; Abo-State et al., 2010; Tiwari et al., 2015a), rice straw (Kang, 2004; Sherief et al., 2010; Singh et al., 2011, 2014; Saritha et al., 2016; Tiwari et al., 2017; Zhao et al., 2018), corn stover (Gao et al., 2008; Xu et al., 2009; Imran et al., 2017; Zhao et al., 2018), grass (Sohail et al., 2009), sugarcane bagasse (Tiwari et al., 2015b), sorghum straw (Mahajan et al., 2016), soybean hull (Ellila et al., 2017), and duckweed (Li et al., 2018) serve as very good substrates for growth as well as cellulase production from different fungi. Ellila et al. (2017) developed a genetically engineered *T. reesei*-based low-cost cellulase production process using soybean hull as substrate, which was far cheaper than the traditional production process using glucose as substrate.

Similarly, a number of agro-industrial by-products provide excellent substrate for fungal cellulase production. Eucalyptus wood chips (Machuca and Ferraz, 2001; Heidorne et al., 2006), pine wood chips (Heidorne et al., 2006), coir waste (Mrudula and Murugammal, 2011), and wheat bran (Jecu, 2000; Chandra et al., 2007; Deswal et al., 2011; Jain et al., 2015, 2017; Saini et al., 2015a; Tiwari et al., 2015a; Chakraborty et al., 2016) provide excellent as well as inexpensive growth and production media for cellulase production from fungi. Chakraborty et al. (2016) used statistical designs to optimize cellulase production under SSF in a cost-effective manner from *Trichoderma* sp. RCK65 grown on wheat bran. The authors calculated the total cost of crude enzyme as INR 5.311/1000 FPase units (Chakraborty et al., 2016).

On the other hand, the feasibility of various fruit and vegetable wastes to act as substrate for microbial cellulase production has also been tested by various research groups. In this regard, Sun et al. (2011) tested the potential of banana peel for cellulase production from *Trichoderma viride* GIM and achieved 5.56, 10.31, and 3.01 U/g of FPase, CMCase, and BG, respectively. Kannahi and Elangeswari (2015) evaluated the production of cellulase by *Aspergillus niger* and *T. viride* on pineapple peel, orange peel, and banana peel. The authors observed pineapple peel as the best substrate among the tested substrates for cellulase production from both the fungi used in the study. However, Philip et al. (2016) isolated cellulose-degrading microorganisms (seven bacterial and four fungi) from banana peel, citrus waste, potato peel, and cucumber peel. The isolated bacterial strains were identified as *Bacillus subtilis, Bacillus licheniformis, Streptococcus, Bacillus smithii, Bacillus firmus, Brevibacillus laterosporus,* and *Pseudomonas chlororaphis,* while the fungal isolates were identified as *A. niger, Penicillium, Aspergillus flavus,* and *Rhizopus* (Philip et al., 2016).

2.5 PRETREATMENTS OF LIGNOCELLULOSIC SUBSTRATE AND ITS IMPORTANCE IN ENHANCING ENZYME TITER

Use of raw biomass for enzyme production has many disadvantages such as low enzyme titer, induction of undesirable enzymes increasing the complexity of enzyme mixture, and binding of secreted cellulase with lignin, which interferes in downstream processing (Sridevi et al., 2008). Hence it has been suggested by various researchers that mild pretreatment of the biomass is always helpful in achieving higher enzyme titer (Table 2.3).

Pretreatment of lignocellulosic substrates helps to bring desirable physical, chemical, and conformational changes in substrates that help the microorganisms to grow profusely and hence give higher enzyme titers. Various physicochemical pretreatments (size reduction, steam pretreatment, steam explosion, alkali, mild acid, AFEX, etc.),

TABLE 2.3 Effect of Different Pretreatments on Cellulase Yield From Different Substrates

Substrate	Pretreatment	Microorganism	Enzyme yields		References
			Without pretreatment	With pretreatment	
56 V Lignocellulose	Steam pretreatment followed by washing	*Trichoderma reesei* C-30	FPase—1 IU/mL	FPase—3.7 IU/mL	(Chahal et al., 1982)
Spruce (*Picea abies*)	Steam pretreatment	*T. reesei* Rut C-30	FPase—0.48 IU/mL	FPase—0.79 IU/mL	(Szengyel et al., 2000)
Soybean hull	Steam pretreatment	*T. reesei*	FPase—0.75 IU/g-ds, CMCase—45 IU/g-ds	FPase—4 IU/g-ds, CMCase—7.29 IU/g-ds	(Brijwani and Vadlani, 2011)
Groundnut husk	2 M NaOH for 1 h	*Aspergillus niger + Trichoderma viride*	FPase—0.12 IU/mL/min, CMCase—0.06 IU/mL/min	FPase—0.04 IU/mL/min, CMCase—0.03 IU/mL/min	(Aliyu et al., 2017)

CMCase, Carboxymethyl cellulase.

biological pretreatments (fungi, bacteria, actinomycetes), and green solvents like ionic liquids have been reported and reviewed by various researchers that can be utilized for pretreatment of lignocellulosic biomass (Saritha and Arora, 2012; Den et al., 2018). Use of mild alkaline treatment, H_2O_2 treatment, or steam pretreatment positively affects the microbial growth and hence higher enzyme yields (Amin et al., 2017). However, pretreatment generates many inhibitory compounds like furfurals, hydroxymethyl, acetic acid, etc., which decreases the microbial growth and hence lowers enzyme titer. It has been suggested by different workers to wash the pretreated biomass before using it as a substrate for producing cellulase.

In this regard, Chahal et al. (1982) investigated the cellulase production potential of *T. reesei* Rut C-30 on glucose, pure cellulose, and a lignocellulosic wood material (56 V) and concluded that use of steam-pretreated lignocellulosic material gave the same productivity of cellulase as that of costly pure cellulose. Szengyel et al. (2000) studied cellulase production from *T. reesei* Rut C-30 using steam-pretreated spruce and reported that washed steam-pretreated spruce gave higher titer of FPase (0.79 IU/mL) compared to without washing (50% filtrate retention) of biomass (0.48 IU/mL).

Sridevi et al. (2008) reported increased cellulase production by *A. niger* on pretreated lignocellulosics (sawdust, wheat straw, sugarcane bagasse, and rice bran) than on the untreated substrates. Wang et al. (2010) used de-ashed and alkali-treated paper mill sludge for the production of a cellulase cocktail from *T. reesei* Rut C-30 and reported a high yield (307 FPU/g glucan of de-ashed sludge) of enzyme.

Interestingly, Brijwani and Vadlani (2011) observed that mild acid, alkali, and steam pretreatment of soybean hull resulted in its increase in crystallinity and bed porosity without disturbing the holocellulosic content. The authors found that the pretreated substrate when fermented with *T. reesei* gave 4 FPU/g, 0.6 IU/g BG, and 45 IU/g endocellulase, whereas the untreated one produced 0.75 FPU/g, 0.06 IU/g β-glucosidase, and 7.29 IU/g-ds endocellulase. Similarly, the authors grew *Aspergillus oryzae* on steam-pretreated soybean hulls and obtained 47.10 IU/g endocellulase compared to 30.82 IU/g in untreated soybean hulls (Brijwani and Vadlani, 2011). However, Aliyu et al. (2017) reported higher FPase and CMCase activities in biologically pretreated groundnut husk compared to alkali as well as untreated husk.

2.6 CELLULASE PRODUCTION METHODS

Microbial enzymes can be produced by either of the two broad fermentation techniques: SSF and SmF. SSF is the microbial growth solid support in the absence of any free-flowing water. SmF is a process involving the growth of microorganisms in liquid medium. Although most of the commercial enzyme-producing industries use the established SmF technology, SSF is also gaining importance due to its cost effectiveness (lower capital investment and operating expenses) and higher product yields, especially in the case of fungal fermentations. Production strategies of both SSF and SmF are depicted in Fig. 2.2, while the overall process of cellulase production is diagrammatically explained in Fig. 2.3.

SSF processes mimic the natural habitats of cellulose utilizing aerobic microorganisms (ascomycetes, basidiomycetes, and deuteromycetes) and as a result give better growth and metabolite production conditions to the organism

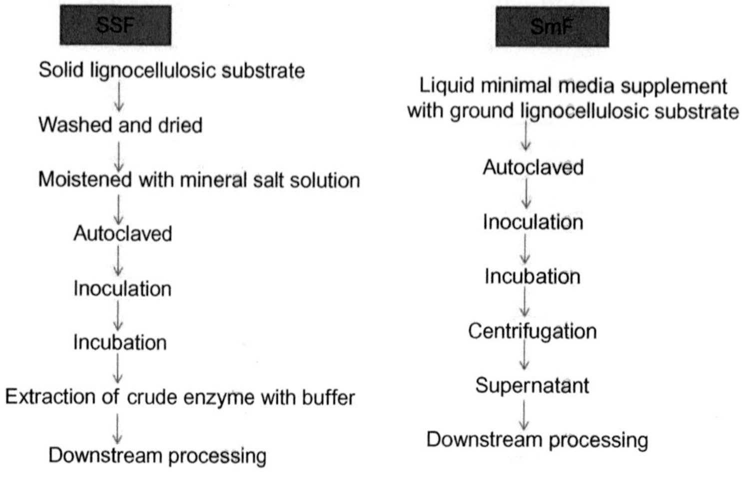

FIG. 2.2 Enzyme production from cellulosic feedstocks using solid substrate fermentation and submerged fermentation technology.

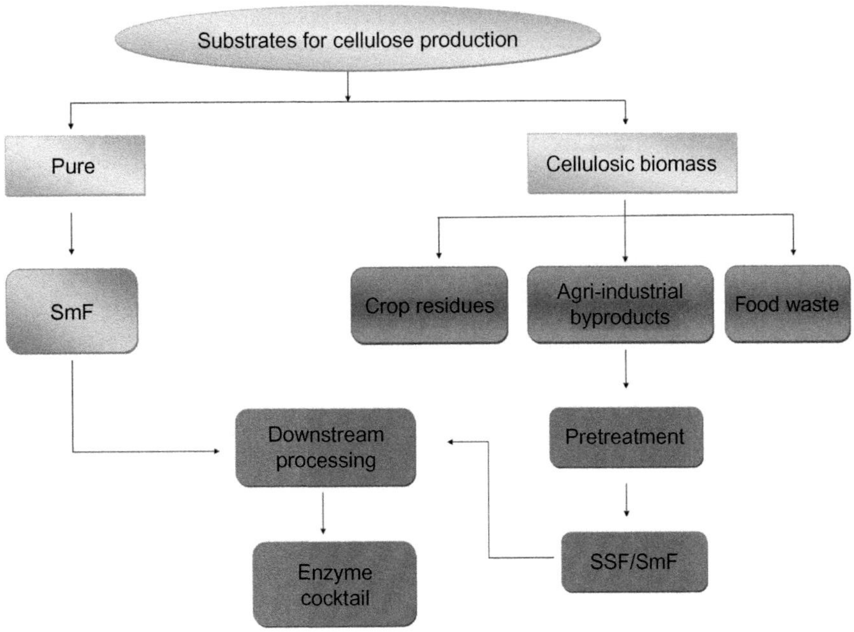

FIG. 2.3 Overall process of cellulase production.

(Kuhad et al., 2016). Since cellulase is usually produced from aerobic fungi on a commercial scale, SSF can prove to be a better technology if scaled up appropriately. The success of SSF is dependent on various factors as well as their interaction. These include selection of a suitable microbe and its substrate for growth, and optimizing culture conditions followed by downstream processing to transfer laboratory-scale fermentation to commercial bioreactors (Hansen et al., 2015). The commonly used bioreactors such as tray, packed bed, fluidized, and horizontal drum bioreactors work well in SSF but bear issues of heat transfer. Many recent designs of bioreactor have been developed to improve the scale-up of SSF technology. A new design of SSF reactor based on air pressure pulsation and internal circulation was found to improve heat and mass transfers (Hongzhang et al., 2002). Various bioreactor designs for SSF have been very well reviewed (Durand, 2003). The problem of heat transfer can also be solved by mathematical modeling, which helps to optimize temperature-to-moisture ratio relevant to microbial growth as well as product formation (Mitchell and Meien, 2000; Philippidis et al., 1992; Sangsurasak and Mitchell, 1998). Cellulase production under both SSF and SmF by various research groups is listed in Table 2.4.

TABLE 2.4 Cellulase Production Under Submerged Fermentation (SmF) and Solid Substrate Fermentation (SSF)

Microbial source	Substrate	Production method	Incubation time	Enzyme activity	References
Aspergillus niger	Wheat bran and wheat straw	SSF	96 h	CMCase—14.8 IU/g	(Jecu, 2000)
Trametes versicolor	Eucalyptus wood chips	SSF	15 days	FPase—4 U/culture, β-glucosidase—3.8 U/culture	(Machuca and Ferraz, 2001)
A. niger KK2	Rice straw	SSF	4 days	CMCase—130 IU/g, FPase—19.5 IU/g	(Kang, 2004)
Ceriporopsis subvermispora	Eucalyptus wood chips	SSF	30 days	CMCase—32.1 U/kg, β-glucosidase— 43.6 U/kg	(Heidorne et al., 2006)
C. subvermispora	Pine wood chips	SSF	30 days	CMCase—15.3 U/kg, β-glucosidase—53 U/kg	(Heidorne et al., 2006)
Trichoderma reesei NRRL 11460	Sugarcane bagasse	SSF	72 h	FPase—154.6 U/g	(Singhania et al., 2006)
A. niger	Wheat bran	SSF	3 days	FPase—2.09 U/g, CMCase—1.36 U/g	(Chandra et al., 2007)
Aspergillus terreus M11	Corn stover	SSF	96 h	CMCase—581 U/g, FPase—243 U/g, β-glucosidase—128 U/g	(Gao et al., 2008)

Continued

TABLE 2.4 Cellulase Production Under Submerged Fermentation (SmF) and Solid Substrate Fermentation (SSF)—cont'd

Microbial source	Substrate	Production method	Incubation time	Enzyme activity	References
Phanerochaete chrysosporium	Corn fiber	SSF	3 days	CMCase—3.42 (μg glucose) (mg protein/min), FPase—0.23 (μg glucose) (mg protein/min)	(Shrestha et al., 2008)
Aspergillus heteromorphus	Wheat straw	SmF	7 days	FPase—13.5 IU/mL	(Singh et al., 2009)
A. niger MS82	Grass	SmF	6 days	CMCase—6.8 U/mL	(Sohail et al., 2009)
Irpex lacteus	Corn stover	SSF	5 days	CBH—69.3 U/g culture, CMCase—2.5 U/g culture, β-glucosidase—11.1 U/g culture	(Xu et al., 2009)
Aspergillus sp. NAM-F35	Wheat straw	SSF	7 days	CMCase—487 U/mL, FPase—79 U/mL	(Abo-State et al., 2010)
Aspergillus fumigatus	Rice straw and wheat bran	SSF	4 days	CMCase—14.7 IU/g, β-glucosidase—8.51 IU/g, FPase—0.93 U/g	(Sherief et al., 2010)
Fomitopsis sp.	Wheat bran	SSF	11–16 days	CMCase—71.69 IU/g, FPase—3.492 IU/g, β-glucosidase—53.67 IU/g	(Deswal et al., 2011)
Rhizopus oryzae	Water hyacinth	SmF	48 h	CMCase—495 U/mL, β-glucosidase—137.32 U/mL	(Karmakar and Ray, 2011)
A. niger USMA1	Sugarcane bagasse and palm kernel cake	SSF	6 days	FPase—3.4 U/g	(Lee et al., 2011)
A. niger	Coir waste	SSF	72 h	CMCase—8.89 U/g, FPase—3.56 U/g	(Mrudula and Murugammal, 2011)
Aspergillus awamori F18	Paddy straw	SSF	8 days	CMCase—87.56 IU/g, FPase—14.59 IU/g	(Singh et al., 2011)
Streptomyces sp. ssr-198	Paddy straw (1.8%)	SmF	96 h	Xylanase—100.3 U/mL, endoglucanase—1.56 U/mL	(Singh et al., 2014)
Thermoascus auranticus RCKK	Wheat bran	SSF	72 h	CMCase—88 IU/g, FPase—15.8 IU/g, β-glucosidase—25.3 IU/g	(Jain et al., 2015)
A. niger	Pineapple peel	SmF	48 h	0.92 IU/mL cellulase	(Kannahi and Elangeswari, 2015)
Trichoderma viride	Pineapple peel	SmF	48 h	0.87 IU/mL cellulase	(Kannahi and Elangeswari, 2015)
A. niger SH3	Wheat bran:wheat straw (1.64:1.36)	SSF	7 days	CMCase—655.94 IU/g, β-glucosidase—540.57 IU/g, FPase—31.48 IU/g, and xylanase—1951.63 IU/g	(Tiwari et al., 2015a)
Pseudomonas lutea BG8	Sugarcane bagasse (2.99%)	SmF	96 h	β-Glucosidase—23.29 IU/g	(Tiwari et al., 2015b)
Penicillium oxalicum	Wheat bran	SmF	4 days	FPase—1.2 U/mL	(Saini et al., 2015a)
Trichoderma atroviride	Vegetable waste	SSF	5 days	CMCase—178.92 U/g, FPase—98.89 U/g	(Bairagi, 2016)
Trichoderma sp. RCK65	Wheat bran	SSF	72 h	CMCase—154 IU/g, FPase—46.9 IU/g, β-glucosidase—144.1 U/g	(Chakraborty et al., 2016)
Malbranchea cinnamomea	Sorghum straw (5 g/mL)	SSF	3–7 days	CMCase—304 U/g-ds, β-glucosidase—234.18 U/g-ds	(Mahajan et al., 2016)
A. terreus CM20	Paddy straw (1.42 g/L) and wheat straw (1.58 g/L)	SSF	7 days	CMCase—365.0318 IU/g, FPase—161.48 IU/g, xylanase—9627.79 IU/g	(Saritha et al., 2016)
Banana peel	*T. viride* GIM	SSF	144 h	CMCase—10.31 IU/g, FPase—5.56 IU/g, β-glucosidase—3.01 U/g	(Sun et al., 2011)

TABLE 2.4 Cellulase Production Under Submerged Fermentation (SmF) and Solid Substrate Fermentation (SSF)—cont'd

Microbial source	Substrate	Production method	Incubation time	Enzyme activity	References
T. reesei	Soybean hull	SSF	96 h	19.3 g/L	(Ellila et al., 2017)
Aspergillus tubingenesis	Corn stover	SSF	4 days	Cellulase—112 μg/mL/min	(Imran et al., 2017)
T. auranticus RCKK	Wheat bran	SSF	72 h	CMCase—102 IU/g, FPase—20.24 IU/g, β-glucosidase—27.32 IU/g	(Jain et al., 2017)
Bacillus subtilis RA 10	Paddy straw (1.24%)	SmF	72 h	β-Glucosidase—57.31 IU/g	(Tiwari et al., 2017)
Mixed culture of *T. reesei* and *A. niger*	NaOH-treated corn stover NaOH-treated rice straw	SmF SmF	5 days	FPase—3.63 U/mL FPase—2.56 U/mL	(Zhao et al., 2018)

CMC, carboxy methyl cellulase; CMCase, Carboxymethyl cellulase.

2.7 PROTEOMIC APPROACHES FOR STUDYING CELLULASE PRODUCTION FROM DIFFERENT FEEDSTOCKS

It is a well-known fact that the biological transformation of plant biomass to soluble sugars is the major blockage in the development of lignocellulose bioenergy, being robustly associated with economical and efficient cellulase production. The conventional colorimetric quantification of cellulose-degrading enzymes in the secretome of potential cellulolytic microbes endures several limitations such as detection sensitivity and reagent cross-reactivity (Manavalan et al., 2011). Alternatively, transcriptome profiling and proteomic techniques have made it possible to recognize a varied set of proteins from composite biological samples. In this regard, Manavalan et al. (2011) quantified the secretome of *P. chrysosporium* grown in cellulose, lignin, and a mixture of cellulose and lignin with isobaric tag for relative and absolute quantification (iTRAQ)-based proteomics using liquid chromatography tandem mass spectrometry. The authors quantified 117 enzymes consisting of cellulose-hydrolyzing, hemicellulose-hydrolyzing, pectin-degrading, and lignin-degrading enzymes. It was found that in culture conditions of cellulose and cellulose + lignin, enzymes such as endoglucanases, BGs, and CBHs were appreciably upregulated, and the iTRAQ data proposed oxidative and hydrolytic degradation of cellulose. However, when only lignin was used in the culture medium, enzymes such as copper radical oxidases, oxidoreductases, alcohol dehydrogenases, and aryl alcohol oxidases were significantly upregulated. The study suggests the role of substrate in inducing the expression of particular enzymes (Manavalan et al., 2011). Similarly, Adav et al. (2011) observed considerable induction in the expression of some enzymes in the tested medium (cellulosic) compared to the noncellulosic medium using iTRAQ-based quantitative proteomic analysis of the secretome of *Thermobifida fusca*. The authors concluded the existence of metabolic pathways induced by cellulose during its utilization (Adav et al., 2011). Furthermore, the secretome of *T. fusca* in SSF of different lignocellulosic substrates (corn stover, hay, sawdust, sugarcane bagasse, wood chips, and undried green plants) was studied by Adav et al. (2012). The study observed increased quantitative expression of 29 glycoside hydrolases when different lignocellulosic substrates were used as growth substrates for the fungus when compared with cellulose as the growth medium. Interestingly, Bischof et al. (2013) compared the transcriptome of *T. reesei* grown on wheat straw and lactose as carbon sources. The authors observed enhanced gene expression on wheat straw compared to that on lactose, with 1619 genes induced only on wheat straw and not on lactose. These genes comprised of 30% CAZome (carbohydrate-active enzymes). Likewise, Liu et al. (2013) used iTRAQ-based quantitative proteomics to study the secretome of *Aspergillus fumigatus* Z5 and concluded that most of the cellulases, hemicellulases, and glycoside hydrolases are considerably upregulated when wheat straw and avicel are used as carbon sources. Peciulyte et al. (2014) studied the mass spectrometry-based protein profiling of crude cellulase extract of *T. reesei* Rut C-30 grown on five different cellulosic substrates with minute differences in their chemical composition and observed considerable differences in the enzyme production profiles on different feedstocks. While, Singh et al. (2015) characterized the secretome of a novel *Streptomyces* sp. grown on paddy straw and reported a total of 80 proteins: 17 glycoside hydrolases, 17 proteases, 3 polysaccharide lyases, 2 esterases, and 14 hypothetical proteins. In the same way, Saritha et al. (2016) identified 63 proteins in the secretome of *Aspergillus terreus* CM20 grown on paddy straw under SSF conditions. On the other hand, holocellulase from *A. niger* SH3 was characterized by

Kumar et al. (2017) and it was found to contain 125 proteins, including cellulases (26), hemicellulases (21), chitinases (10), esterases (6), amylases (4), and hypothetical proteins (32). Ogunmolu et al. (2018) studied spatiotemporal variation in the secretome of *Penicillium funiculosum* when grown on four different cellulosic polymers: avicel, wheat bran, ammonium-treated wheat straw and avicel, and glucose during early as well as late log phases of growth. The most abundant proteins present in the secretome for avicel and wheat bran were CBHI, CBHII, BG, arabinofuranosidase, and β-xylosidase. It was also found that the proteins were secreted in two phases, so that the early ones hydrolyze composite substrates before the late ones act on specific substrates in the culture medium.

Keeping the foregoing in mind, it can be summarized that microorganisms secrete unique extracellular proteins for the consumption of utilizable carbon in the growth medium. Therefore the selection of suitable substrate for cellulose production is the major deciding factor in securing high enzyme yields from the potential microbe.

2.8 COMMERCIAL FORMULATION OF CELLULASES IN THE MARKET

Cellulases being an important hydrolytic enzyme are produced commercially by a number of companies listed in Table 2.5. Various formulations of cellulase are available in the market, for example, Novozyme 188, Accelerase 1000, Accelerase 1500 from Sigma Aldrich, Cellic CTech2, Cellic CTech3 from Novozymes, and others such as Celluclast produced from *T. reesei*. Use of these enzymes confers several advantages such as greater process and substrate versatility, scope to optimize biomass-to-sugar conversion processes such as increasing saccharification efficiency, lower enzyme loading, increasing substrate loading, reduction in time of conversion, and also help in reducing severity of different pretreatment methods.

2.9 CONCLUSION

Depletion of fossil fuels resulting in greater than ever petrol prices as well as the pollution caused by their burning has led to an escalating demand for cellulases to be used for biofuel and biochemical production. However, the need for more vigorous enzymes and the cost of cellulases are the weakest links in the commercialization of future sustainable biorefineries. The market trend is also following this change with companies like Merck and Novozyme providing enzymatic cocktails for biomass degradation. Apart from modern technologies to genetically engineer the fermenting biocatalyst, enzyme cost can alternatively be reduced by choosing inexpensive substrates that maximize enzyme production. Studies on cellulase production using pure cellulosic substrates, which are very expensive and do not support production of the full cocktail of enzymes for lignocellulose degradation, are mainly

TABLE 2.5 Commercial Formulations of Cellulases

			Cellulase activities	
Commercial enzyme	**Company**	**Microorganism**	**FPU/mL**	**FPU/mg**
Accelerase 1500	Dupont (Genencor) (Rochester, NY, USA)	*Trichoderma reesei*	57	0.50
Cellic Ctec2	Merck, NJ, USA	–	119	0.46
Cellic Ctec3	Novozymes (Franklinton, NC, USA)	–	196	0.75
Cellulase	Worthington Biochemical Corporation, NJ, USA	*T. reesei*	–	45
Cellulase	MP Biomedicals, CA, USA	*Aspergillus niger*	–	60
Cellulase	Creative Enzymes, NY, USA	*A. niger*		0.3
Cellulase	Creative Enzymes, NY, USA	*Trichoderma* sp.	–	5
Cellulase (Onozuka R-10)	Bioworld, Dublin, USA	*Trichoderma viride*	–	1.0
Cellulase AP3	Amano Enzyme Inc., Japan	–	–	1.2–1.8
Cellulase DS	Amano Enzyme, USA	*A. niger*	–	
Cytolase CL	DSM (Seclin, France)	–	117	0.82
Novozyme 188	Novozymes (Franklinton, NC, USA)	*A. niger*	487	–

aimed at studying the aspects of the production mechanism of cellulolytic enzymes and other enzyme properties. Nevertheless, mass production of fungal cellulases in a cost-effective manner is possible only when inexpensive and abundant lignocellulosic biomass is used as substrate in SSF processes. However, designing of efficient SSF bioreactors coupled with mathematical modeling remains a major challenge to further enhance the cost effectiveness of enzyme production processes. The future of enzyme technology lies in full integration of educational, political, and industrial efforts as the key drivers to apply and develop new fundamental knowledge to upscale the production process.

References

Abo-State, M.A.M., Hammad, A.I., Swelim, M., Gannam, R.B., 2010. Enhanced production of cellulase(s) by *Aspergillus* spp. isolated from agriculture wastes by solid state fermentation. Environ. Sci. 9.

Adav, S.S., Ng, C.S., Sze, S.K., 2011. iTRAQ-based quantitative proteomic analysis of *Thermobifida fusca* reveals metabolic pathways of cellulose utilization. J. Proteome 74, 2112–2122.

Adav, S.S., Cheow, E.S.H., Ravinandran, A., Dutta, B., Sze, S.K., 2012. Label free quantitative proteomic analysis of secretome by *Thermobifida fusca* on different lignocellulosic biomass. J. Proteome 75, 3694–3706.

Adsul, M., Ghule, J., Singh, R., Shaikh, H., Bastawde, K., Gokhale, D., Varma, A., 2004. Polysaccharides from bagasse: applications in cellulase and xylanase production. Carbohydr. Polym. 57, 67–72. https://doi.org/10.1016/j.carbpol.2004.04.001.

Ahmed, S., Bashir, A., Saleem, H., Saadia, M., Jamil, A., 2009. Production and purification of cellulose-degrading enzymes from a filamentous fungus *Trichoderma harzianum* 9. Pak. J. Bot. 41 (3), 1411–1419.

Ali, B.R.S., Zhou, L., Graves, F.M., Freedman, R.B., Black, G.W., Gilbert, H.J., Hazlewood, G.P., 1995. Cellulases and hemicellulases of the anaerobic fungus *Piromyces* constitute a multiprotein cellulose-binding complex and are encoded by multigene families. FEMS Microbiol. Lett. 125, 15–22.

Aliyu, N., Kassim, H., Aliyu, H., Abu, R., Adamu, Z., 2017. Comparative production of cellulase from chemical and biological pretreated groundnut husk. Adv. Res. 10, 1–7.

Amin, F.R., Khalid, H., Zhang, H., Rahman, S.U., Zhang, R., Liu, G., Chen, C., 2017. Pretreatment methods of lignocellulosic biomass for anaerobic digestion. AMB Express 7, 72.

Asem, D., Leo, V.V., Passari, A.K., Tonsing, M.V., Joshi, J.B., Uthandi, S., Hashem, A., Abd Allah, E.F., Singh, B.P., 2017. Evaluation of gastrointestinal bacterial population for the production of holocellulose enzymes for biomass deconstruction. PLoS One 12, e0186355.

Bairagi, S., 2016. Optimization of Cellulase enzyme from vegetable waste by using *Trichoderma atroviride* in solid State fermentation. IOSR J. Environ. Sci. Toxicol. Food Technol. 10, 68–73.

Bayer, E.A., Chanzy, H., Lamed, R., Shoham, Y., 1998. Cellulose, cellulases and cellulosomes. Curr. Opin. Struct. Biol. 8, 548–557.

Bayer, E.A., Belaich, J.-P., Shoham, Y., Lamed, R., 2004. The cellulosomes: multienzyme machines for degradation of plant cell wall polysaccharides. Annu. Rev. Microbiol. 58, 521–554.

Bischof, R., Fourtis, L., Limbeck, A., Garnauf, C., Seiboth, B., Kubicek, C.P., 2013. Comparative analysis of the *Trichoderma reesei* transcriptome during growth on the cellulase inducing substrates wheat straw and lactose. Biotechnol. Biofuels 6, 127.

Brijwani, K., Vadlani, P.V., 2011. Cellulolytic enzyme production via solid state fermentation: effect of pretraetament methods on physicochemical characteristics of substrate. Enzyme Res. 2011.

Cantarel, B.L., Coutinho, P.M., Rancurel, C., Bernard, T., Lombard, V., Henrissat, B., 2009. The Carbohydrate-Active EnZymes database (CAZy): an expert resource for glycogenomics. Nucleic Acids Res. 37, D233–D238.

Chahal, D.S., McGuire, S., Pikor, H., Noble, G., 1982. Production of cellulase complex by *Trichoderma reesei* rut-C30 on lignocellulose and its hydrolytic potential. Biomass 2, 127–137.

Chakraborty, S., Gupta, R., Jain, K.K., Kuhad, R.C., 2016. Cost-effective production of cellulose hydrolysing enzymes from Trichoderma sp. RCK65 under SSF and its evaluation in saccharification of cellulosic substrates. Bioprocess Biosyst. Eng. 39, 1659–1670.

Chandra, M.S., Viswanath, B., Reddy, B.R., 2007. Cellulolytic enzymes on lignocellulosic substrates in solid state fermentation by *Aspergillus niger*. Indian J. Microbiol. 47, 323–328.

Choudhary, J., Singh, S., Nain, L., 2016. Thermotolerant fermenting yeasts for simultaneous saccharification fermentation of lignocellulosic biomass. Electron. J. Biotechnol. 21, 82–92.

Davies, G., Henrissat, B., 1995. Structures and mechanisms of glycosyl hydrolases. Structure 3, 853–859.

Den, W., Sharma, V.K., Lee, M., Nadadur, G., Varma, R.S., 2018. Lignocellulosic biomass transformations via greener pretreatment processes: access to energy and value-added chemicals. Front. Chem. 6, 141.

Deswal, D., Khasa, Y.P., Kuhad, R.C., 2011. Optimization of cellulase production by a brown rot fungus *Fomitopsis* sp. RCK2010 under solid state fermentation. Bioresour. Technol. 102, 6065–6072.

Dhingra, D., Chopra, S., Rai, D.R., 2012. Stabilization of raw rice bran using ohmic heating. Agribiol. Res. 1, 392–398.

Durand, A., 2003. Bioreactor designs for solid state fermentation. Biochem. Eng. J. 13, 113–125.

Ellila, S., Fonseca, L., Uchima, C., Cota, J., Goldman, G.H., Saloheima, M., Sacon, V., Siika-aho, M., 2017. Development of a low cost cellulase production process using *Trichoderma reesei* for Brazilian biorefineries. Biotechnol. Biofuels 10, 30.

El-Tayeb, T.S., Abdelhafez, A.A., Ali, S.H., Ramadan, E.M., 2012. Effect of acid hydrolysis and fungal biotreatment on agro-industrial wastes for obtainment of free sugars for bioethanol production. Braz. J. Microbiol. 43, 1523–1535.

Gao, J., Weng, H., Zhu, D., Yuan, M., Guan, F., Xi, Y., 2008. Production and characterization of cellulolytic enzymes from the thermoacidophilic fungal *Aspergillus terreus* M11 under solid-state cultivation of corn stover. Bioresour. Technol. 99, 7623–7629.

Ghanem, N.B., Yusef, H.H., Mahrouse, H.K., 2000. Production of *Aspergillus terreus* xylanase in solid-state cultures: application of the Plackett–Burman experimental design to evaluate nutritional requirements. Bioresour. Technol. 73, 113–121.

Gupta, R., Lee, Y.Y., 2009. Mechanism of cellulase reaction on pure cellulosic substrates. Biotechnol. Bioeng. 102, 1570–1581.

Hansen, G.H., Lübeck, M., Frisvad, J.C., Lübeck, P.S., Andersen, B., 2015. Production of cellulolytic enzymes from ascomycetes: comparison of solid state and submerged fermentation. Process Biochem. 50, 1327–1341.

Heidorne, F.O., Magalhães, P.O., Ferraz, A.L., Milagres, A.M.F., 2006. Characterization of hemicellulases and cellulases produced by *Ceriporiopsis subvermispora* grown on wood under biopulping conditions. Enzym. Microb. Technol. 38, 436–442.

Hishida, A., Suzuki, T., Iijima, T., Higaki, M., 1999. Production of extracellular cellulolytic enzymes of the brown-rot fungus, *Tyromyces palustris* under illuminated condition: cellulase and beta-D-glucosidase. J. Agric. Sci. Tokyo Nogyo Daigaku Jpn 43, 186–196.

Hodrová, B., Kopecný, J., Kás, J., 1998. Cellulolytic enzymes of rumen anaerobic fungi *Orpinomyces joyonii* and *Caecomyces communis*. Res. Microbiol. 149, 417–427.

Hongzhang, C., Fujian, X., Zhonghou, T., Zuohu, L., 2002. A novel industrial-level reactor with two dynamic changes of air for solid-state fermentation. J. Biosci. Bioeng. 93, 211–214.

Howard, R.L., Abotsi, E., Jansen, v.R.E.L., Howard, S., 2003. Lignocellulose biotechnology: issues of bioconversion and enzyme production. Afr. J. Biotechnol. 2, 602–619.

Hulme, M., Stranks, D., 1974. Influence of carbohydrate accessibility on cellulase production by *Lenzites trabea* and *Polyporus versicolor*. Aust. J. Biol. Sci. 27, 457.

Imran, M., Anwar, Z., Irshad, M., Javid, A., Hussain, A., Ali, S., 2017. Optimization of cellulase production from a novel strain of *Aspergillus Tubingensis* IMMIS2 through response surface methodology. Biocatal. Agric. Biotechnol. 12, 191–198.

Jain, K.K., Bhanja, T.D., Kumar, S., Kuhad, R.C., 2015. Production of thermostable hydrolases (cellulases and xylanase) from *Thermoascus auranticus* RCKK: a potential fungus. Bioprocess Biosyst. Eng. 38, 787–796.

Jain, K.K., Kumar, S., Deswal, D., Kuhad, R.C., 2017. Improved production of thermostable cellulase from Thermoascus auranticus RCKK by fermentation biopreocessing and its application in the hydrolysis of office waste paper, algal pulp and biologically treated wheat straw. Appl. Biochem. Biotechnol. 181, 784–800.

Jecu, L., 2000. Solid state fermentation of agricultural wastes for endoglucanase production. Ind. Crop. Prod. 11, 1–5.

Johnson, E., 2016. Integrated enzyme production lowers the cost of cellulosic ethanol: integrated enzyme production lowers the cost of cellulosic ethanol. Biofuels Bioprod. Biorefin. 10, 164–174.

Kang, S., 2004. Production of cellulases and hemicellulases by *Aspergillus niger* KK2 from lignocellulosic biomass. Bioresour. Technol. 91, 153–156.

Kannahi, M., Elangeswari, S., 2015. Enhanced production of cellulase on different fruit Peel under submerged fermentation. Int. J. Pharm. Sci. Rev. Res. 32, 161–165.

Karmakar, M., Ray, R.R., 2011. A statistical approach for optimization of simultaneous production of β-glucosidase and endoglucanase by *Rhizopus oryzae* from solid-state fermentation of water hyacinth using central composite design. Biotechnol. Res. Int. 2011, 1–7.

Kim, D., 2018. Physico-chemical conversion of lignocellulose: inhibitor effects and detoxification strategies: a mini review. Molecules 23, 309.

Ko, C., Chen, W., Tsai, C., Jane, W., Liu, C., Tu, J., 2007. *Paenibacillus campinasensis* BL11: a wood material-utilizing bacterial strain isolated from black liquor. Bioresour. Technol. 98, 2727–2733.

Kuhad, R.C., Kumar, M., Singh, A., 1994. A hypercellulolytic mutant of *Fusarium oxysporum*. Lett. Appl. Microbiol. 19, 397–400.

Kuhad, R.C., Singh, A., Eriksson, K.-E.L., 1997. Microorganisms and enzymes involved in the degradation of plant fiber cell walls. In: Biotechnology in the Pulp and Paper Industry, Advances in Biochemical Engineering/Biotechnology. Springer, Berlin, Heidelberg, pp. 45–125.

Kuhad, R.C., Gupta, R., Khasa, Y.P., Singh, A., Zhang, Y.-H.P., 2011. Bioethanol production from pentose sugars: current status and future prospects. Renew. Sust. Energ. Rev. 15, 4950–4962.

Kuhad, R.C., Deswal, D., Sharma, S., Bhattacharya, A., Jain, K.K., Kaur, A., Pletschke, B.I., Singh, A., Karp, M., 2016. Revisiting cellulase production and redefining current strategies based on major challenges. Renew. Sust. Energ. Rev. 55, 249–272.

Kumar, A., Singh, S., Tiwari, R., Goel, R., Nain, L., 2017. Immobilization of indigeneous holocellulase on iron oxide (Fe_2O_3) nanoparticles enhanced hydrolysis of alkali pretreated paddy straw. Int. J. Biol. Macromol. 96, 538–549.

Lee, C.K., Darah, I., Ibrahim, C.O., 2011. Production and optimization of cellulase enzyme using *Aspergillus niger* USM AI 1 and comparison with *Trichoderma reesei* via solid state fermentation system. Biotechnol. Res. Int. 2011, 1–6.

Leis, B., Held, C., Bergkemper, F., Dennemarck, K., Steinbauer, R., Reiter, A., Mechelke, M., Moerch, M., Graubner, S., Liebl, W., Schwarz, W.H., Zverlov, V.V., 2017. Comparative characterization of all cellulosomal cellulases from *Clostridium thermocellum* reveals high diversity in endoglucanase product formation essential for complex activity. Biotechnol. Biofuels 10.

Li, C., Feng, J., Bai, X., Chen, S., Zhang, D., He, R., Li, D., 2018. Duckweed (*Lemna minor*) is a novel natural inducer of cellulase production in *Trichoderma reesi*. Peer J. Preprints 6, e26529v1.

Li, D.C., Li, A.N., Papageorgiou, A.C., 2011. Cellulases from thermophilic fungi: recent insights and biotechnological potential. Enzyme Res. 2011, 1–9.

Liguori, R., Ventorino, V., Pepe, O., Faraco, V., 2016. Bioreactors for lignocellulose conversion into fermentable sugars for production of high added value products. Appl. Microbiol. Biotechnol. 100, 597–611.

Lina, F., Ballesteros, J.A., Teixeira, S.I., Mussatto, 2014. Chemical, functional, and structural properties of spent coffee grounds and coffee silverskin. Food Bioprocess Technol. 7, 3493–3503.

Liu, D., Li, J., Zhao, S., Zhang, R., Wang, M., Miao, Y., Shen, Y., Shen, Q., 2013. Secretome diversity and quantitative analysis of cellulolytic *Aspergillus fumigates* Z5 in the presence of different carbon sources. Biotechnol. Biofuels 149.

Lo, Y.-C., Lu, W.-C., Chen, C.-Y., Chen, W.-M., Chang, J.-S., 2010. Characterization and high-level production of xylanase from an indigenous cellulolytic bacterium *Acinetobacter junii* F6-02 from southern Taiwan soil. Biochem. Eng. J. 53, 77–84.

Machuca, A., Ferraz, A., 2001. Hydrolytic and oxidative enzymes produced by white- and brown-rot fungi during *Eucalyptus grandis* decay in solid medium. Enzym. Microb. Technol. 29, 386–391.

Mahajan, C., Basotra, N., Singh, S., Di Falco, M., Tsang, A., Chadha, B.S., 2016. *Malbranbranchea cinnamomea*: a thermophilic fungal source of catalytically efficient lignocellulolytic glycosyl hydrolases and metal dependent enzymes. Bioiresour. Technol. 200, 53–63.

Maki, M., Leung, K.T., Qin, W., 2009. The prospects of cellulase-producing bacteria for the bioconversion of lignocellulosic biomass. Int. J. Biol. Sci. 500–516.

Manavalan, A., Adav, S.S., Sze, S.K., 2011. iTRAQ-based quantitative secretome analysis of *Phanerochaete chrysosporium*. J. Proteome 75, 642–654.

Mathew, G.M., Sukumaran, R.K., Singhania, R.R., Pandey, A., 2008. Progress in research on fungal cellulases for lignocellulose degradation. J. Sci. Ind. Res. 67, 898–907.

Mejia-Castillo, T., Hidalgo-Lara, M.E., Brieba, L.G., Ortega-Lopez, J., 2008. Purification, characterization and modular organization of a cellulose-binding protein, CBP105, a processive β-1,4-endoglucanase from *Cellulomonas flavigena*. Biotechnol. Lett. 30, 681–687.

Mitchell, D.A., Meien, O.F.V., 2000. Mathematical modeling as a tool to investigate the design and operation of the zymotis packed-bed bioreactor for solid-state fermentation. Biotechnol. Bioeng. 68, 127–135.

Mrudula, S., Murugammal, R., 2011. Production of cellulase by *Aspergillus niger* under submerged and solid state fermentation using coir waste as a substrate. Braz. J. Microbiol. 42, 1119–1127.

Muñoz, I.G., Ubhayasekera, W., Henriksson, H., Szabó, I., Pettersson, G., Johansson, G., Mowbray, S.L., Ståhlberg, J., 2001. Family 7 cellobiohydrolases from *Phanerochaete chrysosporium*: crystal structure of the catalytic module of Cel7D (CBH58) at 1.32 A resolution and homology models of the isozymes. J. Mol. Biol. 314, 1097–1111.

Murray, P.G., Grassick, A., Laffey, C.D., Cuffe, M.M., Higgins, T., Savage, A.V., Planas, A., Tuohy, M.G., 2001. Isolation and characterization of a thermostable endo-β-glucanase active on 1,3-1,4-β-d-glucans from the aerobic fungus *Talaromyces emersonii* CBS 814.70. Enzym. Microb. Technol. 29, 90–98.

Neethu, K., Rubeena, M., Sajith, S., Sreedevi, S., Priji, P., Unni, K.N., Josh, M.K.S., Jisha, V.N., Pradeep, S., Benjamin, S., 2012. A novel strain of *Trichoderma viride* shows complete lignocellulolytic activities. Adv. Biosci. Biotechnol. 03, 1160–1166.

Nigam, P.S., Gupta, N., Anthwal, A., 2009. Pre-treatment of agro-industrial residues. In: Nigam, P.S., Pandey, A. (Eds.), Biotechnology for Agro-Industrial Residues Utilization. 1st edn. Springer, Netherlands, pp. 13–33.

Nyonyo, T., Shinkai, T., Mitsumori, M., 2014. Improved culturability of cellulolytic rumen bacteria and phylogenetic diversity of culturable cellulolytic and xylanolytic bacteria newly isolated from the bovine rumen. FEMS Microbiol. Ecol. 88, 528–537.

Ogunmolu, F.E., Kaur, I., Pasari, N., Gupta, M., Yazdani, S.S., 2018. Qunatitaive multiplexed profiling of *Penicillium funiculosum* secretome grown on polymeric cellulase inducers and glucose. J. Proteome 179, 150–160.

Paepatung, N., Nopharatana, A., Songkasiri, W., 2009. Bio-methane potential of biological solid materials and agricultural wastes. Asian J. Energy Environ. 10, 19–27.

Pandiyan, K., Singh, A., Singh, S., Saxena, A.K., Nain, L., 2019. Technological interventions for utilization of crop residues and weedy biomass for second generation bio-ethanol production. Renew. Energy 132, 723–741.

Pang, J., Liu, Z.-Y., Hao, M., Zhang, Y.-F., Qi, Q.-S., 2017. An isolated cellulolytic *Escherichia coli* from bovine rumen produces ethanol and hydrogen from corn straw. Biotechnol. Biofuels, 10.

Peciulyte, A., Anasontzis, G., Karlstrom, K., Larsson, P., Olsson, L., 2014. Morphology and enzyme production of *Trichoderma reesei* Rut C-30 are affected by the physical and structural characteristics of cellulosic substrates. Fungal Genet. Biol. 72, 64–72.

Philip, J., Tanuja, T., Bedi, S., 2016. Occurrence of cellulose degraders in fruit and vegetable decaying wastes. Bioremediat. Biodegrad. 7, 373.

Philippidis, G.P., Spindler, D.D., Wyman, C.E., 1992. Mathematical modeling of cellulose conversion to ethanol by the simultaneous saccharification and fermentation process. Appl. Biochem. Biotechnol. 34–35, 543.

Poulsen, H.V., Willink, F.W., Ingvorsen, K., 2016. Aerobic and anaerobic cellulase production by *Cellulomonas uda*. Arch. Microbiol. 198, 725–735.

Reddy, K.V., Vijayalashmi, T., Ranjit, P., Raju, M.N., 2018. Characterization of some efficient cellulase producing bacteria isolated from pulp and paper mill effluent contaminated soil. Braz. Arch. Biol. Technol. 60.

Riley, R., Salamov, A.A., Brown, D.W., Nagy, L.G., Floudas, D., Held, B.W., Levasseur, A., et al., 2014. Extensive sampling of basidiomycete genomes demonstrates inadequacy of the white-rot/brown-rot paradigm for wood decay fungi. Proc. Natl. Acad. Sci. 111, 9923–9928.

Rivas, B., Torrado, A., Torre, P., Converti, A., Domínguez, J.M., 2008. Submerged citric acid fermentation on orange peel autohydrolysate. J. Agric. Food Chem. 56, 2380–2387.

Rowell, R.M., 1992. Opportunities for lignocellulosic materials and composites. In: Rowell, R.M., Schultz, T.P., Narayan, R. (Eds.), Emerging Technologies for Materials and Chemicals from Biomass. American Chemical Society, Washington, DC, pp. 12–27.

Rubeena, M., Neethu, K., Sajith, S., Sreedevi, S., Priji, P., Unni, K.N., Josh, M.K.S., Jisha, V.N., Pradeep, S., Benjamin, S., 2013. Lignocellulolytic activities of a novel strain of *Trichoderma harzianum*. Adv. Biosci. Biotechnol. 04, 214–221.

Ryu, J.S., Shary, S., Houtman, C.J., Panisko, E.A., Korripally, P., St. John, F.J., Crooks, C., Siika-aho, M., Magnuson, J.K., Hammel, K.E., 2011. Proteomic and functional analysis of the cellulase system expressed by *Postia placenta* during brown rot of solid wood. Appl. Environ. Microbiol. 77, 7933–7941.

Saharay, M., Guo, H., Smith, J.C., 2010. Catalytic mechanism of cellulose degradation by a cellobiohydrolase, CelS. PLoS One 5, e12947.

Saini, J.K., Saini, R., Tewari, L., 2015b. Lignocellulosic agriculture wastes as biomass feedstocks for second-generation bioethanol production: concepts and recent developments. 3 Biotech 5, 337–353.

Saini, R., Saini, J.K., Adsul, M., Patel, A.K., Mathur, A., Tuli, D., Singhania, R.R., 2015a. Enhanced cellulase production by *Penicillium oxalicum* for bioethanol application. Bioresour. Technol. 188, 240–246.

Sajith, S., Sreedevi, S., Priji, P., Unni, K.N., Benjamin, S., 2014. Production and partial purification of cellulase from a novel fungus, , BS1. Ann. Microbiol. 64, 763–771.

Salinas, A., Vega, M., Lienqueo, M.E., Garcia, A., Carmona, R., Salazar, O., 2011. Cloning of novel cellulases from cellulolytic fungi: heterologous expression of a family 5 glycoside hydrolase from *Trametes versicolor* in *Pichia pastoris*. Enzym. Microb. Technol. 49, 485–491.

Sánchez, C., 2009. Lignocellulosic residues: biodegradation and bioconversion by fungi. Biotechnol. Adv. 27, 185–194.

Sandberg, A.-S., Hasselblad, C., Hasselbad, K., 1982. The effect of wheat bran on the absorption of minerals in the small intestine. Braz. J. Nut. 48, 185–191.

Sangsurasak, P., Mitchell, D.A., 1998. Validation of a model describing two-dimensional heat transfer during solid-state fermentation in packed bed bioreactors. Biotechnol. Bioeng. 60, 739–749.

Santa-Rosa, P.S., Souza, A.L., Roque, R.A., Andrade, E.V., Astolfi-Filho, S., Mota, A.J., Nunes-Silva, C.G., 2018. Production of thermostable β-glucosidase and CMCase by *Penicillium* sp. LMI01 isolated from the Amazon region. Electron. J. Biotechnol. 31, 84–92. https://doi.org/10.1016/j.ejbt.2017.11.005.

Saritha, M., Arora, A., 2012. Biological pretreatment of lignocellulosic substrates for enhanced delignification and enzymatic digestibility. Indian J. Microbiol. 52, 122–130.

Saritha, M., Tiwari, R., Singh, S., Nain, P.K., Rana, S., Adak, A., Arora, A., Nain, L., 2016. Glycoside hydrolase production by *Aspergillus terreus* CM20 using mixture design approach for enhanced enzymatic saccharification of alkali pretreated paddy straw. Indian J. Exp. Biol. 54, 518–524.

Schwarz, W.H., 2001. The cellulosome and cellulose degradation by anaerobic bacteria. Appl. Microbiol. Biotechnol. 56, 634–649.

Sethi, S., Datta, A., Gupta, B.L., Gupta, S., 2013. Optimization of cellulase production from bacteria isolated from soil. ISRN Biotechnol. 2013, 1–7.

Sherief, A.A., El-Tanash, A.B., Atia, N., 2010. Cellulase production by *Aspergillus fumigatus* grown on mixed substrate of rice straw and wheat bran. Res. J. Microbiol. 5, 199–211.

Shrestha, P., Rasmussen, M., Khanal, S.K., Pometto, A.L., van Leeuwen, J.H., 2008. Solid-substrate fermentation of corn fiber by *Phanerochaete chrysosporium* and subsequent fermentation of hydrolysate into ethanol. J. Agric. Food Chem. 56, 3918–3924.

Singh, N., Mathur, A.S., Gupta, R.P., Barrow, C.J., Tuli, D., Puri, M., 2018. Enhanced cellulosic ethanol production via consolidated bioprocessing by *Clostridium thermocellum* ATCC 31924. Bioresour. Technol. 250, 860–867.

Singh, R., Kumar, R., Bishnoi, K., Bishnoi, N.R., 2009. Optimization of synergistic parameters for thermostable cellulase activity of *Aspergillus heteromorphus* using response surface methodology. Biochem. Eng. J. 48, 28–35.

Singh, S., Shukla, L., Khare, S., Nain, L., 2011. Detection and characterization of new thermostable endoglucanse from *Aspergiluus awamori* strain F18. J. Mycol. Plant Pathol. 41, 97–103.

Singh, S., Pranaw, K., Singh, B., Tiwari, R., Nain, L., 2014. Production, optimization nad evaluation of multicomponent holocellulase produced by *Streptomyces* sp. ssr-198. J. Taiwan Instt. Chem. Eng. 45, 2379–2386.

Singh, S., Tiwari, R., Renuse, S., Pranaw, K., Nain, L., 2015. Proteomic analysis of *Streptomyces* sp. ssr-198 grown on paddy straw. J. Basic Microbiol. 55, 790–797.

Singhania, R.R., Sukumaran, R.K., Pillai, A., Prema, P., Szakacs, G., Pandey, A., 2006. Solid-state fermentation of lignocellulosic substrates for cellulase production by *Trichoderma reesei* NRRL 11460. Indian J. Biotechnol. 5.

Siqueira, G., Várnai, A., Ferraz, A., Milagres, A.M.F., 2013. Enhancement of cellulose hydrolysis in sugarcane bagasse by the selective removal of lignin with sodium chlorite. In: Appl. Energy, Special Issue on Advances in Sustainable Biofuel Production and Use—XIX International Symposium on Alcohol Fuels—ISAF 102, pp. 399–402.

Sohail, M., Siddiqi, R., Ahmad, A., Khan, S.A., 2009. Cellulase production from *Aspergillus niger* MS82: effect of temperature and pH. New Biotechnol. 25, 437–441.

Sohail, M., Ahmad, A., Khan, S.A., 2016. Production of cellulase from *Aspergillus terreus* MS105 on crude and commercially purified substrates. 3 Biotech 6.

Sreena, C.P., Sebastian, D., 2018. Augmented cellulase production by *Bacillus subtilis* strain MU S1 using different statistical experimental designs. J. Genet. Eng. Biotechnol. 16, 9–16.

Sridevi, A., Narasimha, G., Reddy, R.B., 2008. Production of cellulase by *Aspergillus niger* on natural and pretreated lignocellulosic wastes. Internet J. Microbiol. 7, 1.

Srinivasan, K., Murakami, M., Nakashimada, Y., Nishio, N., 2001. Efficient production of cellulolytic and xylanolytic enzymes by the rumen anaerobic fungus, *Neocallimastix frontalis*, in a repeated batch culture. J. Biosci. Bioeng. 91, 153–158.

Sukumaran, R.K., Singhania, R.R., Mathew, G.M., Pandey, A., 2009. Cellulase production using biomass feed stock and its application in lignocellulose saccharification for bio-ethanol production. Renew. Energy 34, 421–424.

Sun, H.Y., Li, J., Zhao, P., Peng, M., 2011. Banana peel: a novel substrate for cellulase production under solid-state fermentation. Afr. J. Biotechnol. 10.

Szengyel, Z., Zacchi, G., Varga, A., Réczey, K., 2000. Cellulase production of Trichoderma reesei rut C 30 using steam-pretreated spruce. Appl. Biochem. Biotechnol. 84, 679–691.

Takano, M., Hoshino, K., 2018. Bioethanol production from rice straw by simultaneous saccharification and fermentation with statistical optimized cellulase cocktail and fermenting fungus. Bioresour. Bioprocess. 5.

Tirado-Gonzalez, N.D., Jauregui-Rincon, J., Tirado-Estrada, G.G., Martinez-Hernandez, A.p., Guevara-Lara, F., Miranda-Romeo, L.A., 2016. Production of cellulases and xylanases by white-rot fungi cultures in corn stover media for ruminant feed applications. Anim. Feed Sci. Technol. 221, 147–156.

Tiwari, R., Singh, S., Nain, P.K.S., Rana, S., Sharma, A., Pranaw, K., Nain, L., 2013. Harnessing the hydrolytic potential of phytopathogenic fungus *Phoma exigua* ITCC 2049 for saccharification of lignocellulosic biomass. Bioresour. Technol. 150, 228–234.

Tiwari, R., Nain, P.K., Singh, S., Adak, A., Saritha, M., Rana, S., Sharma, A., Nain, L., 2015a. Cold active holocellulase cocktail from *Aspergillus Niger*: process optimization for production and biomass hydrolysis. J. Taiwan Inst. Chem. Eng. 56, 57–66.

Tiwari, R., Pranaw, K., Singh, S., Nain, P.K.S., Shukla, P., Nain, L., 2015b. Two-step statistical optimization for cold active β-glucosidase production from *Pseudomonas lutea* BG8 and its application for improving saccharification of paddy straw. Biotechnol. Appl. Biochem. 63, 659–668.

Tiwari, R., Singh, P.K., Singh, S., Nain, P.K.S., Nain, L., Shulka, P., 2017. Bioprospecting of novel thermostable β-glucosidase from *Bacillus subtilis* RA 10 and its application in biomass hydrolysis. Biotechnol. Biofuels 10, 246.

Ververis, C., Georghiou, K., Christodoulakis, N., Santas, P., Santas, R., 2004. Fiber dimensions, lignin and cellulose content of various plant materials and their suitability for paper production. Ind. Crop. Prod. 19, 245–254. https://doi.org/10.1016/j.indcrop.2003.10.006.

Vocadlo, D.J., Davies, G.J., 2008. Mechanistic insights into glycosidase chemistry. Curr. Opin. Chem. Biol. 12, 539–555.

Wang, W., Kang, L., Lee, Y.Y., 2010. Production of cellulase from kraft paper mill sludge by *Trichoderma reesei* Rut C-30. Appl. Biochem. Biotechnol. 161, 382–394.

Weshahy, A.A., Rao, V.A., 2012. Potato peel as a source of important phytochemical antioxidant nutraceuticals and their role in human health—a review. In: Phytochemicals as Nutraceuticals—Global Approaches to Their Role in Nutrition and Health. Intech Open Limited, London, UK, pp. 207–224.

Wilson, D.B., 2008. Three microbial strategies for plant cell wall degradation. Ann. N. Y. Acad. Sci. 1125, 289–297.

Wood, T.M., 1985. Properties of cellulolytic enzyme systems. Biochem. Soc. Trans. 13, 407–410.

Xu, C., Ma, F., Zhang, X., 2009. Lignocellulose degradation and enzyme production by *Irpex lacteus* CD2 during solid-state fermentation of corn stover. J. Biosci. Bioeng. 108, 372–375.

Zhao, C., Zou, Z., Li, J., Jia, H., Liesche, J., Chen, S., Fang, H., 2018. Efficient bioethanol production from sodium hydroxide pretreated corn stover and rice straw in the context of on-site cellulase production. Renew. Energy 118, 14–24.

Further Reading

Carere, C.R., Sparling, R., Cicek, N., Levin, D.B., 2008. Third generation biofuels via direct cellulose fermentation. Int. J. Mol. Sci. 9, 1342–1360.

Choudhary, J., Singh, S., Nain, L., 2017. Bioprospecting thermotolerant ethanologenic yeasts for simultaneous saccharification and fermentation from diverse environments. J. Biosci. Bioeng. 123, 342–346.

Gupta, R., Mehta, G., Khasa, Y.P., Kuhad, R.C., 2011. Fungal delignification of lignocellulosic biomass improves the saccharification of cellulosics. Biodegradation 22, 797–804.

Saini, A., Aggarwal, N.K., Yadav, A., 2017. Cost-effective cellulase production using *Parthenium hysterophorus* biomass as an unconventional lignocellulosic substrate. 3 Biotech 7.

Schu, M., 2000. Protein engineering of cellulases. Biochim. Biophys. Acta 1543, 239–252.

C H A P T E R

3

Role of Compositional Analysis of Lignocellulosic Biomass for Efficient Biofuel Production

*Neha Srivastava**, *Kajal Mishra**, *Manish Srivastava*†, *Kumar Rohit Srivastava**, *Vijai Kumar Gupta*‡, *P.W. Ramteke*§, *P.K. Mishra**

*Department of Chemical Engineering and Technology, Indian Institute of Technology (Banaras Hindu University), Varanasi, India †Department of Physics and Astrophysics, University of Delhi, Delhi, India ‡Department of Chemistry and Biotechnology, ERA Chair of Green Chemistry, School of Sciences, Tallinn University of Technology, Tallinn, Estonia §Department of Biological Sciences, Sam Higginbottom University of Agriculture Technology & Sciences (Formerly Allahabad Agricultural Institute), Allahabad, India

3.1 INTRODUCTION

The progressive depletion of fossil fuel and world energy resources based on nonrenewable fuel is a primary concern these days. Moreover, long-term economic and environmental concerns associated with the large-scale utilization of fossil fuel due to availability, geographic distribution, and global warming are also rising day by day (Rass-Hansen et al., 2007; Moon et al., 2010; Pandey et al., 2011; Serrano-Ruiz et al., 2010; Zhang, 2010). Also, there are other concerns associated with the expected increase in the global population by approximately 5 billion people by 2050, which will directly increase the need for fuels. One estimate indicated that to meet the growing demand of industrialized countries and the rapid development of emerging economies, world energy consumption would increase by 45% over the next 30 years (Yat et al., 2008; Sivakumar et al., 2010). Therefore the concerns have resulted in a great amount of research over the past couple of decades into alternative renewable resources as fuels to replace fossil fuels (Chovau et al., 2013). Among existing renewable energy options, biofuels are most attractive due to their low-cost production and pollution-free nature. Biofuels are also known as transportation fuels, whose energy is derived from biological resources or through biological processes (Srivastava et al., 2015a,b). Biofuels such as biohydrogen, biomethane, biogas, ethanol, and butanol have a number of advantages. Production of these biofuels from cellulosic biomass is one of the most potent and economic ways for practical implementation of these fuels. Moreover, a major portion of the global potential of bioenergy resources is described in the form of lignocellulosic complexes and energy crops. Presently, biomass accounts for around 15.8% of the world's primary energy use annually, among which 40% is used in modern forms, i.e., liquid biofuel and steam, and the rest (60%) is used traditionally, i.e., domestic heating with an energy density of 15–20 MJ/kg (Yat et al., 2008; Chovau et al., 2013; Zhang and Mielenz, 2011). Lignocellulosic biomass is a cellulose-rich, abundant, massive substrate, and an additional advantage is its eco-friendly nature due to negligible release of greenhouse gases (Valentine et al., 2012; Hill et al., 2006). Apart from biofuel production, lignocellulosic biomass can also be converted into value-added products such as chemicals, renewable biobased polymers, and composites derived to reduce environmental impact (Chandel and Singh, 2011).

From Cellulose to Cellulase: Strategies to Improve Biofuel Production
https://doi.org/10.1016/B978-0-444-64223-3.00003-5

Lignocellulosic biomass is converted to produce biofuel, which is one of the important choices for the exploitation and production of energy sources. Thus the development of plant-based biofuel has gained strategic importance and will enhance economic development (Srivastava et al., 2015a,b). One study indicated that the production of biofuel would increase from 1.9 million barrels per day in 2010 to 5.9 million barrels per day in 2030, which is approximately up to 55% (Zhang, 2010). Thus from the point of cost economy, lignocellulosic biomass is used as a substrate to reduce the cost of biofuel production (Merwe et al., 2013; Kullander, 2010).

From a biofuel production perspective, the cell walls of plant biomass are mainly composed of cellulose, hemicellulose, lignin, ash, extractives/volatiles, and pectin, minerals that are collectively referred as lignocellulose. Lignocellulosic biomass is rich in cellulose, a polymer of glucose, and is a potential source of sugar production for biofuel production. Cellulosic raw material includes rice straw, wheat straw, sugarcane bagasse, corncob, and rice husk. Cellulose is mainly composed of hydrogen-bonded chains of β-1,4-linked glucose coated with hemicellulose. Hemicellulose is a polymer of β-1,4-linked xylose, which may have branches containing other sugars such as glucuronic acid or arabinose, depending on the plant species. Saccharification of cellulose and hemicellulose releases glucose and xylose that in turn can be fermented to bioethanol. Lignin is a complex polymer of methoxylated and hydroxylated phenylpropanoids and cross-link plant secondary cell walls to provide mechanical strength and hydrophobicity, which contribute to defense against pathogens. From cellulose to biofuel production, the compact structure of these biomasses covered by lignin is the main barrier that hinders the total productivity of the process. The total percentage of lignin content in the cell wall varies between plants and is a crucial parameter affecting the decomposition efficiency of the polysaccharides. An analytical review of the accurate compositional characteristics for selecting the best feedstock for a particular conversion pathway enables evaluation of conversion yield and process economics, because the feedstock may have a great amount of innate variability between different biomass types and individual biomass species. Therefore the objective of this chapter is to explore the compositional impact of cellulosic biomass to enhance biofuel production. Details of production and composition of cellulosic biomass are discussed in this chapter along with the application of these biomasses in biofuel production. Additionally, the pretreatment process to remove lignin for smooth assessment of enzymes during the bioconversion process is also provided with existing drawbacks and sustainable approaches.

3.2 LIGNOCELLULOSIC BIOMASS PRODUCTION STATUS AND AVAILABILITY

Due to the tremendous contribution of fossil fuels, issues such as environmental pollution, greenhouse gas emissions, and global warming contribute to the loss of biodiversity. Therefore production of biofuels from biomass will provide an eco-friendly solution to overcome these issues in an environmentally friendly manner. Additionally, biofuel production using these cellulosic biomasses is a potential process since these biomasses are rich in cellulose content. These cellulosic feedstocks can be classified into three groups, which are starchy crops, sugar (sugar crops and by-products of sugar refineries), and lignocellulosic biomass (LCB). All these groups differ considerably from one another with respect to the presence of sugar content (Valentine et al., 2012). Though the production of biofuel from edible crops (sugars and starch) is easier than LCBs, economic and environmental concerns are the major barriers for sustainability of production due to the "food versus fuel" conflict (Hill et al., 2006). Besides commercialization, the gross production of first-generation biofuels is limited and has been estimated to be roughly 3% of the total consumed fuels in the transport sector (Chandel and Singh, 2011). For all these reasons, research efforts have been more focused on LCBs, because they are mostly waste materials, rich in cellulose content (in general 35%–55%, 20%–35%, and 10%–25% of cellulose, hemicellulose, and lignin contents are present, respectively) (Zhu et al., 2009a,b), available with low stable prices, and are noncompetitive with the food chains (Çetinkol et al., 2010). Additionally, lignocellulosic biofuels cause lower net greenhouse gases, helping to mitigate global climate change by reducing emissions (mainly CO_2), thereby reducing environmental pollution (Lee et al., 2009).

Sources of lignocellulosic biomass are promising and abundantly available throughout the world (Table 3.1) because of the ability to obtain numerous harvests from a single planting, which significantly reduces average annual costs for establishing and managing crops in comparison with conventional crops, which are classified into different groups and include energy crops (perennial grasses, e.g., *Miscanthus* spp., *Phalaris arundinacea*, *Panicum virgatum*), forest biomass and wastes (softwoods, e.g., pine, spruce, cedar, and hardwoods, e.g., cottonwood, oak, poplar, sawdust pruning, and bark thinning residues), agricultural residues (cereal straws, stovers, and bagasse), aquatic plants (water hyacinth), and municipal solid wastes (food waste, newspaper, and Kraft paper) (Limayem and Ricke, 2012; Sanchez and Cardona, 2008). The production of biomass globally, where 90% is lignocellulose, amounts to about 200×10^9 t/year, where about 80–120×10^9 t/year remain potentially accessible (Beede and Bloom, 1995; Al-Salem et al., 2009). Maize, rice, sugarcane, and wheat are the four agricultural crops with maximum production as well

TABLE 3.1 Indian Renewable Feedstock Available for Biofuel Production

Crop	Residue	Season	Biomass potential (kt/year)	Crop	Residue	Season	Biomass potential (kt/year)
Arecanut	Fronds, husk	November–February	1000.8	Meshta	Stalks, leaves	June–July, February–March	1645.5
Arhar	Stalks, husk	Annual crop	5734.6	Moong	Stalks, husk	March–June	762.5
Bajra	Stalks, cobs, husk	April–July	15,831.8	Moth	Stalks	Annual crop	17.8
Banana	Residue	April–May	11,936.5	Mustard	Stalks, husk	October–March	8657.1
Barley	Stalks	April–August	563.2	Niger seed	Stalks		94
Barseem	Stalks	October–March	71.6	Oilseeds	Stalks	July–October	1143.1
Black pepper	Stalks	May–June	29.1	Onion	Stalks	July–October	66.5
Cardamom	Stalks	June–July	43.6	Others	Others		0.34
Cashew nut	Stalks, shell	April–May	189.4	Paddy	Straw, husk, stalks	July–October	169,965.1
Castor seed	Stalks, husk	April–June	1698.6	Peas and beans	Stalks	October–March	27.4
Casuarina	Wood	March–July	211.8	Potato	Leaves, stalks	October–January	887.3
Coconut	Fronds, husk, pith, shell	March–June	10,463.6	Pulses	Stalks	July–October, October–March	1390.4
Coffee	Husk, pruning and wastes	June–December	1591	Ragi	Straw	July–October	2630.2
Coriander	Stalks	June–July, October–November	188.3	Rubber	Primary and secondary wood	June–July	2492.2
Cotton	Stalks, husk, bollshell	April–June	52,936.5	Safflower	Stalks	October–March	539.3
Cow gram	Stalks	March–June	48.5	Sunnhemp	Stalks	March–April	14.1
Cumin seed	Stalks	June–July	182.6	Sawan	Stalks		0.22
Dry chilly	Stalks	May–June	268.6	Small millets	Stalks	Annual crop	600.1
Dry ginger	Residue	May–June	5.3	Soybean	Stalks	July–October	9940.2
Eucalyptus	Stalks		162.8	Sugarcane	Tops and leaves	January–March	12,143.9
Gram	Stalks	July–December	5440.6	Sunflower	Stalks	August–February	1407.6
Groundnut	Shell, stalks	March–May	15,120.4	Sweet potato	Stalks	November–January	12.8
Guar	Stalks	March–September	233.3	Tapioca	Stalks	September–October	3959
Horse gram	Stalks	March–May	191.3	Tea	Sticks	Annual crop, March–November	909.8
Jowar	Cobs, stalks, husk	April–May	24,207.8	Til	Stalks		1207.7
Kesar	Stalks	June–September	9.4	Tobacco	Stalks	Annual crop	204.8
Kodo millets	Stalks	June (mid)–July (end)	3.13	Turmeric	Stalks	May–July	32.3
Linseed	Stalks	October–January	86.3	Urad	Stalks, husk	July–October, October–March, summer	924.9
Maize	Stalks, cobs	July–October	26,957.7	Wheat	Stalks, panicle	October–March	112,034
Masoor	Stalks	March–June	600.3	Total			511,041.39

as area under cultivation. These agricultural crops generate the majority of lignocellulosic biomass in the agricultural sector; the rest of the agrowastes constitute a minor proportion of total agrowaste production in the world (Alvira et al., 2010). The global production of major agrowastes and their biofuel production potentials are shown in Table 3.2. In Asia, rice straw and wheat straw are the maximum generated crops, whereas in the United States, corn straw and sugarcane bagasse are mainly produced. Also, it has been anticipated that 442 billion liters of biofuel can be produced each year using LCBs if wasted crops as well as total crop residues are considered (Liu, 2010; Yu et al., 2008). The United States alone generated a total of 1368 MT of biomass for biofuel production out of which 428 MT come from agricultural residue, while 370, 377, 87, 58, and 48 MT come from forestry wastes, energy crops, grains and corn, municipal and industrial wastes, and other wastes, respectively (Beringer et al., 2011). In India, sugarcane, wheat straw, and rice straw are the most grown crops accounting for over 91% of production crops, and the top four states that generate these crops are Uttar Pradesh, Andhra Pradesh, Punjab, and Maharashtra (Byrt et al., 2011). Corn stover is the leftover residue after harvesting corn kernels and mainly comprises leaves, stalks, cobs, and husks. The content of cellulose present is around 38%–40% and annual production is approximately 1 kg/kg or 4 t/acre (Cook and Devoto, 2011; Jordan et al., 2012; Kim and Dale, 2004). Wheat straw with a cellulose content around 33%–38% is generated during wheat grain harvest under rigorous farming conditions and the production rate is 1–3 t/acre

TABLE 3.2 Compositional Variation for Woody, Agricultural, and Municipal Solid Wastes Along With Anatomical Fractions of Corn, Woody Biomass, and Wheat

Feedstock composition	Proximate analysis			Ultimate analysis					Structural analysis		
	V (%)	A (%)	FC (%)	H (%)	C (%)	N (%)	O (%)	S (%)	Cellulose (%)	HC (%)	Lignin (%)
Woody biomass	84 (2.1)[193]	1.3 (0.9)[193]	14.7 (1.6)[193]	6 (0.1)[192]	50.7 (4.71)[192]	0.32 (0.01)[192]	41.9 (1.4)[134]	0.03 (0.01)[135]	51.2 (8.7)[241]	21 (8.7)[241]	26.1 (5.3)[241]
Pine	83.5 (2.5)[46]	0.7 (0.6)[46]	15.7 (1.9)[46]	6.1 (0.1)[45]	51.5 (1)[45]	0.17 (0.12)[45]	41.4 (1)[38]	0.02 (0.01)[39]	47.4 (2.2)[55]	21.9 (4.9)[55]	28.6 (0.7)[55]
Softwood	81.3 (2.9)[18]	2.1 (2)[18]	16.5 (1.6)[18]	6.1 (0.1)[18]	51.8 (0.9)[18]	0.27 (0.21)[18]	39.7 (1.8)[14]	0.03 (0.01)[14]	42.1 (7.1)[26]	25.1 (5.2)[26]	29.1 (1.7)[26]
Hardwood	85.1 (3.0)[11]	1.8 (1.2)[11]	13.1 (1.8)[11]	6.1 (0.1)[11]	50.2 (0.5)[11]	0.55 (0.49)[11]	41.1 (1.6)[10]	0.05 (0.05)[10]	50.8 (6.9)[24]	29.7 (4.3)[24]	19.5 (4.1)[24]
Hybrid poplar	84 (1.3)[41]	1.3 (0.5)[41]	14.6 (0.1)[41]	6 (0.1)[41]	50 (1.1)[41]	0.35 (0.17)[41]	42.8 (1.2)[28]	0.03 (0.01)[14]	42.1 (7.1)[26]	25.1 (5.2)[26]	29.1 (1.7)[26]
Agricultural biomass	79.1 (5.8)[284]	5.5 (3.2)[284]	15.4 (4.0)[284]	5.8 (0.3)[276]	47.4 (1.9)[276]	0.75 (0.49)[276]	41 (2.4)[107]	0.10 (0.32)[107]	32.1 (4.5)[242]	18.6 (3.4)[242]	16.3 (3.3)[242]
Sugarcane bagasse	82.2 (1.9)[48]	3.4 (1.6)[48]	14.4 (1)[48]	6.1 (0.1)[48]	48.8 (0.9)[48]	0.43 (0.20)[48]	–	–	32.1 (3.2)[479]	19.5 (1.9)[479]	16.3 (1.8)[479]
Corn stover	78.1 (5.0)[50]	6.3 (3.5)[50]	15.6 (4.4)[50]	5.7 (0.3)[40]	47.1 (2.3)[40]	0.63 (0.32)[40]	40.3 (2.2)[39]	0.14 (0.53)[39]	34.3 (2.5)[251]	20.7 (2.0)[251]	15.2 (1.6)[251]
Sorghum	77 (3.7)[44]	7.2 (2.6)[44]	15.7 (2.3)[44]	5.7 (0.2)[44]	46.4 (1.3)[44]	1.04 (0.38)[44]	40.3 (0.6)[3]	0.11 (0.01)[3]	28.6 (2.6)[488]	15.4 (1.6)[488]	12.2 (1.9)[488]
Switch grass	82.4 (4.1)[43]	4 (2)[43]	13.6 (3)[43]	5.9 (0.2)[43]	47.1 (1.1)[43]	0.60 (0.26)[43]	42.4 (2.3)[42]	0.06 (0.03)[42]	34.2 (2.7)[348]	21.9 (2.6)[348]	19.2 (1.4)[348]
Miscanthus	82.5 (3.5)[35]	2.6 (1.3)[35]	14.8 (2.9)[35]	5.8 (0.1)[35]	48.9 (1.5)[35]	0.35 (0.17)[35]	42.3 (1.1)[4]	0.04 (0.02)[4]	38.9 (3.2)[274]	20.1 (1.4)[274]	21.1 (1.6)[274]
Mixed grasses	78.6 (2.8)[47]	6.6 (1.7)[47]	14.8 (2.4)[47]	5.8 (0.3)[47]	47.6 (1.1)[47]	1.38 (0.54)[47]	39.5 (0.7)[2]	0.12 (0.02)[2]	28.9 (2.9)[465]	16.7 (3.9)[465]	15.7 (1.7)[465]
Wastes	76.7 (5.5)[21]	6.6 (6.7)[21]	14.8 (5.0)[21]	(5.9) (0.4)[21]	46 (4)[21]	1.3 (1.6)[21]	38.3 (4.2)[7]	0.15 (0.16)[7]	28.4 (13.2)[27]	16.4 (5.5)[27]	12.5 (2.7)[27]
MSW	76.5 (1.1)[11]	11.8 (5.2)[11]	11.2 (5.2)[11]	5.6 (0.4)[11]	43.3 (3.3)[11]	1.52 (1.72)[11]	36.3 (4.8)[4]	0.25 (0.14)[4]	28.4 (13.2)[15]	16.4 (5.5)[15]	12.5 (2.7)[15]
C&D waste	76.5 (3.7)[9]	0.8 (0.4)[9]	18.9 (2.1)[9]	6.2 (0.2)[9]	48.3 (1.2)[9]	1.09 (1.47)[9]	42.4 (0.1)[2]	0.02 (0.01)[2]	–	–	–

TABLE 3.2 Compositional Variation for Woody, Agricultural, and Municipal Solid Wastes Along With Anatomical Fractions of Corn, Woody Biomass, and Wheat—cont'd

ANATOMICAL FRACTIONS				
Structural components	*Cellulose*	*Hemicelulose*	*Lignin*	*Extractives*
Woody biomass (wt%-daf)				
Whole tree	51.2	23.4	25.4	3.0
Leaves	26.5	47.2	26.3	3.7
Twigs	15.4	62.3	22.3	1.6
Bark	22.0	47.0	31	3.3
Wheat (wt%-db)				
Internode 1	34.34	21.30	16.36	16.24
Internode 2	39.04	21.07	18.58	10.98
Internodes 3/4/5	38.92	21.56	19.50	9.67
Corn (wt%-db)				
Corn cob	35.92	30.7	16.44	5.89
Corn husk	37.73	31.18	10.52	5.80
Corn leaves	34.33	22.77	13.09	10.54
Corn internodes	40.21	20.03	17.24	12.29

Average (standard deviation); V = volatiles, A = ash, FC = fixed carbon, H = hydrogen, C = carbon, N = nitrogen, O = oxygen, S = sulfur, HC = hemicellulose, MSW = municipal solid waste, C&D = construction & demolition, daf = dry ash free, db = dry basis; superscript, number of samples.

(Jordan et al., 2012). Rice straw is the leftover of rice production and comprises stems, leaf sheaths, leaf blades, and the remains of the panicle after threshing. It is one of the most abundant lignocellulosic waste materials with a cellulose content of 28%–36%, hemicellulose content of 23%–28%, and lignin content of 12%–14%. Moreover, out of an annual global production of 731 MMT of rice straw, Asia alone produces 667.6 MMT and India produces around 112 MMT. Bagasse is a cheap renewable resource with a cellulose content of 42%–48% and is produced in huge amounts during sugarcane processing (Lal, 2005; Pauly and Keegstra, 2008). Also, it has been anticipated that 442 billion liters of biofuel can be produced each year using LCBs if wasted crops as well as total crop residues are considered (Liu, 2010; Yu et al., 2008). Thus agricultural production of various crops like sugarcane, sorghum, sweet sorghum, pulses, chilly, cotton, oilseed, etc. results in the generation of huge amounts of wastes that do not have any alternative use, are either left in fields, or are burned off. Thus being waste and cellulose rich, these could be used as good alternative resources to generate biofuels in an environmentally friendly manner. Moreover, agricultural residues help in the reduction of deforestation by decreasing the reliance on forest woody biomass. Also, crop residues have a short harvest period that makes them more consistently available for biofuel production (Puri et al., 2012; Reddy et al., 2012; Pauly and Keegstra, 2008). Therefore huge availability and high cellulose content make lignocellulosic biomass an ideal candidate for biofuel production. Exploration of particular feedstocks and their compositional analysis may help to utilize these biomasses effectively.

3.3 BIOMASS COMPOSITIONAL ANALYSIS

The cellulosic cell walls of plants are the most abundant renewable resource on earth with 150–170 × 10^9 t produced annually (Jorgensen et al., 2007). They are mainly composed of cellulose, hemicellulose, lignin, extractives (nonstructural soluble materials such as nitrogenous materials, chlorophyll, nonstructural sugars, and waxes), and ash (Sun and Cheng, 2002). The composition of lignocellulosic biomass materials is complex, heterogeneous, and varies significantly from biomass to biomass, e.g., in rice straw the percentage of cellulose, hemicellulose, and lignin is around 32%–41%, 15%–24%, and 10%–18%, respectively (Perez et al., 2002; Zhang et al., 2008; Bian et al., 2004), whereas in sugarcane bagasse it is around 26%–50%, 24%–34%, and 10%–26% (Zhang et al., 2006; Kuhad et al., 1997; Lal., 2005).

Thus the composition of feedstock is an important parameter that influences the efficiency of the biomass-to-biofuel production process. In lignocellulosic biomass, celluloses are the main structural component and are found in an organized fibrous structure (Zhang et al., 2008). The structure of cellulose is linear and consists of D-glucose subunits linked through β-(1,4)-glycosidic bonds. The supramolecular structure of cellulose showed that the noncrystalline phase, i.e., unorganized cellulose chain and the crystalline phase, intertwines to form the cellulose. When tested by X-ray diffraction, the noncrystalline phase assumes an amorphous state because most hydroxyl groups on glucose are amorphous. Moreover, in the crystalline phase large amounts of hydroxyl groups form many hydrogen bonds as well as van der Waals bonds and these construct a huge network that directly contributes to the compact structure (Vassilev et al., 2012). Also, the ratio of crystalline to amorphous cellulose is said to be a significant factor in glucose production, because enzymes are reported to degrade amorphous cellulose more efficiently (Girio et al., 2010) (Table 3.1). The deconstruction of LCBs with high cellulose content should translate to larger amounts of glucose for subsequent conversion to biofuel (Kuhad et al., 1997).

The second most abundant compound is hemicellulose containing about 20%–35% of lignocellulosic biomass. Hemicellulose is not chemically homogeneous like cellulose but has branches with short lateral chains consisting of different types of sugars, which include pentoses (xylose, rhamnose, and arabinose), hexoses (glucose, mannose, and galactose), and uronic acids (4-*O*-methylglucuronic, D-glucuronic, and D-galacturonic acids) (McMillan, 1993). The backbone of hemicellulose is either heteropolymer or homopolymer with short branches linked by β-(1,4)-glycosidic bonds; they have lower molecular weight compared to cellulose and are easily hydrolyzed because of short lateral chain branches (Saha, 2003; Aspinall, 1980). A hemicellulose "coat" has cellulose fibrils and it has been proposed that 50% of hemicellulose should be removed to significantly increase cellulose digestibility because it provides strength by cross-linking with both cellulose and lignin. Additionally, severity parameters should be significantly optimized to avoid the formation of hemicellulose degradation products such as furfurals and hydroxymethyl furfurals (HMFs), which have been reported to inhibit the fermentation process by ultimately degrading the biofuel yield (Demirbas, 2005). Therefore, for this reason, depending on the pretreatment method, severity conditions are compromised to maximize sugar recovery, and hemicellulose could be obtained either as a combination of solid and liquid fractions or as a solid fraction only (Saha, 2003).

The third most abundant compound is lignin containing about 10%–25% of a lignocellulosic biomass (Aspinall, 1980). It is a complex structure composed of a large molecular structure containing cross-linked polymers of phenolic monomers. These phenolic monomers are linked by alkyl-aryl, alkyl-alkyl, and aryl-aryl ether bonds and their main monomers are coniferyl alcohol (guaiacol propanol), coumaryl alcohol (*p*-hydroxyphenyl propanol), and sinapyl alcohol (syringyl alcohol) (Fig. 3.1). Lignin is generally called a "glue" that binds the different components of lignocellulosic biomass together, thus making it insoluble in water (Demirbas, 2005). They are also said to confer rigidity, impermeability, and resistance to microbial attack and oxidative stress (Mielenz, 2001). Lignin has been identified as a major disincentive for enzymatic hydrolysis because of its close association with cellulose microfibrils (Li et al., 2008). It has been shown that LCB digestibility is increased by decreasing the lignin content (Ladisch et al., 2010). Apart from being a physical barrier, the harmful effects of lignin can be (1) interference with, and nonproductive binding of, cellulolytic enzymes to lignin–carbohydrate complexes, (2) nonspecific adsorption of hydrolytic enzymes to "sticky" lignin, and (3) toxicity of lignin derivatives to microorganisms (INL DOE, 2015). Different feedstocks contain different amounts of lignin such as wheat straw, which contains 16%–24% of lignin, sugarcane bagasse, which contains 23%–32% lignin, and corncobs, which contain 15% lignin, whereas corn stover contains around 18%–22% of lignin that must be removed via pretreatment to enhance biomass digestibility. It is believed that during pretreatment, lignin melts and coalesces upon cooling such that its properties are altered and further can be subsequently precipitated (Li et al., 2008; Ladisch et al., 2010; INL DOE, 2015). Delignification causes biomass swelling, disruption in the lignin structure, increase in internal surface area, and increase in accessibility of cellulolytic enzymes to cellulose fibers (INL DOE, 2015). Therefore pretreated biomass becomes more digestible to further processing in comparison to raw biomass, which directly helps in increasing the biomass yield (Mielenz, 2001).

Extractives/volatiles and ash are the other important compounds of LCBs and they have a major influence on what ends up being the optimal conversion process. Extractives/volatiles may be ethanol soluble (chlorophyll and waxes) and water soluble (nonstructural sugars and proteins). Ash from sulfur, nitrogen, phosphorus, calcium, iron, potassium, magnesium, copper, zinc, sodium, manganese, chlorine, etc. is an indispensable material. When biomass is dried at temperatures of 105–250°C and further processed at 550–750°C in a high-temperature muffle furnace, elements such as carbon, nitrogen, hydrogen, oxygen, sulfur, etc. disappear in the form of gaseous compounds and the residue is ash, which contains many types of mineral elements in the form of oxides. Thus different components present in LCBs have different roles and their analysis plays a significant part in evaluating and improving biofuel production processes (Fig. 3.2).

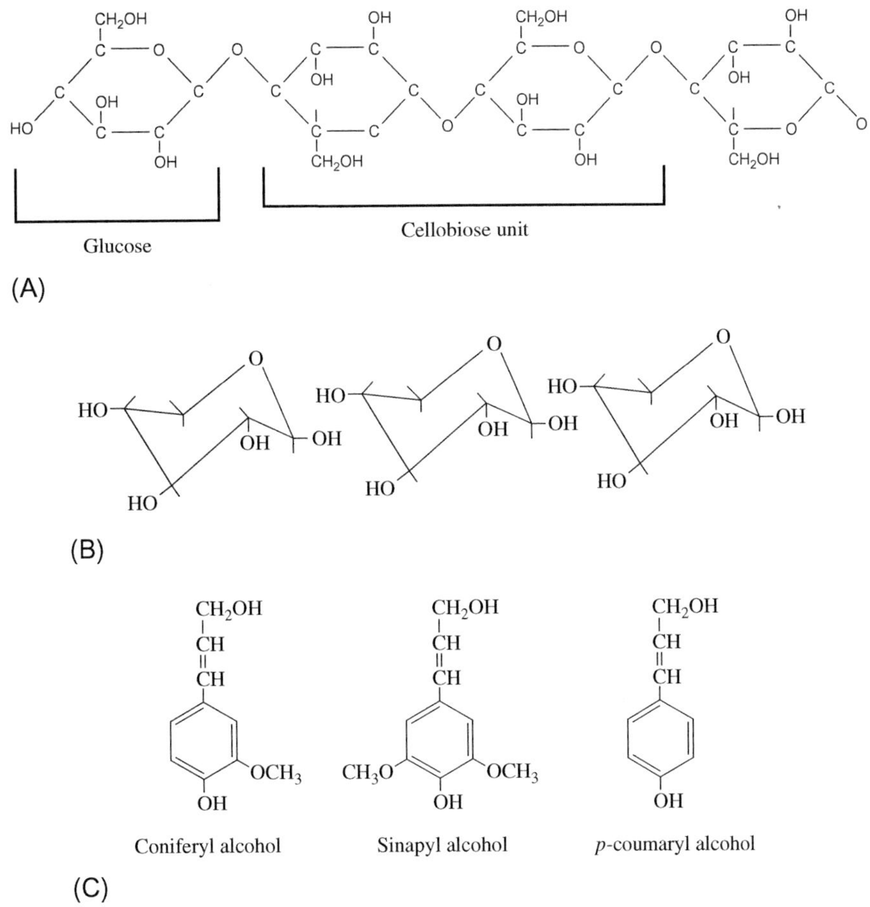

FIG. 3.1 (A) Structural chain of cellulose, (B) structural chain of hemicellulose, (C) basic structural units of lignin. *From Iqbal, H.M.N., Kyazze, G., Keshavarz, T., 2013 Advances in valorization of lignocellulosic materials by biotechnology: an overview. BioResources 8, 3157–3176.*

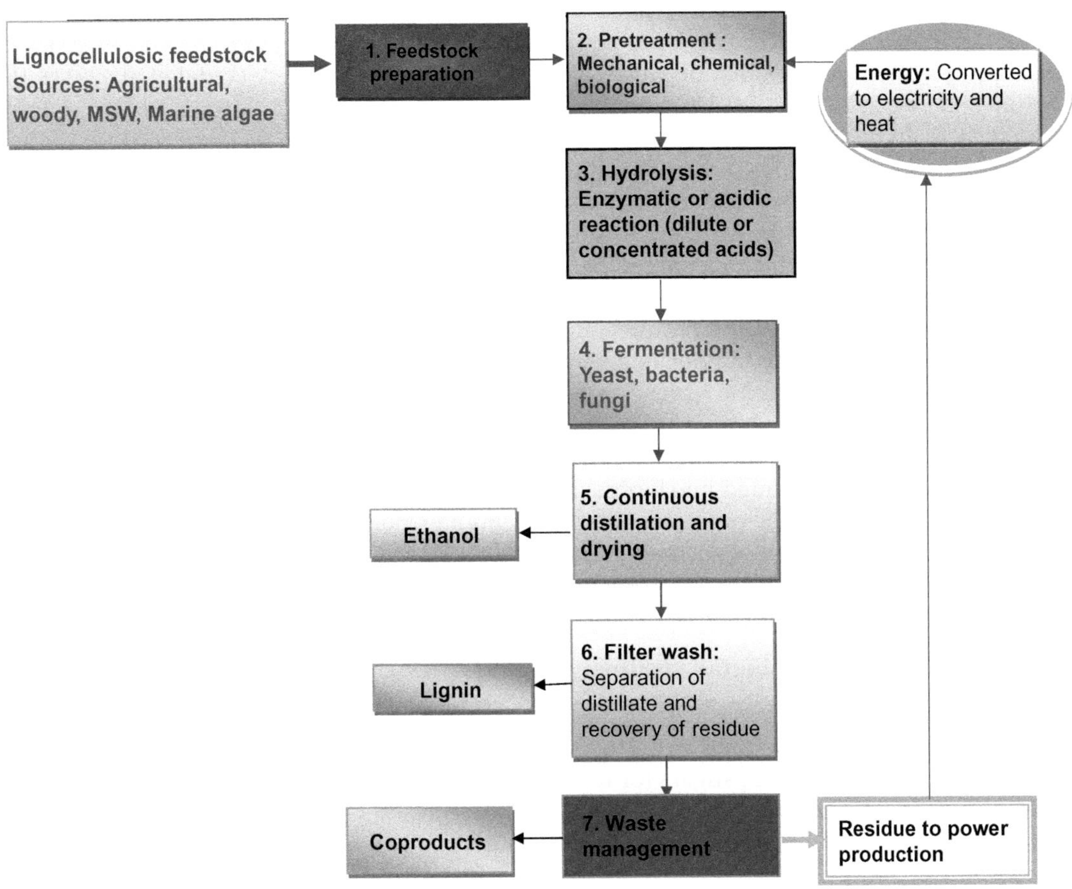

FIG. 3.2 Process flow diagram.

Compositional variability may depend on many factors, including component analysis methods, feedstock types, environmental factors, harvesting practices, storage conditions, and preprocessing techniques. Some of these can be controlled through standardization practices, while other factors can be difficult to control. Significant compositional variation occurs in different biomass types but there is also a lot of variation in a single feedstock as well. Table 3.2 describes the large difference in composition across woody, agricultural, and municipal solid waste and variability within individual feedstock categories along with substantial compositional variability between different anatomical fractions of the same type of biomass.

From the table it is evident that a significant amount of variability exists in the overall compositions of different feedstocks. The differences are so large that the conversion reactors have to be operated under different conditions based on the type of material supplied. Apart from high variability across broad categories such as woody, agricultural, or municipal solid waste, there is also significant variability within individual feedstock categories such that it can be easily seen from the table that the lignocellulosic biomass contains a large fraction of cellulose in the whole tree in comparison to the bark, which contains a high amount of lignin. In corn stove the majority of extractives are in the leaves and internodes, whereas in woody biomass the extractives are evenly distributed. Therefore studying anatomical fractions could allow for greater control over product output; it can also increase the economic viability of a process by extracting only high-value components. All the values reported in the table were collected from the Idaho National Laboratory (ASTM D3172-13, 2013) and stored in the bioenergy feedstock library (Wolfrum and Sluiter, 2009). The values of cellulose, hemicellulose, and lignin reported in the table are a combination of glucose, which represents cellulose fractions, and xylose, galactose, and arabinose, which represent hemicellulose fractions, and are measured using a near-infrared spectroscopy-based predictive model developed at the National Renewable Energy Lab. The values of volatiles were determined by heating samples at 950°C in an inert atmosphere. Therefore broad range compositional analysis may play a crucial role in improving biofuel production processes by generating high sugars.

3.4 OVERVIEW OF THE CONVERSION OF CELLULOSIC FEEDSTOCK TO BIOFUEL

Raw lignocellulosic biomass has a chemical composition high in cellulose, hemicellulose, lignin, volatiles, ash, and moisture content (Demirbas, 2007). This combination does not make lignocellulosic biomass suitable of further processing to convert it into biofuel. The barrier to lignin presenting as a raw biomass may be overcome by a physical and chemical pretreatment process that produces a conversion-ready feedstock (Vidal et al., 2011). Pretreatment alters the structure and size of biomass, and also its chemical composition, so that hydrolysis of the cellulosic fraction into monomeric sugars can be achieved rapidly with higher yields.

After pretreatment of cellulosic biomass, enzymatic hydrolysis is carried out as a second step in the biomass-to-biofuel conversion step. In this step, sugars are released by breaking down the cellulose chains before they are fermented to biofuel. This process can be done through (1) enzymatic hydrolysis or (2) acid hydrolysis. In acid hydrolysis, dilute as well as concentrated acids can be used for decrystallization of a cellulosic mixture of acid and sugars that reacts in the presence of water to release individual sugar molecules (Zhang et al., 2013). This process leads to the formation of toxic degradation products that can interfere with fermentation. Similarly, lignocellulosic waste can be enzymatically hydrolyzed under relatively mild conditions (50°C and pH around 5.0), which enable effective cellulose breakdown without the formation of by-products that would otherwise inhibit enzyme activity (Zhu and Pan, 2010).

The third process involves the fermentation process in which the generated sugar such as glucose, xylose, arabinose, galactose, or mannose is readily fermented into biofuel by yeast or bacteria. When enzymatic hydrolysis and fermentation are conducted in a single step, the processes are called simultaneous saccharification and fermentation and separated saccharification and fermentation (Yang and Wyman, 2008). The next step involves purification, which includes distillation, molecular sieves, and different separation techniques employed to produce biofuels as a final product.

3.4.1 Pretreatment Overview of Lignocellulosic Biomass

Pretreatment of lignocellulosic biomass is regarded as the most expensive step and is the main challenge in biofuel production (Zhu et al., 2009a,b; Mu et al., 2010). In the pretreatment process, the complex structure of cellulose is broken down for enzyme reaction to get free sugars for fermentation. The most concerning factor in cellulose hydrolysis is the accessible surface area (Sánchez and Cardona, 2008). Theoretically, <20% cellulose hydrolysis is done directly through enzymes (absence of pretreatment), while about 90% is done after the pretreatment process (Zheng et al., 2009).

Thus the main objectives of the pretreatment process are removal of lignin and hemicelluloses, enhancement of the porosity of raw substrate, and reduction to a crystalline cellulose complex. Additionally, it improves sugar production and smooths enzymatic hydrolysis, avoiding the formation of inhibitors for hydrolysis and fermentation processes.

Different pretreatment methods, namely physical (milling, grinding), chemical (ozone, alkali, acid, organic solvents), biological, electrical, or a combination of these, have been used to remove the recalcitrant cell wall material of plants, depending on the raw material extracted (Wyman et al., 2005). Considerable advancement in the development of pretreatment processes has been made (Banerjee et al., 2009; Sun and Cheng, 2002; Cardona and Sanchez, 2007; Lynd et al., 2002; Mosier et al., 2005). Table 3.3 gives a brief idea of the pretreatment processes that have been examined over the years. Although some of them remain economically unfeasible due to key technical issues such as the involvement of recovery steps in dilute acid treatment, in the Organosolv pretreatment method organic solvents require recycling, and in some of the pretreatment methods, such as steam explosion, high energy is required for the process to occur. Forest and woody material have high lignin (approximately 29%) and cellulose

TABLE 3.3 Different Types of Pretreatment Methods

Methods	Pretreated materials	Procedure	Advantages	Disadvantages	Remarks
PHYSICAL METHODS					
Mechanical comminution	Corn stover, sugarcane bagasse, *Miscanthus*, woody and forestry wastes (hardwood, alfalfa wood)	Chipping, grinding, milling (<1)	Reduces cellulose crystallinity	Extensive power consumption usually more than inherent biomass energy	Vibratory ball mill (final size: 0.2–2mm), hammer mill (final size: 3–6mm)
Pyrolysis	Corn stover, waste cotton	High temperature > 300°C, cooling and condensing	Produces gas and liquid products	High temperature ends up producing ash	Less volatile products, mild diluted acid hydrolysis (1N H_2SO_4 for 2.5h), 80%–85% reducing sugars
CHEMICAL METHODS					
Ozonolysis	Poplar sawdust, wheat straw, pine, bagasse, cotton straw	Ozone, ambient conditions	Reduces lignin content; no toxic residue produced	Expensive as large amount of ozone required	Identification of different types of compounds such as oxalic acid, formic acid, glycolic acid, melonic acid, etc.; cellulose conversion can be >57%
Dilute acid hydrolysis	Wheat straw, rice straw, oil palm, bagasse	0.75%–5% HNO_3, HCl H_2SO_4, T=121–160°C; high solid load (10–40wt% dry substrate/mixture); T>160°C; low solid load (5–10wt% dry substrate/mixture)	Alters lignin structure; hydrolyzes hemicellulose to xylose and other sugars	Formation of toxic components; high amount of by-product generation	pH neutralization is required, 70%–90% xylose recovery, depolymerization occurs at certain degrees
Concentrated acid hydrolysis	Bagasse	Silo-type system, 170–190°C, 1:1.6S/L ratio, 10%–30% H_2SO_4, 21–60% peracetic acid	Powerful agent for cellulose hydrolysis	High cost, equipment corrosion	Compared to dilute acid, residence time is greater
Alkaline hydrolysis	Sugarcane bagasse, corn stover, wheat straw, rice straw, hardwood	2% H_2O_2 with 1.5% NaOH, 121°C, 15min; 60°C, 24h, diluted NaOH; 4h, $Ca(OH)_2$ 120°C	Increases accessible surface area; removes hemicellulose and lignin	Irrecoverable salts formed and incorporated into biomass; long residence time required	Reactor costs are lower than acid pretreatment
BIOLOGICAL METHODS					
Fungal pretreatment	Wheat straw, corn stover	Brown, white, and soft rot fungi	Low energy requirement; degrades lignin and hemicellulose	Hydrolysis rate is slow	Fungi produces cellulases, hemicellulases, and lignin-degrading enzymes: ligninases, lignin, peroxidases, polyphenoloxidases, laccase, and quinone-reducing enzymes

(approximately 44%) contents, which make them more recalcitrant (Demirbas, 2005). Agricultural residues like rice straw, wheat straw, and corn stover are composed of less lignin (3%–13%) making the texture less resistant, but they contain a higher level of pentose sugar.

The main aim of physical pretreatment is to reduce cellulose crystallinity. In this process, lignocellulosic material is employed by a combination of chipping, grinding, milling, and sieving. Sun and Cheng (2002) concluded that the size of the material is usually 10–30mm after chipping and 0.2–2mm after milling or grinding. Millet et al. (1976) found that vibratory ball milling is more effective than ordinary ball milling because it reduces cellulose crystallinity of aspen chips and spruce chips to improve their digestibility.

Chemical pretreatment methods for conversion of lignocellulosic biomass were started a century ago. The most studied method among all existing pretreatment methods includes the use of different chemical agents such as ozone, acid, peroxide, alkali, and organic solvents. Ozone treatment is used to reduce the lignin content of the lignocellulosic wastes because they increase the in vitro digestibility of the treated materials and do not produce toxic residues. Ozonolysis mainly degrades lignin; hemicellulose is slightly affected but there is no effect on cellulose content in many lignocellulosic materials such as bagasse, peanut, pine (Kilzer and Broido, 1965), wheat straw (Shafizadeh and Bradbury, 1979), cotton straw (Neely, 1984), or sawdust (Ben-Ghedalia and Miron, 1981). Quesada et al. (1999) reported that enzymatic hydrolysis yield increases from 0% to 57% as the percentage of lignin decreased from 29% to 8% after ozonolysis pretreatment of sawdust. In acid hydrolysis, inorganic acids such as H_2SO_4 and HCl (concentrated and dilute) have been mostly used for biomass pretreatment. Acid hydrolysis pretreatment improves the enzymatic hydrolysis of lignocellulosic biomass to release fermentable sugars. Concentrated acids are powerful agents for cellulose hydrolysis but they are toxic, hazardous, and corrosive, and thus require reactors that are resistant to corrosion, which directly makes the process expensive. Additionally, it is important to recover the concentrated acid after pretreatment to make it economically viable (Sun and Cheng, 2002; Ben-Ghedalia and Miron, 1981).

In alkaline pretreatment, some bases can be used for the pretreatment of lignocellulosic materials and their effect depends on the lignin content of the feedstock used (Shafizadeh and Bradbury, 1979; Ben-Ghedalia and Shefet, 1983). They increase the internal surface and degree of polymerization and crystallinity, and break the link between lignin content and other polymers (McMillan, 1994). Alkaline pretreatment processes are utilized at ambient conditions and lower temperatures and pressures, but the time requirement is on the order of hours and days rather than minutes and seconds. When compared with acidic pretreatment, in alkaline pretreatment many of the caustic salts can be recovered and/or regenerated, which cause less sugar degradation. Commonly used alkaline agents are sodium, potassium, calcium, and ammonium hydroxides of which sodium hydroxide has been studied the most (McMillan, 1997; Elshafei et al., 1991; Soto et al., 1994; Fox et al., 1989). Moreover, calcium hydroxide (slake lime) is an effective pretreatment agent and is the least expensive, and there is a possibility of recovering calcium from an aqueous reaction system by neutralizing it with inexpensive carbon dioxide (MacDonald et al., 1983; Kim and Holtzapple, 2006). Chang and Holtzapple (2000) concluded that there is a correlation between enzymatic digestibility and three structural factors, i.e., crystallinity, lignin content, and acetyl content: (1) regardless of acetyl content and crystallinity, extensive delignification is sufficient to obtain high digestibility, (2) crystallinity significantly affects the initial hydrolysis rate but has less effect on ultimate sugar yield, and (3) delignification and deacetylation remove barriers to enzymatic hydrolysis. Fan et al. (1987) reported that using dilute NaOH treatment caused swelling, leading to an increase in internal surface area, decrease in degree of polymerization, decrease in crystallinity, separation of structural linkage between lignin and holocellulose, and delignification. The drawback of using alkaline pretreatment is that it is not feasible for large-scale studies due to the high cost (Lynd et al., 1999).

Most of the pretreatment technologies, particularly physical or thermochemical processes, require expensive instruments or require abundant energy for conversion. The biological process is an environmentally friendly and safe process that uses various types of rot fungi such as soft, white, and brown rot fungi and the process does not require high energy for lignin removal from lignocellulosic biomass, in spite of extensive lignin degradation (Okano et al., 2005). White and soft rot fungi attack both cellulose and lignin, whereas brown rot attacks mainly cellulose. White rot fungi are most effective in lignin degradation through the action of lignin-degrading enzymes such as laccase and peroxide (Lee et al., 2007), and these enzymes are regulated by nitrogen and carbon sources. Delignification of bermudagrass by white rot fungi was studied by Akin et al. (1995), who reported that biodegradation of bermudagrass stem was improved by 29%–32% after 6weeks using *Ceriporiopsis subvermispora* and by 63%–77% using *Cyathus stercoreus*. *Phanerochaete chrysosporium*, white rot, and soft rot fungi attack fungi producing lignin-degrading enzymes, lignin peroxidases, and manganese-dependent peroxidases during the second metabolism, in response to nitrogen and carbon limitation (Boominathan and Reddy, 1992). The evaluation of eight bioagents, including fungi and bacteria, was studied by Singh et al. (2008) to see the pretreatment effects on sugarcane trash, and they concluded that the C/N ratio was important for biomass pretreatment because degradation of lignocellulosic material depends on the material's C/N ratio to degrade each molecule of carbon; a definite proportion of nitrogen is required by the microorganisms, depending on the different kinds of microflora (Singh et al., 2008). The advantages associated with the biological pretreatment method are the low energy requirements and mild environmental conditions, but the low hydrolysis rate makes this process less viable. It is predicted that in next 10 years the cost of ethanol production will be reduced by using these lignocellulosic biomasses along with the improved technologies for the conversion process (Boominathan and Reddy, 1992).

3.4.2 Hydrolysis Overview

For an effective hydrolysis process, a successful pretreatment operation is essential (Gamage et al., 2010). During hydrolysis, the released sugar polymers cellulose and hemicellulose are hydrolyzed into free monomer molecules for fermentation conversion to biofuel (Chandel et al., 2007). Enzymatic hydrolysis can act as an alternative to acid hydrolysis because the process can be performed under mild conditions (temperature <50°C and pH around 5) with lower environmental impact and energy consumption (Nazhad et al., 1995). Additionally, it gives a high yield of glucose with low formation of inhibitors that are favorable for subsequent hydrolysate use in the fermentation processes; also, there are no corrosion problems associated with enzymatic hydrolysis (Harris and Beglinger, 1946). During enzymatic hydrolysis, substrate enzymatic digestibility is improved with increased porosity of the substrate and cellulose accessibility to cellulases. *Trichoderma reesei* is an efficient productive fungus and is used to cleave the β-(1,4)-glycosidic bonds, which are β-glucosidase, endoglucanases, and exoglucanases (Eggeman and Elander, 2005). However, cellulose enzymes exposed to lignin and phenolic-derived lignin have an adverse effect, which has two aspects, including nonproductive adsorption and limitation of the accessibility of cellulose to cellulase. Therefore genetic modification transforms lignin effects, and lignin has been shown to be a potential source of self-sustaining energy and added-value components. Liu et al. (2010) studied the application of metal components, namely Ca(II) and Mg(II), via lignin metal complexation, which reported enhanced enzymatic hydrolysis.

3.5 CURRENT CHALLENGES IN LIGNOCELLULOSIC BIOFUEL PRODUCTION

Lignocellulosic biomass is a potential feedstock for the production of sustainable biofuel. The recalcitrant nature of biomass and complexities is a major technical and economic obstacle in the biomass-to-biofuel production pathway (Liu et al., 2010). The complex composition of these biomasses, which covers cellulose with hemicellulose and lignin, is the main barrier (Zhu and Pan, 2010). Fermentative sugar for biofuel production is obtained from the cellulose fraction of biomass, thus it is mandatory to remove lignin and hemicellulose prior to fermentation. Since these structures are encapsulated with strong hydrogen and covalent bonds, the cell walls are not only compact but limit enzyme access to the surface (Olofsson et al., 2008). Cellulose, a polymer of glucose linked via a β-(1,4)-glycosidic bond, has a crystalline and compact structure, whereas hemicellulose is an amorphous part composed of hexose and pentose sugar linked via a β-(1,4)-xylan linkage. Thus the complexities of the cell wall are mainly due to these varieties of sugar polymer and linkages between molecules, and therefore the hydrolysis process necessitates different cost-prohibitive enzymes to cleave polysaccharides entirely into fermentable sugars. Therefore it is desirable to remove these components, including organic acid and phenolics, to get an efficient cellulose conversion via enzymatic hydrolysis (Keeny and Muller, 2006).

Many pretreatment methods are well known for conversion of these complex biomasses into simpler forms, but selection of the appropriate pretreatment method is itself a challenge because it relies on environmental, technological, and economical factors, including recycling issues, substrate recovery, wastewater, and energy savings along with minimal chemical use and maximal solid loading yield (Berndes, 2002). Acidic pretreatment with dilute acid pretreatment is extensively studied because it enhances the subsequent enzymatic hydrolysis process and it is effective at a low cost (Agbor et al., 2011; Dale and Ong, 2012); however, the main challenge with the process is the requirement of high pressure, high temperature, and low pH, which impose high demands for the material of construction and ultimately contribute to the high cost of pretreatment. Also, concentration of acid, residence time, temperature, and pressure will contribute to the formation of degradation products such as HMF, furfural, and several phenolic lignins, which can be inhibitory for downstream processing, i.e., enzyme hydrolysis and microbial fermentation. Additionally, phenolic components like phenolic hydroxyl groups can influence cellulase enzyme activities, which will directly affect the enzymatic hydrolysis process (Kurian et al., 2013; Sharma et al., 2013). Similarly, in the concentrated acid pretreatment, corrosion-resistant equipment is required, and recovery of the acid used in the process can be a challenge because it can be hazardous and toxic. Furthermore, toxic inhibitors influence the microbial growth rate and volumetric productivity, so selecting a fermentative culture from modified microorganisms will improve the microbial resistance to inhibitors (Agbor et al., 2011). In the alkaline pretreatment process, depolymerization of lignin requires a long reaction time, and low biomass concentration can be a severe obstacle. In spite of the mild environmental conditions and no chemical requirement, the biological pretreatment method has not been an effective commercial process due to its slow process rate and high cost of enzymes (Dale and Ong, 2012). The cost of enzymes is the biggest challenge, which can be resolved by (1) producing novel enzymes using different genetic engineering approaches (Agbor et al., 2011), and (2) identifying multifunctional enzymes that can hydrolyze different polysaccharide linkages (Dale and Ong, 2012). Lignocellulosic biomass varies in its composition as discussed elsewhere, and not all types of lignocellulosic material require the same type of pretreatment processes, therefore due to these heterogeneities it is important to select appropriate pretreatment and downstream strategies (Perez et al., 2002).

3.6 CONCLUSION

This chapter showed the significance of compositions and compositional analysis in sustainable biofuel production. The process of biofuel production via lignocellulosic biomass is directly related to cellulose content present in biomass and the strategies to access cellulose cost effectively. To bring the overall biofuel production costs down, compositional analysis and pretreatment strategies as well as the enzymatic hydrolysis step play crucial roles. Zone-wise identification and selection of cellulose-rich biomass may contribute to bringing down the production cost of biofuels. Moreover, suitable and cheap pretreatment strategies will also aid to develop economic processes.

Acknowledgments

Authors N.S. and P.K.M. acknowledge the Department of Chemical Engineering and Technology, IIT (BHU), Varanasi, India for providing research facilities. N.S. also acknowledges IIT (BHU), Varanasi for providing Institute PDF fellowship. MS acknowledges DST for awarding the DST-INSPIRE Faculty [IFA-13 MS-02] award.

References

Agbor, V.B., Cicek, N., Sparling, R., Berlin, A., Levin, D.B., 2011. Biomass pretreatment: fundamentals toward application. Biotechnol. Adv. 29 (6), 675–685.

Akin, D.E., Rigsby, L.L., Sethuraman, A., Morrison III, W.H., Gamble, G.R., Eriksson, K.E.L., 1995. Alterations in structure, chemistry, and biodegradability of grass lignocellulose treated with the white rot fungi *Ceriporiopsis subvermispora* and *Cyathus stercoreus*. Appl. Environ. Microbiol. 61, 1591–1598.

Al-Salem, S., Lettieri, P., Baeyens, J., 2009. Recycling and recovery routes of plastic solid waste (PSW): a review. Waste Manag. 29, 2625–2643.

Alvira, P., Tomás-Pejó, E., Ballesteros, M., Negro, M.J., 2010. Pretreatment technologies for an efficient bioethanol production process basedon enzymatic hydrolysis: a review. Bioresour. Technol. 101 (13), 4851–4861.

Aspinall, G.O., 1980. Chemistry of cell wall polysaccharides. In: Preiss, J. (Ed.), Thebiochemistry of Plants (a Comprehensive Treatise). Carbohydrates Structureand Function. vol. 3. Academic Press, NY, pp. 473–500.

ASTM D3172-13, 2013. Standard Practice for Proximate Analysis of Coal and Coke. ASTM International, West Conshohocken, PA.

Banerjee, S., Mudliar, S., Sen, R., Balendu, B.G., Chakrabarti, T., Pandy, R.A., 2009. Commercializing lignocellulosic bioethanol: technology bottlenecks and possible remedies. Biofuel Bioprod. Biorg. 4, 77–93.

Beede, D.N., Bloom, D.E., 1995. Economics of the Generation and Management of MSW. NBERWorking Papers 5116, National Bureau of Economic Research, Inc.

Ben-Ghedalia, D., Miron, J., 1981. The effect of combined chemical and enzyme treatment on the saccharification and in vitro digestion rate of wheat straw. Biotechnol. Bioeng. 23, 823–831.

Ben-Ghedalia, D., Shefet, G., 1983. Chemical treatments for increasing the digestibility of cotton straw. J. Agric. Sci. 100, 393–400.

Beringer, T., Lucht, W., Schaphoff, S., 2011. Bioenergy production potential of global biomassplantations under environmental and agricultural constraints. GCB Bioenergy 3, 299–312.

Berndes, G., 2002. Bioenergy and water e the implications of large-scale bioenergy production for water use and supply. Glob. Environ. Chang. 12, 253–271.

Bian, H.Y., Zhou, Z.G., Chen, B.L., Jiang, G.H., 2004. Biological synthesis of cellulose during cotton fiber thickening process. Cotton Sci. 16 (6), 374–378.

Boominathan, K., Reddy, C.A., 1992. cAMP-mediated differential regulation of lignin peroxidase and manganese-dependent peroxidases production in the white-rot basidiomycete *Phanerochaete chrysosporium*. Proc. Natl. Acad. Sci. U. S. A. 89 (12), 5586–5590.

Byrt, C.S., Grof, C.P.L., Furbank, R.T., 2011. C4 plants as biofuel feedstocks: optimising biomass production and feedstock quality from a lignocellulosic perspective. J. Integr. Plant Biol., 153,120–135.

Cardona, C.A., Sanchez, O.J., 2007. Fuel ethanol production: process design trends and integration opportunities. Bioresour. Technol. 98, 2415–2457.

Çetinkol, Ö.P., Dibble, D.C., Cheng, G., Kent, M.S., Knierim, B., Auer, M., Wemmer, D.E., Pelton, J.G., Melnichenko, Y.B., Ralph, J., Simmons, B.A., Holmes, B.M., 2010. Understanding the impact of ionic liquid pretreatment on eucalyptus. Biofuels 1 (1), 33–46.

Chandel, A.K., Singh, V., 2011. Weedy lignocellulosic feedstock and microbial metabolicengineering: advancing the generation of 'biofuel'. Appl. Microbiol. Biotechnol. 89, 1289–1303.

Chandel, A.K., Chan, E., Rudravaram, R., Narasu, M.L., Rao, L.V., Ravindra, P., 2007. Economics and environmental impact of bioethanol production technologies: an appraisal. Biotechnol. Mol. Biol. Rev. 2, 14–32.

Chang, V.S., Holtzapple, M.T., 2000. Fundamental factors affecting biomass enzymatic reactivity. Appl. Biochem. Biotechnol. 84–86, 5–37.

Chovau, S., Degrauwe, S., Bruggen, B.V., 2013. Critical analysis for techno-economic estimate for the production cost of lignocellulosic bioethanol production. Renew. Sust. Energ. Rev. 26, 307–321.

Cook, C., Devoto, A., 2011. Fuel from plant cell walls: recent developments in second generationbioethanol research. J. Sci. Food Agric. 91, 1729–1732.

Dale, B.E., Ong, R.G., 2012. Energy, wealth, and human development: why and how biomass pretreatment research must improve. Biotechnol. Prog. 28 (4), 893–898.

Demirbas, A., 2005. Bioethanol from cellulosic materials: a renewable motor fuel from biomass. Energy Sources 27, 327–337.

Demirbas, A., 2007. Progress and recent trends in biofuels. Prog. Energy Combust. Sci. 33, 1–18.

Eggeman, T., Elander, R.T., 2005. Process and economic analysis of pretreatments technologies. Bioresour. Technol. 96, 2019–2025.

Elshafei, A.M., Vega, J.L., Klasson, K.T., Clausen, E.C., Gaddy, J.L., 1991. The saccharification of corn stover by cellulase from *Penicillin funiculosum*. Bioresour. Technol. 35, 73–80.

Fan, L.T., Gharpuray, M.M., Lee, Y.-H., 1987. Cellulose hydrolysis. In: Biotechnology Monographs. vol. 3. Springer, Berlin, pp. 57.

Fox, D.J., Gray, P.P., Dunn, N.W., Warwick, L.M., 1989. Comparison of alkali and steam (acid) pretreatments of lignocellulosic materials to increase enzymic susceptibility: evaluation under optimized pretreatment conditions. J. Chem. Technol. Biotechnol. 44, 135–146.

Gamage, J., Howard, L., Zisheng, Z., 2010. Bioethanol production from lignocellulosic biomass. J. Biobased Mater. Bioenerg. 4, 3–11.

Girio, F.M., Fonseca, C., Carvalheiro, F., Duarte, L.C., Marques, S., Bogel-Lukasic, R., 2010. Hemicellulose. Bioresour. Technol. 101, 4775–4800.

Harris, E.E., Beglinger, E., 1946. The Madison Wood-Sugar Process. Report no R1617. US Department of Agriculture, Forest Products Laboratory, Madison, Wisconsin.

Hill, J., Nelson, E., Tilman, D., Polasky, S., Tiffany, D., 2006. Environmental, economic, and energeticcosts and benefits of biodiesel and ethanol biofuels. Proc. Natl. Acad. Sci. USA A103, 11206–11210.

INL DOE biomass feedstock library, 2015. *bioenergy.inl.gov*. (Accessed October 3, 2015).

Jordan, D.B., Bowman, M.J., Braker, J.D., Dien, B.S., Hector, R.E., Lee, C.C., Mertens, J.A., Wagschal, K., 2012. Plant cell walls to ethanol. Biochem. J. 442, 241–252.

Jorgensen, H., Kristensen, J.B., Felby, C., 2007. Enzymatic conversion of lignocellulose into fermentable sugars: challenges and opportunities. Biofuels Bioprod. Biorefin. 1, 119–134.

Keeny, D., Muller, M., 2006. Waste Use by Ethanol Plants Potential Challenges. 7 Institute for Agriculture and Trade Policy, Minneapolis, MN.

Kilzer, F.J., Broido, A., 1965. Speculations on the nature of cellulose pyrolysis. Pyrodynamics 2, 151–163.

Kim, S., Dale, B.E., 2004. Global potential bioethanol production from wasted crops and cropresidues. Biomass Bioenergy 26, 361–375.

Kim, S., Holtzapple, M.T., 2006. Effect of structural features on enzyme digestibility of corn stover. Bioresour. Technol. 97, 583–591.

Kuhad, R.C., Singh, A., Eriksson, K.E., 1997. Microorganisms and enzymes involved in degradation of plant fiber cell walls. Adv. Biochem. Eng. Biotechnol. 57, 45–125.

Kullander, S., 2010. Food security: crops for people not for cars. Ambio 39, 249–256.

Kurian, J.K., Nair, G.R., Hussain, A., Raghavan, G.S.V., 2013. Feedstocks, logistics and pre-treatment processes for sustainable lignocellulosic biorefineries: a comprehensive review. Renew. Sust. Energ. Rev. 25, 205–219.

Ladisch, M.R., Mosier, N.S., Kim, Y., Ximenes, E., Hogsett, D., 2010. Converting cellulose tobiofuels. SBE special supplement biofuels. CEP 106 (3), 56–63.

Lal, R., 2005. World crop residues production and implications of its use as a biofuel. Environ. Int. 31, 575–584.

Lee, J.W., Gwak, K.S., Park, J.Y., Park, M.J., Choi, D.H., Kwon, M., Choi, I.G., 2007. Biological pretreatment of softwood *Pinus densiflora* by three white rot fungi. J. Microbiol. 45 (6), 485–491.

Lee, S.H., Doherty, T.V., Linhardt, R.J., Dordick, J.S., 2009. Ionic liquid-mediated selective extraction of lignin from wood leading to enhanced enzymatic cellulose hydrolysis. Biotechnol. Bioeng. 102 (5), 1368–1376.

Li, X., Weng, J.K., Chapple, C., 2008. Improvement of biomass through lignin modification. Plant J. 54, 569–581.

Limayem, A., Ricke, S.C., 2012. Lignocellulosic biomass for bioethanol production: current perspectives, potential issues and future prospects. Prog. Energy Combust. Sci. 38, 449–467.

Liu, H., Zhu, J.Y., Fu, S., 2010. Effects of lignin-metal complexation on enzymatic hydrolysis of cellulose. J. Agric. Food Chem. 58, 7233–7238.

Liu, S., 2010. Woody biomass: niche position as a source of sustainable renewable chemicals and energy and kinetics of hot-water extraction/ hydrolysis. Biotechnol. Adv. 28 (5), 563–582.

Lynd, L.R., Wyman, C.E., Gerngross, T.U., 1999. Biocommodity engineering. Biotechnol. Prog. 15, 777–793.

Lynd, L.R., Weimer, P.J., van Zyl, W.H., Pretorius, I.S., 2002. Microbial cellulose utilization: fundamentals and biotechnology. Microbiol. Mol. Biol. Rev. 66, 506–577.

MacDonald, D.G., Bakhshi, N.N., Mathews, J.F., Roychowdhury, A., Bajpai, P., Moo-Young, M., 1983. Alkali treatment of corn stover to improve sugar production by enzymatic hydrolysis. Biotechnol. Bioeng. 25, 2067–2076.

McMillan, J.D., 1993. Pretreatment of lignocellulosic biomass. In: Himmel, M.E., Baker, J.O., Overend, R.P. (Eds.), Enzymatic Conversion of Biomass for Fuel Production. American Chemical Society, Washington, DC, pp. 292–323.

McMillan, J.D., 1994. Pretreatment of lignocellulosic biomass. In: Himmel, M.E., Baker, J.O., Overend, R.P. (Eds.), Enzymatic Conversion of Biomass for Fuels Production. American Chemical Society, Washington, DC, pp. 292–324.

McMillan, J.D., 1997. Bioethanol production: status and prospectus. Renew. Energy 10, 295–302.

Merwe, A.B.V.D., Cheng, H., Gorgens, J.F., Knoetze, J.H., 2013. Comparison of energy efficiency and economics of process designs for biobutanol production from sugarcane molasses. Fuel 105, 451–458. https://doi.org/10.1016/j.fuel.2012.06.058.

Mielenz, J.R., 2001. Ethanol production from biomass: technology and commercializationstatus. Curr. Opin. Microbiol. 4, 324–325.

Millet, M.A., Baker, A.J., Scatter, L.D., 1976. Physical and chemical pretreatment for enhancing cellulose saccharification. Biotechnol. Bioeng. Symp. 6, 125–153.

Moon, H.C., Song, I.S., Kim, D.H., 2010. Concentrated acid impregnation, hydrolysis, and fermentation of reed to ethanol. Int. J. Green Energy 7, 5.

Mosier, N., Wyman, C., Dale, B.E., Elander, R., Lee, Y.Y., Holtzapple, M., 2005. Features of promising technologies for pretreatment of lignocellulosic biomass. Bioresour. Technol. 96, 673–686.

Mu, D., Seager, T., Suresh Rao, P., Zhao, F., 2010. Comparative life cycle assessment of lignocellulosic ethanol production: biochemical versus thermochemical conversion. Environ. Manag. 46, 565–578.

Nazhad, M.M., Ramos, L.P., Paszner, L., Saddler, J.N., 1995. Structural constraints affecting the initial enzymatic hydrolysis of recycled paper. Enzyme Microb. Technol. 17, 66–74.

Neely, W.C., 1984. Factors affecting the pretreatment of biomass with gaseous ozone. Biotechnol. Bioeng. 26, 59–65.

Okano, K., Kitagaw, M., Sasaki, Y., Watanabe, T., 2005. Conversion of Japanese red cedar (*Cryptomeria japonica*) into a feed for ruminants by white-rot basidiomycetes. Anim. Feed Sci. Technol. 120, 235–243.

Olofsson, K., Bertlisson, M., Liden, G., 2008. A short review on SSF e an interesting process option from lignocellulosic feedstocks. Biotechnol. Biofuels 1, 1–7.

Pandey, A., Srivastava, N., Sinha, P., 2011. Optimization of hydrogen production by Rhodobacter sphroides NMBL-01. Biomass Bioenergy 37, 251–256.

Pauly, M., Keegstra, K., 2008. Cell wall carbohydrates and their modification as a resource for biofuels. Plant J. 54, 559–568.

Perez, J., Dorado, J.M., Rubia, T.D., Martinez, J., 2002. Biodegradation and biological treatment of cellulose, hemicellulose and lignin: an overview. Int. Microbiol. 5, 53–63.

Puri, M., Abraham, R.E., Colin, I., Barrow, J., 2012. Biofuel production: prospects, challenges andfeedstock in Australia. Renew. Sust. Energ. Rev. 16, 6022–6031.

Quesada, J., Rubio, M., Gomez, D., 1999. Ozonation of lignin rich solid fractions from corn stalks. J. Wood Chem. Technol. 19, 115–137.

Rass-Hansen, J.H., Falsig, B., Jorgensen, Christensen, C.H., 2007. Bioethanol: fuel or feedstock. J. Chem. Technol. Biotechnol. 82, 329–333.

Reddy, K.O., Maheswari, C.U., Shukla, M., Rajulu, A.V., 2012. Chemical composition and structuralcharacterization of Napier grass fibers. Mater. Lett. 67, 35–38.

Saha, B.D., 2003. Hemicellulose bioconversion. J. Ind. Microbiol. Biotechnol. 30, 279–291.

Sanchez, O.J., Cardona, C.A., 2008. Trends in biotechnological production of fuel ethanol fromdifferent feedstocks. Bioresour. Technol. 99, 5270–5295.

Sánchez, Ó.J., Cardona, C.A., 2008. Trends in biotechnological production of fuel ethanol from different feedstocks. Bioresour. Technol. 99, 5270–5295.

Serrano-Ruiz, J.C., West, R.M., Dumesic, J., 2010. Catalytic conversion of renewable biomass resources to fuels and chemicals. Chem. Biomol. Eng. Ann. Rev. 1, 79–100.

Shafizadeh, F., Bradbury, A.G.W., 1979. Thermal degradation of cellulose in air and nitrogen at low temperatures. J. Appl. Polym. Sci. 23, 1431–1442.

Sharma, B., Ingalls, R.G., Jones, C.L., Khanchi, A., 2013. Biomass supply chain design and analysis: basis, overview, modeling, challenges, and future. Renew. Sust. Energ. Rev. 24, 608–627.

Singh, P., Suman, A., Tiwari, P., Arya, N., Gaur, A., Shrivastava, A.K., 2008. Biological pretreatment of sugarcane trash for its conversion to fermentable sugars. World J. Microbiol. Biotechnol. 24, 667–673.

Sivakumar, G., Vali, D.R., Xu, J., Burner, D.M., Jr, J.O.L., Ge, X., Weathers, P.J., 2010. Bioethanol and biodiesel: alternative liquid fuels for future generations. Eng. Life Sci. 10 (1), 8–18.

Soto, M.L., Dominguez, H., Nunez, M.J., Lema, J.M., 1994. Enzymatic saccharification of alkali-treated sunflower hulls. Bioresour. Technol. 49, 53–59.

Srivastava, N., Rawat, R., Oberoi, H.S., Ramteke, P.W., 2015a. A review on fuel ethanol production from lignocellulosic biomass. Int. J. Green Energy 12, 949–960. https://doi.org/10.1080/15435075.2014.890104.

Srivastava, N., Singh, J., Srivastava, M., Ramteke, P.W., Mishra, P.K., 2015b. Improved production of reducing sugars from rice straw using crude cellulase activated with Fe_3O_4/alginate nanocomposite. Bioresour. Technol. 183, 262–266. https://doi.org/10.1016/j.biortech.2015.02.059.

Sun, Y., Cheng, J., 2002. Hydrolysis of lignocellulosic materials for ethanol production: a review. Bioresour. Technol. 83, 1–11.

Valentine, J., Clifton-Brown, J., Hastings, A., Robson, P., Allison, G., Smith, P., 2012. Food vs. fuel: the use of land for lignocellulosic 'next generation' energy crops that minimize competition with primary food production. GCB Bioenergy 4, 1–19.

Vassilev, S.V., Baxter, D., Andersen, L.K., Vassileva, C.G., Morgan, T.J., 2012. An overview of the organic and inorganic phase composition of biomass. Fuel 94, 1–33. https://doi.org/10.1016/j.fuel.2011.09.030.

Vidal, B.C., Dien, B.S., Ting, K.C., Singh, V., 2011. Influence of feedstock particle size on lignocellulose conversion. Appl. Biochem. Biotechnol. 164, 1405–1421.

Wolfrum, E.J., Sluiter, A.D., 2009. Improved multivariate calibration models for corn stover feedstock and dilute-acid pretreated corn stover. Cellulose 16 (4), 567–576.

Wyman, C.E., Dale, B.E., Elander, R.T., Holtz apple, M., Ladisch, M.R., Lee, Y.Y., 2005. Coordinated development of leading biomass pretreatment technologies. Bioresour. Technol. 96, 1959–1966.

Yang, B., Wyman, C.E., 2008. The key to unlocking low-cost cellulosic ethanol. Biofuels Bioprod. Biorefin. 2, 26–40.

Yat, S.C., Berger, A., Shonnard, D.R., 2008. Kinetic characterization of dilute surface acid hydrolysis of timber varieties and switchgrass. Bioresour. Technol. 99, 3855–3863.

Yu, Y., Lou, X., Wu, H.W., 2008. Some recent advances in hydrolysis of biomass in hot-compressed water and its comparisons with other hydrolysis methods. Energy Fuel 22 (1), 46–60.

Zhang, P., Hu, H.R., Shi, S.L., 2006. Application of hemicellulose. Tianjin Pap Mak 2, 16–18.

Zhang, J.Q., Lin, L., Sun, Y., Mitchell, G., Liu, S.J., 2008. Advance of studies on structure and decrystallization of cellulose. Chem. Ind. For. Prod. 28 (6), 109–114.

Zhang, Q.P., Pei, Z.J., Wang, D., 2013. Relationships between cellulosic biomass particle size and enzymatic hydrolysis sugar yield: analysis of inconsistent reports in the literature. Renew. Energy 60, 127–136.

Zhang, Y.-H.P., 2010. Renewable carbohydrates are a potential high-density hydrogen carrier. Int. J. Hydrog. Energy 35, 10334–10342.

Zhang, Y.-H.P., Mielenz, J.R., 2011. Renewable hydrogen carrier-carbohydrate: constructing the carbon-neutral carbohydrate economy. Energies 4, 254–275.

Zheng, Y., Pan, Z., Zhang, R., 2009. Overview of biomass pretreatment for cellulosic ethanol production. Int. J. Agric. Biol. Eng. 2, 51–68.

Zhu, J.Y., Pan, H.J., 2010. Woody biomass treatment for cellulosic ethanol production: technology and energy consumption evaluation. Bioresour. Technol. 101, 4992–5002.

Zhu, J.Y., Wang, G.S., Pan, X.J., Gleisner, R., 2009a. Specific surface to evaluate the efficiencies of milling and pretreatment of wood for enzymatic saccharification. Chem. Eng. Sci. 64, 474–485.

Zhu, Z.G., Sathitsuksanoh, N., Vinzant, T., Schell, D.J., McMillan, J.D., Zhang, Y.H.P., 2009b. Comparative study of corn stover pretreated by dilute acid and cellulose solvent-based lignocellulose fractionation: enzymatic hydrolysis, supramolecular structure, and substrate accessibility. Biotechnol. Bioeng. 103 (4), 715–724.

Further Reading

Carroll, A., Somerville, C., 2009. Cellulosic biofuels. Annu. Rev. Plant Biol. 60, 165–182.

Jiang, T.D., 2001. Lignin. Chemical Industry Press, Beijing.

Wichern, F., Miiller, T., Joergensen, R.G., Buerkert, A., 2004. Effect of manure quality and application forms on soil C and N turnover of a subtropical oasis soil under laboratory conditions. Biol. Fertil. Soils 39, 165–171.

CHAPTER

4

Microorganisms as an Efficient Tool for Cellulase Production: Availability, Diversity, and Efficiency

Vincent Vineeth Leo[*,†], *Zothanpuia*[*], *Lalthafala*[*], *N. Ramesh*[†,‡], *Bhim Pratap Singh*[*]

[*]Molecular Microbiology and Systematic Laboratory, Department of Biotechnology, Mizoram University, Aizawl, India [†]Department of Biotechnology, J.J College for Arts and Science, Pudukkottai, India [‡]PG and Research Department of Botany, Govt. Arts College for Men, Krishnagiri, India

4.1 INTRODUCTION

Microorganisms from diverse natural habitats have adapted to develop pathways for the production of several extracellular enzyme systems for the conversion of cellulosic substrates to simpler sugars and allied products, which may be further converted to useful energy sources (Payne et al., 2015). The complexity of arrangement of celluloses entangled with hemicellulose and lignin makes the action of cellulase enzymes complex. Hence the naturally occurring biomass-degrading microbes produce a cluster of hydrolytic enzymes of synergistic effects to tackle the complex-natured biomasses (Sukumaran et al., 2009; Leo et al., 2016). Two major paradigms of cellulase have been in action for cellulose depolymerization: the "free" enzyme paradigm and the cellulosomal paradigm. Cellulase secreted by aerobes tends to produce cellulase individually or "free" enzymes, which have been categorized into exocellulases or exoglucanases or cellobiohydrolases (which cleave the reducing and nonreducing ends of cellulosic polymers), endocellulases or endoglucanases (which cleave open the internal linkages within the polymeric celluloses or the amorphous cellulose), and β-glucosidases (which help to convert the endo- and exocellulase product to monomeric sugar units). Cellulase produced by anaerobes for enzymatic cellulose depolymerization employs another paradigm, "cellulosome," which is a large complex of enzymes of cellulolytic properties that are noncovalently bound by cohesion-dockerin interactions in close proximity on large protein scaffolds. Multimodular hydrolytic enzymes with multiple catalytic domains within the individual enzymes have also been identified, which mark the way for another paradigm of cellulase action (Payne et al., 2015).

Cellulases have been well documented based on their catalytic- and carbohydrate-binding properties in an exclusive database of carbohydrate-active enzymes (CAZymes) (http://www.cazy.org/). The majority of microbial cellulases are based on their catalytic activities, and hydrolytic properties have been grouped within glycoside hydrolases (GHs) GH1, GH3, GH5, GH6, GH8, GH9, GH12, GH18, GH19, GH44, GH45, GH48, GH51, and GH74. The carbohydrate binding modules (CBMs) of cellulases that bind to celluloses along with hemicelluloses are CBM1, CBM2, CBM3, CBM6, CBM8, CBM30, and CBM44 (López-Mondéjar et al., 2016).

Several fungi and bacteria that are known to reside over a vast range of environmental niches produce cellulases that are extremely resistant to environmental stresses. Though the most studied microbes capable of cellulase

From Cellulose to Cellulase: Strategies to Improve Biofuel Production
https://doi.org/10.1016/B978-0-444-64223-3.00004-7

production are from mesophilic conditions, industrial needs and the need for more enhanced activity enzymes have prompted the focus on microbial extremophiles. Hence the microbes residing in thermophilic, psychrophilic, alkaliphilic, acidophilic, etc. have been explored for their capability of producing extreme conditions tolerant enzymes. The stability of these cellulases could ensure efficiency for the bioconversion process, which in turn enhances rates of catalytic hydrolysis, fermentation, and subsequent product recovery (Maki et al., 2009).

Most of the commercially available cellulase has aerobic microbial origin from either fungi or bacteria. From an application perspective, though cellulase finds its primary usage in the pretreatment of plant biomass feedstocks, it also has other industrial usages ranging from textile, detergent, paper, food, and even animal feed industries. Given the ever-growing demand for fast-acting and more durable cellulases, various improvement techniques have been involved in the enhancement of its activity (Sukumaran et al., 2009). In this perspective we discuss in detail the availability of microbes capable of cellulase production, the diversity of their habitats, and the enhancement techniques used to improve the activities of the cellulases.

4.2 MICROBIAL AVAILABILITY FOR CELLULASE PRODUCTION

Microorganisms are the most prolific source for the production of hydrolytic enzymes like cellulases. The major group of microorganisms having the ability to produce cellulase are fungi, actinobacteria, eubacteria, and archaea, which are explained in brief in the sections that follow. Microbial cellulase availability, sources, and forms are represented in Fig. 4.1.

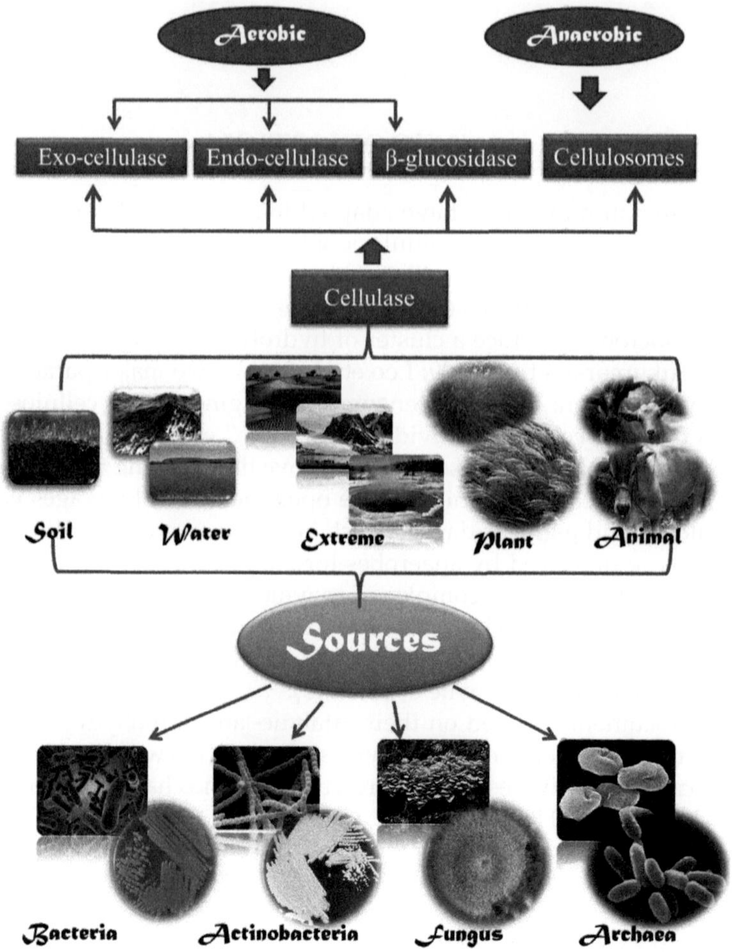

FIG. 4.1 Microbial cellulase sources, diversity, and forms.

4.2.1 Fungal Cellulase Producers

Among the microorganisms, the eukaryotic filamentous fungi are the most dynamic cellulose decomposer and account for the majority of the cellulosic components within the naturally occurring lignocellulosic biomasses of the world. Fungal cellulases are mainly seen among phyla Ascomycota and Basidiomycota, respectively, with a few reported from other phyla. The most effective and explored fungal cellulases belong to the species of *Trichoderma, Aspergillus, Penicillium, Chaetomium, Fusarium, Cladosporium, Myrothecium, Humicola*, etc. (Helal, 2005).

4.2.2 Bacterial Cellulase Producers

Bacteria are also renowned cellulose degraders by producing cellulases. Though a few of the bacterial cultures are capable of synthesis the complete spectra of cellulolytic enzymes needed for hydrolyzing crystalline cellulose are found in nature. Among them, anaerobic genera of *Clostridium* and *Acetivibrio* and aerobic genera of *Bacillus* and *Pseudomonas* are most extensively studied (Sadhu and Maiti, 2013; Kuhad et al., 2011).

4.2.3 Actinobacterial Cellulase Producers

Phylum Actinobacteria is mainly aerobic and the cellulases produced are very noncomplexed (Anderson et al., 2012). This has particularly been of huge interest from an industrial perspective, because actinobacteria are capable of surviving in harsher conditions than many microbes. The few known anaerobic cellulase-producing actinobacteria are *Actinomycosis bovis* and *Cellulomonas flavigena*. Among the actinobacterial strains studied, *Cellulomonas fimi, Thermobifida fusca, Thermomonospora curvata*, and *Xylanimonas cellulosilytica* have been reported as cellulase producers (Anderson et al., 2012; Saini et al., 2015; Lynd et al., 2002).

4.2.4 Archaeal Cellulase Producers

Among the few Archaea cellulases reported, most are from extreme habitats. *Desulfurococcus fermentans* and *Thermogladius cellulolyticus* have been reported for certain cellulase genes, while hyperthermophilic Archaea are *Pyrococcus furiosus, Pyrococcus horikoshii*, and *Sulfolobus solfataricus* (Wilson, 2016; Graham et al., 2011; Bauer et al., 1999; Ando et al., 2002; Huang et al., 2005).

4.3 MICROBIAL DIVERSITY FOR CELLULASE PRODUCTION

Given the capability of the microorganisms (fungi, bacteria, actinobacteria, and archaea) to exist in diverse habitats the versatility of cellulase produced is also enormous. The diversity of microorganisms that produce cellulase and their properties is described in the following and represented in Table 4.1.

4.3.1 Soil-Derived Microbial Cellulase

Soil with its immense repository of microorganisms has been the most exploited for obtaining cellulosic-degrading cellulase enzymes. Plant litter is the major source for cellulosic substrates; its natural decomposers might reside within such ecosystems. Hence many studies have been carried out with forest soil and forest litter as the major source for obtaining microbial cellulases. However, polluted soil, agriculture land soil, and extreme climate soil have all been explored for this purpose (Lin and Stutzenberger, 1995; Da-Vinha et al., 2011; Prasad et al., 2013; Carrasco et al., 2016). Fungi are believed to be the chief decomposer of naturally occurring plant biomasses like forest litters. Among the fungal genera *Trichoderma, Penicillium, Aspergillus, Humicola, Acremonium, Fusarium, Paecilomyces, Chaetomium, Mortierella, Coprinopsis, Trichosporon, Cryptococcus*, etc. are some of the well-known cellulase producers (Avellaneda-Torres et al., 2014; Stursová et al., 2012; Reanprayoon and Pathomsiriwong, 2012). However, bacterial cultures as a source for cellulases have also made niches for themselves, being fast and easy producers of cellulase. Numerous bacterial cultures have been isolated from soil ecosystems with major studies carried out in *Bacillus, Pseudomonas*, and *Clostridium* with other genera like *Thermoanaerobacterium, Serratia, Stenotrophomonas, Paenibacillus, Variovorax, Janthinobacterium*, etc. (López-Mondéjar et al., 2016; Sethi et al., 2013; Stursová et al., 2012). Among the actinobacterial cellulase-producing genera *Streptomyces, Thermomonospora, Cellulomonas, Rhodococcus, Nocardia, Actinomadura, Micromonospora, Actinopolyspora, Streptosporangium*, etc. are the most studied and explored from soil ecosystems (Saini et al., 2016; Stursová et al., 2012; López-Mondéjar et al., 2016).

TABLE 4.1 An Outlook Into Microbial Cellulase Diversities and Their Catalytic Properties

Microbial species	Source	Enzyme	Features/actions	References
FUNGI				
Aureobasidium pullulans	Marine	CMCase	pH5.6 and 40°C	Rong et al. (2015)
Cladosporium sphaerospermum		CMCase, FPase	pH4 and 25°C	Trivedi et al. (2015)
Talaromyces emersonii	Soil	Endoglucanase (endo-β-glucanase) with cellobiohydrolase and FPase	pH4.8 and 80°C Thermostable	Busk and Lange (2013); Murray et al. (2001); Brienzo et al. (2008)
Chaetomium thermophilum		FPase	70°C	Busk and Lange (2013)
Aspergillus wentii		Cellulase	60–65°C	Kvesitadze (2017)
Aspergillus versicolor				
Aspergillus terreus			pH2.8–6.0	Sohail et al. (2016)
Penicillium citrinum		Endoglucanase	pH8.0	Dutta et al. (2008)
Rhodotorula glacialis	Soil samples of sub-Antarctic region	CMCase	4–22°C	Carrasco et al. (2016)
Mrakia blollopis				
Penicillium roqueforti	Antarctic wood	Endoglucanase	Below 15°C	Duncan et al. (2006)
Cadophora malorum				
Trichoderma sp. IS-05	Coastal plains	Cellulase	pH3.0	Andrade et al. (2011)
Trichoderma atroviride 102C1	Amazon rainforest soil		pH3.6 and 66°C	De-Oliveira et al. (2016)
Aspergillus fumigatus	Sugarcane bagasse	CMCase	pH1.0 and 65°C	Grigorevski-Lima et al. (2009)
A. terreus AUMC 10138	Soda lake	FPase, CMCase, β-glucosidase	pH9.0–11.0.	Isaac and Abu-Tahon (2015)
Aspergillus nigri (niger)	Endophytic tissues of Madagascar periwinkle	CMCase	pH5.0 and 70°C	Yopi et al. (2017)
BACTERIA				
Bacillus subtilis subsp. *subtilis*	Marine/marine sediments	CMCase, FPase, cellobiase	50°C	Kim et al. (2009)
Bacillus halodurans		Cellulase	pH12 and 80°C	Annamalai et al. (2013)
Pseudomonas sp.		CMCase (Cel5M)	10–30°C	Yang and Dang (2011)
Rhodothermus marinus		Cel 2A (GH10) and Cel A (GH12)	Thermostable	Hamed et al. (2017)
Alteromonas haloplanktis	Antarctic region	CMCase	5–20°C	Buchon et al. (2000)
Clostridium sp.	Psychrophilic climate	Endoglucanase	15°C	Akila and Chandra (2003)
Clostridium thermocellum	Compost	Cellulase	60°C	Lv and Yu (2013)
Bacillus licheniformis	Geothermal wells	CMCase	Thermostable (60–80°C)	Kiio et al. (2016)
Aquifex aeolicus	Hot spring	Cellulase	95°C	Deckert et al. (1998)
Thermotoga sp.	Geothermally heated areas	Exo- and endocellulase	100–106°C	Ruttersmith and Daniel (1991)
Alicyclobacillus acidocaldarius	Acidic hot spring	CMCase	pH4.0 and 80°C	Eckert and Schneider (2003)
Bacillus sp.	Gastrointestinal tract of Tibetan pig	CMCase and avicelase	pH4.0 and 60°C	Ma et al. (2015)

TABLE 4.1 An Outlook Into Microbial Cellulase Diversities and Their Catalytic Properties—cont'd

Microbial species	Source	Enzyme	Features/actions	References
Marinimicrobium sp.	Marine solar saltern	CMCase	pH 11.0	Zhao et al. (2012)
Gracilibacillus sp.	Salt lake	Exo-/endocellulase and β-glucosidase	pH 6.0–10.0 (alkali stable)	Yu and Li (2015)
Serratia rubidaea DBT4 *Aneurinibacillus aneurinilyticus* DBT87	Goat gut/swine gut	Cellulase (CMCase and FPase)	pH 8 and 55°C	Asem et al. (2017)
Burkholderia cepacia BPS-GB3 *Alcaligenes faecalis* BPS-GB5 *Enterobacter hormaechei* BPS-GB8	Perennial grass verities (*Nayraudia reynaudiana* and *Sacharrum arundinaceum*)	β-Glucosidase (5.36–12.34 IU), FPase (0.26–0.53 IU), CMCase (2.31–4.65 IU)	Triculture study resulted in 55%–70% loss in dry weight on perennial biomass feedstock	Leo et al. (2016)
Bacillus amyloliquefaciens	Rhinoceros dung	CMCase	Activity optima 0.079 U/mL within 48 h	Singh et al. (2013)
Ochrobactrum sp.	Gastrointestinal tract of giant African snail	CMCase and avicelase	CMCase activity of 501.75 IU/mL (K38) extract and avicelase activity of 3116.92 IU/mL extract (Cf60)	Dar et al. (2015)
ACTINOBACTERIA				
Micromonospora sp.	Fresh water	Cellulase	Native cellulose-like cotton degradation	De-Menezes et al. (2008)
Actinoalloteichus sp.	Marine		Cellulase activity of 14.379 IU/mL	Gobalakrishnan and Kannan (2017)
Arthrobacter sulfonivorans *Brevibacterium antarcticum* *Streptomyces beijiangensis* *Streptomyces cellulosae* *Rhodococcus* sp. *Streptomyces thermospinosisporus*	South Shetland Islands, Antarctica		Cold stable cellulases	Lamilla et al. (2017)
Acidothermus cellulolyticus	Acidic hot springs	Avicelase, FPase, CMCase, and β-glucosidase	pH 4.0–6.0 and thermostable	Ding et al. (2003); Tucker et al. (1989)
Streptomyces omiyaensis	Goat rumen	Avicelase	Thermostable	Alam et al. (2004)
Thermomonospora curvata	Solid waste compost	CMCase	70°C	Lin and Stutzenberger (1995)
Thermomonospora sp.			pH 7.0–10.0 and 50°C	George et al. (2001)
Streptomyces viridobrunneus	Sugarcane plantation soil	CMCase	pH 4.9 and 50°C	Da-Vinha et al. (2011)
Streptomyces misionensis		Endoglucanase	pH 3.0–3.6 and 62–70°C	Franco-Cirigliano et al. (2013)
Streptomyces griseoaurantiacus	Municipal waste	Endoglucanase and β-glucosidase	pH up to 11.0 and 45°C	Prasad et al. (2013); Kumar (2015)
Amycolatopsis cihanbeyliensis	Salt mine	CMCase	pH 8 (activity even at 6% addition of NaCl)	Adıgüzel and Tunçer (2017)

Continued

TABLE 4.1 An Outlook Into Microbial Cellulase Diversities and Their Catalytic Properties—cont'd

Microbial species	Source	Enzyme	Features/actions	References
Kocuria sp. *Streptomyces thermocarboxydus* *Streptomyces mutabilis* *Leifsoniaxyli* sp. *Streptomyces olivaceus* *Microbacterium testaceum*	Endophytic tissues of *Rhynchotechumellipticum*	CMCase	Enzyme activity ranged between 54.6 and 74.2 U/mL	Passari et al. (2017)
ARCHAEON				
Pyrococcus furiosus	Geothermally heated marine sediments	Endoglucanase activity (GH12)	112°C	Bauer et al. (1999)
Pyrococcus horikoshii	Hydrothermal fluid		97°C	Ando et al. (2002)
Sulfolobus solfataricus	Sulfur-rich hot springs		pH 1.8 and 80°C	Huang et al. (2005)

4.3.2 Water Source-Derived Microbial Cellulase

Cellulase from microbes originating from water sources has also gained momentum over the years, mainly due to the high demand for enhanced cellulolytic active and stable enzyme requirements for the industry. Among them, marine habitats are the most explored for such enzymes with variable environments, especially with respect to pH and temperature. Although freshwater lakes and hot water springs as sources for microbial cellulases have also been investigated.

Rong et al. (2015) reported a marine yeast *Aureobasidium pullulans* 98 capable of producing high yields of carboxymethyl cellulase (CMCase) with an activity of 4.51 U (mg protein)$^{-1}$ at an optimal pH of 5.6 and temperature of 40°C. Trivedi et al. (2015) worked on the marine fungal culture *Cladosporium sphaerospermum* to utilize its cellulase production capabilities and to act on the seaweed *Ulva fasciata* as a potent biomass. The study revealed that the cellulase produced at 25°C and pH 4 showed CMCase and a filter paper activity (FPase) assay of 10.20 ± 0.40 U/g and 9.60 ± 0.64 U/g on a dry weight basis, respectively. Similarly, Xue et al. (2012) reported the action of another marine fungus *Aspergillus niger* capable of producing cellulase (17.80 U/g dry weight) on another seaweed *Eichhornia crassipes*. Zare-Maivan and Shearer (1988) reported cellulase-producing fungi belonging to Ascomycetes, Deuteromycetes and an Oomycete, showing mesophilic cellulase production and great wood-decaying activity.

Among the marine bacterial cultures, Muñoz et al. (2014) reported microbes belonging to four bacterial genera of *Aeromonas*, *Pseudomonas*, *Chryseobacterium*, and *Raoultella* having cellulase (endoglucanase) activity. These microbial cellulases were used to successfully implement enzymatic pretreatment of microalgal biomass for biogas production. The *Bacillus* genus is one of the most reported bacterial cellulase producers of marine habitats. Kim et al. (2009) reported *Bacillus subtilis* subsp. *subtilis* isolated from sea water having cellulase activity. This enzyme on purification showed enhanced activity against carboxymethyl cellulose (CMC), cellobiose, and filter paper at 50°C. Annamalai et al. (2013) reported *Bacillus halodurans* CAS isolated from marine sediments collected from the Parangipettai coast, India, having extracellular cellulase productivity, that were thermostable (80°C) and haloalkaline (pH 12.0). Yu and Li (2015) similarly reported another alkali-stable cellulase obtained from *Gracilibacillus* sp. SK1 isolated from a marine source. This alkali-stable cellulase was then used in saccharification of corn stover and rice straw for bioethanol production. Leschine and Canale-Parola (1983) reported various mesophilic *Clostridia* strains (*Clostridium cellobioparum*) isolated from freshwater environments, capable of producing cellulase at an optimal temperature of 40°C.

Freshwater actinobacteria have also been reported to produce cellulases. De-Menezes et al. (2008) reported isolation of a few *Micromonospora* strains from freshwater habitats having cellulase production capabilities. The study revealed a new method using scanning electron microscopy (SEM) analysis to find whether the *Micromonospora* strains have cellulase production capabilities. The inner penetrations caused by *Micromonospora* strains on "cotton" were correlated to the extent of degradation on cellulosic substrate, which was imaged using SEM. An actinobacterial marine cellulase producer was reported by Gobalakrishnan and Kannan (2017), where a marine actinobacterium *Actinoalloteichus* sp. MHA15 was isolated that showed cellulase activity of 14.379 IU/mL. Veiga et al. (1983), in one of the older reports in this regard, screened around 36 *Streptomyces* spp. isolated from marine sediments for their cellulase production. Out of the 36 screened, almost half of the *Streptomyces* spp. showed cellulase activity.

4.3.3 Extremophilic Microbial Cellulase

Microbes capable of surviving in harsher environmental conditions like extreme heat, cold, saline, acidic, etc. have been of great interest in recent times. Enzymes that might be derived from such extremophiles might also mimic such properties, hence their exploration has resulted in more competent and stable enzymes at extreme conditions. A few such cellulases derived from extremophiles are discussed next.

4.3.3.1 *Psychrophilic Microbial Cellulase*

Quite a major portion of the earth is covered in water bodies, so the study of and exploration for psychrophilic or psychrotolerant microorganisms (10°C to −20°C) has resulted in obtaining cold active enzymes produced by such microbes. The benefit of such enzymes like cold active cellulases is that these are more capable of being active at colder temperatures and are specific in their activity. Such enzymes find their applications in the production of fragrances, cosmetics, and juices, where cold-based juice clarifications are required (Pulicherla et al., 2011).

A number of psychrophilic fungal varieties have also been reported to show cellulase productivity. Isolates of *Penicillium roqueforti* and *Cadophora malorum* having the capability of producing endoglucanase at temperatures below 15°C were one of the earlier reports on psychrophilic fungal varieties (Duncan et al., 2006). Da-Qing et al. (2016) reported production of endoglucanase, cellobiohydrolase, and β-glucosidase from a fungal variety *Cladosporium* sp. isolated from a hypothermal litter area, with an average soil temperature of 0°C. The study reported filter paper degradation of 30% within 15 days at 10°C with an FPase activity of 0.51 U/mL. Carrasco et al. (2016) in their study identified a few yeast varieties that were psychrotolerant in nature with cellulose-degrading capabilities. Among them, cold active (4–22°C) cellulase (CMCase) productivity was shown by *Rhodotorula glacialis* and *Mrakia blollopis* sp., respectively. Extracellular hydrolase production such as cellulases from microfungi like *Geomyces* sp. and *Pseudogymnoascus* sp. isolated from Arctic and Antarctic regions has been reported by Krishnan et al. (2018).

Amid the bacterial psychrophiles that produce cold-tolerant cellulases, most have been restricted to a few cultures like *Pseudoalteromonas haloplanktis* (*Alteromonas haloplanktis*). Buchon et al. (2000) used the *A. haloplanktis* strain collected by Feller et al. (1994) from the Antarctic region to study the potential of this bacterium to produce cold active cellulase, which proved that the CMCase activity was stable at 5–20°C. This study was followed up by Garsoux et al. (2004), Zeng et al. (2006), and Wang et al. (2015), who worked on *P. haloplanktis*, *Pseudoalteromonas* sp. DY3, and *Pseudoalteromonas* sp. NJ64, respectively. These works helped to understand the potential of the cellulases produced by these bacteria and also in characterizing cold active cellulase genes like CelX and CelG. Akila and Chandra (2003) reported another psychrophilic bacterium *Clostridium* strain PXYL1 having endoglucanase activity at around 15°C, with an optimal activity shown at 20°C. Yang and Dang (2011) worked on a novel *endo*-β-1,4-glucanase (Cel5M) belonging to GH5 from *Pseudomonas* sp. MM15, a deep-sea psychrophile. Though the optimal CMCase activity of this endoglucanase obtained was at 30°C, its activity remained at 10°C, opening avenues into cloning of this particular endoglucanase and its overexpression into *Escherichia coli* DH 5a/pCEL. Dutta et al. (2016) worked on another psychrophilic strain *B. subtilis*, whose cellulase was immobilized by magnesium nanoparticles onto graphene oxide nanosupport material by glutaraldehyde as cross linker. The enzyme that was immobilized showed an increased activity of 2.98-fold at 8°C.

Lamilla et al. (2017) isolated psychrophilic actinobacterial cultures from South Shetland Islands, Antarctica, among which cold-stable cellulases were exhibited by *Arthrobacter sulfonivorans*, *Brevibacterium antarcticum*, *Rhodococcus* sp., *Streptomyces thermospinosisporus*, *Streptomyces* sp., *Streptomyces cellulosae*, and *Streptomyces beijiangensis*.

4.3.3.2 *Thermophilic Microbial Cellulase*

Thermophilic microbes have been found from naturally hot environments like deserts, hot springs, volcanic areas, etc., as well as from compost, animal wastes, or even wood chips (Busk and Lange, 2013). The main advantage of thermophilic microbes is that they produce thermostable enzymes like thermophilic cellulases, which in turn reduce or eliminate the separate thermal pretreatment steps needed at times for biomass conversion.

Thermostable cellulases have been reported previously from a small number of fungal varieties like *Myceliophthora thermophila*, *Thielavia terrestris*, *Talaromyces emersonii*, *Chaetomium thermophilum*, and *Acremonium thermophilum* with the most extensively studied fungal cultures being *Humicola insolens* and *Humicola grisea* (Maheshwari et al., 2000; Hayashida and Yoshioka, 2014; Xu et al., 2015). Busk and Lange (2013) did a screening study for cellulase producers from 16 thermophilic fungi belonging to orders of Sordariales, Eurotiales, Onygenales, and Mucorales of which *T. emersonii* and *C. thermophilum* were found to be the most efficient in thermophilic conditions. *C. thermophilum* interestingly showed more stable FPase activity than endoglucanase activity at 70°C, whereas enhanced thermostable cellulase was observed in *T. emersonii*. Kvesitadze (2017) reported cellulases from *Aspergillus wentii*, *Aspergillus versicolor*, and *C. thermophilum* at a temperature range of 60–65°C.

Lv and Yu (2013) reported the isolation of thermophilic bacteria *Clostridium thermocellum* by using the enrichment method, which showed optimal cellulase activity at 60°C and thermostablity. The culture mostly grew on cellulosic substrates like filter paper and cellobiose. The genera of *Caldicellulosiruptor* (*Caldicellulosiruptor bescii*, *Caldicellulosiruptor saccharolyticus*, and *Anaerocellum thermophilum*) that are anaerobic in nature are renowned for their production of thermophilic cellulases at temperature ranges between 70 and 80°C (Blumer-Schuette et al., 2008; Yang et al., 2009). Among the aerobic thermophilic bacteria *Rhodothermus marinus* has been studied quite extensively for its thermostable cellulases called Cel 2A (GH10) and Cel A (GH12). This Cel A was overexpressed in *Streptomyces lividans* TK24 that showed substantial cellulase productivity and activity (Hamed et al., 2017). Among the *Bacillus* species reported, thermostable cellulases with both avicelase and CMCase activity were reported in *Bacillus licheniformis* isolated from geothermal wells in the Kenyan rift valley (Kiio et al., 2016), which was similarly found in another *B. licheniformis* 380 by De-Marco et al. (2017). Among the aerobic bacteria the highest thermo-tolerant cellulase has been reported in *Aquifex aeolicus*, which was isolated from the Aeolic Islands in Sicily, and produced cellulase tolerant at 95°C (Deckert et al., 1998). Bok et al. (1998) and Ruttersmith and Daniel (1991) reported *Thermotoga* sp. to be capable of having cellulase activities (exo- and endocellulase) at higher temperatures of 100–106°C.

Leo et al. (2018) in their comprehensive review of holocellulose-degrading actinobacteria pointed out that many actinobacteria are capable of producing thermostable cellulases. *Acidothermus cellulolyticus* is one of the most studied hyperthermophilic actinobacterial varieties, which is capable of producing thermostable cellulases especially for avicelase and endoglucanases (Ding et al., 2003). Tucker et al. (1989) reported the production of FPase, CMCase, and β-glucosidase at 75, 83, and 75°C, respectively, from *A. cellulolyticus*. Among the genus *Streptomyces*, Park et al. (2001) reported slightly thermophilic exo- and endocellulase at 60°C by *Streptomyces* sp., while *Streptomyces omiyaensis* was reported to be capable of producing thermostable avicelase at 45°C (Alam et al., 2004). Lin and Stutzenberger (1995) studied *T. curvata* that showed thermostable CMCase activity at 70°C. Kim et al. (2005) showed the application of thermophilic cellulases from *T. fusca* on plant sources for its bioconversion in biofuel production.

Among the archaeons, *P. furiosus* and *P. horikoshii* are renowned hyperthermophilic cellulase producers with endoglucanase activity (GH12) being intact at 112 and 97°C, respectively (Bauer et al., 1999; Ando et al., 2002). *S. solfataricus* is another such archaeon that produces thermostable (half-life of 8h) endoglucanases (GH12) at 80°C (Huang et al., 2005).

4.3.3.3 *Acidophilic Microbial Cellulase*

Acid-tolerant cellulases have applications within the food industry, especially in the hydrolysis of coffee residues, along with direct application of bleaching in paper recycle plants (Kuhad et al., 2011). The need for such cellulases has opened avenues for exploring more acidophilic microbes capable of producing acid-stable cellulases.

Three fungal cultures *Aspergillus terreus*, *Myrothecium roridum*, and *Myrothecium verrucaria* produced cellulase at acidic pH ranges from 2.8 to 6.0 according to Gross and Robbins (2000) and Sohail et al. (2016). Andrade et al. (2011) reported *Trichoderma* sp. IS-05, isolated from sandy coastal plains of northeast Brazil, having an optimal cellulase production at a pH of 3.0. Another *Trichoderma* mutant species *Trichoderma atroviride* 102C1 also showed optimal endoglucanase production at pH3.6 and thermophilic conditions (De-Oliveira et al., 2016). However, Grigorevski-Lima et al. (2009) studied *Aspergillus fumigatus* and revealed that the optimal endoglucanase (CMCase) activity was observed at an extreme pH condition of pH1.0 and at thermophilic conditions. Nevertheless, the most studied acidophilic fungi have been *T. emersonii*, whose *endo*-β-glucanase activity has been extensively explored, though it also produces endoglucanase, cellobiohydrolase, and even FPase at acidic pH (Brienzo et al., 2008). This fungal variety showed its endo-β-glucanase production at pH4.8 and 80°C and cellulase activity of 142.9IU/mg, with structural stability at pH2.8 (Murray et al., 2001).

Many acidophilic bacteria have also been identified to have cellulase activities. *Alicyclobacillus acidocaldarius* have been reported to produce CMCase at an acidic pH of 4.0 at thermophilic temperatures of 80°C (Eckert and Schneider, 2003). Zhu et al. (2011) isolated *B. subtilis* strain BEC-1 from a forest soil ecosystem, which exhibited CMCase activity, whose gene *egl173* was cloned and overexpressed on *E. coli*. The subsequent enzyme produced was stable at pH4.0, with 85% activity maintained within the acidic pH range. Rawat and Tewari (2012) worked on another *Bacillus* sp. (*B. subtilis* strain LFS3), whose CMCase produced was purified by $(NH_4)_2SO_4$ precipitation, ion exchange, and gel filtration chromatography to obtain CMCase with activity at an acidic pH of 4.0. Ma et al. (2015) reported yet another *Bacillus* sp. with cellulase activity of 8.61U/mL when fermented on wheat bran. This enzyme on purification showed both CMCase and avicelase activity at pH optima of 4.0 and 60°C.

A. cellulolyticus is one of the most studied actinobacterial acidophilic cellulase producers. It was initially isolated from the acidic hot springs in Yellowstone National Park and, given its capability to survive in harsh temperatures and acidic conditions, its potential was explored in detail (Mohagheghi et al., 1986). *A. cellulolyticus* is renowned for its various cellulases having an acid tolerance range of pH 4–6, whose endoglucanase E1 was purified and has even been overexpressed in plants. Such plants that were modified made them suitable for feedstock improvements in biomass conversions and subsequent cellulosic ethanol conversions (Ding et al., 2003; Thomas et al., 1995; Sticklen, 2008). *Streptomyces viridobrunneus* SCPE-09 isolated from sugarcane plantation soil in Brazil was also reported to produce acidic (CMCase) cellulase (pH 4.9) when grown on wheat bran and corn steep liquid as biomass feedstock (Da-Vinha et al., 2011). Similarly, it was found that *Streptomyces misionensis* PESB-25 showed thermos-acidophilic endoglucanase production at a pH range of 3.0–6.0 (Franco-Cirigliano et al., 2013).

Among the archaeon, *S. solfataricus*, which is also thermophilic, was reported to produce *endo*-β-glucanase types such as SSO1354, SSO1949, and SSO2534, with SSO1949 enzyme showing enzyme activity at pH 1.8 and 80°C, which was deemed capable of hydrolyzing CMC, cellobiose, and cellotriose (Huang et al., 2005).

4.3.3.4 Alkaliphilic Microbial Cellulase

The early reports on alkali-tolerant (higher pH of 10) fungal cellulase producers was carried out by Solovyeva et al. (1999). In that study a large number of fungal genera like *Acremonium*, *Alternaria*, *Cylindrocarpon*, *Fusarium*, *Penicillium*, *Verticillium*, *Myrothecium*, *Ulocladium*, *Gliocladium*, *Chaetomium*, *Phoma*, *Microsphaeropsis*, and *Aposphaeria* were reported to be capable of producing alkali-stable cellulases. Vega et al. (2012), in their pursuit for obtaining alkaline cellulase producers, found three fungal varieties (*Aspergillus* sp. LMHP32, *Penicillium* sp. LM-HP33, and *Penicillium* sp. LMHP37) showing alkaline cellulase production under industrial conditions, which may find immense usage in textile industries. Dutta et al. (2008) showed that filamentous fungi *Penicillium citrinum* had potential to produce alkali-tolerant endoglucanase at pH 8.0. Halophilic fungal varieties have also been reported with recent works reported on *Aspergillus* sp. in Isaac and Abu-Tahon (2015)s study of thermohalophilic *A. terreus* AUMC 10138, revealing an enhancement in properties of alkaline cellulase production (FPase, CMCase, and β-glucosidase) at a pH range of 9.0–11.0. Similarly, Gunny et al. (2015) in their work on pretreated lignocelluloses (in situ saccharification using ionic liquids) by *A. terreus* UniMAP AA-6 showed the capability to produce halophilic cellulases.

Zhao et al. (2012) isolated a marine bacterium *Marinimicrobium* sp. LS-A18 that produced novel halotolerant and alkali-stable CMC with the enzyme retaining almost 84% of its activity at alkaline pH ranges up to pH 11. Yu and Li (2015) worked on a halophilic bacteria *Gracilibacillus* SK1 isolated from Yuncheng Salt Lake, which produced a multicomponent cellulolytic system (exo-/endocellulase and β-glucosidase) with a stable activity over broad alkaline pH (6.0–10.0). Shikata et al. (1990), in one of the early studies on bacterial alkaliphiles, isolated and screened a few *Bacillus* spp. for their alkaline cellulase productions (pH 8.5–9.5) and tried their application as an additive to detergent. Studies carried out by De-Marco et al. (2017) on *B. licheniformis* 380 isolated from compost found that the cellulase produced had stable activities at alkaline pH 7.0–10.0.

George et al. (2001), in one of the early studies, found that an alkali-thermophilic actinomycete *Thermomonospora* sp. had CMCase alkali stability at pH ranges of 7.0–10.0. Prasad et al. (2013) characterized *Streptomyces griseoaurantiacus* for its cellulase production under alkaline conditions, which was followed up by another study by Kumar (2015) on UV mutagenesis of *S. griseoaurantiacus* that resulted in enhanced productivity of endoglucanase and β-glucosidase at alkaline pH buffers (up to pH 11.0). Adıgüzel and Tunçer (2017) studied the halotolerant actinobacteria *Amycolatopsis cihanbeyliensis* Mut43, mutated by UV, to enhance its endoglucanase production. The work revealed that endoglucanase activity enhanced with 6% addition of NaCl performed well in alkaline pH ranges in commercially available laundry detergent solutions as an additive.

4.3.4 Plant-Derived Microbial Cellulase

Microbes that reside within the tissues of plants were thought to be beneficial in their purpose with mutualistic relations. Recent exploration of these microbes revealed that they are also capable of producing hydrolytic enzymes like cellulase for their defense mechanisms or mutual help. These microbial enzymes like cellulases have the uniqueness of being extremely substrate specific.

Endophytic fungal varieties have also exhibited cellulase production capabilities. Yopi et al. (2017) reported three fungal verities (Class Coelomycetes, *Acremonium* sp., and *A. niger*) isolated from endophytic tissues of some medicinal plants (ironwood tree, pin flower tree, Madagascar periwinkle) capable of producing cellulases having an activity range of 0.26–0.43 U/mL.

Bibi et al. (2017) reported bacterial endophytic bacterial cultures (*B. licheniformis*, *Erwinia toletana*, *Vibrio antiquaries*, *Aidingimonas halophila*, *Halomonas anticariensis*, *Microbulbifer celer*, *Halomonas* sp., *Alteromonas macleodii*, *Marinobacter xestospongiae*, *Chromohalobacter israelensis*, and *Microbulbifer halophilus*) having halophilic cellulases isolated from mangrove plants situated within Saudi Arabia. Leo et al. (2016) reported bacterial (*Burkholderia cepacia* BPS-GB3,

Alcaligenes faecalis BPS-GB5, and *Enterobacter hormaechei* BPS-GB8) endophytes isolated from *Nayraudia reynaudiana* and *Saccharum arundinaceum*, which were used as tricultured consortia that produced FPase, CMCase, and β-glucosidase.

Castro et al. (2014) reported actinobacterial (*Curtobacterium oceanosedimentum* MBR2.21, *Curtobacterium flaccumfaciens* MBR2.22, *Microbacterium arborescens* MBR 2.54, and *Microbacterium oleivorans* MBR 2.52) cellulases isolated from endophytic tissues of Brazilian mangroves, while endoglucanase was reported from seven actinobacterial endophytes of which two *Streptomyces* sp. (DBT 210 and DBT 204) showed maximum enzyme activity isolated from inner tissues of *Solanumly copersicum* (Passari et al., 2016). This work was followed by another report by Passari et al. (2017) where again a few actinobacterial cultures (*Kocuria* sp., *Microbacterium testaceum*, *Streptomyces olivaceus*, *Streptomyces* sp., *S. cellulosae*, *Leifsoniaxyli* sp., *Streptomyces mutabilis*, and *Streptomyces thermocarboxydus*) were reported to be cellulase (CMCase) producers that were isolated tissues samples from *Rhynchotechum ellipticum*.

4.3.5 Animal/Insect Gut-Derived Microbial Cellulase

Ruminant animals, small insects, termites, snails, and even the gut of aquatic fishes have all been reported as a repository of microbial cellulases, which were mostly bacterial and actinobacterial cultures.

Among the early reports, Oyeleke and Okusanmi (2008) isolated numerous microbes from the rumen of three animals (goat, sheep, and cow) that exhibited cellulolytic properties. Among the microorganisms obtained were fungal cultures like the genera *Penicillium, Aspergillus, Mucor, and Fusarium*, while the bacterial species included *Pseudomonas aeruginosa*, *Streptococcus* sp., and *Bacillus* sp. Anand and Sripathi (2004) reported eight bacterial isolates isolated from Indian flying fox (*Pteropus giganteus*) having cellulase activities. Russell et al. (2009) in a comprehensive review of ruminant-dominant bacteria in cow reflected upon the presence of cellulase producer bacteria as *Bacteroides succinogenes*, *Ruminococcus albus*, and *Ruminococcus flavefaciens*. Ko et al. (2011) in their study worked on novel *endo-* and exocellulase from anaerobic *C. thermocellum* that was ruminal in its origin. This very enzyme CelEdx16 when overexpressed on *E. coli* was capable of showing endo- and exocellulase activities of 15.9 and 3.6×10^{-2} U/mg, respectively. Han et al. (2015) did a comprehensive study on rumen bacterial diversity of goats for a period of time. The study revealed that four cellulase-producing bacteria *R. flavefaciens*, *Fibrobacter succinogenes*, *Butyrivibrio fibrisolvens*, and *Clostridium* sp. were dominant in the 110-day-old goat rumen. Yang et al. (2014) and Ma et al. (2015) isolated *Bacillus* sp., namely *B. subtilis* BY-2 and *B. subtilis* BY-4, respectively, from the gastrointestinal portion of Tibetan pig, which showed cellulase productivity. Asem et al. (2017) isolated bacterial cultures from the ruminal and gastrointestinal portions of an indigenous swine (Zovawk) and a domesticated goat (Black Bengal) of which *Serratia rubidaea* DBT4 and *Aneurinibacillus aneurinilyticus* DBT87 showed cellulase activities.

Wenzel et al. (2002) isolated cellulase-positive bacteria and actinobacteria from the gut of the termite *Zootermopsis angusticollis*. These anaerobic and facultative aerobic bacteria belong to the genera *Cellulomonas*, *Microbacterium*, *Kocuria*, *Bacillus*, *Brevibacillus*, *Paenibacillus*, *Afipia*, *Brucella*, *Ochrobactrum*, *Pseudomonas*, and *Sphinomonas*, respectively.

There are even reports of cellulase producers from bacteria isolated from the dung of rhinoceros and cow like *Bacillus amyloliquefaciens* SS35 and *B. subtilis*, respectively (Singh et al., 2013; Bai et al., 2012).

Dar et al. (2015) worked on cellulase-producing bacteria *Ochrobactrum* sp. K38 and *B. subtilis* Cf60 from the gastrointestinal tract of the giant African snail *Achatina fulica* that exhibited CMCase activity of 501.75 IU/mL extract (K38) and avicelase activity of 3116.92 IU/mL extract (Cf60).

Among the aquatic animals, Ray et al. (2007) reported two bacterial cellulase producers, *B. subtilis* CY5 (26 U) and *Bacillus circulans* TP3 (20.2 U), isolated from the gastrointestinal tracts of common carp *Cyprinus carpio* L. and Mozambique tilapia *Oreochromis mossambicus*, while *Bacillus* sp. AC1 isolated from the gastric juices of a mollusca *Ampullaria crosseans* exhibited endoglucanase activity (*Ba*-EGA) of a specific activity of 35.0 IU/mg (Li et al., 2006).

4.4 MICROBIAL EFFICIENCY FOR CELLULASE PRODUCTION

Given the plethora of microbial cellulase available, it is important to optimize its activity to the specifications required for its actions in various industrial usages. In this scenario the technological advancements made have to be put into usage for enhancing the efficiency of the obtained microbial cellulases. Various improvements and technological inputs for enhancing the understanding and enhancement of microbial cellulases are hereby reviewed.

4.4.1 Celluosome/Secretome Study and Efficiency

Cellulosomes are deemed to be multienzyme complexes secreted by anaerobic bacteria, though a few anaerobic fungal cellulases have been reported from *Neocallimastix*, *Piromyces*, and *Orpinomyces* (Maki et al., 2009). "Cel A," a cellulase cocktail isolated from the thermophilic anaerobic firmicutes *C. bescii*, has gained immense interest, with its cellulosome-like cellulase action on hydrolyzing crystalline cellulose as well as low crystallinity substrates. Additionally, Cel A was tried on a variety of biomass substances, and cavity formations by the enzymatic actions within the crystalline structures of the biomasses were observed (Brunecky et al., 2017). Secretome profiling (total enzyme secretions of a microbe) has recently aroused immense interest, which in turn provides innumerable additions into the CAZymes databases. Borin et al. (2015) worked on the most studied fungal cellulase producers *A. niger* and *T. reesei* secretomes and compared them for their actions on sugarcane biomass. The secretomes were filtered, purified, and their proteins were profiled using mass spectrometry and subjected to CAZYmes classifications. The work reported 45 CAZymes in *T. reesei* and 89 in *A. niger*, which was further correlated with the cell wall degradation capability of the secretome to that of monosaccharides released. The success of the hydrolysis effect revealed the effectiveness of secretome system capability.

4.4.2 Mutants Efficiency

The thermohalophilic fungal variety *A. terreus* AUMC 10138 when mutated with physical and chemical mutants produced an enhanced mutant *A. terreus* M17, capable of producing alkali-stable cellulases. Under their optimized growth conditions, it produced FPase (729U/g), CMCase (1783U/g), and β-glucosidase (342U/g), which was 1.85, 1.97-, and 2.31-fold higher than its wild-type strain. The work reported high potential for these strains to be active on biomass-derived edible corn such as corn stover (Isaac and Abu-Tahon, 2015). Zheng et al. (2018) worked on enhancing the activity of a novel GH5 cellulase GtCel5 from *Gloeophyllum trabeum* by site-directed mutation on loop 6 of the enzyme. The specific activity for CMCase was 1117 ± 43U/mg at pH4.0 and 70°C, which was far better than its wild type.

4.4.3 Recombinants Efficiency

Qian et al. (2017) worked on the most popular cellulase producer microbe *T. reesei* to optimize its activity. They used pyrG marker to overexpress certain cellulolytic genes like EG2 and BGL1 (endoglucanase2 and β-glucosidase1) that resulted in an 11.6-fold enhancement in total cellulase activity. Thus the newly engineered *T. reesei* QEB4 had the potential to hydrolyze almost complete cellulosic residues of the delignified corn cob residues, as expressed by 94% conversion capability.

Another development was overexpression of microbial cellulolytic genes within plants themselves. GH5 endocellulase from *A. cellulolyticus* (AcCel5A), which is hyperthermostable in nature (81°C), has been characterized and opted to be overexpressed in a variety of plants like tobacco, potato, rice, maize, and even *Arabidopsis thaliana* as part of an effort to understand biomass digestibility (Donohoe et al., 2017; Jiang et al., 2017).

4.4.4 Co- and Triculture Studies and Efficiency

Maki et al. (2009), on their review of cellulase-producing bacteria and their lignocellulose conversion capability, observed that the coculture system employed using a few bacterial cultures was efficient in utilizing cellulosic substrates. *C. thermocellum*, an anaerobic bacterium capable of fermenting pentose sugars to ethanol, was mixed with other anaerobic, thermophilic bacteria such as *Thermoanaerobacterium saccharolyticum*, *Clostridium thermohydrosulfuricum*, and *Thermoanaerobium brockii* (Venkateswaren and Demain, 1986; Ng et al., 1981; Lamed and Zeikus, 1980). The system caused a syntrophic relation among the parent microbe *C. thermocellum* and the secondary microbes, wherein the cellulases and hemicellulases are equally shared to hydrolyze cellulose to cellobiose and cellodextrans, while hemicelluloses are converted to xylobiose, arabinoxylans, and xylooligosaccharides. Thus the primary organism helped to utilize the broken-down cellulosic simple sugars to ethanol, while the secondary microbe utilizes the hemicellulosic hydrolytic by-products to be converted to ethanol.

Leo et al. (2016) reported a triculture approach to efficiently utilize the lignocellulosic nonedible biomass for conversion to simple sugars by an endophytic-derived microbial triculture consortia. They used *B. cepacia* BPS-GB3, *A. faecalis* BPS-GB5, and *E. hormaechei* BPS-GB8 whose consortia were capable of producing β-glucosidase, CMCase, and FPase along with xylanase and laccase during 12 days of submerged cultivation on *N. reynaudiana* BPS-G109. This was justified with a 70% loss in dry weight, along with analysis using Fourier-transform infrared spectroscopy and SEM, which showed disruption of the surface structure and chemical composition of polymeric lignocellulosic units.

4.4.5 Consolidated Bioprocessing and Efficiency

Consolidated bioprocessing (CBP) helps in combining the effects of cellulases and hemicellulases, along with fermentation of hydrolyzed products (small sugars) to end products like ethanol, all in one step. Lee et al. (2017) combined various efficient cellulases, including *C. thermophilum* cellobiohydrolase (CtCBH1), *C. thermophilum* cellobiohydrolase (ClCBH2), *T. reesei* endoglucanase (TrEGL2), and *Saccharomycopsi fibuligera* β-glucosidase (SfBGL1), which were individually overexpressed in the yeast *Saccharomyces cerevisiae*. Thus 0.6–2.0 g/L of combined cellulases were expressed within the same fermentation media and were found to be effective against avicel and many other cellulosic biomasses. This CBP technique of using four cellulase-induced yeast varieties produced approximately 14 g/L ethanol from the pretreated rice straw containing 35 g/L glucan. Thus this work reported a threefold productivity retrospect to normal fermentation. This cost-effective technique surely will be a boon for bioethanol production.

4.4.6 Nanotechnology Usage and Efficiency

Nanotechnology is one method that could help to immobilize enzymes, especially cellulases, by using various nanomaterials like nanotubes, nanopores, nanocomposites, etc., which are then termed nanocatalysts. The major benefit of this technology is enhancement in effectiveness. This nanosized material helps to ensure that the enzyme has enough surface area to act on substrates. Also, immobilized enzymes on nanomaterials ensure higher enzyme activity retention and reusage (Leo and Singh, 2018).

Various nanomaterial-based cellulases have been used in applications. Cellulases from well-known fungal cellulase producers like *Aspergillus* sp. and *Trichoderma* sp. immobilized onto nanocarriers have shown immense potential. Abraham et al. (2014) reported 93% hydrolysis of the hemp hurd biomass by the action of cellulase derived from *T. reesei* immobilized on magnetic nanoparticles (zinc + magnetite), having 50% enzyme activity retention after five reusages at 80°C. Honda et al. (2015) worked on endoglucanase and β-glucosidase sourced from *C. thermocellum* that was immobilized onto a magnetic nanoparticle–magnetosome system as a nanomaterial of 75 nm size, which exhibited 70% activity even after five cycles of hydrolysis. Lima et al. (2016) reported 69% retention of activity even after eight cycles of usage for *T. terrestris*-derived endoglucanase that was immobilized onto poly(methyl methacrylate)-Fe_3O_4 nanomaterial of 150 nm size.

4.4.7 Metagenomic Approaches and Its Efficiency

Metagenomic approaches have been used to discover many novel CAZymes from soil (Nacke et al., 2012), cow rumens (Pope et al., 2010), sediments (Klippel et al., 2014), and aquatic environments (Martin et al., 2014). Garg et al. (2016) during their metagenomic analysis of soil samples and subsequent studies revealed the presence of a novel cellulase Cel5R with high halotolerant and thermal stability. Given the potential of this cellulase it was recombined onto *E. coli* Rosetta (DE3). The applicability of this enzyme in various industries will be substantial. Kolinko et al. (2018) reported a very interesting approach for consortia-based cellulase detection and usage. They worked on a thermophilic (60°C) cellulose-degrading bacterial community that was grown in aerobic conditions isolated from compost from Vacaville, USA. The durability of this consortia was measured and sampled for 3 years scaled to a 15–300 L bioreactor for bioprocess studies. Though consortia-based studies have previously been carried out, the metagenomics profiling addition to this work has enhanced the understanding of the repertoire of microbes responsible for cellulolytic activities. The metagenomics profiling helped reveal that cellulases (multidomain glycoside hydrolases) responsible for cellulose degradation came from *R. marinus* (45%), *Thermus thermophilus* (29%), and *Thermobispora bispora* (14%), and hence renamed the consortia *Candidatus Reconcilibacillus cellulovorans*.

4.5 CONCLUSION

The majority of the commercially available cellulases have their origins in microorganisms. Cellulases as such have found numerous applications other than their major purpose in the conversion of native cellulosic biomass into usable simple sugars. The enzyme has found applications at the industrial scale, especially during the production of paper, textiles, detergents, food, beverages, etc., and in many other biotechnological products. Given the advancements made and required in every field, it is natural that more emphasis has to be put into developing and exploring the known or novel cellulases. It is in this aspect that microbes that produce cellulases and their diverse foray of habitats come to the forefront, especially those that are extremophiles and ruminal gut-derived enzymes. Such cellulases are capable of being tolerant and stable during harsh conditions, while the gut-derived cellulases help in understanding the specificity and mechanism of cohesive cellulolytic actions, given that aerobic and anaerobic microbial cellulases are in mutualistic action there. Recent advancements in metagenomics has helped to understand the idea of total cellulolytic gene profiles available within the site and their action as part of a community, whereas proteomics work has helped to simplify the complex actions of the enzymatic secretomes and cellulosomes, which thereby helped to design more robust recombinant models of cellulases. Thus such advancements in understanding microbial cellulase availability and their diversities have resulted in enhancing more effective cellulases meant for application in various industrial purposes. The search for novel microbial cellulases and their modifications is still an ongoing and interesting field of study, given the fact that almost 80%–90% of the total microbial communities in this world are yet to be explored in full.

Acknowledgment

The authors are thankful to the DBT-sponsored NER-Twinning project (No. BT/323/NE/TBP/2012) that the Government of India sanctioned to BPS.

References

Abraham, R.E., Verma, M.L., Barrow, C.J., Puri, M., 2014. Suitability of magnetic nanoparticle immobilised cellulases in enhancing enzymatic saccharification of pretreated hemp biomass. Biotechnol. Biofuels 7, 90. https://doi.org/10.1186/1754-6834-7-90.

Adıgüzel, A.O., Tunçer, M., 2017. Production, purification, characterization and usage of a detergent additive of endoglucanase from isolated halotolerant *Amycolatopsis cihanbeyliensis* mutated strain Mut43. Biocatal. Biotransformation 35 (3), 197–204. https://doi.org/10.1080/10242422.2017.1315106.

Akila, G., Chandra, T.S., 2003. A novel cold-tolerant *Clostridium* strain PXYL1 isolated from a psychrophilic cattle manure digester that secretes thermolabile xylanase and cellulase. FEMS Microbiol. Lett. 219 (1), 63–67. https://doi.org/10.1016/S0378-1097(02)01196-5.

Alam, M.Z., Sultana, M., Anwar, M.N., 2004. Isolation, purification, characterization of cellulolytic enzymes produced by the isolate *Streptomyces omiyaensis*. Pak. J. Biol. Sci. 7 (10), 1647–1653. https://doi.org/10.3923/pjbs.2004.1647.1653.

Anand, A.A.P., Sripathi, K., 2004. Digestion of cellulose and xylan by symbiotic bacteria in the intestine of the Indian flying fox (*Pteropus giganteus*). Comp. Biochem. Phys. A Mol. Integr. Physiol. 139 (1), 65–69. https://doi.org/10.1016/j.cbpb.2004.07.006.

Anderson, I., Abt, B., Lykidis, A., Klenk, H.P., Kyrpides, N., Ivanova, N., 2012. Genomics of aerobic cellulose utilization systems in actinobacteria. PLoS One 7 (6), 1–10. https://doi.org/10.1371/journal.pone.0039331.e39331.

Ando, S., Ishida, H., Kosugi, Y., Ishikawa, K., 2002. Hyperthermostable endoglucanase from *Pyrococcus horikoshii*. Appl. Environ. Microbiol. 68, 430–433.

Andrade, J.P., Bispo, A.S.R., Marbach, P.A.S., Do-Nascimento, P.R., 2011. Production and partial characterization of cellulases from *Trichoderma* sp. IS-05 isolated from sandy coastal plains of Northeast Brazil. Enzyme Res. 2011 (167248), 1–7. https://doi.org/10.4061/2011/167248.

Annamalai, N., Rajeswari, M.V., Elayaraja, S., Balasubramanian, T., 2013. Thermostable, haloalkaline cellulase from *Bacillus halodurans* CAS 1 by conversion of lignocellulosic wastes. Carbohydr. Polym. 94 (1), 409–415. https://doi.org/10.1016/j.carbpol.2013.01.066.

Asem, D., Leo, V.V., Passari, A.K., Tonsing, M.V., Joshi, J.B., Uthandi, S., et al., 2017. Evaluation of gastrointestinal bacterial population for the production of holocellulose enzymes for biomass deconstruction. PLoS One 12 (10), 1–17. https://doi.org/10.1371/journal.pone.0186355.

Avellaneda-Torres, L.M., Pulido, C.P.G., Rojas, E.T., 2014. Assessment of cellulolytic microorganisms in soils of Nevados Park, Colombia. Braz. J. Microbiol. 45 (4), 1211–1220.

Bai, S., Kumar, R.M., Kumar, D.J.M., Balashanmugam, P., Kumaran, M.D.B., Kalaichelvan, P.T., 2012. Cellulase production by *Bacillus subtilis* isolated from cow dung. Arch. Appl. Sci. Res. 4 (1), 269–279.

Bauer, M.W., Driskill, L.E., Callen, W., Snead, M.A., Mathur, E.J., Kelly, R.M., 1999. An endoglucanase, EglA, from the hyperthermophilic archaeon *Pyrococcus furiosus* hydrolyzes β-1, 4 bonds in mixed-linkage (1-3)-(1-4)-β-D-glucans and cellulose. J. Bacteriol. 181, 284–290.

Bibi, F., Ullah, I., Alvi, S.A., Bakhsh, S.A., Yasir, M., Al-Ghamdi, A.A.K., et al., 2017. Isolation, diversity, and biotechnological potential of rhizo- and endophytic bacteria associated with mangrove plants from Saudi Arabia. Genet. Mol. Res. 16 (2), 1–12. https://doi.org/10.4238/gmr16029657.

Blumer-Schuette, S.E., Kataeva, I., Westpheling, J., Adams, M.W., Kelly, R.M., 2008. Extremely thermophilic microorganisms for biomass conversion: status and prospects. Curr. Opin. Biotechnol. 19 (3), 210–217. https://doi.org/10.1016/j.copbio.2008.04.007.

Bok, J.D., Yernool, D.A., Eveleigh, D.E., 1998. Purification, characterization, and molecular analysis of thermostable cellulases CelA and CelB from *Thermotoganea politana*. Appl. Environ. Microbiol. 64, 4774–4781.

Borin, G.P., Sanchez, C.C., de Souza, A.P., de Santana, E.S., de Souza, A.T., Leme, A.F.P., et al., 2015. Comparative secretome analysis of *Trichoderma reesei* and *Aspergillus niger* during growth on sugarcane biomass. PLoS One 10 (6), e0129275. https://doi.org/10.1371/journal.pone.0129275.

Brienzo, M., Arantes, V., Milagres, A.M.F., 2008. Enzymology of the thermophilic ascomycetous fungus *Thermoascus aurantiacus*. Fungal Biol. Rev. 22 (3–4), 120–130. https://doi.org/10.1016/j.fbr.2009.04.001.

Brunecky, R., Donohoe, B.S., Yarbrough, J.M., Mittal, A., Scott, B.R., Ding, H., et al., 2017. The multi domain *Caldicellulosiruptor bescii* CelA cellulase excels at the hydrolysis of crystalline cellulose. Sci. Rep. 7, 9622. https://doi.org/10.1038/s41598-017-08985-w.

Buchon, L., Laurent, P., Gounot, A.M., Guespin-Michel, J.F., 2000. Temperature dependence of extracellular enzymes production by psychrotrophic and psychrophilic bacteria. Biotechnol. Lett. 22 (19), 1577–1581. https://doi.org/10.1023/A:1005641119076.

Busk, P.K., Lange, L., 2013. Cellulolytic potential of thermophilic species from four fungal orders. AMB Express 3 (1), 47. https://doi.org/10.1186/2191-0855-3-47.

Carrasco, M., Villarreal, P., Barahona, S., Alcaíno, J., Cifuentes, V., Baeza, M., 2016. Screening and characterization of amylase and cellulase activities in psychrotolerant yeasts. BMC Microbiol. 16 (21), 1–9. https://doi.org/10.1186/s12866-016-0640-8.

Castro, R.A., Quecine, M.C., Lacava, P.T., Batista, B.D., Luvizotto, D.M., Marcon, J., et al., 2014. Isolation and enzyme bioprospection of endophytic bacteria associated with plants of Brazilian mangrove ecosystem. Springerplus 3 (382), 1–9. https://doi.org/10.1186/2193-1801-3-382.

Da-Qing, W., Wen-Ran, J., Tai-Peng, S., Yu-Tian, M., Wei, Z., Hong-Yan, W., 2016. Screening psychrophilic fungi of cellulose degradation and characteristic of enzyme production. J. Northeast. Agric. Univ. 23 (2), 20–27. https://doi.org/10.1016/S1006-8104(16)30044-7.

Dar, M.A., Pawar, K.D., Jadhav, J.P., Pandit, R.S., 2015. Isolation of cellulolytic bacteria from the gastro-intestinal tract of *Achatina fulica* (Gastropoda: Pulmonata) and their evaluation for cellulose biodegradation. Int. Biodeterior. Biodegrad. 98, 73–80. https://doi.org/10.1016/j.ibiod.2014.11.016.

Da-Vinha, F.N.M., Gravina-Oliveira, M.P., Franco, M.N., Macrae, A., Bon, E.P.S., Nascimento, R.P., et al., 2011. Cellulase production by *Streptomyces viridobrunneus* SCPE-09 using lignocellulosic biomass as inducer substrate. Appl. Biochem. Biotechnol. 164 (3), 256–267. https://doi.org/10.1007/s12010-010-9132-8.

Deckert, G., Warren, P.V., Gaasterland, T., Young, W.G., Lenox, A.L., Graham, D.E., et al., 1998. The complete genome of the hyperthermophilic bacterium *Aquifex aeolicus*. Nature 392 (6674), 353–358. https://doi.org/10.1038/32831.

De-Marco, É.G., Heck, K., Martos, E.T., Vander-Sand, S.T., 2017. Purification and characterization of a thermostable alkaline cellulase produced by *Bacillus licheniformis* 380 isolated from compost.*An*. Acad. Bras. Cienc. 89 (3), 2359–2370. https://doi.org/10.1590/0001-3765201720170408.

De-Menezes, A.B., Lockhart, R.J., Cox, M.J., Allison, H.E., McCarthy, A.J., 2008. Cellulose degradation by *Micromonospora*'s recovered from freshwater lakes and classification of these actinomycetes by DNA gyrase B gene sequencing. Appl. Environ. Microbiol. 74, 7080–7084. https://doi.org/10.1128/AEM.01092-08.

De-Oliveira, M.M.Q., De-Lima, A.L.G., Bon, E.P.S., Coelho, R.R.R., Do-Nascimento, R.P., 2016. Production of thermophilic and acidophilic endoglucanases by mutant *Trichoderma atroviride* 102C1 using agro-industrial by-products. Afr. J. Biotechnol. 15 (11), 423–430. https://doi.org/10.5897/AJB2015.15007.

Ding, S.-Y., Adney, W.S., Vinzant, T.B., Decker, S.R., Baker, J.O., Thomas, S.R., et al., 2003. Glycoside hydrolase gene cluster of *Acidothermus cellulolyticus*. In: Mansfield, S., Saddler, J.N. (Eds.), Applications of Enzymes to Lignocellulosics. ACS Symposium Series 855 (332–360). American Chemical Society, Washington, DC.

Donohoe, B.S., Wei, H., Mittal, A., Shollenberger, T., Lunin, V.V., Himmel, M.E., et al., 2017. Towards an understanding of enhanced biomass digestibility by in planta expression of a family 5 glycoside hydrolase. Sci. Rep. 7 (1), 4389. https://doi.org/10.1038/s41598-017-04502-1.

Duncan, S.M., Farrell, R.L., Thwaites, J.M., Held, B.W., Arenz, B.E., Jurgens, J.A., Blanchette, R.A., 2006. Endoglucanase-producing fungi isolated from Cape Evans historic expedition hut on Ross Island, Antarctica. Environ. Microbiol. 8 (7), 1212–1219. https://doi.org/10.1111/j.1462-2920.2006.01013.x.

Dutta, T., Sahoo, R., Sengupta, R., Ray, S.S., Bhattacharjee, A., Ghosh, S., 2008. Novel cellulases from an extremophilic filamentous fungi *Penicillium citrinum*: production and characterization. J. Ind. Microbiol. Biotechnol. 35 (4), 275–282. https://doi.org/10.1007/s10295-008-0304-2.

Dutta, N., Biswas, S., Saha, M.K., 2016. Biophysical characterization and activity analysis of nano-magnesium supplemented cellulase obtained from a psychrobacterium following graphene oxide immobilization. Enzyme Microb. Technol. 95, 248–258. https://doi.org/10.1016/j.enzmictec.2016.04.012.

Eckert, K., Schneider, E., 2003. A thermoacidophilic endoglucanase (CelB) from *Alicyclobacillus acidocaldarius* displays high sequence similarity to arabinofuranosidases belonging to family 51 of glycoside hydrolases. Eur. J. Biochem. 270, 3593–3602. https://doi.org/10.1046/j.1432-1033.2003.03744.x.

Feller, G., Narinx, E., Arpigny, J.L., Zekhnini, Z., Swings, J., Gerday, C., 1994. Temperature dependence of growth, enzyme secretion and activity of psychrophilic Antarctic bacteria. Appl. Microbiol. Biotechnol. 41 (4), 477–479. https://doi.org/10.1007/BF01982539.

Franco-Cirigliano, M.N., Rezende, R.C., Gravina-Oliveira, M.P., Pereira, P.H., Nascimento, R.P., Bon, E.P., et al., 2013. *Streptomyces misionensis* PESB-25 produces a thermoacidophilic endoglucanase using sugarcane bagasse and corn steep liquor as the sole organic substrates. Biomed. Res. Int. 2013 (584207), 1–9. https://doi.org/10.1155/2013/584207.

Garg, R., Srivastava, R., Brahma, V., Verma, L., Karthikeyan, S., Sahni, G., 2016. Biochemical and structural characterization of a novel halotolerant cellulase from soil metagenome. Sci. Rep. 6, 39634. https://doi.org/10.1038/srep39634.

Garsoux, G., Lamotte, J., Gerday, C., Feller, G., 2004. Kinetic and structural optimization to catalysis at low temperatures in a psychrophilic cellulase from the Antarctic bacterium *Pseudoalteromonas haloplanktis*. Biochem. J. 384 (2), 247–253. https://doi.org/10.1042/BJ20040325.

George, S.P., Ahmad, A., Rao, M.B., 2001. Studies on carboxymethyl cellulase produced by alkalothermophilic actinomycetes. Bioresour. Technol. 77 (2), 171–175. https://doi.org/10.1016/S0960-8524(00)00150-4.

Gobalakrishnan, R., Kannan, S., 2017. Systematic characterization of potential cellulolytic marine actinobacteria *Actinoalloteichus* sp. MHA15. Biotechnol. Rep. 13, 30–36. https://doi.org/10.1016/j.btre.2016.12.003.

Graham, J.E., Clark, M.E., Nadler, D.C., Huffer, S., Chokhawala, H.A., Rowland, S.E., et al., 2011. Identification and characterization of a multidomain hyperthermophilic cellulase from an archaeal enrichment. Nat. Commun. 2, 375. https://doi.org/10.1038/ncomms1373.

Grigorevski-Lima, A.L., Da-Vinha, F.N.M., Souza, D.T., Bispo, A.S.R., Bon, E.P.S., Coelho, R.R.R., et al., 2009. *Aspergillus fumigatus*thermophilic and acidophilic endoglucanases. Appl. Biochem. Biotechnol. 155 (1–3), 18–26. https://doi.org/10.1007/s12010-008-8482-y.

Gross, S., Robbins, E.I., 2000. Acidophilic and acid-tolerant fungi and yeasts. Hydrobiologia 433 (1–3), 91–109. https://doi.org/10.1023/A:1004014603333.

Gunny, A.A.N., Arbain, D., Jamal, P., Gumba, R.E., 2015. Improvement of halophilic cellulase production from locally isolated fungal strain. Saudi J. Biol. Sci. 22 (4), 476–483. https://doi.org/10.1016/j.sjbs.2014.11.021.

Hamed, M.B., Karamanou, S., Ólafsdottir, S., Basílio, J.S.M., Simoens, K., Tsolis, K.C., et al., 2017. Large-scale production of a thermostable *Rhodothermus marinus* cellulase by heterologous secretion from *Streptomyces lividans*. Microb. Cell Factories 16 (1), 232. https://doi.org/10.1186/s12934-017-0847-x.

Han, X., Yang, Y., Yan, H., Wang, X., Qu, L., Chen, Y., 2015. Rumen bacterial diversity of 80 to 110-day-old goats using 16S rRNA sequencing. PLoS One 10 (2), e0117811. https://doi.org/10.1371/journal.pone.0117811.

Hayashida, S., Yoshioka, H., 2014. Production and purification of thermostable cellulases from *Humicola insolens* YH-8. Agric. Biol. Chem. 4 (8), 1721–1728. https://doi.org/10.1080/00021369.1980.10864207.

Helal, G.A., 2005. Bioconversion of straw into improved fodder: fungal flora decomposing rice straw. Mycobiology 33, 150–157. https://doi.org/10.4489/MYCO.2005.33.3.150.

Honda, T., Tanaka, T., Yoshino, T., 2015. Stoichiometrically controlled immobilization of multiple enzymes on magnetic nanoparticles by the magnetosome display system for efficient cellulose hydrolysis. Biomacromolecules 16 (12), 3863–3868. https://doi.org/10.1021/acs.biomac.5b01174.

Huang, Y., Krauss, G., Cottaz, S., Driguez, H., Lipps, G., 2005. A highly acid-stable and thermostable endo-β-glucanase from the thermoacidophilic archaeon *Sulfolobus solfataricus*. Biochem. J. 385, 581–588. https://doi.org/10.1042/BJ20041388.

Isaac, G.S., Abu-Tahon, M.A., 2015. Enhanced alkaline cellulases production by the thermohalophilic *Aspergillus terreus* AUMC 10138 mutated by physical and chemical mutagens using corn stover as substrate. Braz. J. Microbiol. 46 (4), 1269–1277. https://doi.org/10.1590/S1517-838246420140958.

Jiang, X., Zhou, X., Liu, Q., Zheng, L., Yu, N., Li, W., 2017. Expression of *Acidothermus cellulolyticus* thermostable cellulases in tobacco and rice plants. Biotechnol. Biotechnol. Equip. 31 (1), 23–28. https://doi.org/10.1080/13102818.2016.1236671.

Kiio, I.K., Jackim, M.F., Munyali, W.B., Muge, E.K., 2016. Isolation and characterization of a thermostable cellulase from *Bacillus licheniformis*strain Vic isolated from geothermal wells in the Kenyan rift valley. Open Biotechnol. J. 10, 198–207. https://doi.org/10.2174/1874070701610010198.

Kim, B., Lee, B., Lee, Y., Jin, I., Chung, C., Lee, J., 2009. Purification and characterization of carboxymethyl cellulase isolated from a marine bacterium, *Bacillus subtilis* subsp. *subtilis* A-53. Enzyme Microb. Technol. 44, 411–416. https://doi.org/10.1016/j.enzmictec.2009.02.005.

Kim, Y.J., Kim, D.O., Chun, O.K., Shin, D.H., Jung, H., Lee, C.Y., et al., 2005. Phenolic extraction from apple peel by cellulases from *Thermobifida fusca*. J. Agric. Food Chem. 53, 9560–9565. https://doi.org/10.1021/jf052052j.

Klippel, B., Sahm, K., Basner, A., Wiebusch, S., John, P., Lorenz, U., et al., 2014. Carbohydrate-active enzymes identified by metagenomic analysis of deep-sea sediment bacteria. Extremophiles 18 (5), 853–863. https://doi.org/10.1007/s00792-014-0676-3.

Ko, K.C., Han, Y., Choi, J.H., Kim, G.J., Lee, S.G., Song, J.J., 2011. A novel bifunctional endo–/exo-type cellulase from an anaerobic ruminal bacterium. Appl. Microbiol. Biotechnol. 89 (5), 1453–1462. https://doi.org/10.1007/s00253-010-2949-9.

Kolinko, S., Wu, Y.W., Tachea, F., Denzel, E., Hiras, J., Gabriel, R., et al., 2018. A bacterial pioneer produces cellulase complexes that persist through community succession. Nat. Microbiol. 3, 99–107. https://doi.org/10.1038/s41564-017-0052-z.

Krishnan, A., Convey, P., Gonzalez, M., Smykla, J., Alias, S.A., 2018. Effects of temperature on extracellular hydrolase enzymes from soil microfungi. Polar Biol. 41 (3), 537–551. https://doi.org/10.1007/s00300-017-2215-z.

Kuhad, R.C., Gupta, R., Singh, A., 2011. Microbial cellulases and their industrial applications. Enzym. Res. 2011, 280696. https://doi.org/10.4061/2011/280696.

Kumar, A.K., 2015. UV mutagenesis treatment for improved production of endoglucanase and β-glucosidase from newly isolated thermotolerant actinomycetes, *Streptomyces griseoaurantiacus*. Bioresour. Bioprocess. 2, 22. https://doi.org/10.1186/s40643-015-0052-x.

Kvesitadze, G., 2017. Cellulases from extremophiles. Curr. Trends Biomed. Eng. Biosci. 4 (3), 1–6. https://doi.org/10.19080/CTBEB.2017.04.555640.

Lamed, R., Zeikus, J.G., 1980. Ethanol production by thermophilic bacteria: relationship between fermentation product yields of and catabolic activities in *Clostridium thermocellum* and *Thermoanaerobium brockii*. J. Bacteriol. 144 (2), 569–578.

Lamilla, C., Pavez, M., Santos, A., Hermosilla, A., Llanquinao, V., Barrientos, L., 2017. Bioprospecting for extracellular enzymes from culturable Actinobacteria from the South Shetland Islands, Antarctica. Polar Biol. 40, 719–726. https://doi.org/10.1007/s00300-016-1977-z.

Lee, C.R., Sung, B.H., Lim, K.M., Kim, M.J., Sohn, M.J., Bae, J.H., et al., 2017. Co fermentation using recombinant*Saccharomyces cerevisiae* yeast strains hyper secreting different cellulases for the production of cellulosic bioethanol. Sci. Rep. 7 (1), 4428. https://doi.org/10.1038/s41598-017-04815-1.

Leo, V.V., Singh, B.P., 2018. Prospectus of nanotechnology in bioethanol productions. In: Srivastava, N., Srivastava, M., Pandey, H., Mishra, P., Ramteke, P. (Eds.), Green Nanotechnology for Biofuel Production. Biofuel and Biorefinery Technologies. vol. 5. Springer, Cham, pp. 129–139. https://doi.org/10.1007/978-3-319-75052-1_9.

Leo, V.V., Passari, A.K., Joshi, B., Mishra, V.K., Uthandi, S., Ramesh, N., et al., 2016. A novel triculture system (CC3) for simultaneous enzyme production and hydrolysis of common grasses through submerged fermentation. Front. Microbiol. 7 (447), 1–13. https://doi.org/10.3389/fmicb.2016.00447.

Leo, V.V., Asem, D., Zothanpuia and Singh. B. P., 2018. Actinobacteria: a highly potent source for holocellulose degrading enzymes. In: Singh, B.P., Gupta, V.K., Passari, A.K. (Eds.), New and Future Developments in Microbial Biotechnology and Bioengineering Actinobacteria: Diversity and Biotechnological Applications. Elsevier, The Boulevard, United Kingdom, pp. 191–205. https://doi.org/10.1016/B978-0-444-63994-3.00013-8.

Leschine, S.B., Canale-Parola, E., 1983. Mesophilic cellulolytic Clostridia from freshwater environments. Appl. Environ. Microbiol. 46 (3), 728–737.

Li, Y.H., Ding, M., Wang, J., Xu, G.J., Zhao, F., 2006. A novel thermoacidophilic endoglucanase, Ba-EGA, from a new cellulose-degrading bacterium, *Bacillus* sp.AC-1. Appl. Microbiol. Biotechnol. 70 (4), 430–436. https://doi.org/10.1007/s00253-005-0075-x.

Lima, J.S., Araújo, P.H.H., Sayer, C., Souza, A.A.U., Viegas, A.C., de-Oliveira, D., 2016. Cellulase immobilization on magnetic nanoparticles encapsulated in polymer nanospheres. Bioprocess Biosyst. Eng. 40 (4), 511–518. https://doi.org/10.1007/s00449-016-1716-4.

Lin, S.B., Stutzenberger, F.J., 1995. Purification and characterization of the major β-1, 4-endoglucanase from *Thermomonospora curvata*. J. Appl. Microbiol. 79, 447–453. https://doi.org/10.1111/j.1365-2672.1995.tb03160.x.

López-Mondéjar, R., Zühlke, D., Becher, D., Riedel, K., Baldrian, P., 2016. Cellulose and hemicellulose decomposition by forest soil bacteria proceeds by the action of structurally variable enzymatic systems. Sci. Rep. 6, 25279. https://doi.org/10.1038/srep25279.

Lv, W., Yu, Z., 2013. Isolation and characterization of two thermophilic cellulolytic strains of *Clostridium thermocellum* from a compost sample. J. Appl. Microbiol. 114 (4), 1001–1007. https://doi.org/10.1111/jam.12112.

Lynd, L.R., Weimer, P.J., Van Zyl, W.H., Pretorius, I.S., 2002. Microbial cellulose utilization: fundamentals and biotechnology. Microbiol. Mol. Biol. R 66 (3), 506–577. https://doi.org/10.1128/MMBR.66.3.506-577.2002.

Ma, L., Yang, W., Meng, F., Ji, S., Xin, H., Cao, B., 2015. Characterization of an acidic cellulase produced by *Bacillus subtilis* BY-4 isolated from gastrointestinal tract of Tibetan pig. J. Taiwan Inst. Chem. Eng. 56, 67–72. https://doi.org/10.1016/j.jtice.2015.04.025.

Maheshwari, R., Bharadwaj, G., Bhat, M.K., 2000. Thermophilic fungi: their physiology and enzymes. Microbiol. Mol. Biol. Rev. 64 (3), 461–488. https://doi.org/10.1128/MMBR.64.3.461-488.2000.

Maki, M., Leung, K.T., Qin, W.S., 2009. The prospects of cellulase-producing bacteria for the bioconversion of lignocellulosic biomass. Int. J. Biol. Sci. 5 (5), 500–516. https://doi.org/10.7150/ijbs.5.500.

Martin, M., Biver, S., Steels, S., Barbeyron, T., Jam, M., Portetelle, D., et al., 2014. Identification and characterization of a halotolerant, cold-active marine endo-β-1,4-glucanase by using functional metagenomics of seaweed-associated microbiota. Appl. Environ. Microbiol. 80 (16), 4958–4967. https://doi.org/10.1128/AEM.01194-14.

Mohagheghi, A., Grohmann, K., Himmel, M., Leighton, L., Updegraff, D.M., 1986. Isolation and characterization of *Acidothermus cellulolyticus* gen. nov., sp. nov., a new genus of thermophilic, acidophilic, cellulolytic bacteria. Int. J. Syst. Bacteriol. 36, 435–443. https://doi.org/10.1099/00207713-36-3-435.

Muñoz, C., Hidalgo, C., Zapata, M., Jeison, D., Riquelme, C., Rivas, M., 2014. Use of cellulolytic marine bacteria for enzymatic pretreatment in microalgal biogas production. Appl. Environ. Microbiol. 80 (14), 4199–4206. https://doi.org/10.1128/AEM.00827-14.

Murray, P.G., Grassick, A., Laffey, C.D., Cuffe, M.M., Higgins, T., Savage, A.V., et al., 2001. Isolation and characterization of a thermostable endo-β-glucanase active on 1,3-1,4-β-D-glucans from the aerobic fungus *Talaromyces emersonii* CBS 814.70. Enzym. Microb. Technol. 29, 90–98. https://doi.org/10.1016/S0141-0229(01)00354-4.

Nacke, H., Engelhaupt, M., Brady, S., Fischer, C., Tautzt, J., Daniel, R., 2012. Identification and characterization of novel cellulolytic and hemicellulolytic genes and enzymes derived from German grassland soil metagenomes. Biotechnol. Lett. 34 (4), 663–675. https://doi.org/10.1007/s10529-011-0830-2.

Ng, T.K., Ben-Bassat, A., Zeikus, J.G., 1981. Ethanol production by thermophilic bacteria: fermentation of cellulosic substrates by cocultures of *Clostridium thermocellum* and *Clostridium thermohydrosulfuricum*. Appl. Environ. Microbiol. 41 (6), 1337–1343.

Oyeleke, S.B., Okusanmi, T.A., 2008. Isolation and characterization of cellulose hydrolyzing microorganism from the rumen of ruminants. Afr. J. Biotechnol. 7 (10), 1503–1504. https://doi.org/10.5897/AJB08.142.

Park, C., Kawaguchi, T., Sumitani, J., Arai, M., 2001. Purification and characterization of cellulases (CBH I and EGL 1) produced by thermophilic microorganism *Streptomyces* sp. M23. Appl. Biol. Sci. 7, 27–35.

Passari, A.K., Chandra, P., Zothanpuia, M.V.K., Leo, V.V., Gupta, V.K., et al., 2016. Detection of biosynthetic gene and phytohormone production by endophytic actinobacteria associated with *Solanum lycopersicum* and their plant growth-promoting effect. Res. Microbiol. 167 (8), 692–705. https://doi.org/10.1016/j.resmic.2016.07.001.

Passari, A.K., Mishra, V.K., Singh, G., Singh, P., Kumar, B., Gupta, V.K., et al., 2017. Insights into the functionality of endophytic actinobacteria with a focus on their biosynthetic potential and secondary metabolites production. Sci. Rep. 7 (11809), 1–17. https://doi.org/10.1038/s41598-017-12235-4.

Payne, C.M., Knott, B.C., Mayes, H.B., Hansson, H., Himmel, M.E., Sandgren, M., et al., 2015. Fungal cellulases. Chem. Rev. 115 (3), 1308–1448. https://doi.org/10.1021/cr500351c.

Pope, P.B., Denman, S.E., Jones, M., Tringe, S.G., Barry, K., Malfatti, S.A., et al., 2010. Adaptation to herbivory by the Tammar wallaby includes bacterial and glycoside hydrolase profiles different from other herbivores. Proc. Natl. Acad. Sci. U. S. A. 107 (33), 14793–14798. https://doi.org/10.1073/pnas.1005297107.

Prasad, P., Singh, T., Bedi, S., 2013. Characterization of the cellulolytic enzyme produced by *Streptomyces griseorubens* (Accession No. AB184139) isolated from Indian soil. J. King. Saud. Univ. Sci. 25 (3), 245–250. https://doi.org/10.1016/j.jksus.2013.03.003.

Pulicherla, K.K., Ghosh, M., Kumar, P.S., Rao, K.R.S.S., 2011. Psychrozymes—the next generation industrial enzymes. J. Marine Sci. Res. Dev. 1 (1), 1–7. https://doi.org/10.4172/2155-9910.1000102.

Qian, Y., Zhong, L., Gao, J., Sun, N., Wang, Y., Sun, G., et al., 2017. Production of highly efficient cellulase mixtures by genetically exploiting the potentials of *Trichoderma reesei* endogenous cellulases for hydrolysis of corncob residues. Microb. Cell Factories 16, 207. https://doi.org/10.1186/s12934-017-0825-3.

Rawat, R., Tewari, L., 2012. Purification and characterization of an acidothermophilic cellulase enzyme produced by *Bacillus subtilis* strain LFS3. Extremophiles 16 (4), 637–644. https://doi.org/10.1007/s00792-012-0463-y.

Ray, A.K., Bairagi, A., Ghosh, S.K., Sen, S.K., 2007. Optimization of fermentation conditions for cellulase production by *Bacillus subtilis* CY5 and *Bacillus circulans* TP3 isolated from fish gut. Acta Ichthyol. Piscat. 37 (1), 47–53. https://doi.org/10.3750/AIP2007.37.1.07.

Reanprayoon, P., Pathomsiriwong, W., 2012. Tropical soil fungi producing cellulase and related enzymes in biodegradation. J. Appl. Sci. 12 (18), 1909–1916. https://doi.org/10.3923/jas.2012.1909.1916.

Rong, Y., Zhang, L., Chi, Z., Wang, X., 2015. A carboxymethyl cellulase from a marine yeast (*Aureobasidium pullulans* 98): its purification, characterization, gene cloning and carboxymethyl cellulose digestion. J. Ocean Univ. China 14 (5), 913–921. https://doi.org/10.1007/s11802-015-2574-4.

Russell, J.B., Muck, R.E., Weimer, P.J., 2009. Quantitative analysis of cellulose degradation and growth of cellulolytic bacteria in the rumen. FEMS Microbiol. Ecol. 67 (2), 183–197. https://doi.org/10.1111/j.1574-6941.2008.00633.x.

Ruttersmith, L.D., Daniel, R.M., 1991. Thermostable cellobiohydrolase from the thermophilic eubacterium *Thermotoga* sp. strain FjSS3-B1. Biochem. J. 277, 887–890. https://doi.org/10.1042/bj2770887.

Sadhu, S., Maiti, T.K., 2013. Cellulase production by bacteria: a review. Br. Microbiol. Res. J. 3, 235–258. https://doi.org/10.9734/BMRJ/2013/2367.

Saini, A., Aggarwal, N.K., Sharma, A., Yadav, A., 2015. Actinomycetes: a source of lignocellulolytic enzymes. Enzyme Res. 15, 279381. https://doi.org/10.1155/2015/279381.

Saini, A., Aggarwal, N.K., Yadav, A., 2016. Cellulolytic potential of actinomycetes isolated from different habitats. Bioeng. Biosci. 4 (5), 88–94. https://doi.org/10.13189/bb.2016.040503.

Sethi, S., Datta, A., Gupta, B.L., Gupta, S., 2013. Optimization of cellulase production from bacteria isolated from soil. ISRN Biotechnol. 2013 (985685), 1–7. https://doi.org/10.5402/2013/985685.

Shikata, S., Saeki, K., Okoshi, H., Yoshimatsu, T., Ozaki, K., Kawai, S., et al., 1990. Alkaline cellulase for laundry detergents: production by alkalophilic strains of *Bacillus* and some properties of crude enzyme. Agric. Biol. Chem. 54 (1), 91–96. https://doi.org/10.1080/00021369.1990.10869922.

Singh, S., Moholkar, V.S., Goyal, A., 2013. Isolation, identification, and characterization of a cellulolytic *Bacillus amyloliquefaciens*strain SS35 from rhinoceros dung. ISRN Microbiol. 2013 (728134), 1–7. https://doi.org/10.1155/2013/728134.

Sohail, M., Ahmad, A., Khan, S.A., 2016. Production of cellulase from *Aspergillus terreus* MS105 on crude and commercially purified substrates. 3 Biotech 6 (1), 103. https://doi.org/10.1007/s13205-016-0420-z.

Solovyeva, I.V., Okunev, O.N., Kryukova, E.G., Kochkina, G.A., 1999. Occurrence of neutral and alkaline cellulases among alkali-tolerant *Micromycetes*. Syst. Appl. Microbiol. 22 (4), 546–550. https://doi.org/10.1016/S0723-2020(99)80007-1.

Sticklen, M.B., 2008. Plant genetic engineering for biofuel production: towards affordable cellulosic ethanol. Nat. Rev. Genet. 9, 433–443. https://doi.org/10.1038/nrg2777.

Stursová, M., Zifčáková, L., Leigh, M.B., Burgess, R., Baldrian, P., 2012. Cellulose utilization in forest litter and soil: identification of bacterial and fungal decomposers. FEMS Microbiol. Ecol. 80 (3), 735–746. https://doi.org/10.1111/j.1574-6941.2012.01343.x.

Sukumaran, R.K., Singhania, R.R., Mathew, G.M., Pandey, A., 2009. Cellulase production using biomass feed stock and its application in lignocellulose saccharification for bio-ethanol production. Renew. Energy 34, 421–424. https://doi.org/10.1016/j.renene.2008.05.008.

Thomas, S.R., Laymon, R.A., Chou, Y.C., Tucker, M.P., Vinzant, T.B., Adney, W.S., et al., 1995. Initial approaches to artificial cellulase systems for conversion of biomass to ethanol. In: Saddler, J.N., Penner, M.H. (Eds.), Enzymatic Degradation of Insoluble Polysaccharides, ACS Series. vol. 618. American Chemical Society, Washington, DC, pp. 208–236. https://doi.org/10.1021/bk-1995-0618.ch014.

Trivedi, N., Reddy, C.R.K., Radulovich, R., Jha, B., 2015. Solid state fermentation (SSF)-derived cellulase for saccharification of the green seaweed *Ulva* for bioethanol production. Algal Res. 9, 48–54. https://doi.org/10.1016/j.algal.2015.02.025.

Tucker, M.P., Mohagheghi, A., Grohmann, K., Himmel, M.E., 1989. Ultra-thermostable cellulases from *Acidothermus cellulolyticus*: comparison of temperature optima with previously reported cellulases. Bio/Technology 7, 817–820. https://doi.org/10.1038/nbt0889-817.

Vega, K., Villena, G.K., Sarmiento, V.H., Ludeña, Y., Vera, N., Gutiérrez-Correa, M., 2012. Production of alkaline cellulase by fungi isolated from an undisturbed rain forest of Peru. Biotechnol. Res. Int. 2012 (934325), 1–7. https://doi.org/10.1155/2012/934325.

Veiga, M., Esparis, A., Fabregas, J., 1983. Isolation of cellulolytic actinomycetes from marine sediments. Appl. Environ. Microbiol. 46 (1), 286–287.

Venkateswaren, S., Demain, A.L., 1986. The *Clostridium thermocellum-Clostridium thermosaccharolyticum* ethanol production process: nutritional studies and scale-down. Chem. Eng. Commun. 45 (1–6), 53–60. https://doi.org/10.1080/00986448608911371.

Wang, Y.W., Gao, C., Zheng, Z., Liu, F.M., Zang, J.Y., Miao, J.L., 2015. Immobilization of cold-active cellulase from antarctic bacterium and its use for kelp cellulose ethanol fermentation. Bioresources 10 (1), 1757–1772. https://doi.org/10.15376/biores.10.1.1757-1772.

Wenzel, M., Schönig, I., Berchtold, M., Kämpfer, P., König, H., 2002. Aerobic and facultatively anaerobic cellulolytic bacteria from the gut of the termite *Zootermopsis angusticollis*. J. Appl. Microbiol. 92 (1), 32–40. https://doi.org/10.1046/j.1365-2672.2002.01502.x.

Wilson, D.B., 2016. Microbial diversity and cellulase production. In: Gupta, V.K. (Ed.), New and Future Developments in Microbial Biotechnology and Bioengineering-Microbial Cellulase System Properties and Applications. Elsevier, The Boulevard, United Kingdom, pp. 43–48. https://doi.org/10.1016/B978-0-444-63507-5.00005-8.

Xu, X., Li, J., Zhang, W., Huang, H., Shi, P., Luo, H., et al., 2015. A neutral thermostable β-1,4-glucanase from *Humicola insolens* Y1 with potential for applications in various industries. PLoS One 10 (4), e0124925. https://doi.org/10.1371/journal.pone.0124925.

Xue, D.S., Chen, H.Y., Lin, D.Q., Guan, Y.X., Yao, S.J., 2012. Optimization of a natural medium for cellulase by a marine *Aspergillus niger* using response surface methodology. Appl. Biochem. Biotechnol. 167, 1963–1972.

Yang, J., Dang, H., 2011. Cloning and characterization of a novel cold-active endoglucanase establishing a new subfamily of glycosyl hydrolase family 5 from a psychrophilic deep-sea bacterium. FEMS Microbiol. Lett. 325 (1), 71–76. https://doi.org/10.1111/j.1574-6968.2011.02413.x.

Yang, S.J., Kataeva, I., Hamilton-Brehm, S.D., Engle, N.L., Tschaplinski, T.J., Doeppke, C., et al., 2009. Efficient degradation of lignocellulosic plant biomass, without pretreatment, by the thermophilic anaerobe *Anaerocellum thermophilum* DSM 6725. Appl. Environ. Microbiol. 75 (14), 4762–4769. https://doi.org/10.1128/AEM.00236-09.

Yang, W., Meng, F., Peng, J., Han, P., Fang, F., Ma, L., et al., 2014. Isolation and identification of a cellulolytic bacterium from the Tibetan pig's intestine and investigation of its cellulase production. Electron. J. Biotechnol. 17 (6), 262–267. https://doi.org/10.1016/j.ejbt.2014.08.002.

Yopi, Y., Tasia, W., Melliawati, R., 2017. Cellulase and xylanase production from three isolates of indigenous endophytic fungi. IOP Conf. Ser.: Earth Environ. Sci. 101 (1), 012035. https://doi.org/10.1088/1755-1315/101/1/012035.

Yu, H.-Y., Li, X., 2015. Alkali-stable cellulase from a halophilic isolate, *Gracilibacillus* sp. SK1 and its application in lignocellulosic saccharification for ethanol production. Biomass Bioenergy 81, 19–25. https://doi.org/10.1016/j.biombioe.2015.05.020.

Zare-Maivan, H., Shearer, C.A., 1988. Wood decay activity and cellulase production by freshwater lignicolous fungi. Int. Biodeterior. 24 (6), 459–474. https://doi.org/10.1016/0265-3036(88)90034-6.

Zeng, R., Xiong, P., Wen, J., 2006. Characterization and gene cloning of a cold-active cellulase from a deep-sea psychrotrophic bacterium *Pseudoalteromonas* sp. DY3. Extremophiles 10 (1), 79–82. https://doi.org/10.1007/s00792-005-0475-y.

Zhao, K., Guo, L.Z., Lu, W.D., 2012. Extracellular production of novel halotolerant, thermostable, and alkali-stable carboxymethyl cellulase by marine bacterium *Marinimicrobium* sp. LS-A18. Appl. Biochem. Biotechnol. 168 (3), 550–567. https://doi.org/10.1007/s12010-012-9796-3.

Zheng, F., Tu, T., Wang, X., Wang, Y., Ma, R., Su, X., et al., 2018. Enhancing the catalytic activity of a novel GH5 cellulase GtCel5 from *Gloeophyllum trabeum* CBS 900.73 by site-directed mutagenesis on loop 6. Biotechnol. Biofuels 11, 76. https://doi.org/10.1186/s13068-018-1080-5.

Zhu, C., Xu, Z., Song, R., 2011. The endoglucanase from *Bacillus subtilis* BEC-1 bears halo-tolerant, acidophilic and dithiothreitol-stimulated enzyme activity. World J. Microbiol. Biotechnol. 27 (12), 2863–2871. https://doi.org/10.1007/s11274-011-0767-6.

CHAPTER

5

Role of Bioprocess Parameters to Improve Cellulase Production: Part I

Misbah Ghazanfar, Muhammad Irfan*, Muhammad Nadeem†, Quratulain Syed†*

*Department of Biotechnology, University of Sargodha, Sargodha, Pakistan †Food & Biotechnology Research Center, PCSIR Labs Complex, Lahore, Pakistan

5.1 INTRODUCTION

Cellulose is the most abundant organic polymer, and is supposed to be an everlasting source of raw material for various commodities, representing about 1.5×10^{12} tons of the biomass produced through photosynthesis annually. It is the most incumbent waste matter and inexhaustible biopolymer from agriculture and is abundantly found in nature. According to Wang et al. (2016), cellulose is the main outcome of photosynthesis, and approximately 100 billion tons of dry lignocellulosic biomass are formed annually in nature. Cellulose is also produced by some animals and plants, other than bacteria and fungi. It is frequently found in nature because it is the main constituent of a plant's cell wall. Generally, cellulose is found in association with lignin, hemicellulose, and occasionally with silica (e.g., rice husk, rice straw) in the lignocellulosic biomass structure. Moreover, it constitutes nearly 35%–50% of dry weight of plants, although hemicelluloses cover nearly 20%–35%, and lignin accounts for 5%–30% (Zabeda et al., 2016). Connected units of monomers of glucose are linked by a β-1,4-glycosidic bond to form a linear polysaccharide known as cellulose. The cellulase enzyme is required to conduct enzymatic hydrolysis to liberate these monomeric molecules. Fermentable sugars, especially small chains of glucose and cellobiose molecules, are obtained through conversion of the insoluble polymer of cellulose present in the lignocellulosic material (Taherzadeh and Karimi, 2007). Cellulase converts cellulosic substrate into monomeric sugars, thus playing an important role in the degradation of cellulosic material. The cellulase enzyme system is composed of three main subcomponents, which are endoglucanases, exoglucanases or cellobiohydrolases, and β-glucosidases (Thota et al., 2017), as mentioned in Fig. 5.1. For efficient degradation of cellulosic matter these three kinds of constituent enzymes are required to work together. Exoglucanases produce.

cellobiose by attacking the crystalline ends of cellulosic substrate, while the glycosidic bonds found in the amorphous part of cellulosic substrate are cleaved by endoglucanases (Yoon et al., 2014). β-Glucosidases degrade the liberated cellobiose and glucose molecules (Wang et al., 2013), as mentioned in Fig. 5.2. In nature, cellulase enzymes are commonly found and fungi are also considered to be a powerful producer of cellulase. Production of high yields of enzyme and the secretory pathways are the principal benefits of employing fungi for the yield of cellulases. Fungi like *Chrysosporium*, *Acremonium*, and *Penicillium* are being studied for their potential of cellulase production.

Researchers are investigating the new cellulase-producing systems for economic viability, enhanced yield, and also for spreading the employment of cellulase-producing systems by proceeding toward increasingly industrially adaptable fungi or bacterial production systems. Moreover, several enzymatic hydrolysis studies are now focusing

From Cellulose to Cellulase: Strategies to Improve Biofuel Production
https://doi.org/10.1016/B978-0-444-64223-3.00005-9

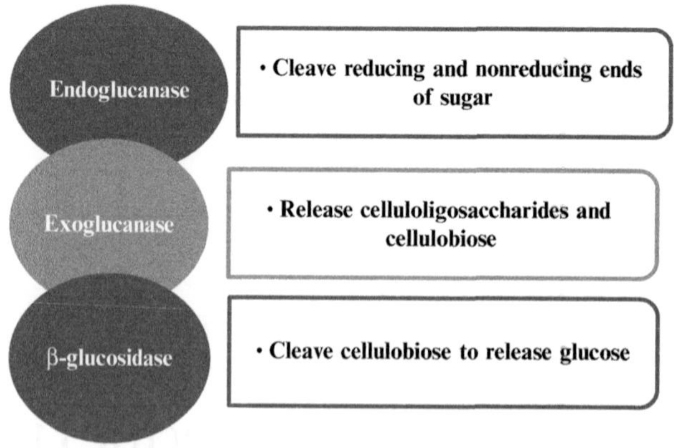

FIG. 5.1 Components of the cellulase system and their function.

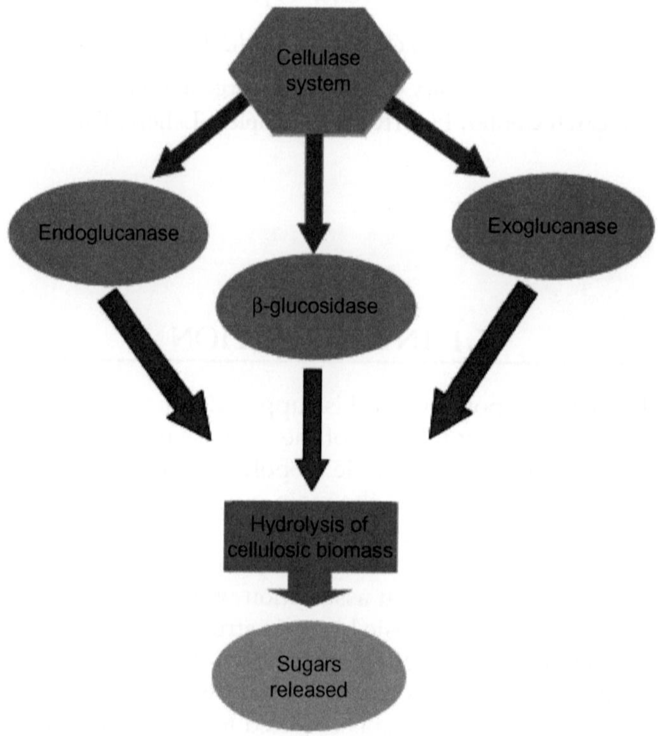

FIG. 5.2 Mode of action of the cellulase system.

on producing cellulase with high activity of enzymes for advancement of the ongoing technology used for the production of cellulase enzyme (Chandel et al., 2012; Garvey et al., 2013). Cellulases have numerous applications in different fields, including the textile industry, the beer and wine industry, the. Paper and pulp industry, the food industry, biofuel production (Gao et al., 2008; Ibrahim et al., 2015), household laundry, agriculture, etc. (Wilson, 2009). Cellulose is mostly degenerated by multicomplex enzymes called cellulases (Pandey et al., 2014). Lee and Koo (2001) reported that when microorganisms are grown on cellulosic material, they produce cellulase enzymes.

Cellulase is among the most competent industrial enzymes. It can be formed by white rot fungi such as *Phanerochaete chrysosporium, Sporotrichum thermophile, Agaricus arvensis, Trametes versicolor, Pleurotus ostreatus, Phlebia gigantean*; soft rot fungi such as *Aspergillus niger, Aspergillus oryzae, Aspergillus nidulans, Aspergillus terreus, Fusarium oxysporum, Fusarium solani, Melanocarpus albomyces, Humicola insolens, Humicola grisea, Penicillium occitanis, Penicillium brasilianum, Penicillium decumbans, Trichoderma longibrachiatum, Trichoderma harzianum, Trichoderma reesei, Neurospora crassa, Chaetomium thermophilum, Chaetomium cellulyticum, Thermoascus aurantiacus, Penicillium fumigosum, Penicillium janthinellum, Mucor circinelloides, Trichoderma atroviride, Paecilomyces inflatus, Penicillium echinulatum*; brown rot fungi such as *Tyromyces palustris, Coniophora puteana, Lanzites trabeum, Poria placenta, Fomitopsis* sp.; anaerobic bacteria such as

Butyrivibrio fibrisolvens, Acetivibrio cellulolyticus, Clostridium acetobutylium, Clostridium cellulolyticum, Clostridium thermocellum, Clostridium papyrosolvens, Ruminococcus albus, Fibrobacter succinogenes; actinomycetes such as *Cellulomonas uda, Cellulomonas fimi, Cellulomonas bioazotea, Streptomyces lividans, Streptomyces drozdowiczii, Thermomonospora curvata, Thermomonospora fusca*; and aerobic bacteria such as *Acinetobacter amitratus, Acinetobacter junii, Acidothermus cellulolyticus, Anoxybacillus* sp., *Bacillus pumilus, Bacillus subtilis, Bacillus amyloliquefaciens, Bacillus circulans, Bacillus flexus, Bacillus licheniformis, Bacteriodes* sp., *Eubacterium cellulosolvens, Cellulomonas biazotea, Cellvibrio gilvus, Microbispora bispora, Paenibacillus curdlanolyticus, Geobacillus* sp., *Pseudomonas cellulosa, Salinivibrio* sp., and *Rhodothermus marinus* (Kuhad et al., 2011).

However, fungi are commonly used for the production of cellulases. They are most usual producer of cellulases. The slow growth rate of fungi and the substrates used in production are the main causes of the high cost of cellulase production. Bacteria are considered potentially good for cellulase production because of the high rate of growth as compared to fungi. However, employment of bacteria is not common for the production of cellulase. Cellulase from bacteria is generally deprived of filter paper activity (FPase), one of the three cellulase activities. Yet bacterial cellulases are usually more efficacious catalysts. Because they face less feedback inhibition they are generally less influenced by the existence of already degraded material. The most significant advantage of using bacteria is that they can be easily genetically modified. This is especially required to improve the production of cellulase (Abdullah, 2004).

However, due to high penetration ability and versatile substrate utilization, fungi are always favored over bacteria for efficient cellulase production. It is very unusual that a single fungus has all the elements for efficient degradation of biomass in spite of the fact that fungi are proficient for the desirable production of cellulase. Moreover, different fungal strains are employed through submerged fermentation (SmF) for the production of commercial cellulase to establish a cellulase system containing its three principal coherent enzymes: exocellulases, endocellulases, and β-glucosidase. However, SmF is an expensive process because of the supplementary purification steps and less yield of the end products (Singhania et al., 2010; Yoon et al., 2014). Genetically modified organisms are being used by some commercial companies nowadays to produce cellulase enzymes, but the developing process is on a small scale due to high production cost (Bhalla et al., 2013). Therefore to construct the process of biofuel production that is economical, there is a need to produce cost-effective cellulase. In this scenario, recent research has been conducted on thermophilic/thermotolerant, hyperthermophilic, and extreme thermophilic fungi as potential producers of thermostable and thermoactive enzymes (Indira et al., 2016). The whole process of the production of enzymes can be made more cost effective by these thermoactive enzymes because under various prevailing conditions it is observed that they are extremely substantial, and they have maximum substrate solubility with increased diffusion rates, a lengthy half-life, and high specific activities (Wu and Arnold, 2013). The cellulase enzymes formed by these microorganisms that decompose cellulose have become more attractive due to the multiplicity of their implementation. The consequential industrial application of cellulases as mentioned in Fig. 5.3 are in biofinishing and biopolishing of fabrics, biostoning of denims, biopolishing in the textile industry, grain alcohol fermentation, malting and brewing, starch

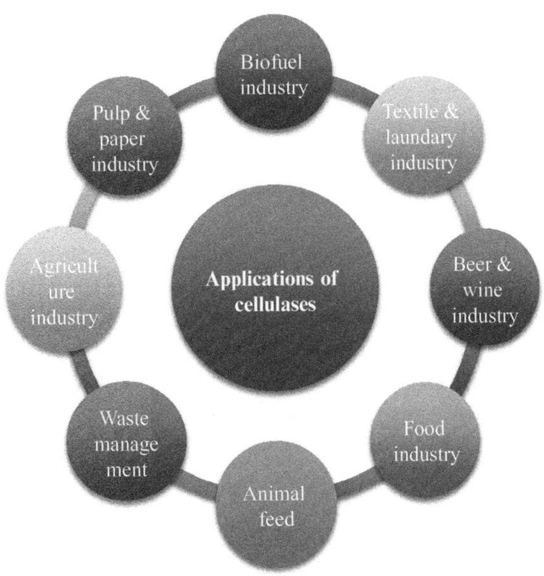

FIG. 5.3 Applications of cellulases in different industries.

processing in the beer and wine industry and paper and pulp industry, extraction and processing of vegetables and fruit juices in the food industry (Gao et al., 2008; Ibrahim et al., 2015), detergents for improving the brightness and softness of fabrics in household laundry, as well as controlling diseases and plant pathogen in agriculture, etc. (Wilson, 2009). The wide applications of cellulase have gradually increased its demand in industries. However, for bioethanol production, the available literature has reported considerably different costs, including \$0.10/gal (Aden and Foust, 2009), \$0.30/gal (Lynd et al., 2008), \$0.40/gal (Kazi et al., 2010), \$0.32/gal (Dutta et al., 2010), and \$0.35/gal (Klein et al., 2010).

There are several difficulties attached to the commercial processes of production of biofuels and technoeconomic analysis because of contradictions regarding the cost of enzymes, specifically for applications in biofuel. For the economical production of cellulase, many factors, such as high substrate loading, low enzyme loading, and a short hydrolysis period, are crucial. According to Klein et al. (2012), these factors can considerably decrease the production cost of enzymes on a commercial platform. At the commercial scale, there are a number of industries involved in the production of cellulase, whereas at the global stage, two main companies, namely Genencor and Novozymes, are known for the industrial production of cellulase. These companies have significantly contributed to substantially bring down the cost of cellulase. Genencor has introduced a cellulase complex called Accelerase1500, specifically for lignocellulosic biomass processing industries, which is further considered to be more cost effective than its previously available predecessor Accelerase1000 for bioethanol production industries. Moreover, Accelerase1500 contains higher levels of β-glucosidase enzyme activity than the other commercial cellulases and thus it seems to be efficient for the conversion of cellobiose into glucose (Singhania et al., 2010). There are many potential cellulases that can hydrolyze biomass along with β-glucosidase. Furthermore, Novozymes also represent various ranges of cellulase preparations depending on their application. In addition, Novozymes also have cellulase preparation for the hydrolysis of biomass. In addition, Amano Enzyme Inc., Japan and MAP's India are the other enzyme producers that have actively participated in the production of cellulase. Although many cellulase-producing companies around the world have participated in its production and marketing, only a few of them have developed cellulase for the conversion of biomass.

The appropriate medium components such as nitrogen, carbon, minerals, etc. and the parameters of fermentation such as pH, temperature, agitation speed, etc. must be determined and optimized accordingly for designing a production medium. Hence maximum product can be obtained by optimizing these parameters (Gupte and Kulkarni, 2003; Franco-Lara et al., 2006; Wang et al., 2011). The overall cost of the product and production costs are reduced by an increase in productivity. Generally, increased productivity can be obtained either by the optimization of the process parameters or by improving the strain. However, both optimization and strain improvement are "Catch-22" situations because it is difficult to select a leading strain until we have the best medium, and it is also impossible to choose a prime medium until we have a leading strain. Hence most researchers around the globe are clinging to one integrant at a time to avoid this jeopardy. This approach, however, does not mean that by using another medium the leading strain will yield maximum enzymes. To avoid the contradictory scenario, several novel methods have been explored, so that both strain improvement and medium design can be done concurrently (Singh et al., 2017). Alam et al. (2004) reported that the nature of cellulosic material, the origin of cellulase enzymes, and the fermentation conditions mainly affect the successful bioconversion of cellulosic biomass into valuable items. The production of cellulase is positively influenced by several factors such as temperature, air circulation, carbon sources, cellulose quality, incubation period, pH, and composition of the medium (Immanuel et al., 2006). Production of cellulase is a desirable area of research around the world for the cost-effective production of biofuels and therefore many researchers are consistently working in this field. Low production and high cost have always been major constraints, which must be overcome by adopting novel and versatile approaches. In this context, use of inexpensive raw materials as substrates, use of genetically modified microorganisms, and use of efficient crude thermostable/thermophilic enzymes are some of the key factors that can enhance cellulase production significantly (Srivastava et al., 2015a).

For improved productivity, reproducibility, and effectiveness, optimization and development of bioprocesses are mainly dependent on accurate, real-time control of physical and chemical process variables. Already existing sensor technologies for some of the important parameters, including pH, temperature, and dissolved oxygen, which have an effect on the process, are being used commonly for process control (Madsen et al., 2011; Yin et al., 2013). Generally, these bioprocess parameters are divided into two categories as shown in Fig. 5.4. For overproduction of the desired product it is required to make nutritional and environmental conditions suitable for the microorganism, leading to the development of a successful fermentation process.

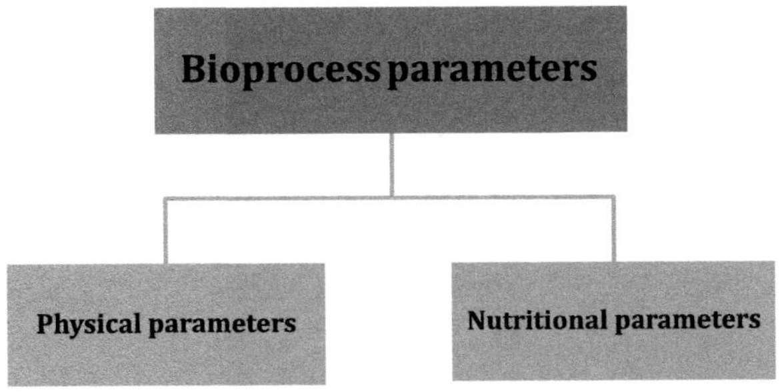

FIG. 5.4 Types of parameters.

5.2 PHYSICAL PARAMETERS

The microbial production of enzymes is influenced by physicochemical parameters, genetic nature of the organism, and components of the medium of fermentation and their concentrations. Therefore optimization of bioprocess parameters is essential to produce a plausible bioprocess framework and to acquire maximum yields for industrial applications. Hence a sensible assemblage of the bioprocess parameters can considerably increase the yield of enzymes. Increase in enzyme yield upon optimization of the bioprocess parameters has been reported by several authors (Iyer and Singhal, 2010). Some of the important physical parameters as described in Fig. 5.5 affect cellulase production and are discussed next.

5.2.1 Effect of pH

By prompting morphological changes in microbes and enzyme production, the pH of the growth medium plays a vital role in various physical parameters. Product stability in the medium also depends on the pH of media during the growth of microbes (Gupta et al., 2003). To obtain higher production of cellulase enzymes, the optimum pH observed was 6.0 during propagation in SmF and solid-state fermentation (SSF) (Mrudula and Murugammal, 2011). Enzyme production is heavily influenced by the initial pH of the medium. With the addition of NaOH or HCl, the pH values can be adjusted to assess their effects on cellulase synthesis in substrate. It is reported that for cellulase activity the optimum pH is generally 6.5. By increasing the pH to the optimum, enzyme activity gradually increased followed by a gradual fall in activity. The activity of cellulase is also observed to be stable at a pH range of 5.0–8.0 (Gautam et al., 2011). For regulation of the three-dimensional structure of the enzyme-active site an optimum pH is essential and the alteration in pH denatures the enzyme by destroying its functional structure and changing the ionic bonding (Lugani et al., 2015).

Fungal cellulase stability and activity was investigated at different pH values to check the effect of pH. At pH 5.0 the cellulase exhibited its complete stability by maintaining its full activity, whereas it secured its half-life at pH 7.5.

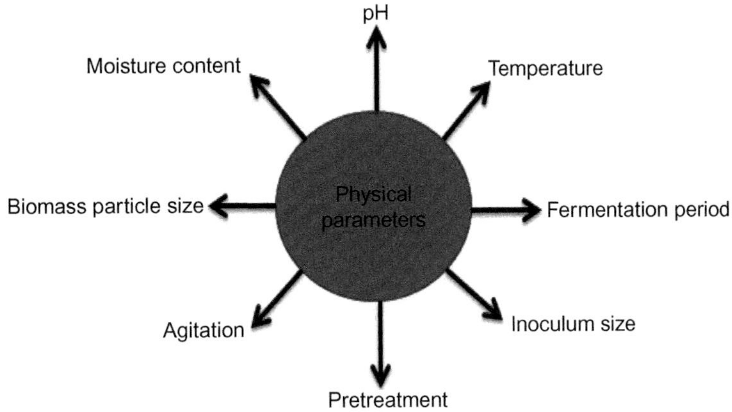

FIG. 5.5 Different physical parameters that affect cellulase production.

At this point, it is also necessary to state that the cellulase displayed its utilitarian stability across a wide range of pH values (4.5–7.5). However, from a stability point of view, a notable decline in pH is seen from pH 6.0, and a distinct decline is observed from pH 8.0–10. Thus the reality that the fungi propagate rapidly in acidic medium supports this phenomenon (Bajpai, 1997). Additionally, the pH of the medium is one of the most significant parameters that regulates growth and enzyme production, and also participates in the movement of various components across the cell membrane. Higher pH influences the activity of cellulase activity due to the reaction catalysis on the dispersion of enzyme molecules, charge and substrate, and dependence of substrate binding.

5.2.2 Effect of Temperature

Temperature has a key part in the physiology, growth, and enzymatic activity of microbes (Kanmani et al., 2011). Fermentation temperature is a principal parameter in the regulation of normal metabolic activities of a microorganism. Generally, the temperature regulated in the SSF system ranges from 25 to 35°C and relies on the rate of growth of the microbes engaged and the factors affecting it instead of the enzyme produced (Mrudula and Murugammal, 2011). Above the optimum values, thermal denaturation of the enzyme occurs and results in the loss of enzyme activity, thus above 45°C low enzyme activity is observed (Lugani et al., 2015). *Bacillus cereus* was grown at different temperatures, including 20, 30, 35, 40, 45, and 50°C. The maximum cellulase activity was observed at 35°C, and low enzyme activity was observed above 35°C. The enzyme activity decreased as the temperature increased because of the thermal denaturation of the enzyme. It has been investigated that the cellulase enzyme from *Bacillus pumilis* showed a higher activity at 35°C (Kanmani et al., 2011).

According to Juturu and Wu (2014), fermentation temperature is very important for optimal enzyme production because temperature fluctuations cause changes in microbial protein structure and properties. At low or high optimum temperatures, metabolic activities are decreased and eventually inhibit the propagation and synthesis of enzymes (Ray et al., 2007). To attain high enzyme yields, it is desirable to obtain the optimum temperature for microorganism growth because the cultivation temperature also influences enzyme production. For thermostable enzymes, thermostable microorganisms are the potential sources. Because of the presence of hydrophobic, electrostatic, and disulfide interaction, thermophiles can tolerate high temperatures better than nonthermotolerant organisms by using increased interaction (Kumar and Nussinov, 2001). Either in a flask or a fermentor, many researchers have investigated different temperatures for maximum cellulase production. It is reported that the strain variation of the microorganism also influences optimal temperature for cellulase production (Murao et al., 1988; Lu et al., 2003). Thermostable enzymes have diverse applications in biotechnology and industries. They are more suitable for the severe conditions because of their thermal stability. An enzyme can be considered as thermostable if it is active and stable at high temperatures. For example, exoglucanases with molecular weights of 40–70 kDa are thermostable enzymes showing maximum activity at 50–70°C and are glycoproteins in nature, while endoglucanases originating from fungi are thermostable having molecular weights of 30–100 kDa with a carbohydrate content of 2%–50% and exhibiting maximum activity at a temperature range of 55–80°C. Thermostable enzymes can be obtained from both mesophilic and thermophilic sources as well (Srivastava et al., 2015b). Moreover, the molecular weights of thermophilic cellulases vary from 30 to 250 kDa with different ranges of carbohydrate from 2% to 50%. Additionally, thermophilic cellulases are substantial at 60°C with extended half-lives at 70, 80, and 90°C, thus exhibiting excellent thermostability (Li et al., 2011). However, some investigations revealed that finding factors that improve or support the thermostable and thermophilic enzymes is difficult. Furthermore, there is comparatively little understanding of the function and nature of the thermostable and thermophilic enzymes obtained from fungi compared to bacteria. Hence for a better understanding of the role of amino acid residues that affect the thermophilic nature and thermal sustainability of cellulases from fungi in the improvement of the thermostability of fungal cellulases there is a need for comprehensive studies and characterization of these amino acid residues. Because the degradation of lignocellulosic matter by using thermostable cellulases enhances the rate of reaction, solubility of substrate, and the rate of bioavailability of organic compounds, the diffusion coefficient simultaneously reduces the contamination, the viscosity, and the cost, therefore thermostable cellulases are counted as potential enzymes for bioprocessing industries. Thermostable cellulases are remarkable in aiding complete hydrolysis reactions (Khelila and Cheba, 2014). Furthermore, thermostable cellulases are also marked as the model system for investigating enzymatic activity and temperature stability because of these characteristics, which would definitely open a new horizon for protein engineering (Haki and Rakshit, 2003). According to Srivastava et al. (2014) there is a demand for viable improvements and modifications in the enzyme-based technology for economic feasibility of biofuel production processes. By employing various possible approaches such as increasing the recycling ability or thermal stability and efficiency of cellulase enzymes, cellulase-assisted biomass hydrolysis can be enhanced. In this respect, thermostability and the efficiency of cellulase enzymes can be enhanced by nanotechnology. Srivastava et al. (2015a) discussed the use of thermostable enzymes using various microorganisms for partial/complete hydrolysis of cellulose. These authors have systematically discussed the utility of thermostable cellulase over cellulose for effective hydrolysis. Ang et al. (2013) reported thermostable cellulase production from *Aspergillus fumigatus* SK1. This thermostable cellulase exhibited higher filter paper unit activity with effective hydrolysis for a shorter time period. It has also been reported that the thermostable enzyme can reduce the hydrolysis time (Dutta et al., 2014; Srivastava et al., 2015b).

5.2.3 Effect of Moisture Content

In SSF processes, moisture content is a crucial parameter because this factor effects the growth, biosynthesis, and production of enzymes (Mrudula and Murugammal, 2011). In SSF the initial moisture content of culture media is a principal factor. Kalogeris et al. (2003) reported that the microorganism employed the nature of the biomass, and the type of end product suggested the optimum moisture content. Kim et al. (2013) reported that the optimum initial moisture content for *Penicillium* sp. GDX01 was 40%–50% for the highest cellulases. Particularly at 50% moisture content, CMCase and FPase activities were maximum, whereas that of β-glucosidase was at 40% (Kim et al., 2013). El-Naggar et al. (2011) found that at 50% moisture level all the purposed activities were maximum and reduced on decreasing or increasing the moisture level. These results are collaborated with the findings of Lu et al. (2003); using dry koji at 40%–50% moisture they achieved high enzyme production. It is considered that enzyme yield is reduced by increasing moisture due to lowering the oxygen transfer, decreasing substrate porosity, increasing soluble protein, and changing particle structure. On the other hand, the decline in nutrient solubility and low level of swelling of substrate are observed in low moisture contents (Gervais and Molin, 2003).

5.2.4 Effect of Biomass Particle Size

Nandakumar et al. (1994) investigated that the degradation of nutrient is a feature of both extracellular enzymes and cell-bound enzymes because in bioprocesses usually the nutrient sources provide physical support for the growth of microorganisms. According to Pandey (2003) and Viniegra-Gonza'lez et al. (2003) the physical properties of the materials, such as the amorphous or crystalline nature, porosity, surface area, accessible area, and particularly particle size, influence the enzyme action on the substrate. Different researchers investigated the influence of particle size on microorganism growth and formation of enzyme in SSF (Zadrazil and Puniya, 1995; Reddy et al., 2003). Enzyme production in SSF could also be affected by biomass particle size. Due to a surface/volume ratio, smaller-sized particles lead to better absorption of nutrient, but a finely ground biomass may counter heat transfer and gas exchange. Consequently, for maximum enzyme production, an optimum limit of particle size of biomass must be ascertained (Kim et al., 2013). To illustrate the relation between particle size of biomass and the production of cellulase in SSF, Kim et al. in 2013 investigated two different particle sizes: 2 and 5 mm. They observed maximum cellulase activity with the particle size of 2 mm. These findings match the results of Bahrin et al. (2011), which revealed that by using biomass with small-sized particles, maximum CMCase and FPase activities can be achieved.

5.2.5 Effect of Pretreatment of Substrate (Lignocellulosic Waste)

To get the efficient bioconversion lignocellulosic biomass into fermentable sugars is a huge task for biotechnology. Because of the presence of lignin and the crystalline structure of the cellulose, untreated lignocellulose is difficult to hydrolyze (Gianni and Lisbeth, 2007). By using various pretreatment techniques, the hard and rigid structure of lignocellulosic biomass could be degraded. Different types of pretreatment techniques involve physical, chemical, biological, and combinations of all these techniques. Every pretreatment technique has remarkable influence on the conversion of lignocellulosic biomass (Monte et al., 2011; Sun et al., 2004).

By changing the chemical and physical structure, lignocellulosic biomass pretreatment enhances the rate of enzymatic action. Pretreatment also facilitates microorganisms to have easier access to biomass (Kim et al., 2013). Pretreatment is responsible for quick and effective conversion of carbohydrates into sugars by changing the chemical and structural configuration of the lignocellulosic biomass (Chang and Holtzapple, 2000).

To enhance the accessibility of enzymes to cellulosic fibers, various physical (grinding, hydrothermolysis), chemical (acid, alkali, solvents, ozone), physicochemical (ammonia fiber explosion, steam explosion), and biological pretreatment techniques have been employed (Ghazanfar et al., 2018). Effect of pretreatment on the lignocellulosic biomass is schematically represented in Fig. 5.6.

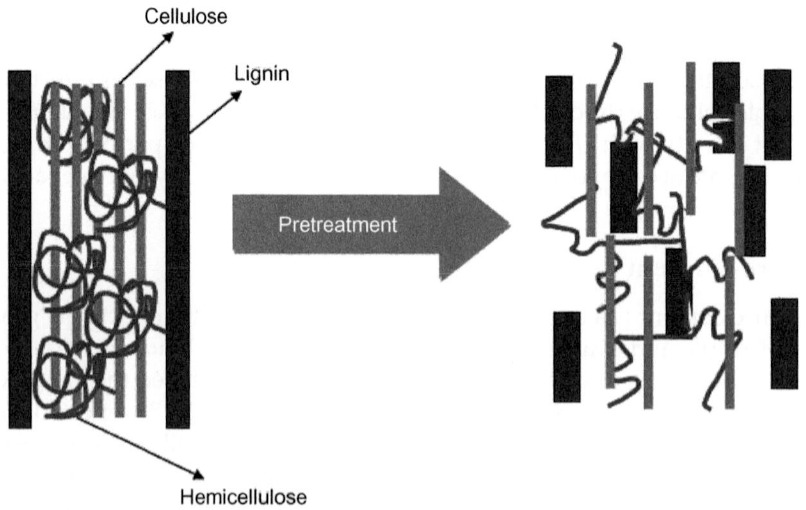

FIG. 5.6 Diagrammatic representation of the effect of pretreatment on biomass.

5.2.6 Effect of the Fermentation Period

Time of fermentation has an important role in product formation. Enzyme production is associated with time of incubation (Gautam et al., 2011). To a certain extent, the production of enzymes and other metabolic activities directly depends on the incubation period (Kang et al., 2004). Enzyme production is reduced by increasing the optimal incubation time. It may be caused by suppressed physiology of microbes due to exhaustion of nutrients in the medium resulting in the inactivation of the enzymes' secretory machinery (Nochure et al., 1993). The cellulase produced from *B. cereus* and isolated from the Persian Gulf showed a higher cellulase activity after 48 h of incubation time. It has been shown that (Samira et al., 2011) production of enzymes could be made inexpensive if microbes require a short period of incubation (Sonjoy et al., 1995). Muniswaran and Charyulu (1994) have also reported a similar trend in cellulase production using *Trichoderma viride*. However, according to the characteristics and culturing conditions the optimum fermentation period may vary from strain to strain. This influence is due to the exhaustion of nutrients or gathering of other by-products in the fermentation media over time, which causes reduced cellulase activity (Bajaj et al., 2009).

5.2.7 Effect of Inoculum Size

Enzyme activity is distinctly decreased on altering the optimal inoculum concentration, because microbial growth was reduced when inoculum size increases due to raised competition for nutrients and space among cells. The length of the stationary phase is also influenced by these factors, which are followed by loss of activity of enzymes due to the assemblage of secondary metabolites and toxic products (Lugani et al., 2015). Enzyme yield reduced by increasing inoculum size due to the limitation of nutrients. While high inoculum amount causes rapid multiplication of microbial biomass and low inoculum, it may require a longer time for microbial proliferation and substrate utilization to produce the desired enzymes. So, as suggested by Ramachandran et al. (2004), a balance between the growing biomass and substrate consumption will yield maximum enzymes.

5.2.8 Effect of Agitation

To fulfill the need for oxygen, distribution of nutrients, and uniform mixing during the fermentation process, agitation is usually required. Increase in agitation lessens hostility to the transfer of oxygen into the medium and then into the microbial cells (Bartholomew et al., 1950), increasing the rate of oxygen transfer in the fermentation broth by supplying a large gas/liquid interface surface area (Das et al., 2013). Studies clearly revealed that higher agitation speed inhibited enzyme activity. It is reported that agitation has an influence on the mixing of the nutrients and level of aeration in the fermentation medium (Giavasis et al., 2006). Many investigations revealed maximum cellulase production at 50 rpm (Mmango-Kaseke et al., 2016).

5.3 NUTRITIONAL PARAMETERS

We already know that cellulase systems contain three subcomponents: endoglucanases, exoglucanases, and β-glucosidase; the conversion of cellulose into glucose is possible by the synergy of all these enzymes (Lynd et al., 2002). Mg et al. (2015) reported that nutrient sources are the critical parameters for the production of cellulase, because carbon is believed to be the primary nutrient for the bacteria. Some important nutritional parameters that affect cellulase production are mentioned in Fig. 5.7 and are discussed next.

5.3.1 Effect of Carbon Sources

Carbon plays a key part in the propagation of microbes and production of cellulase, thus it is the most significant component in the media and also an energy source for microorganisms. The rate of metabolism of the carbon source often influences the formation of biomass and desired enzymes (Marwick et al., 1999).

The existence of substrate that acts as an inducer for the production of cellulase also affects cellulase production, which is an inducible enzyme (Lugani et al., 2015). A variety of lignocellulosic biomasses such as rice husk, tea production waste, sugarcane bagasse, wheat bran, rice bran, coconut coir pith, etc. have been engaged for the production of cellulase by employing different microorganisms (Pandey et al., 1999). However, above the optimal concentration of substrate, cell concentration is affected adversely because it also contains minerals in addition to carbon that may act as a nourishing supplement (Omojasola and Jilani, 2008). For bacteria, it is believed that carbon is the chief nutrient; different sources of carbon such as fructose, glucose, starch, maltose, sucrose, lactose, and mannitol are also being used (Bai et al., 2012). Medium components and raw biomasses are the principal parts of the cost of product in fermentation processes. Therefore the assemblage of these components is a crucial assignment for the production of cellulase. The type of carbon source also influences the nature and quantity of the production of enzymes along with the speed of integration of sources of carbon. Single-cell protein production of ethanol is an example of where the raw substances constitute nearly 60%–77% of the production cost. Thus it is necessary to estimate the role of the cost as well as the dynamics of the carbon source in fermentation processes (Singh et al., 2017).

5.3.2 Effect of Substrate Concentration

In the SSF process, particularly when using trays or flasks during the fermentation process, substrate concentration plays an important part (Satyanarayana, 1994). The high market demand of cellulase and its immense potential in bioenergy and biobased industries have motivated researchers to look for new and efficient cellulases. The cost of cellulase is one major impediment to commercialization of enzymatic cellulose hydrolysis. Utilization of

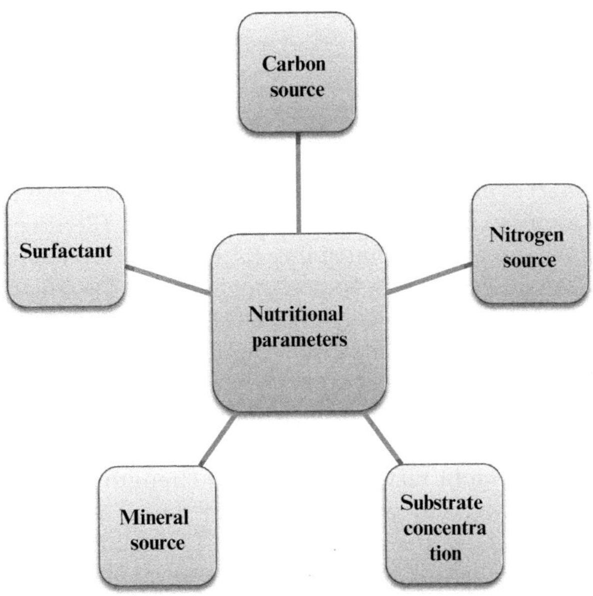

FIG. 5.7 Different nutritional parameters that affect production of cellulase.

lignocellulosic biomass residue for cellulase production not only reduces the production cost, but also leads to waste management (Sreena and Sebastian, 2015). Substrate concentration affects many culture fermentation factors such as depth of substrate, moisture level, transfer of heat, inoculum amount, aeration, etc. (Hansen et al., 1993). Type and particle size of substrate and employed organism also influence the concentration of substrate (Fadel, 2000). It is also reported that moderate concentration led to high enzyme yield and cellulolytic enzyme production is negatively affected by lowering or raising the concentration of substrate (El-Naggar et al., 2011). These findings are also collaborated with the research conducted by Fadel (2000).

5.3.3 Effect of Nitrogen Source

The choice and concentration of nitrogen source in the media also play a critical role (Marwick et al., 1999; Singh et al., 2007). The basic need is to fulfill the demand for optimal growth provision of utilizable forms of nitrogen source to organisms (Jyotsna et al., 2015). Usually, nitrogen is imparted in combination with substrate and is used in inorganic or organic form (Singhania et al., 2006). However, many researchers have revealed that organic sources of nitrogen cause high production of cellulases as compared to inorganic sources of nitrogen (Jeya et al., 2010; Deswal et al., 2011). During extracellular enzyme production the existence of external nitrogen sources is necessary in the fermentation media for efficient use of soluble carbohydrates. For high cellulase production the use of organic nitrogen is a more acceptable source as compared to inorganic sources (Ray et al., 2007; Ariffin et al., 2008). Nitrogen is an important constituent of proteins essential in cell metabolism. Biomass production increases with the incorporation of extra nitrogen but diminishes cellulase production (Masaoka et al., 1993). Cellulase productivity can be optimized using different nitrogen supplementation media such as ammonium nitrate, peptone, yeast extract, sodium nitrate, and ammonium sulfate. In SSF, maximum cellulase production was noticed in rice bran for a nitrogen source in ammonium nitrate, and SmF maximum cellulase production was noticed in rice bran for a nitrogen source in ammonium nitrate, respectively (Shobana and Maheswari, 2013). In 2013 Kim et al. studied the impact of various organic nitrogen sources on the production of cellulases by using wheat bran, yeast extract, rape seed meal, and peptone. Among these nitrogen sources, highest cellulase production was exhibited by yeast extract and the second highest cellulase production was shown by wheat bran, whereas peptone showed limited activity. It was observed that rape seed meal showed the same activity as yeast extract; rape seed meal is cheaper than yeast extract so is a reasonable substitute and its use as a nitrogen source is economical. To estimate the optimum concentration of yeast extract for the production of cellulase, different concentrations of yeast extract were used, and it was concluded that a concentration of 5% of yeast extract caused the maximum activity of FPase. *Penicillium* sp. GDX01 cannot produce cellulase without yeast extract, revealing that nitrogen is necessary for the production of cellulase. However, with 10% concentration of yeast extract, *Penicillium* sp. GDX01 cellulase production decreased with 10% yeast extract due to growth inhibition. The highest activities of CMCase and β-glucosidase were obtained by using 5% concentration of yeast extract. These findings distinctly show that an optimized concentration of nitrogen source must be determined for enhanced cellulase production.

5.3.4 Effect of Surfactants

Many studies have reported the stimulatory effect of surfactants on cellulolytic enzyme production (Okeke and Obi, 1993; Menon et al., 1994; Kuhad et al., 1994). Evidently, the main effects are the release of cell-bound enzyme (Reese and Manguire, 1969), an increase in cell membrane permeability (Reese et al., 1969), and a diminution in mycelial growth caused by a decrease in oxygen availability (Hulme and Stranks, 1970). The effect of surfactant on cellulase production is familiar (Pardo, 1996). Tween 80 does not denature the enzymes, thus its use is advantageous. Incorporation of the surfactant enhances cellulose conversion because the first phase of saccharification is affected by enzyme adsorption into the insoluble cellulosic material. Alignment and adsorption of the molecules of surfactant at the liquid/solid interface must make the substrate quickly wettable using the solution of enzyme, bringing the substrate quickly into interaction with the enzyme and permitting the enzymes to access the substrate. Otherwise, reachable site in biomass to enzyme is very poor. Moreover, interactions between the substrates, the molecules of surfactant, and the adsorbed enzymes might result in a reduced strength, in which the enzymes are kept on the substrates. Castanon and Wilke (2004) observed that when the enzyme was in the liquid phase, the activity of endoglucanase was kept for a longer period. Singh et al. (2007) reported that the surfactants are compounds that have both hydrophilic and hydrophobic parts, which can decrease interfacial and surface tensions by assembling at the immiscible fluids' interface and increasing the bioavailability, mobility, solubility, and subsequent biodegradation of organic compounds, which are insoluble or hydrophobic in nature.

5.3.5 Effect of Mineral Source

According to Weinhouse and Benziman (1974) minerals play a significant role in the growth and production of microbial cellulases regardless of existing in small concentrations. Martin et al. in 1990 and Wong in 1993 reported that elements like magnesium, calcium, potassium, iron, and sodium are nutrients that have a vital role because they are enzymatic cofactors. Among other minerals in low concentration, soluble phosphates with calcium, sodium, potassium, iron, copper, magnesium, manganese, and cobalt are used as sources of phosphorus (Lima et al., 2001). Ferric ions are cofactor oxygenases that help in the formation of phosphoric complexes with high energy such as adenosine triphosphate (Nohata and Kurane, 1997). Sulfur is an important constituent of vitamins, amino acids, and prosthetic groups from several significant proteins in redox reactions. The sulfates and amino acids are also used by the bacteria (Trabulsi and Alterthum, 1999). García et al. (1974) investigated the roles of magnesium in the production of different lipophilic components from the enzymatic system. For cellulase stability the effect of different types of metal ions has been investigated. They observed that cellulase is almost stable in the existence of most bivalent metal ions (Mg^{2+}, Fe^{2+}, Mn^{2+}, Cu^{2+}, and Hg^{2+}) except Ca^{2+} because the enzyme retains nearly 80% of relative activity. Moreover, in the existence of Fe^{3+} and Mn^{2+} ions, the enzyme attains its complete relative activity. Metal ions have a catalytic impact on the enzyme because these ions bind to the active site of the enzyme and boost its activity, hence enhancing enzyme stability. The unsupportive protein association with metal ions may cause the decline in enzyme stability in the presence of Ca^{2+} (Olajuyigbe and Ogunyewo, 2016). Increase in the activity of cellulase in the existence of Fe^{3+} ions is noticed, and improved activity of cellulase in the existence of iron oxide nanoparticles has also been reported in one of the recent investigations (Srivastava et al., 2015b).

References

Abdullah, N., 2004. Strategies for expanded bed purification of recombinant protein. PhD Thesis, University of Cambridge, United Kingdom.

Aden, A., Foust, T., 2009. Technoeconomic analysis of the dilute sulfuric acid and enzymatic hydrolysis process for the conversion of corn Stover to ethanol. Cellulose 16, 535–545.

Alam, M.Z., Manchur, M.A., Anwar, M.N., 2004. Isolation, purification, characterization of cellulolytic enzymes produced by the isolate Streptomyces omiyaensis. Pak. J. Biol. Sci. 7, 1647–1653.

Ang, T.N., Ngoh, G.C., Chua, A.S.M., 2013. Development of a novel inoculum preparation method for solid-state fermentation: cellophane film culture (CFC) technique. Ind. Crop. Prod. 43, 774–777.

Ariffin, H., Hassan, M.A., Shah, U.K., Abdullah, N., Ghazali, F.M., Shirai, Y., 2008. Production of bacterial endoglucanase from pretreated oil palm empty fruit bunch by *Bacillus pumilus* EB3. J. Biosci. Bioeng. 106, 231–236.

Bahrin, E.K., Seng, P.Y., Abd-Aziz, S., 2011. Effect of oil palm empty fruit bunch particle size on cellulase production by *Botryosphaeria* sp. under solid state fermentation. Aust. J. Basic Appl. Sci. 5, 276–280.

Bai, S., Kumar, M.R., Kumar, D.J.M., Balashanmugam, P., Kumaran, M.D.B., Kalaichelvan, P.T., 2012. Cellulase production by *Bacillus subtilis* isolated from cow dung. Arch. Appl. Sci. Res. 4, 269–279.

Bajaj, B.K., Pangotra, H., Wani, M.A., Sharma, P., Sharma, A., 2009. Partial purification and characterization of a highly thermostable and pH stable endoglucanase from a newly isolated *Bacillus* strain M-9. Indian J. Chem. Technol. 16, 382–387.

Bajpai, P., 1997. Microbial xylanolytic enzyme system properties and applications. Appl. Environ. Microbiol. 43, 141–189.

Bartholomew, W.H., Karrow, E.O., Sfat, M.R., Wilhelm, R.H., 1950. Oxygen transfer and agitation in submerged fermentations. Effect of air flow and agitation rates upon fermentation of *Penicilium chrysogenum* and *Streptomyces griseus*. Ind. Eng. Chem. 42, 1801–1809.

Bhalla, A., Bansal, N., Kumar, S., Bischoff, K.M., Sani, R.K., 2013. Improved lignocelluloses conversion to biofuels with thermophilic bacteria and thermostable enzymes. Bioresour. Technol. 128, 751–759.

Castanon, M., Wilke, C.R., 2004. Effects of the surfactant Tween 80 onenzymatic hydrolysis of newspaper. Biotechnol. Bioeng. 23, 1365–1372.

Chandel, A.K., Chandrasekhar, G., Silva, M.B., Silva, S.S.D., 2012. The realm of cellulases in biorefinerydevelopment. Crit. Rev. Biotechnol. 32, 187–202.

Chang, V., Holtzapple, M., 2000. Fundamental factors affecting biomass enzymatic reactivity. Appl. Biochem. Biotechnol. 84, 5–37.

Das, A., Paul, T., Halder, S.K., Maity, C., Mohapatra, P.K.D., Pati, B.R., Mondal, K.C., 2013. Study on regulation of growth and biosynthesis of cellulolytic enzymes from newly isolated *Aspergillus fumigatus* ABK9. Pol. J. Microbiol. 62, 31–43.

Deswal, D., Khasa, Y.P., Kuhad, R.C., 2011. Optimization of cellulase production by a brown rot fungus *Fomitopsis sp.* RCK2010 under solid state fermentation. Bioresour. Technol. 102, 6065–6072.

Dutta, A., Dowe, N., Ibsen, K.N., Schell, D.J., Aden, A., 2010. An economic comparison of different fermentation configurations to convert corn Stover to ethanol using *Z. mobilis* and *Saccharomyces*. Biotechnol. Prog. 26, 64–72.

Dutta, N., Mukhopadhyay, A., Dasgupta, A.K., Chakrabarti, K., 2014. Improved production of reducing sugars from rice husk and rice straw using bacterial cellulase and xylanase activated with hydroxyapatite nano particles. Bioresour. Technol. 153, 269–277.

El-Naggar, N.E.-A., Sherief, A.A., Hamza, S.S., 2011. Bioconversion process of rice straw by thermotolerant cellulolytic *Streptomyces viridiochromogenes*under solid-state fermentation conditions for bioethanol production. Afr. J. Biotechnol. 10, 11998–12011.

Fadel, M., 2000. Production physiology of cellulases and β-glucosidase enzymes of *Aspergillus niger*grown under solid state fermentation conditions. Online J. Biol. Sci. 1.401–411.

Franco-Lara, E., Link, H., Weuster-Botz, D., 2006. Evaluation of artificial neural networks for modelling and optimization of medium composition with a genetic algorithm. Process Biochem. 41, 2200–2206.

Gao, J., Weng, H., Zhu, D., Yuan, M., Guan, F., Xi, Y., 2008. Production and characterization of cellulolytic enzymes from the thermoacidophilic fungal *Aspergillus terreus* M11 under solid-state cultivation of corn stover. Bioresour. Technol. 99, 7623–7629.

García, R.C., Recondo, E., Dankert, M., 1974. Polysaccharide biosynthesis in *Acetobacter xylinum*: enzymatic synthesis of lipid diphosphate and monophospate sugars. Eur. J. Biochem. 43, 93–105.

Garvey, M., Klose, H., Fischer, R., Lambertz, C., Commandeur, U., 2013. Cellulases for biomass degradation: comparing recombinant cellulase expression platforms. Trends Biotechnol. 31 (10), 581–593.

Gautam, S.P., Bundela, P.S., Pandey, A.K., Khan, J., Awasthi, M.K., Sarsaiya, S., 2011. Optimization for the production of cellulase enzyme from municipal solid waste residue by two novel cellulolytic fungi. Biotechnol. Res. Int. 2011. Article ID 810425, 8 pages. https://doi.org/10.4061/2011/810425.

Gervais, P., Molin, P., 2003. The role of water in solid-state fermentation. Biochem. Eng. J. 13, 85–101.

Ghazanfar, M., Irfan, M., Nadeem, M., 2018. Statistical modeling and optimization of pretreatment of *Bombax ceiba* with KOH through Box-Behnken design of response surface methodology. Energy Sources A Recov. Util. Environ. Effects 40 (9), 1114–1124.

Gianni, P., Lisbeth, O., 2007. Effect of compounds released during pretreatment of wheat straw on microbial growth and enzymatic hydrolysis rates. Biotechnol. Bioeng. 96, 250–258.

Giavasis, I., Harvey, L.M., McNeil, B., 2006. The effect of agitation and aeration on the synthesis and molecular weight of gellan in batch cultures of *Sphingomonas paucimobilis*. Enzym. Microb. Technol. 38, 101–108.

Gupta, R., Gigras, P., Mohapatra, H., Goswami, V.K., Chauhan, B., 2003. Microbial-amylases. A biotechnological perspective. Process Biochem. 38, 1599–1616.

Gupte, M., Kulkarni, P., 2003. A study of antifungal antibiotic production by *Thermomonospora* sp MTCC3340 using full factorial design. J. Chem. Technol. Biotechnol. 78, 605–610.

Haki, G.D., Rakshit, S.K., 2003. Developments in industrially important thermostable enzymes: a review. Bioresour. Technol. 89, 17–34.

Hansen, R.C., Keener, H.M., Marugg, C., Dick, W.A., Hoitink, H.A.J., 1993. Composting of poultry manure. In: Hoitink, H.A.J. (Ed.), Science and Engineering of Composting: design, Environmental, Microbiological and Utilization Aspects. Renaissance Publications, Ohio, pp. 131–153.

Hulme, M.A., Stranks, D.W., 1970. Induction and the regulation of production of cellulase by fungi. Nature 226, 469–470.

Ibrahim, M.F., Abd-Aziz, S., Yusoff, M.E.M., Phang, L.Y., Hassan, M.A., 2015. Simultaneous enzymatic saccharification and ABE fermentation using pretreated oil palm empty fruit bunch as substrate to produce butanol and hydrogen as biofuel. Renew. Energy 77, 447–455.

Immanuel, G., Dhanusha, R., Prema, P., Palavesam, A., 2006. Effect of different growth parameters on endoglucanase enzyme activity by bacteria isolated from coir retting effluents of estuarine environment. Int. J. Environ. Sci. Technol. 3, 25–34.

Indira, D., Sharmila, D., Balasubramanian, P., Thirugnanam, A., Jayabalan, R., 2016. Utilization of sea water based media for the production and characterization of cellulase by *Fusarium subglutinans* MTCC 11891. Biocatal. Agric. Biotechnol. 7, 187–192.

Iyer, P.V., Singhal, R.S., 2010. Glutaminase production using *Zygosaccharomyces rouxii* NRRL-Y 2547: effect of aeration, agitation regimes and feeding strategies. Chem. Eng. Technol. 33, 52–62.

Jeya, M., Joo, A., Lee, K.M., Sim, W.I., Oh, D.K., Kim, Y.S., Kim, I.W., Lee, J.K., 2010. Charaterization of endo-β-1,4-glucanase from 1 novel strain of *Penicillium pinophilum* KMJ601. Appl. Microbiol. Biotechnol. 85, 1005–1014.

Juturu, V., Wu, J.C., 2014. Microbial cellulases: engineering, production and applications. Renew. Sust. Energ. Rev. 33, 188–203.

Jyotsna, K.P., Rao, A.R., Devaki, K., 2015. Effect of nutritional factors on cellulase production by *Streptomyces albaduncus* from the gut of earthworm, *Eiseniafoetida*. Pest Manag, Hortic. Ecosyst. 21, 75–80.

Kalogeris, E., Christakopoulos, P., Katapodis, P., Alexiou, A., Vlachou, S., Kekos, D., Macris, B.J., 2003. Production and characterization of cellulolytic enzymes from the thermophilic fungus *Thermoascus aurantiacus* under solid state cultivation of agricultural wastes. Process Biochem. 38, 1099–1104.

Kang, S.W., Park, Y.S., Lee, J.S., Hong, S.I., Kim, S.W., 2004. Production of cellulases and hemicellulases by *Aspergillus niger* KK2 from lignocellulosic biomass. Bioresour. Technol. 91, 153–156.

Kanmani, R., Vijayabaskar, P., Jayalakshmi, S., 2011. Saccharification of banana-agro waste and clarification of apple juice by cellulase enzyme produced from *Bacillus pumilis*. World Appl. Sci. J. 12, 2120–2128.

Kazi, F.K., Fortman, J.A., Anex, R.P., Hsu, D.D., Aden, A., Dutta, A., Kothandaraman, G., 2010. Technoeconomic comparison of process technologies for biochemical ethanol production from corn stover. Fuel 89, S20–S28.

Khelila, O., Cheba, B., 2014. Thermophilic cellulolytic microorganisms from western Algerian sources: promising isolates for cellulosic biomass recycling. Procedia Technol. 12, 519–528.

Kim, D.M., Cho, E.J., Kim, J.W., Lee, Y.W., Chung, H.G., 2013. Production of cellulases by *Penicillium* sp. in a solid-state fermentation of oil palm empty fruit bunch. Afr. J. Biotechnol. 13, 145–155.

Klein, M.D., Oleskowicz, P.P., Simmons, B.A., Blanch, H.W., 2010. Technoeconomic analysis of biofuels: a wiki-based platform for lignocellulosic biorefineries. Biomass Bioenergy 34, 1914–1921.

Klein, M.D., Oleskowicz, P.P., Simmons, B.A., Blanch, H.W., 2012. The challenge of enzyme cost in the production of lignocellulosic biofuels. Biotechnol. Bioeng. 109, 1083–1087.

Kuhad, R.C., Kumar, M., Singh, S., 1994. A hypercellulolytic mutant of *Fusarium oxysporum*. Lett. Appl. Microbiol. 19, 397–400.

Kuhad, R.C., Gupta, R., Singh, A., 2011. Microbial cellulases and their industrial applications. Enzyme Res. 2011. Article ID 280696, 10 pages. https://doi.org/10.4061/2011/280696.

Kumar, S., Nussinov, R., 2001. How does thermophilic protein deal with heat. Cell. Mol. Life Sci. 58, 1216–1233.

Lee, S.M., Koo, Y.M., 2001. Pilot-scale production of cellulase using *Trichoderma reesei* Rut C-30 in fedbatch mode. J. Microbiol. Biotechnol. 11, 229–233.

Li, D.C., Li, A.N., Papageorgiou, A.C., 2011. Cellulases from thermophilic fungi: recent insights and biotechnological potential. Enzym. Res. 2011, 1–9.

Lima, U.A., Aquarone, E., Borzani, W., Schmidell, W., 2001. Biotecnologia Industrial: Processos Fermentativos e Enzimáticos. Edgard Blücher, São Paulo. 593 pp.

Lu, W., Li, D., Wu, Y., 2003. Influence of water activity and temperature on xylanase biosynthesis in pilot-scale solid-state fermentation by *Aspergillus sulphureus*. Enzym. Microb. Technol. 32, 305–311.

Lugani, Y., Singla, R., Sooch, B.S., 2015. Optimization of cellulase production from newly isolated *Bacillus sp.* Y3. J. Bioprocess. Biotechnol. 5, 264. https://doi.org/10.4172/2155-9821.1000264.

Lynd, L.R., Weimer, P.J., van Zyl, W.H., Pretorius, I.S., 2002. Microbial cellulose utilization: fundamentals and biotechnology. Microbiol. Mol. Biol. Rev. 66, 506–577.

Lynd, L.R., Laser, M.S., Bransby, D., Dale, B.E., Davison, B., Hamilton, R., Himmel, M., Keller, M., McMillan, J.D., Sheehan, J., Wyman, C.E., 2008. How biotech can transform biofuels. Nat. Biotechnol. 26, 169–172.

Madsen, M., Holm-Nielsen, J.B., Esbensen, K.H., 2011. Monitoring of anaerobic digestion processes: a review perspective. Renew. Sust. Energ. Rev. 15, 3141–3155.

Marwick, J.D., Wright, P.C., Burgess, J.G., 1999. Bioprocess intensification for production of novel marine bacterial antibiotics through bioreactor operation and design. Mar. Biotechnol. 1, 495–507.

Masaoka, S., Ohe, T., Sakota, N., 1993. Production of cellulose from glucose by *Acetobacter xylinum*. J. Ferment. Bioeng. 75, 18–22.

Menon, K., Rao, K.K., Pushalkar, S., 1994. Production of β-glucosidase by *Penicillium rubrum* O. Stoll. Indian J. Exp. Biol. 32, 706–709.

Mg, Z.L.M., Than, W.M., Myint, M., 2015. Study on the cellulase enzyme producing activity of bacteria isolated from manure waste and degrading soil. Int. J. Tech Res. Appl. 3, 165–169.

Mmango-Kaseke, Z., Okaiyeto, K., Nwodo, U.U., Mabinya, L.V., Okoh, A.I., 2016. Optimization of cellulase and xylanase production by *Micrococcus* species under submerged fermentation. Sustainability 8, 1168. https://doi.org/10.3390/su8111168.

Monte, J.R., Brienzo, M., Milagres, A.M.F., 2011. Utilization of pineapple stem juice to enhance enzyme-hydrolytic efficiency for sugarcane bagasse after an optimized pre-treatment with alkaline peroxide. Appl. Energy 88, 403–408.

Mrudula, S., Murugammal, R., 2011. Production of cellulase by *Aspergillus niger* under submerged and solid state fermentation using coir waste as a substrate. Braz. J. Microbiol. 42, 1119–1127.

Muniswaran, P.K.A., Charyulu, N.C.L.N., 1994. Solid state fermentation of coconut coir pith for cellulase production. Enzym. Microb. Technol. 16 (5), 436–446.

Murao, S., Sakamoto, R., Arai, M., 1988. Cellulase of *Aspergillus aculeatus*. In: Wood, W.A., Kellog, S.T. (Eds.), Methods in Enzymology. vol. 160. Academic Press, London, UK, pp. 275–284.

Nandakumar, M.P., Thakur, M.S., Raghavarao, K.S.M.S., Ghildyal, N.P., 1994. Mechanism of solid particle degradation by *Aspergillus niger* in solid substrate fermentation. Process Biochem. 29, 545–551.

Nochure, S.V., Roberts, M.F., Demain, A.I., 1993. True cellulases production by *Clostridium thermocellum* grown on different carbon sources. Biotechnol. Lett. 15, 641–646.

Nohata, Y., Kurane, R., 1997. Complete defined medium for largescale production of polysaccharide bio absorbent from *Alcaligenes latus* B-16. J. Ferment. Bioeng. 83, 116–117.

Okeke, B.C., Obi, S.K.C., 1993. Production of cellulolytic and xylanolytic enzymes by an *Arthrographis* sp. World J. Microbiol. Biotechnol. 9, 345–349.

Olajuyigbe, F.M., Ogunyewo, O.A., 2016. Enhanced production and physicochemical properties of thermostable crude cellulase from *Sporothrix carnis* grown on corn cob. Biocatal. Agric. Biotechnol. 7, 110–117.

Omojasola, P.F., Jilani, O.P., 2008. Cellulose production by *Trichoderma longi*, *Aspergillus niger*, and *Saccharomyces cerevisae* cultured on waste materials from orange. Pak. J. Biol. Sci. 11, 2382–2388.

Pandey, A., 2003. Solid-state fermentation. Biochem. Eng. J. 13, 81–84.

Pandey, A., Selvakumar, P., Soccol, C.R., Nigam, P., 1999. Solid state fermentation for the production of industrial enzymes. Curr. Sci. 77, 149–162.

Pandey, S., Shahid, M., Srivastava, M.M., Singh, S., Sharma, A., Kumar, V., Srivastava, Y.K., 2014. Effect of various physiological parameters and different carbon sources on cellulase and xylanase induction by different strains of *Trichoderma* species. Enzym. Eng. 1, 1–5.

Pardo, A.G., 1996. Effect of surfactants on cellulase production by *Nectriacatalinensis*. Curr. Microbiol. 33, 275–278.

Ramachandran, S., Patel, A.K., Nampoothiri, K.M., Francis, F., Nagy, V., Szakacs, G., Pandey, A., 2004. Coconut oil cake-a potential raw material for the production of α-amylase. Bioresour. Technol. 93, 169–174.

Ray, A.K., Bairagi, A., Ghosh, K.S., Sen, S.K., 2007. Optimization of fermentation conditions for cellulase production by *Bacillus subtilis* CY5 and *Bacillus circulans* TP3 isolated from fish gut. Actalchthyol. et Piscat. 37, 47–53.

Reddy, G.V., Ravindra Babu, P., Komaraiah, P., Roy, K.R.R.M., Kothari, I.L., 2003. Utilization of banana waste for the production of lignolytic and cellulolytic enzymes by solid substrate fermentation using two Pleurotus species *(P. ostreatusand P. sajor. caju)*. Process Biochem. 38, 1457–1462.

Reese, E.T., Manguire, A., 1969. Surfactants as stimulants of enzyme production by microorganisms. Appl. Microbiol. 17, 242–245.

Reese, E.T., Lola, J.E., Parrish, F.W., 1969. Modified substrates and modified products as inducers of carbohydrases. J. Bacteriol. 100, 1151–1154.

Samira, M., Mohammad, R., Gholamreza, G., 2011. Carboxymethyl-cellulose and filter-paperase activity of new strains isolated from Persian gulf. Microbiol. J. 1, 8–16.

Satyanarayana, T., 1994. Production of bacterial extracellular enzymes by solid state fermentation. In: Pandey, A. (Ed.), Solid State Fermentation. Wiley Eastern Limited, New Delhi, pp. 122–129.

Shobana, P., Maheswari, N.U., 2013. Production of cellulase from *Aspergillus fumigates* under submerged and solid state fermentation using agricultural waste. Int. J. Adv. Pharm, Biol. Chem. 2 (4).

Singh, A., Hamme, V.J., Ward, O., 2007. Surfactants in microbiology and biotechnology. Appl. Aspects Biotechnol. 25, 99–121.

Singh, V., Haque, S., Niwas, R., Srivastava, A., Pasupuleti, M., Tripathi, C.K.M., 2017. Strategies for fermentation medium optimization: an in-depth review. Front. Microbiol. 7, 2087.

Singhania, R.R., Sukumaran, R.K., Pillai, A., Szakacs, G., Pandey, A., 2006. Solid-state fermentation of lignocellulosic substrates for cellulase production by *Trichoderma reesei* NRRL 11460. Indian J. Biotechnol. 5, 332–336.

Singhania, R.R., Sukumaran, R.K., Patel, A.K., Larroche, C., Pandey, A., 2010. Advancement and comparative profiles in the production technologies using solid-state and submerged fermentation for microbial cellulases. Enzym. Microb. Technol. 46, 541–549.

Sonjoy, S.B., Bex, K., Honston, H., 1995. Cellulase activity of *Trichoderma reesei* (RUT-C30) on municipal solid waste. Appl. Biochem. Biotechnol. 51-52 (1), 145–153.

Sreena, C.P., Sebastian, D., 2015. Cost effective cellulase production by *Bacillus subtilis* MU S1 using lignocellulosic biomass residues. In: Biodiversity & Evaluation: Perspectives and Paradigm Shifts. Cochin University of Science and Technology, Kalady, India. https://doi.org/10.13140/RG.2.1.2833.1604.

Srivastava, N., Rawat, R., Sharma, R., Oberoi, H.S., Srivastava, M., Singh, J., 2014. Effect of nickel-cobaltite nanoparticles on production and thermostability of cellulases from newly isolated thermotolerant *Aspergillus fumigatus* NS (class: eurotiomycetes). Appl. Biochem. Biotechnol. 174, 1092–1103.

Srivastava, N., Rawat, R., Oberoi, H.S., Ramteke, P.W., 2015a. A review on fuel ethanol production from lignocellulosic biomass. Int. J. Green Energy 12, 949–960.

Srivastava, N., Singh, J., Ramteke, P.W., Mishra, P.K., Srivastava, M., 2015b. Improved production of reducing sugars from rice straw using crude cellulase activated with Fe_3O_4/alginate nanocomposite. Bioresour. Technol. 183, 262–266.

Sun, J.X., Sun, X.F., Sun, R.C., Su, Y.Q., 2004. Fractional extraction and structural characterization of sugarcane bagasse hemicelluloses. Carbohydr. Polym. 56, 195–204.

Taherzadeh, M.J., Karimi, K., 2007. Enzyme-based hydrolysis processes for ethanol from lignocellulosic materials: a review. Bioresources 2, 707–738.

Thota, S.P., Badiya, P.K., Yerram, S., Vadlani, P.V., Pandey, M., Golakoti, N.R., Belliraj, S.K., Dandamudi, R.B., Ramamurthy, S.S., 2017. Macro-micro fungal cultures synergy for innovative cellulase enzymes production and biomass structural analyses. Renew. Energy 103, 766–773.

Trabulsi, L.R., Alterthum, F., 1999. Microbiologia, third ed. Atheneu, São Paulo. 586 pp.

Viniegra-Gonza'lez, G., Favela-Torres, E., Aguilar, C.N., Romero-Go'mez, S., Dı'az-Godı'nez, G., Augur, C., 2003. Advantages of fungal enzyme production in solid state over liquid fermentation systems. Biotechnol. Eng. J. 13, 157–167.

Wang, M., Liu, K., Dai, L., Zhang, J., Fang, X., 2013. The structural and biochemical basis for cellulose biodegradation. J. Chem. Technol. Biotechnol. 88, 491–500.

Wang, Y., Fang, X., An, F., Wang, G., Zhang, X., 2011. Improvement of antibiotic activity of *Xenorhabdus bovienii* by medium optimization using response surface methodology. Microb. Cell Factories 10, 1–15.

Wang, Y., Fan, C., Hu, H., Li, Y., Suna, D., Wang, Y., Peng, L., 2016. Genetic modification of plant cellwalls to enhance biomass yield and biofuel production in bioenergy crops. Biotechnol. Adv. 34, 997–1017.

Weinhouse, H., Benziman, M., 1974. Regulatoin of hexose phosphate metabolism in *Acetobacter xylinum*. J. Biochem. 138, 537–542.

Wilson, D.B., 2009. Cellulases and biofuels. Curr. Opin. Biotechnol. 20, 295–299.

Wu, I., Arnold, F.H., 2013. Engineered thermostable fungal Cel 6A and Cel 7A cellobiohydrolases hydrolyze cellulose efficiently at elevated temperatures. Biotechnol. Bioeng. 110, 1874–1883.

Yin, S., Ding, S.X., Abandan Sari, A.H., Hao, H., 2013. Data-driven monitoring for stochastic systems and its application on batch process. Int. J. Syst. Sci. 44, 1366–1376.

Yoon, L.W., Ang, T.N., Ngoh, G.C., Chua, A.S.M., 2014. Fungal solid-state fermentation and various methods of enhancement in cellulase production. Biomass Bioenergy 67, 319–338.

Zabeda, H., Sahu, J.N., Boyce, A.N., Faruq, G., 2016. Fuel ethanol production from lignocellulosic biomass: an overview on feed-stocks and technological approaches. Renew. Sust. Energ. Rev. 66, 751–774.

Zadrazil, F., Puniya, A.K., 1995. Studies on the effect of particle size on solid-state fermentation of sugarcane bagasse into animal feed using white-rot fungi. Bioresour. Technol. 54, 85–87.

Further Reading

Martins, L.O., Brito, L.C., Sá-Correia, I., 1990. Roles of Mn^{2+}, Mg^{2+} and Ca^{2+} on alginate biosynthesis by *Pseudomonas aeruginosa*. Enzym. Microbial. Technol. 12, 794–799.

Wong, T.Y., 1993. Effects of calcium on sugar transport in *Azotobacter vinelandii*. Appl. Environ. Microbiol. 59, 89–92.

CHAPTER

6

Role of Bioprocess Parameters to Improve Cellulase Production: Part II

*Sreedevi Sarsan**, *Ramchander Merugu*†

*Department of Microbiology, St. Pious X Degree & P.G. College, Hyderabad, India †Department of Biochemistry, Mahatma Gandhi University, Nalgonda, India

6.1 INTRODUCTION

Biotechnological processes are employed in the cultivation of microbial cells and production of metabolites such as enzymes. These processes involve various factors such as medium composition, fermenter design, and different controlling parameters to obtain the desired product of high quality and quantity. Bioprocess technology aims at employing the enzymes in different purposes and also producing them under different circumstances. Different processes such as submerged fermentation (SmF), solid-state fermentation (SSF), and immobilized cells are employed in the microbial production of enzymes. Though the cost of production is high, SmF is preferred to SSF, because it can be better monitored and handled. In these fermentation processes, several important parameters such as sterility of bioreactor and air, agitation speed, pH, temperature, and also nutrient requirements are prominent for enhanced/improved growth of microorganisms and their productivity.

A number of microbial products such as enzymes, vaccines, drugs, etc. are being manufactured and produced through biotechnological processes. A number of factors affect and are involved in the production of the desired product by biotechnological processes. Bioreactor design, medium composition, and different physical parameters influence the product quality and quantity. In bioprocess technology, different enzymes are produced from microorganisms using various modes of cultivation: batch cultivation, fed-batch cultivation, and continuous cultivation. In laboratory and industrial fermentations, bacteria can be cultivated in batch mode mostly by SmF or SSF and immobilized cells, and in continuous mode by SmF or with immobilized cells.

The basic approach in the production process of industrial enzymes involves (Fig. 6.1):

- *Enzyme selection*: Among a plethora of industrial enzymes, the enzyme of interest that needs to be produced by industrial fermentation process needs to be decided.
- *Production strain selection*: The discovery of the enzyme producing a microorganism is the starting point for a fermentation process. The master culture of the producing microbe is then prepared and preserved, which is consistent in growth rate and metabolite yield in all successive fermentations.
- *Improving the production strain*: The selected microorganism may be improved with regard to its growth as well as product formation, either genetically or by providing the optimum fermentation conditions.
- *Fermenter design and operation*: Fermenter design and operation under sterile conditions (inoculation, aeration, stirring, sampling) is an important consideration for successful fermentation. Efficient and economic downstream processing is also important for a successful fermentation.
- *Optimizing fermentation media and bioprocess parameters*: To improve productivity in a fermentation process, different strategies can be employed, but generally it is achieved by modifying the various physicochemical and

From Cellulose to Cellulase: Strategies to Improve Biofuel Production
https://doi.org/10.1016/B978-0-444-64223-3.00006-0

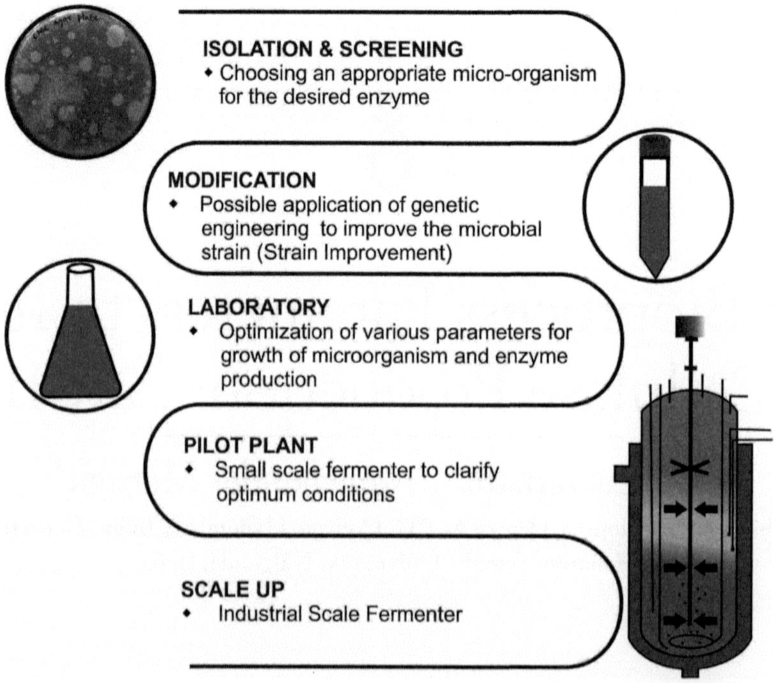

FIG. 6.1 Basic process of industrial enzyme production.

nutritional parameters. Enzyme production is affected by many important parameters such as pH, incubation temperature, moisture level, agitation, aeration, incubation period, inoculum size, and medium components, especially carbon and nitrogen sources. Hence it is crucial to optimize the fermentation process parameters to develop feasible technologies for maximum and cost-effective production of cellulase enzymes. First, optimization studies should be carried out at shake flask level and should then be extended to a larger scale in a fermenter.

- *Scale-up*: Once the fermentation process has been standardized on the pilot scale in the laboratory, scale-up studies should be undertaken for commercial production at the industrial level.

6.2 BIOPROCESS PARAMETERS

Monitoring, controlling, and measuring of the bioprocess parameters decide the quantity and quality of the product in an industrial fermentation process. A large number of parameters such as physical, chemical, and/or biological parameters can be measured during fermentation and bioprocessing for data analysis and appropriate control. Computers are used in industrial biotechnology for data acquisition, data analysis, and developing fermentation models. The computer can be used to develop mathematical models of fermentation processes. These models in turn will be useful for better control over fermentation systems with high productivity in a cost-effective manner.

Important parameters that are measured during bioprocessing are:

- Physical parameters: temperature, pressure, flow rates, viscosity, turbidity, power consumption.
- Chemical parameters: pH, substrate concentration, product concentration, O_2 concentration, waste gas concentration, ionic strength.
- Biological parameters: activities of specific enzymes, protein concentration, energetics (ATP concentration), DNA/RNA content.

Computers can be used for data acquisition, data analysis, and developing fermentation models. Sensors and computers in fermentation systems can be used to know the concentration of O_2 and CO_2, pH, temperature, pressure, viscosity, turbidity, aeration rate, etc. The data collected on a computer can be used for various calculations, e.g., rate of substrate utilization, rate of product formation, rates of O_2 uptake and CO_2 formation, heat balance, and respiratory quotient. Oxygen electrodes and CO_2 electrodes can be used to measure O_2 and CO_2 concentrations, respectively. O_2 and CO_2 can be measured, respectively, by the magnetic property of O_2 and the infrared absorption of CO_2. The mass spectrometer can be used to measure the concentrations of N_2, NH_3, ethanol, and methanol simultaneously. The measurement of dissolved gases can be carried out accurately by using gas-permeable membranes. The measurement of these parameters can be performed by various monitoring systems using special sensors (Fig. 6.2). The measurements of the parameters can be done either directly in the bioreactor or in the laboratory. Through computer data analysis, it is possible to arrive at the optimal productivity for a given fermentation system.

Fermentation media and process conditions are the most significant parts of any industrial fermentation process because they tend to influence product formation, its concentration, and yield, thus affecting the overall production cost. In nature, the original microorganisms may produce low levels of enzymes. Therefore once an organism producing cellulase is isolated and identified, an extensive effort must be expended to increase the fermentation yield and minimize the production cost. Appropriate and suitable fermentation media and process conditions are the key factors that improve the production of metabolites in a fermentation process, thus enhancing microbial efficiency (Hegde et al., 2013). Different strategies can be employed for lowering the production cost of enzymes such as recombination, genetic engineering, metabolic engineering, etc. Greater yields of enzymes with minimal production cost may be best attained by optimization of various bioprocess parameters. Process optimization refers to cost-effective and efficient production of a desirable product by modifying or improving the biochemical process.

Traditionally, optimization of fermentation parameters can be brought about by controlling and monitoring the effect of a single independent variable factor at any specified time in an experiment while fixing other parameters. This method is the most popular and widely used method for optimization of process parameters in fermentation technology because it is easy and convenient. However, this method involves many experiments causing wastage of time, labor, and reagents and thus is considered to be a laborious, expensive, and time-consuming process. Another disadvantage is that the interactive effects between and among the different variables under study are excluded from this method (Lundstedt et al., 1998). Therefore statistical methods of optimization were developed in which many

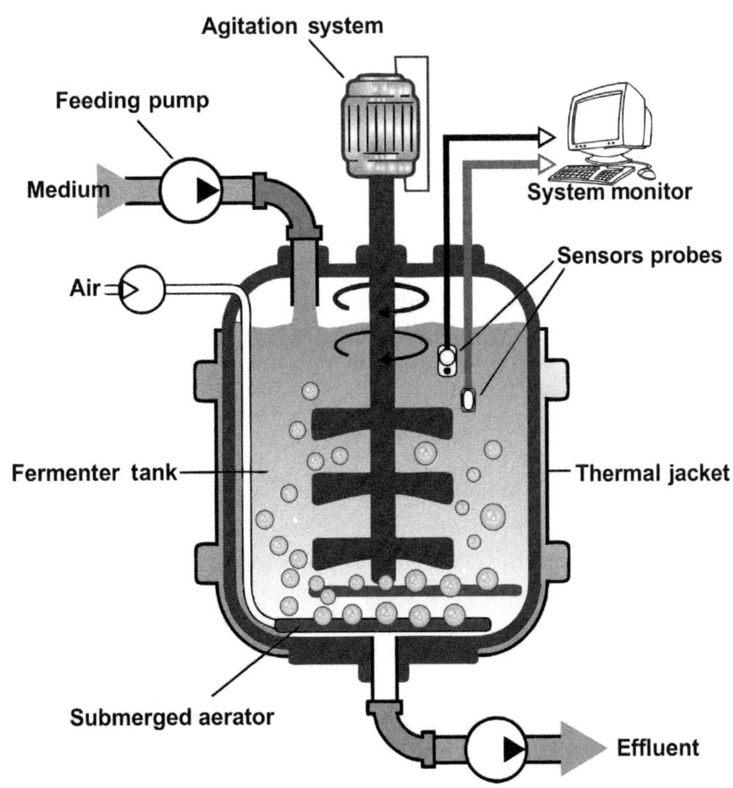

FIG. 6.2 Bioprocess parameters in an industrial fermenter. *From https://commons.wikimedia.org/wiki/File:Bioreactor_principle.svg.*

factors and their interactions can be monitored and assessed at a time with maximum efficiency (Akhtar et al., 2015). Among the various techniques developed to optimize a process parameter that is influenced by several independent variables, response surface methodology (RSM) is the most accepted and popular method. RSM is an amazing statistical tool that uses various statistical, mathematical, and graphical techniques to develop empirical models aimed at optimizing process parameters in an experiment (Panda et al., 2007). RSM is practically used in optimization studies of various products such as ethanol production, bacteriocin production, dye degradation, spore production, and the production of enzymes like chitinase, alkaline protease, and cellulase (Shankar and Isaiarasu, 2011, 2012; Zambare and Christopher, 2011; Deka et al., 2011; Anuradha Jabasingh, 2012; Saravanan et al., 2012; Sandhu et al., 2013; Singh et al., 2014a,b; Aanchal et al., 2016).

6.3 BIOPROCESS PARAMETERS INFLUENCING CELLULASE PRODUCTION

Cellulose, the most abundant carbohydrate, is primarily found in the cell walls of plants. It is a polymer of glucose molecules connected by β-1,4-linkages. Abundant cellulose-containing wastes are dumped into the environment from industries or agriculture. Every year large amounts of biomass are generated on earth in terrestrial and marine ecosystems. A large portion of this biomass generated is lignocellulosic and about 30%–60% of these lignocelluloses is cellulose. There is a need to develop and employ various methods of degradation of the cellulosic biomass, and the biological degradation of cellulose by microbial cellulase enzyme has gained much prominence. Cellulase is an enzyme that causes the hydrolysis of β-1,4-glycosidic linkages and produces glucose molecules from the cellulose (Sethi et al., 2013). Cellulase enzyme has applications in many sectors such as food, textiles, leather, alcoholic beverages, paper, pharmaceuticals, starch processing, biofuel production, etc. (Linko, 1977; Bisaria and Ghose, 1981; Bhat, 2000; Makoi and Ndakidemi, 2008). There are several reports of cellulose-degrading bacteria such as *Bacillus, Brevibacillus, Cellovibrio, Cellulomonas, Fibrobacter, Geobacillus, Micrococcus, Paenibacillus, Pseudomonas, Ruminococcus, Streptomyces*, and *Sporosphytophaga* spp. (Bagnara-Tardif et al., 1992; Weimer, 1996; Mawadza et al., 2000; Lynd et al., 2002; Immanuel et al., 2006; Gautam et al., 2010; Rastogi et al., 2009). The cellulases are also produced by various fungal species like *Trichoderma reesei, Trichoderma viride, Trichoderma longibrachiatum, Aspergillus niger, Neocallimastix frontalis, Penicillium purpurogenum* (Hurst et al., 1977; Mountfort and Asher, 1985; Steiner et al., 1994; Duff and Murray, 1996).

Large amounts of microbial enzymes are being produced by industrial fermentation processes and their production has to be closely monitored, controlled, and modified to improve their productivity. Many environmental as well as nutritional factors have pronounced effects on microbial enzyme production and hence need to establish the optimum conditions required in an individual fermentation process. The costs of production of cellulase enzymes are linked to the enzyme-producing microorganisms and their productivity (Gautam et al., 2011). Any variation in the process parameters will affect the microbial metabolic activity and thus the production yield, which also differs widely among various species. Hence it is necessary to optimize the various parameters in a fermentation process for maximum production of cellulase enzyme at lower cost. Enzyme cellulase production and yield are markedly affected by various interrelated parameters such as the pH of media, temperature, incubation time, agitation/stationary conditions, moisture level, and inoculum size. The various media components such as carbon sources, nitrogen sources, trace elements, presence of inducers, surfactants, and medium additives will also influence the production and yield.

6.3.1 Production Media

The selection of raw materials and the appropriate amounts required for the preparation of media are significant factors for the accomplishment of a successful fermentation process. Cellulases are considered to be inducible enzymes and their production and activity depend on the kind of substrate present in the media. Nutrient source and concentration greatly affect the growth of microorganisms and synthesis of enzymes. The most important factor to be considered in the process of enzyme production is the selection of an ideal substrate. Sometimes no substrate is ideally suitable because the nutrients may be absent or available only in suboptimal concentrations. Therefore it becomes imperative to supplement substrate externally with deficit nutrients for maximum yield of cellulase. An adequate amount of carbon, nitrogen, P, S, trace minerals, and inducers are required that favor the growth of microorganisms to obtain high yields in a fermentation process. Nitrogen sources supply essential elements as nutrients and carbon serves as a cofactor of enzyme, the concentrations of which will influence the fermentation process. A nutrient medium provided with utilizable carbon and nitrogen sources, trace minerals, and other essential elements is a significant feature in improving the production of fermentation. Among the various nutritional parameters affecting the enzyme cellulase yield, the components of media, especially the carbon and nitrogen sources, will have a major influence on the production and yield.

6.3.1.1 *Influence of Carbon Sources on Cellulase Production*

Large-scale production of cellulases requires the proper understanding, monitoring, and controlling of microbial growth and their abilities to produce enzymes. The formulation of media for a fermentation process is of great concern because no single media can favor the growth and cellulase production in all microbes and thus varies with the microbe. The cost of the media, especially the carbon resource, is the major contribution to cellulase production cost. Carbon sources have a significant role in cell metabolism as well as in the biosynthesis of cellulase enzymes. Various carbonaceous compounds such as cellulose, carboxymethyl cellulose (CMC), starch, glucose, lactose, maltose, fructose, and sucrose influence the production of enzymes, which is exemplified in the following studies. Golovchenko et al. (1993) used medium containing glucose, avicel, cellulose, cellobiose, and CMC as a carbon source for the growth of thermophilic *Clostridium thermocellum*. The production of cellulases by employing media containing cellulose was studied in *Aspergillus* spp. (Paroda and Mishra, 1984; Pushalkar et al., 1995). Ali and Sayed (1992) used media containing cellobiose, glucose, and xylose for the production of cellulase enzymes. Cellulose, cellobiose, and a mixture of CMC and cellobiose induced maximal endoglucanase production by *Trametes* in a liquid synthetic medium (Levin and Forchiassin, 1995). Jang and Chen (2003) reported that 1% CMC was better for the production of enzyme cellulase by *Streptomyces* spp. than many other carbon sources such as glucose, salicin, and avicel. Several other studies reported the maximum enzyme activity on medium supplemented with CMC, a synthetic carbon source (Narasimha et al., 2006; Das et al., 2010; Sreedevi et al., 2013; Singh et al., 2014a,b; Lugani et al., 2015). Thus only a few soluble sugars are recommended to be added in the production media for optimum enzyme production.

The production of cellulase enzymes is regulated or controlled by activation repression mechanisms (Schrempt and Walter, 1995; Ilmen et al., 1997; Sukumaran et al., 2010). The production of cellulase enzymes is transcriptionally regulated by carbon sources. For instance, the genes encoding cellulase enzymes are repressed by glucose molecules. CMC is a strong inducer of cellulase and fewer cellulase activities are obtained in the presence of glucose. The growth of microorganisms was maximum with the inclusion of glucose in the medium but it resulted in less cellulase yield (Narasimha et al., 2006; Padmavathi et al., 2012). Similarly, studies by Ruijter and Visser (1997) and Suto and Tomita (2001) have shown that smaller amounts of cellulase enzyme produced in many organisms in the presence of glucose are due to repression of the synthesis of cellulase enzymes. However, sometimes this repression is not observed due to mutations and may result in higher cellulase yields even in media with glucose (Kotchoni and Shonukan, 2002). Cellulase is considered to be an inducible enzyme and its synthesis depends on the substrate in the media, which might also induce the production of enzymes (Lugani et al., 2015). Several inducers were reported in the *T. reesei* cellulase system (Vaheri et al., 1979; Kubicek et al., 2009). Increased cellulase activity is produced on insoluble and amorphous cellulose sources (Stewart and Parry, 1981; Padmavathi et al., 2012). Godden et al., 1989reported cellotriose and cellotetraose, and Levin and Forchiassin (1995) reported cellobiose and crystalline cellulose as inducers for cellulase production in *Streptomyces* sp. and *Trametes trogii*, respectively.

Addition of relevant nutrient supplements in optimal concentrations is desirable and provides favorable conditions for the microbial isolates to increase the enzyme yield. Pure cellulosic substrates like Solka-Floc and Avicel have been employed for the production of cellulase enzymes (Esterbauer et al., 1991). However, the majority of fermentations employ lignocellulosic biomass as carbon sources during commercial production of cellulases. They include rice straw, rice bran, wheat bran, wheat straw, spent hulls of cereals and pulses, bagasse, dairy manure, and many other lignocellulosic substrates (Szabó et al., 1996; Reczey et al., 1996; Romero et al., 1999; Heck et al., 2002; Wen et al., 2005). Different concentrations of various carbon sources ranging from 0.5% to 5% (w/v) also affected enzyme activity. The effect of varied concentrations of the substrate can be studied by adding 0.5%–5.0% of each carbon source to different flasks containing broth and incubating the flasks after inoculating with the culture. The enzyme supernatant is collected after fermentation and then estimated for enzyme activity. Many studies have shown an increase in cellulase production with increase in concentration of sugar initially from 1.0% to 1.5%, and then the cellulase yield reduced slightly with further increase in sugar concentration (Gautam et al., 2011; Lugani et al., 2015). The high yield of cellulase at low substrate concentrations may be due to derepression of enzyme synthesis (Clarke and Lilly, 1969) and the low yield of cellulase at higher concentrations of the substrate may be because of the absorption of enzymes by substrate (Stewart and Parry, 1981).

6.3.1.2 Agro-Based Waste Material as Substrates

The cost of the medium is a key feature to be considered when formulating and developing a fermentation medium. Purified cellulosic substances are too expensive to be used as substrates and thus are not feasible for extensive production of cellulases on a larger scale. To minimize the production costs, agroindustrial residues can be employed as substrates for both growth of microorganisms and enzyme production. Padmavathi et al. (2012) in their studies have shown higher production of cellulase on cellulosic wastes than on sugar substrates in *Aspergillus terreus* and *Mucor plumbeus* sp. Cellulases are used for saccharification of agrowastes and the extent of saccharification can be assayed on the basis of sugars released. Lignocellulosic materials are considered to be the best substrates for bioconversion because they are the most abundantly available raw materials and usually free of cost. Due to their increased bioconversion efficiency of lignocellulosic wastes into sugars, cellulases, particularly microbial cellulases, have gained importance. Moreover, the advantage of biochemical conversion of biomass is that the original carbohydrate structure is preserved in the form of monomeric sugars (Xiros et al., 2009). Several factors that influence the biomass conversion processes are the physical state of cellulosic substrate, stability of enzyme, product binding to the cellulose, and product inhibition (Heinzelman et al., 2009).

Agricultural biomass substances are the chief source of renewable and inexpensive lignocellulosic wastes. Jecu (2000) suggested that these lignocellulosic wastes, particularly agricultural residues, are the best carbon source for the production of microbial enzymes. In *T. reesei* NRRL 11460, agro-industrial residues and wastes are used for the production of cellulases (Singhania et al., 2013). Demain et al. (2005) utilized the lignocellulosic wastes for the generation of biofuels using cellulase enzymes. Similarly, Ingram et al. (1999) have shown ethanol production from nonedible parts of plants, wood, and grasses. Many such studies were conducted to discover the suitability of agricultural biomass waste as alternative substrates for production of cellulase enzymes (Van Wyk, 2001; Kirk et al., 2002; Lynd et al., 2002; Adsul et al., 2004; Pothiraj et al., 2006). Different substrates were incorporated into the growth medium and the enzyme activity was measured after enzyme production. Thus several lignocellulosic substrates such as rice bran, wheat bran, rice straw, wheat straw, water hyacinth, peanut shell, banana peel, groundnut cake, coconut cake, soy cake, okara, corn cobs, bagasse, etc. are being used as the main component of media used for the production of cellulases (Table 6.1).

TABLE 6.1 Agrowastes Used as Substrates in Cellulase Production

Agrowaste substrate	Organism	References
Wheat bran	*Trichoderma reesei, Trichoderma harzianum, Aspergillus niger, Sporotrichum pulverulentum, Botrytis* sp.	Deschamps et al. (1985), Sharada et al. (2013)
Rice bran	*A. niger, T. reesei, Trichoderma viride, Scopulariopsis*	Ikram-ul-Haq et al. (2005a), Latifian et al. (2007), Kodali and Ravindra (2006)
Sugarcane bagasse	*T. reesei* QM9414, *Aspergillus terreus* SUK-1, *Penicillium echinulatum, Humicola insolens* TAS-13, *Aspergillus phoenicis*	Massadeh et al. (2001), Muthuvelayudham and Viruthagiri (2006), Sharada et al. (2013), Tengerdy and Szakacs (2003)
Corn cobs	*A. niger, Aspergillus flavus*	Khan and Singh (2011), Ojumu et al. (2003)
Water hyacinth	*T. reesei* SEMCC	Zhao et al. (2011)
Sawdust	*Trichoderma koningii, A. flavus, Aspergillus candidus, A. niger*	Ojumu et al. (2003), Acharya et al. (2008), Sharada et al. (2013)
Wheat straw	*T. reesei* QMY-1, *T. harzianum, Neurospora crassa, Aspergillus heteromorphous, T. viride, A. niger, Lentinus edodus, A. fumigatus*	Deschamps et al. (1985), Singh et al. (2009), Ikram-Ul-Haq et al. (2005b), Umar Dahot and Harif Noomrio (1996), Sharada et al. (2013), Romero et al. (1999), Thygesen et al. (2003)
Rice straw	*Trichoderma* sp., *Botrytis* sp., *T. viride, A. niger, A. heteromorphous, A. niger* KK2, *T. reesei, T. viride* ZY-01	Ikram-Ul-Haq et al. (2005b), Muthuvelayudham and Viruthagiri (2006), Qi et al. (2007), Sharada et al. (2013)
Sweet sorghum silage	*T. reesei, A. niger, Gliocladium* sp., *Trichoderma* sp., *Penicillium* sp.	Sharada et al. (2013)
Fruit waste (orange, pineapple waste)	*Trichoderma longibrachiatum, A. niger, Saccharomyces cerevisiae, A. flavus, Nigrospora* sp.	Omajasola et al. (2008), Attyia and Ashour (2002)

The amount of cellulase enzyme production on different lignocelluloses varies among different organisms and is attributed to various factors such as varied amounts of cellulose in different plant sources, structural heterogeneity, and differences in microbial cellulolytic abilities (Narasimha et al., 2006). The type of substrate and its raw material content can affect the production and yield of cellulase enzymes. Sometimes, the content of the substrates will be different even though the raw material may be from the same plant species, being affected by several factors such as geographical area, climate, soil type, plant type, pests, etc. Enzyme production and activity are affected mainly by the amount of cellulose content in the substrate because it serves as the source of carbon for microbial growth. The presence of hemicellulose and lignin contents, which are highly resistant to enzymatic degradation, in the substrate also influence the activity of enzymes. For example, the amount of cellulose is 39.68% in corn cob, 38.26% in rice straw, and 29.54% in water hyacinth. The cellulose content is thus lower in water hyacinth than in rice straw or corn cobs, but still the enzyme activity was found to be higher in water hyacinth substrate because of its low amount of lignin (7.46%) when compared to rice straw (10.28%) and corn cobs (13.43%) (Abdullah et al., 2018).

Pretreatment of lignocellulosic biomass is considered to be one of the most important preventive factors in biomass hydrolysis by cellulases and hemicellulases (Dijkerman et al., 1997; Várnai et al., 2010). Zhang and Lynd (2004) have shown that the hydrolysis of native lignocellulose using cellulase enzyme produced smaller amounts of glucose than that produced from cellulose substrates. Eggeman and Elander (2005) have also shown that the cost of using untreated biomass for enzymatic hydrolysis is much higher when compared to the pretreated substrate. Thus the pretreatment of lignocellulosic biomass is a crucial step for effective hydrolysis of substrates. A large number of pretreatment methods, physical, chemical, thermal, biological, and their combinations, are used for treating the lignocellulsoic biomass (Sun and Cheng, 2002).

Cellulases produced and developed from one microbial species may not be very effective in hydrolyzing a particular biomass feedstock. Therefore it is desirable to develop efficient cellulases either by artificially constructing or genetically engineering the cellulase-producing organisms to express the required set of cellulase enzymes (Schülein, 2000). Thus a cocktail of enzymes has been prepared and employed successfully for conversion of lignocellulosic biomass (Zhang et al., 2006; Park et al., 2009; Singh et al., 2009; Sukumaran et al., 2009). A cellulase cocktail for biomass conversion consists of a combination of endocellulase, exocellulase, β-glucosidase, xylanase, pectinase, etc. and varies based on the differences in the composition of biomass (Xin et al., 1993; Berlin et al., 2007).

6.3.1.3 *Influence of Nitrogen Sources on Cellulase Production*

Nitrogen (N_2) is an essential requirement in the fermentation media for effective utilization of soluble carbohydrates during enzyme production. The natural substrates may be supplemented with inorganic nitrogen for an optimal C:N ratio for optimal growth and enzyme production. Nitrogen sources, both organic and inorganic compounds, have a marked effect on cellulase enzyme production. Some of the organic nitrogenous compounds include urea, peptone, yeast extract, and beef extract, and the inorganic compounds include ammonium nitrate, ammonium sulfate, sodium nitrate, and ammonium chloride. Nitrogen is one of the chief components of cell proteins, and increased enzymatic activity in the presence of these nitrogenous compounds is attributed to their direct entry into the cell and their role in protein synthesis (Mandels, 1975; Sethi et al., 2013). The effect of N_2 sources on cellulase production can be checked by adding different nitrogen compounds at the 0.5%–5.0% level in the media and investigating the production and yield. Several studies have shown that the nitrogen compounds enhanced production and caused an increase in enzyme activity (Stewart and Parry, 1981; Debing et al., 2007; Lugani et al., 2015).

The organic nitrogen sources are more suitable than inorganic sources and are used for high production and yield of cellulases (Ray et al., 2007; Ariffin et al., 2008; Lugani et al., 2015). The organic N_2 sources such as peptone, yeast extract, and urea are considered to be good for cellulase production (Enari and Markkanen, 1977; Padmavathi et al., 2012; Jha et al., 1995). Soumita et al. (2011) also suggested peptone as the best nitrogen source for production of cellulase enzymes in *Rhizopus oryzae*. The organic nitrogen sources caused an increase in growth of microbes and production of enzymes, but were not considered to be an effective substitute for inorganic N_2 sources because they were costly (Debing et al., 2007; Sharada et al., 2013). Among the various inorganic salts, ammonium compounds are considered to be the best nitrogen sources for the production of cellulase enzymes.

6.3.1.4 *Influence of Metal Additives on Cellulase Production*

The production of cellulase enzymes is influenced by the addition of metal ions to the production media during the fermentation process. Various metal ions such as Mg^{2+}, Cu^{2+}, Ca^{2+}, Ni^{2+}, K^{+}, Na^{+}, Zn^{2+}, Fe^{3+}, and Co^{2+} have influence on the production of cellulase. Studies relating to the influence of metal additives on the industrial production of cellulases are very important. Mawadza et al. (2000) reported that there was little effect with certain metal ions such as Mg^{2+}, Cu^{2+}, Ca^{2+}, Ni^{2+}, K^{+}, Na^{+}, Zn^{2+}, and Fe^{3+}, whereas Co^{2+} had a significant effect on cellulase activity when used at 1 mmol/L concentration. According to Saha (2004), in *Streptomyces drozdowiczii* there was a 30% inhibition in the activity of cellulase in the presence of copper but many ions like barium, potassium, magnesium, iron, and sodium showed induced enzyme activity. Siddiqui et al. (1997) found that the cellulase enzyme in *Cellulomonas biazotea* was activated by Mn^{2+}. Mandels and Reese (1963) have shown that metal ions like silver, chromium, lead, mercury, and zinc salts inhibited cellulase activity. The metal salts of definite concentrations in the production medium will support the growth of microorganisms, but may or may not stimulate the physiological environment needed for the extracellular synthesis of cellulase enzymes. Some metal ions such as K^+, Na^+, Mg^{2+}, and NH_4^+ have a stimulatory effect and some ions such as Cu^{2+}, Co^{2+}, Zn^{2+}, and Pb^{2+} have an inhibitory effect on the production of cellulase.

6.3.1.5 *Influence of Surfactants on Cellulase Production*

Addition of one or more surfactants to the production media will enhance cellulase production (Singh and Hayashi, 1995). The detection of cellulase activity in the presence of surfactants is possible because they tend to minimize the binding of cellulases to the substrate (Chandra et al., 2007). Surfactants improve the catalytic hydrolysis of cellulolytic materials used in various industrial and agricultural applications. Nevertheless, the enhanced hydrolysis of the biomass substrates depends on the type and concentration of the surfactant (Castanon and Wilke, 1981; Kim et al., 1982; Ooshima et al., 1986; Helle et al., 1993; Eriksson et al., 2002; Alkasrawi et al., 2003). The commonly used surfactants are Tween 20, Tween 80, Triton X-100, and polyethylene glycol. The mechanism of action of these surfactants to enhance enzyme activity may be understood by examining their effect on the functional regions, i.e., catalytic and cellulose-binding domains of the cellulase enzyme (Bayer et al., 1998).

6.3.2 Bioreactors

A bioreactor is the main portion of all biochemical processes where microbes are used for manufacturing important biologically active substances or products like enzymes. Some of the important functions of the bioreactor as defined by Freed and Vunjak-Novakovic (2000) are:

- Inoculation of uniform concentrations of microbial cells.
- Supplying sufficient metabolites and nutrients to cells.
- Monitoring and controlling physiological conditions like temperature, pH, oxygen levels, etc. in the cell culture medium.

The chief purpose of an appropriately designed bioreactor is to provide controlled physiological environment conditions to promote and attain optimum growth of microorganisms as well as maximum product formation in the fermentation process. The design of the bioreactor has a significant impact on the performance of the fermentation process to a great extent. The selection of suitable and specific microorganisms for use in the fermentation process has a direct influence on the design and performance of the bioreactor. However, the choice of bioreactor also depends on other factors such as those that affect microbe growth (growth rate, temperature, induction type, duration of growth) and those that affect the product (chemical and physical nature of the product, kinetics of product synthesis). Knowledge of reaction kinetics such as mass and energy balances, mixing, mass transfer, and heat transfer is also important for understanding the functioning/operation of a bioreactor. The static flask is the simplest type of bioreactor, while another is a stirred-flask bioreactor. Additional shear stresses generated in the stirred-tank bioreactors (STBs) due to stirring of the media cause enhanced microbial cell growth when compared to that under static incubation conditions (Barron et al., 2003).

Fermentation by organisms growing in bioreactors is brought about by SmF or SSF. The advantage of SmF is the relatively high enzyme yield, vast experience with SmF bioreactors and process control, and the possibility for modeling cellulase production and regulating biosynthetic pathways for enhancing a desired enzyme component. The disadvantage of SmF is the high production cost and the difficulty of adjusting the system to produce an optimal multienzyme complex for hydrolyzing a particular substrate. An alternative technology for cellulase production is SSF. SSF is a low-technology, cost-effective way for producing crude enzyme complexes, including cellulases, hemicellulases, pectinases, and lignases (Pandey et al., 1999). SSF may be a viable alternative to SmF when the crude enzyme product is used for bioconversion directly or for enzyme enrichment of a natural substrate (e.g., in the production of biofuels and animal feed supplements). Although SSF has been widely practiced in some countries for some time, it is not universally accepted as an alternative technology because of the difficulty of process control and the extensive labor intensity involved in the SSF system (Pandey, 1992; Pandey et al., 2000). The basic reactor for cellulase production by SSF is the koji-type tray reactor that is widely used for fungal fermentations in industrial-scale reactors in East Asia (Japan, Taiwan, and China). Much effort needs to be expended to develop new SSF reactors that can overcome the problem of heat and moisture control and that can allow online control of the fermentation.

Cellulases are mostly produced commercially in SmF systems and they employ two types of bioreactors: conventional STBs or nonconventional pneumatic-type bioreactors such as airlift and bubble column bioreactors. In industries, the more widely used bioreactors are the STBs, whereas the pneumatic-type bioreactors are employed at laboratory scale for the production of cellulase enzymes. The chief advantage of STBs is that they are versatile in operation and the bioprocess parameters can be easily controlled and monitored. The disadvantage of the system is that it demands an input of high energy to maintain and sustain the homogeneity of the media. Apart from low power consumption and high O_2 transfer rates, the main advantages of pneumatic bioreactors are that they are easy to construct, operate, and maintain (Cerri and Badino, 2010). In pneumatic bioreactors, the air flow generated during agitation induces complete movement of particles leading to lower shear stress and promoting the growth of microbes. However, these are less flexible than the STBs because aeration and mixing are both caused by the injection of air into the pneumatic bioreactors. Both types of bioreactors have been evaluated for cellulase production in SmF. A comparative study of different types of bioreactors used for the production of cellulase enzymes by different *Aspergillus* species has been reported (Wase et al., 1985; Kim et al., 1997; Siedenberg et al., 1997; Michelin et al., 2011, 2013; Cunha et al., 2012). For example, Kim et al. (1997) reported 9.7 FPU/hL for cellulase productivity in *A. niger* on rice straw in a bubble column bioreactor, and Umikalsom et al. (1998) reported 20.8 FPU/hL of cellulase production on oil palm empty fruit bunch fiber by *Chaetomium globosum* 414 in an STB.

There are three modes of operation for these SmF bioreactors, which include batch mode, fed-batch mode, and continuous mode. A closed type of fermentation in which a sterilized medium is introduced into the fermenter and then inoculated for the process to occur with no further inflow or outflow is called a batch process. In continuous fermentation, the media nutrients as well as the products of the fermentation are continuously added and removed from the fermenter at a constant rate. In fed-batch cultivation, a few nutrients of the media are added during the fermentation process but the final products are removed only after the process is completely finished. The most common mode of operation is batch fermentation, because it is the most suitable mode for highly viscous slurry of a lignocellulosic substrate. However, in filamentous fungi, a number of studies have shown higher rates of cellulase production in fed-batch cultivation than in batch fermentation. For example, Hendy et al. (1984) have shown that when *T. reesei* RUT C30 was grown using cellulose as substrate in a 14 L capacity bioreactor, the amount of cellulase produced was 4.2 FPU/mL in batch culture, whereas it was 22.4 FPU/mL in fed-batch culture. Ahamed and Vermette (2008) also used mixed cultures of *Trichoderma* and *Aspergillus* and showed that there was enhanced cellulase production in fed-batch fermentation in a 3 L capacity STB. In fed-batch cultivation, more cellulase is produced because of minimized agitation and aeration difficulties and also the reduced repression effects caused by accumulation of reducing sugar in the medium. Thus fed-batch fermentation increases both yield and productivity in comparison to batch fermentation, mainly because of the optimal growth conditions (fresh substrate, no lag period for growth, dilution of inhibitory product). Continuous fermentation has not been successful for cellulase production, most probably because of the handling problems with the highly viscous slurry.

SmFs can employ either suspended cells or immobilized cells. Suspension bioreactors can operate at a larger scale and can also utilize a wider variety of organisms, since special attachment surfaces are not needed. However, sometimes the organisms may be removed from the bioreactor along with the effluent such as in the continuous fermentation process. In such cases, immobilization is useful because the organisms will not be removed with the effluent. The main limitation is that it is employed at a smaller scale because the microbes are only present on the surfaces of the vessel. Immobilization is a common term that describes the attachment or entrapment of a wide variety of particles such as enzymes, microbial, animal, and plant cells, or their organelles. The immobilization technique keeps enzymes localized during a desired reaction and will enable separation of the immobilized enzymes from reactants, products, and reaction media. Immobilization limits changes to the three-dimensional structure of enzymes, restricting folding and unfolding. Furthermore, interactions of the enzymes' functional groups with support will alter their interactions with substrates. An early attempt at recovery and reuse of cellulase was tested by immobilizing cellulase on polysaccharides (Woodward and Zachry, 1982). This was achieved by adsorption of cellulase to the carbohydrate-binding protein concanavalin A, which was immobilized on Sepharose 4B, a cross-linked and beaded polysaccharide. Initial recovery of cellulase using this method was only 6%. However, this method of immobilization showed positive results in terms of high retention of enzyme activity. This indicates that the active sites of the enzymes were accessible for cellulose substrates. Furthermore, these immobilized enzymes showed relatively higher efficiency than free cellulase for converting cellulosic biomass to glucose during enzymatic hydrolysis. Cellulase has also been recycled by reusing unhydrolyzed residue from a previous hydrolysis cycle that would have enzymes attached to them (Gregg and Saddler, 1996).

The operational behavior of a bioreactor in industries is primarily dependent on measurement and analysis of various bioprocess parameters. Physiological conditions such as pH, temperature, oxygen concentration, and amount of nutrients inside the bioreactor affect the growth and productivity of the organisms. Different sensors or probes are essentially always present in a bioreactor for monitoring and control purposes. A cooling jacket or coils enveloped around all or part of the vessel maintain the appropriate temperature required inside the fermenter. The pH of the medium is monitored, measured, and adjusted with the addition of small amounts of acid or base depending on the fermentation (acidic or alkaline fermentation). Oxygen must be supplied to the fermentations especially for aerobic and some anaerobic fermentations. Oxygen is relatively insoluble in water, therefore air must be supplied continuously and further agitation helps in the transfer of oxygen and mixing of nutrients, thus making the fermentation homogeneous. Nutrients may be added to the fermenter at the start of fermentation as in batch fermentation, or continuously added to the fermenter as in a fed-batch system or continuous fermentations.

6.3.3 Temperature

One of the successful strategies to increase cellulase production by high-yielding cellulolytic organisms is to optimize the culture conditions. It is very important to understand the variables/factors that control cellulase production. Temperature is one such important physiological parameter that influences microbial growth and metabolism. The temperature of incubation during the fermentation process is one of the prominent factors influencing enzyme production. Every microorganism has an optimum temperature at which growth proceeds with greatest rapidity and maximum activity, and their activity decreases when grown at higher or lower temperatures. This was explained by Raju and Divakar (2013) in their studies where they showed that at low temperatures, cell metabolism would be at a slower rate and higher temperatures would damage the enzymes and cell growth, and thus eventually kill the cells. Therefore an optimum temperature must be provided during the production of enzymes. The influence of temperature on cellulase enzyme production can be checked by inoculating and incubating the media flasks at various temperatures between 25° and 100°C. The production of cellulolytic enzymes was found to be maximum at 45–50°C in many thermophilic fungi (Ogundero, 1982; Maheshwari et al., 2000; Gao et al., 2008). Rajoka (2004) explained that at low temperatures, the product yields are lower due to suppressed substrate transport across the cells. Similarly, at higher temperatures the decrease in the production of enzymes may be because of the thermal denaturation of enzymes. However, thermophilic organisms that produce the cellulase enzymes at higher temperatures are also reported, e.g., the optimum temperature for cellulase production is 100°C in *Bacillus licheniformis* C108 as reported by Gaur and Tiwari (2015) and 80°C in *Fusarium subglutinans* MTCC 11891 as reported by Sharmila (2014). Several such studies carried out by many researchers revealed that different organisms have different temperature optima for maximum cellulase production (Table 6.2).

6.3.4 pH

All enzymes have an optimum pH where they have maximum activity and their activity decreases with increases or decreases in pH. The pH of the fermentation medium is considered to be one of the significant factors for the production of cellulase. Initial pH of the medium exerts a strong influence on cell growth, cell membrane transport processes, and enzyme production. The effect of pH on cellulase can be examined by using production media of different pH values in a fermentation process. Initial pH of the medium has a strong influence on cellulase production and was reported in several research works. It was observed that the production of cellulase enzymes gradually increased with an increase in pH value and reached its maximum at an optimum pH and then decreased. Ye et al. (2017) studied the fermentation conditions for production of cellulase in *Bacillus amyloliquefaciens* and reported that the enzyme was stable between pH 5 and 6, and pH 6 was optimum for maximum enzyme activity; also the enzyme yield decreased with higher pH values. Sohail et al. (2011) reported that pH 4 and 35°C were suitable for high cellulase production in *A. niger* MS82 using SSF. In *Cellulomonas flavigena*, optimum cellulase activities were found to be at pH 6 (Pérez-Avalos et al., 2008). pH 6.5 was reported in *Trichoderma koningii* for optimum cellulase production (Li et al., 2004). *B. licheniformis*, *B. amyloliquefaciens*, and *Bacillus subtilis* are reported to produce high amounts of enzyme at pH 7 (Tanyildizi et al., 2005). Enzymological studies of cellulases from *Bacillus* species indicated that the enzymes showed wider pH adaptability and were most active in the pH range of 3–7 with an optimum pH of 4 (Mawadza et al., 2000; Lin et al., 2009). The optimal pH for cellulolytic activity on coir fiber was 7 in *Cellulomonas*, *Micrococcus*, and *Bacillus* species (Immanuel et al., 2006). Thus different organisms have varied pH optima for maximum cellulase production, which are tabulated in Table 6.3.

TABLE 6.2 Optimum Temperature of Cellulase Production by Microorganisms

Cellulase source	Temperature	References
Trichoderma longibrachiatum	45°C	Omojasola and Jilani (2008)
Aspergillus niger	40°C	Gautam et al. (2011)
Trichoderma sp.	45°C	Gautam et al. (2011)
A. niger MS82	35°C	Sohail et al. (2009)
Bacillus amyloliquefaciens	50°C	Lee et al. (2008)
Trichoderma koningii AS3.4262	25–35°C	Lee et al. (2008)
Trichoderma reesei QM 9414	100°C	Aiello et al. (1996)
Streptomyces drozdowiczii	50°C	de Lima et al. (2005)
Fusarium subglutinans MTCC 11891	80°C	Sharmila (2014)
Bacillus vallismortis RG-07	65°C	Gaur and Tiwari (2015)
Bacillus licheniformis C108	100°C	Gaur and Tiwari (2015)
Aspergillus terreus M11	45°C	Gao et al. (2008)
T. reesei RUT C30	30°C	Singhania et al. (2007)
Bacillus subtilis CBTK 106	35°C	Krishna (1999)
A. niger USM AI	30°C	Lee et al. (2011)
B. subtilis subsp. *subtilis* A-53	30°C	Lee et al. (2010)
Anoxybacillus flavithermus EHP2	75°C	Ibrahim and El-diwany (2007)
B. licheniformis strain Vic	60°C	Irene et al. (2016)
Bacillus sp. (alkalophilic) 1139	92°C	Fukumori et al. (1985)
Bacillus sp. (neutrophilic) KSM-522	50°C	Kawai et al. (1988)
Bacillus sp. AC-1	70°C	Li et al. (2006b)
Bacillus sp. LFC15	50°C	Korpole et al. (2011)
B. subtilis	37°C	Nakamura et al. (1991)
Bacillus circulans	40°C	Ray et al. (2007)
Bacillus pumilus EWBCM1	37°C	Shankar and Isaiarasu (2011)

6.3.5 Inoculum Size

An important requirement for successful cellulase production is the availability of a reliable starter culture of the producing organisms. An initial microbial load to a medium also affects metabolite production. Metabolic status of the organism was also one of the important factors for metabolic activity of any microbial strain and effective productivity, and will vary with the fermentation time. The effect of inoculum size on cellulase production can be checked by seeding fermentation media with varied concentrations of inoculums ranging from 1.0% to 15.0% inoculum and incubating under standard conditions. Studies by several workers have shown that the production of cellulase increased with increase in the inoculum concentration ranging from 1% to 10%, and a decrease in enzyme production was observed with a further increase in inoculum size (Garg and Neelakantan, 1981; Alam et al., 2008; Iqbal et al., 2010; Shankar and Isaiarasu, 2011; Gomathi et al., 2012; Azzaz et al., 2012; Srilakshmi et al., 2017; Domingues et al., 2000). Pandey et al. (1999) reported that size/load of inoculum can affect the growth of microbes and production of enzymes and found 10^5–10^6 spores/mL of inoculum was efficient for maximum cellulase activity. Ray et al. (2007) reported in *B. subtilis* CY5 and *Bacillus circulans* TP3 that an inoculum of 9.75×10^7 viable cells/mL culture broth is required for optimum cellulase production.

TABLE 6.3 Optimum pH of Cellulase Production by Microorganisms

Cellulase source	pH	References
Aspergillus niger	6–7	Gautam et al. (2011), Akiba et al. (1995)
Trichoderma sp.	6.5	Gautam et al. (2011)
Trichoderma viride	5.5	Gautam et al. (2010)
Bacillus amyloliquefaciens	6.0	Ye et al. (2013)
A. niger	3.0–4.0	Acharya et al. (2008)
A. niger MS82	4.0	Sohail et al. (2011)
Fomitopsis sp. RCK2010	5.5	Deswal et al. (2011)
B. amyloliquefaciens	7.0	Lee et al. (2008)
Streptomyces drozdowiczii	5.0	de Lima et al. (2005)
Aspergillus terreus M11	3.0	Gao et al. (2008)
Bacillus	4.0	Mawadza et al., 2000
B. subtilis sub sp. *subtilis* A-53	6.8	Lee et al. (2010)
A. niger NCIM 548	4.6	Kumar et al. (2011)
Anoxybacillus flavithermus EHP2	7.5	
Bacillus licheniformis	6.1	Irene et al. (2016)
Bacillus sp. (alkalophilic) 1139	9.0	Fukumori et al. (1985)
Bacillus sp. (neutrophilic) KSM-522	7.0–10.0	Kawai et al. (1988)
Bacillus sp. AC-1	4.5–6.5	Li et al. (2006a,b)
Bacillus sp. LFC15	9–10	Amore et al. (2013)
B. subtilis	7.0	Nakamura et al. (1991)
Bacillus pumilus EWBCM1	6.0	Shankar and Isaiarasu (2011)

Lower inoculum size will yield a lower amount of cellulase because of the lack of cells to use in the fermentation medium. Enzyme production will reach a maximum at an optimum concentration of inoculum. Then enzyme activity is reduced with an increase in inoculum size because of decreased microbial growth, which is attributed to increased competition among cells for space and nutrients. Sharma et al. (1995) reasoned that the decrease in enzyme production due to low inoculum size is because it controls and increases the initial lag phase, whereas the decreased activity due to larger inoculum size is because of increased moisture content. The decrease in cellulase production due to addition of a larger volume of inoculum could also be due to reduced sugar and O_2 uptake and enzyme release resulting from the clumping of cells, and also because of autolysis of cells resulting from nutritional imbalance caused by tremendous growth (Haq et al., 1993; Omojasola and Jilani, 2008).

6.3.6 Incubation Time

Incubation time for enzyme production is a significant parameter of an industrial fermentation process. Production of enzymes and other metabolic products is directly proportional to the incubation period up to a certain extent. Optimization studies of incubation time can be made by incubating the fermentation media flasks inoculated with cultures for different time periods ranging from 1 to 10 days and measuring the rate of cellulase production after every 1 day interval. Most of the studies revealed that the production of cellulase increased as the incubation time increased, and the optimum enzyme recovery period was found to be from 4 to 7 days (Bhalla and Joshi, 1993; Narasimha et al., 2006; Gautam et al., 2010; Srilakshmi et al., 2017). Prolonged incubation will result in a decline in cellulase production, which may be due to nutrient depletion or by-product accumulation in the fermentation media resulting in a decrease in microbial growth and consequently reduced cellulase activity. It may also be because of enzyme denaturation at various pH values during fermentation or due to proteolytic digestion and cell death (Kang et al., 2004; Dhillon et al., 2004; Haq et al., 1993; Sandhya et al., 2005).

6.3.7 Agitation

Agitation speed is one of the significant culture parameters that maintains the homogeneous environment inside a fermenter, disperses dissolved oxygen, and mixes the contents leading to enhanced microbial activity and improved substrate conversion (Singh et al., 2000). Thus the shaking speed increases the dissolved oxygen content and uniformly distributes the contents in the culture media necessary for growth of microbes and production of enzymes. The influence of agitation rate on cellulase production can be assessed by varying the agitation speed in a shaker incubator from 0 to 400 rpm at 50 rpm intervals. Many studies by researchers indicated that specific cellulase activity was higher in the bioreactor than in flask fermentation. It was also found that there was an increase in cellulase activity when aeration and agitation speeds increased from 0 to 400 rpm, above which cellular growth and enzyme activity decreased. The optimum agitation speed required for the production of cellulase was reported to be 121 rpm in *Bacillus* spp. (Rastogi et al., 2009, 2010; Deka et al., 2013). Kaseke et al. (2016) in their studies have shown that the high production of cellulase was obtained at 50 rpm, and further decrease in activity was observed with an increase in agitation speed up to 200 rpm. Wase et al. (1985) checked for the effects on cellulase production when fermentation was carried out in disc turbine-agitated vessels and airlift fermenters in *Aspergillus fumigatus*. A decline in the cellulase yield was seen in the agitated vessels, whereas in airlift fermenters there was approximately a 20% increase in enzyme yield when the agitation rate was increased. This was because no shear effects were created in the airlift fermenters as compared to those in agitated fermenters. The optimum speed of agitation for microbial growth and cellulase production was found to be 400 and 300 rpm, respectively. Many other studies have reported a decline in the rate of cellulase production with increase in agitation speed as in *B. amyloliquefaciens* (Jo et al., 2008), *T. reesei* (Lejeune and Baron, 1995), and *Thermomyces lanuginosus* (Singh et al., 2000). This decline in enzyme production with agitation speed is ascribed to increased shear stress on the cells and reduced cellular growth (Deng and Fong, 2010).

6.3.8 Aeration

Aeration is a very important parameter of optimizing cell culture, especially aerobic microorganisms in a fermentation process. The amount of oxygen delivered to the cells and the bioreactor culture during the entire process forms an important aspect of the fermentation environment. The rate of oxygen demand will increase as the cell mass increases and thus the cells need a continuous supply of oxygen when in the bioreactor. Because of the low solubility of oxygen in the liquid phase and increasing metabolic consumption by the cells with time, it needs to be supplied continuously to the cell culture. Oxygen supply is carefully controlled for optimal cell growth by manipulating bioreactor parameters. k_La defines a theoretical maximum cell density that could be achieved in cell culture and needs to be kept constant. The conventional scale-up of bioprocesses is based on physicochemical and geometric similarity. During the comparison, k_La is utilized as a target performance metric when a process is transferred from one bioreactor platform to another design. Bioreactor hardware design such as stirrer geometry and aeration sparger option, and running parameters such as gas flow rate or power input, are altered to achieve a similar k_La, providing a similar cell density.

k_La can be represented as:

$$k_La = k_L \times a$$

where k_La is the mass transfer coefficient from gas to liquid phase given in s^{-1}, k_L is the liquid side mass transfer coefficient (resistance in gas side film can be neglected), and a is the bubble surface (available for diffusion).

Any change to process and engineering parameters or to physical characteristics will have an impact on k_La and should be considered when evaluating bioreactor platforms and performing scaling calculations. In a bioreactor, mixing is used to eliminate gradients of concentration of cell, gas, medium, nutrient, temperature, and other properties. Mixing efficiency widely characterized by mixing time is one of the significant factors affecting both performance and scale-up in a bioreactor. Gas bubble size and residency time are highly dependent on three mixing conditions: impeller type, speed, and location. k_La values generally increase as tip speed increases but also cause increased shear forces that can lead to cell death. Bioreactors therefore are designed with different impeller types, combinations, and locations to achieve target k_La values without creating these shear forces. Generally, k_La values are closely associated with impeller design, with Rushton typically higher than paddle, which is typically higher than marine and pitched impeller.

k_La is impacted by multiple factors and its values will vary widely with sparger characteristics, including number, pore size, and surface area, because these factors affect bubble size, gas velocity, and flow rates. When gas bubble size decreases, surface area and gas residency time increase, causing bubbles to stay in the culture longer. Thus there is a greater opportunity for oxygen to release mass transfer into the cell culture medium. An increase in this oxygen residence time improves k_La. The production of smaller amounts of enzyme during fungal growth under static conditions is mostly due to oxygen limitation caused by less diffusion of oxygen. Moreover, agitated and aerated conditions favor not only oxygen transfer but also heat transfer and CO_2 dissipation promoting growth and enzyme production (Pirota et al., 2013). Higher oxygen availability increases k_La drive. Increasing oxygen supply to a bioreactor drives this availability and can be controlled by modifying concentration (air vs. O_2 enrichment) and volumetric flow. Apart from attaining high k_La values, it is important to take care of the operating conditions and implications to increase cell viability and decrease the process costs.

6.3.9 Moisture

The biosynthesis and secretion of cellulase enzymes is hugely affected by the moisture content of the production media. Water is a source of hydrogen and oxygen elements, which are essential for the synthesis of various biomolecules in the cell. Water has several important functions in the metabolism of organisms such as enzymatic hydrolysis processes and membrane transport processes of nutrients and metabolites. An increase in moisture content increases enzyme activity. Thus a positive influence of moisture content was observed on cellulase production. The impact of moisture level on microbial growth and cellulase formation is attributed to its effect on physical properties of the solid substrates (Shah and Madamwar, 2005).

SSF employs insoluble substrate that contains enough water for microbial growth and thus affects cellulase production. The success of a fermentation process is determined by the amount of moisture content of solid substrate employed. The water-binding characteristics of the substrate will determine the moisture level at which free moisture occurs. An optimum moisture level is desirable for enzyme production. If the water content is insufficient in a substrate, the solubility reduces and diffusion of nutrients decreases, causing an insufficient supply of nutrients for microbial growth and disruption of enzymes, leading to lower cellulase production. Decreased growth and enzyme production due to low moisture may also be because of reduced solubility of substrate causing high water stress (Poorna and Prema, 2007). A higher content of moisture than the optimum level also results in decreased cellulase production because of reduced porosity and gas volume, lower oxygen transfer, and increased contamination (Narahara et al., 1982; Abdullah et al., 2018). An additional barrier for diffusion of molecules is imposed along with the solid substrate because of the free excess liquid in case of high moisture level causing a decrease in growth of microorganisms and production of enzymes (Muniswaran and Charyulu, 1994). Moreover, the high moisture level causes accumulation of the substrate particles to the fermenter vessel making the substrate more exposed to microbial contamination (Lonsane et al., 1985).

The optimum water content is usually 50%–80% in an SSF. Many works reported 70% initial moisture content in the substrate as favorable for cellulase production. An optimum moisture of 70% is required by *B. subtilis* CBTK 106 in SSF to obtain a 12-fold higher increase in cellulase production when compared to SmF (Krishna, 1999). Lee et al. (2011) reported that *A. niger* USM AI required 70% moisture content to grow on a 1:1 (w/w) ratio of sugarcane bagasse and palm kernel cake substrate. The optimum moisture content was 70% even in *T. reesei* QM9414 as reported by Latifian et al. (2007). Other organisms were reported to have different optimal moisture contents for cellulase production:

- 37.56% in *T. reesei* RUT C30 (Singhania et al., 2007).
- 50% in some *Aspergillus* sp. (da Silva Delabona et al., 2013).
- 74% in *Trichoderma harzianum* (Deschamps et al., 1985).
- 80% in *Aspergillus terreus* M11 (Gao et al., 2008), *T. reesi* LM-UC4, and *Aspergillus phoenicis* QM 329 (Duenas et al., 1995).

Bansal et al. (2012) studied the cellulase production system in *A. niger* NS-2. Various substrates sufficiently moistened with water like corn cobs, rice husk, wheat bran, wheat straw, fruit and vegetable peelings (carrot, orange, pineapple, potato), composite, grass, leaves, sugarcane bagasse, sawdust, etc. were compatible for microbial growth and cellulase production in SSF. Among all, wheat bran was considered to be the preferred substrate because it possessed better water-absorbing ability and had the largest moisture content. The maximum water adsorption for wheat bran was approximately around 70% and any additional water made the substrate inappropriate for SSF.

6.3.10 Extraction Solvent

Different compounds are obtained from soil sediments, polymers, plants, bacteria, algae, fungi, and other sources using the solvent extraction technique (El Hattab et al., 2007; Plaza et al., 2010). For optimum recovery of desirable biological compounds, the most important step of this technique is their extraction from natural materials by using different organic solvents (Li et al., 2006a,b; Dunford and Edwards, 2010). Among the various solvents used in the extraction procedures, the most extensively used solvents are hexane, chloroform, ether, benzene, acetonitrile, and ethanol (Starmans and Nijhuis, 1996; Vatai et al., 2009; Plaza et al., 2010; Tokuoka et al., 2010). Different ratios of these organic solvents and water are used for the extraction of phenols, alkaloids, aromatic hydrocarbons, organochlorine pesticides, fatty acids and oils, and several other polar and nonpolar organic compounds (Li et al., 2006a,b; El Hattab et al., 2007). The solvent extraction technique is more advantageous over other methods because of its ease of operation and low processing cost. However, this technique has some disadvantages, e.g., it uses toxic solvents, involves an evaporation step for recovery, and generally requires large quantities of solvents and more time. Also, due to the high temperature solvents used and long extraction times, there is a possibility of thermal degradation of biological compounds. Hence to obtain better yields the solvent extraction technique has been improved by means of other methods such as Soxhlet, ultrasound, or microwave extraction (Szentmihalyi et al., 2002).

The optimum conditions required for operation of the enzyme extraction process under SSF are dependent on various factors. The different variables that mostly affect the extraction of enzymes are the solvent used, solid-to-liquid ratio, temperature, contact time, and agitation type (Castilho et al., 2000). Although there are many studies reporting the various strategies employed for improving the production of fungal cellulases under SSF, there are only a few studies relating to the enzyme recovery or extraction step. The influence of the composition of the extraction buffer on the extraction and recovery of cellulase enzyme was studied by Rezaei et al. (2011). The extraction of the cellulase produced under SSF and evaluation of each individual parameter separately on the recovery of the enzyme were studied. The efficiency of cellulase recovery by performing sequential extractions was also determined (Chandra et al., 2010).

6.4 CONCLUSION

Cellulase enzyme is considered to be one of the most significant among all the industrially produced enzymes because it uses lignocelluloses as substrate, the largest renewable resource on earth. The high price of the enzyme cellulase is still one of the chief limiting factors in the bioprocesses of conversion of lignocellulosic substrates using cellulase. This can be overcome either by selection and improvement of the best cellulolytic organisms or by judicial choice of fermentation technology used in industrial production. In future, the chief goals of cellulase research would be to improve the overall performance of cellulases making them more effective, and to radically decrease the production cost. The commercial or industrial production of cellulase enzymes at a larger scale requires further improvement for more efficient and cost-effective cellulases. Thus identification of cheaper substrates, developing appropriate media, and standardizing bioprocess parameters for microbial growth and cellulase enzyme production remain a great challenge for future microbiologists.

References

Aanchal, A., Kanika, N., Goyal, D., Goyal, A., 2016. Response surface methodology for optimization of microbial cellulase production. Rom. Biotechnol. Lett. 21 (5), 11832–11841.

Abdullah, B., Maftukhah, S., Listyaningrum, E., Faradhiba, F., 2018. Effect of some variable in cellulase production by *Aspergillus niger* ITBCC L74 using solid state fermentation. IOP Conf. Ser. Mater. Sci. Eng. 316 (1), 012066. IOP Publishing.

Acharya, P.B., Acharya, D.K., Modi, H.A., 2008. Optimization for cellulase production by *Aspergillus niger* using saw dust as substrate. African J. Biotechnol. 7 (22), 4147–4152.

Adsul, M.G., Ghule, J.E., Singh, R., Shaikh, H., Bastawde, K.B., Gokhale, D.V., Varma, A.J., 2004. Polysaccharides from bagasse: applications in cellulase and xylanase production. Carbohydr. Polym. 57 (1), 67–72.

Ahamed, A., Vermette, P., 2008. Culture-based strategies to enhance cellulase enzyme production from *Trichoderma reesei* RUT-C30 in bioreactor culture conditions. Biochem. Eng. J. 40 (3), 399–407.

Aiello, C., Ferrer, A., Ledesma, A., 1996. Effect of alkaline treatments at various temperatures on cellulase and biomass production using submerged sugarcane bagasse fermentation with *Trichoderma reesei* QM 9414. Bioresour. Technol. 57 (1), 13–18.

Akhtar, N., Goyal, D., Goyal, A., 2015. Simplification and optimization of media ingredients for enhanced production of CMCase by newly isolated *Bacillus subtilis* NA15. Environ. Prog. Sustain. Energy 34 (2), 533–541.

Akiba, S., Kimura, Y., Yamamoto, K., Kumagai, H., 1995. Purification and characterization of a protease-resistant cellulase from *Aspergillus niger*. J. Ferment. Bioeng. 79 (2), 125–130.

Alam, M.Z., Muyibi, S.A., Wahid, R., 2008. Statistical optimization of process conditions for cellulase production by liquid state bioconversion of domestic wastewater sludge. Bioresour. Technol. 99 (11), 4709–4716.

Ali, S., Sayed, A., 1992. Regulation of the cellulase biosynthesis in *Aspergillus terreus*. World J. Microbiol. Biotechnol. 8 (1), 73–75.

Alkasrawi, M., Eriksson, T., Börjesson, J., Wingren, A., Galbe, M., Tjerneld, F., Zacchi, G., 2003. The effect of Tween-20 on simultaneous saccharification and fermentation of softwood to ethanol. Enzym. Microb. Technol. 33 (1), 71–78.

Amore, A., Pepe, O., Ventorino, V., Aliberti, A., Faraco, V., 2013. Cellulolytic *Bacillus* strains from natural habitats—a review. Chim. Oggi 31, 49–52.

Anuradha Jabasingh, S., 2012. Response surface methodology for the evaluation and comparison of cellulase production by *Aspergillus nidulans* SU04 and *Aspergillus nidulans* MTCC344 cultivated on pretreated sugarcane bagasse. Chem. Biochem. Eng. Q. 25 (4), 501–511.

Ariffin, H., Hassan, M.A., Shah, U.K.M., Abdullah, N., Ghazali, F.M., Shirai, Y., 2008. Production of bacterial endoglucanase from pretreated oil palm empty fruit bunch by *Bacillus pumilus* EB3. J. Biosci. Bioeng. 106 (3), 231–236.

Attyia, S.H., Ashour, S.M., 2002. Biodegradation of agro industrial orange waste solid state fermentation and natural environmental conditions. Egypt. J. Biol. 4, 23–30.

Azzaz, H.H., Murad, H.A., Kholif, A.M., Hanfy, M.A., Gawad, M.A., 2012. Optimization of culture conditions affecting fungal cellulase production. Res. J. Microbiol. 7 (1), 23–31.

Bagnara-Tardif, C., Gaudin, C., Belaich, A., Hoest, P., Citard, T., Belaich, J.P., 1992. Sequence analysis of a gene cluster encoding cellulases from *Clostridium cellulolyticum*. Gene 119 (1), 17–28.

Bansal, N., Tewari, R., Soni, R., Soni, S.K., 2012. Production of cellulases from *Aspergillus niger* NS-2 in solid state fermentation on agricultural and kitchen waste residues. Waste Manag. 32 (7), 1341–1346.

Barron, J., Rockström, J., Gichuki, F., Hatibu, N., 2003. Dry spell analysis and maize yields for two semi-arid locations in east Africa. Agric. For. Meteorol. 117 (1-2), 23–37.

Bayer, E.A., Chanzy, H., Lamed, R., Shoham, Y., 1998. Cellulose, cellulases and cellulosomes. Curr. Opin. Struct. Biol. 8 (5), 548–557.

Berlin, A., Maximenko, V., Gilkes, N., Saddler, J., 2007. Optimization of enzyme complexes for lignocellulose hydrolysis. Biotechnol. Bioeng. 97 (2), 287–296.

Bhalla, T.C., Joshi, M., 1993. Production of cellulase and xylanase by *Trichoderma viride* and *Aspergillus* sp. on apple pomace. Indian J. Microbiol. 33 (4), 253–255.

Bhat, M.K., 2000. Cellulases and related enzymes in biotechnology. Biotechnol. Adv. 18 (5), 355–383.

Bisaria, V.S., Ghose, T.K., 1981. Biodegradation of cellulosic materials: substrates, microorganisms, enzymes and products. Enzym. Microb. Technol. 3 (2), 90–104.

Castanon, M., Wilke, C.R., 1981. Effects of the surfactant Tween 80 on enzymatic hydrolysis of newspaper. Biotechnol. Bioeng. 23 (6), 1365–1372.

Castilho, L.R., Medronho, R.A., Alves, T.L., 2000. Production and extraction of pectinases obtained by solid state fermentation of agroindustrial residues with *Aspergillus niger*. Bioresour. Technol. 71 (1), 45–50.

Cerri, M.O., Badino, A.C., 2010. Oxygen transfer in three scales of concentric tube airlift bioreactors. Biochem. Eng. J. 51 (1), 40–47.

Chandra, R.P., Bura, R., Mabee, W.E., Berlin, D.A., Pan, X., Saddler, J.N., 2007. Substrate pretreatment: the key to effective enzymatic hydrolysis of lignocellulosics? In: Biofuels. Springer, Berlin, Heidelberg, pp. 67–93.

Chandra, M., Kalra, A., Sharma, P.K., Kumar, H., Sangwan, R.S., 2010. Optimization of cellulases production by *Trichoderma citrinoviride* on marc of *Artemisia annua* and its application for bioconversion process. Biomass Bioenergy 34 (5), 805–811.

Clarke, P.H., Lilly, M.D., 1969. The regulation of enzyme synthesis during growth. Symp. Soc. Gen. Microbiol. 19, 113–159.

Cunha, F.M., Esperanca, M.N., Zangirolami, T.C., Badino, A.C., Farinas, C.S., 2012. Sequential solid-state and submerged cultivation of *Aspergillus niger* on sugarcane bagasse for the production of cellulase. Bioresour. Technol. 112, 270–274.

da Silva Delabona, P., Pirota, R.D.P.B., Codima, C.A., Tremacoldi, C.R., Rodrigues, A., et al., 2013. Effect of initial moisture content on two Amazon rainforest Aspergillus strains cultivated on agro-industrial residues: Biomass-degrading enzymes production and characterization. Ind. Crop. Prod. 42, 236–242.

Das, A., Bhattacharya, S., Murali, L., 2010. Production of cellulase from a thermophilic *Bacillus* sp. isolated from cow dung. Am. Eurasian J. Agric. Environ. Sci. 8, 685–691.

de Lima, A.L.G., Nascimento, R., da Silva Bon, E.P., Coelho, R., 2005. *Streptomyces drozdowiczii* cellulase production using agro-industrial by-products and its potential use in the detergent and textile industries. Enzyme Microb. Technol. 37, 272–277.

Debing, J., Li, P., Xiong, X.Z., Wang, L., 2007. Optimization of cellulase complex formulation for peashrub biomass hydrolysis. Appl. Microbiol. Biotechnol. 75 (4), 793–800.

Deka, D., Bhargavi, P., Sharma, A., Goyal, D., Jawed, M., Goyal, A., 2011. Enhancement of cellulase activity from a new strain of *Bacillus subtilis* by medium optimization and analysis with various cellulosic substrates. Enzyme Res. 2011 (1), 151656.

Deka, D., Das, S.P., Sahoo, N., Das, D., Jawed, M., Goyal, D., Goyal, A., 2013. Enhanced cellulase production from *Bacillus subtilis* by optimizing physical parameters for bioethanol production. ISRN Biotechnol. 2013, 965310.

Demain, A.L., Newcomb, M., Wu, J.D., 2005. Cellulase, clostridia, and ethanol. Microbiol. Mol. Biol. Rev. 69 (1), 124–154.

Deng, Y., Fong, S.S., 2010. Influence of culture aeration on the cellulase activity of *Thermobifida fusca*. Appl. Microbiol. Biotechnol. 85 (4), 965–974.

Deschamps, F., Giuliano, C., Asther, M., Huet, M.C., Roussos, S., 1985. Cellulase production by Trichoderma harzianum in static and mixed solid-state fermentation reactors under nonaseptic conditions. Biotechnol. Bioeng. 27 (9), 1385–1388.

Deswal, D., Khasa, Y.P., Kuhad, R.C., 2011. Optimization of cellulase production by a brown rot fungus *Fomitopsis* sp. RCK2010 under solid state fermentation. Bioresour. Technol. 102 (10), 6065–6072.

Dhillon, S.S., Gill, R.R., Gill, S.S., Singn, M., 2004. Studies on the utilization of citrus feel for pectinase production using fungus *Aspergillus niger*. Int. J. Environ. Stud. 61 (2), 199–210.

Dijkerman, R., den Camp, H.J.O., Van der Drift, C., Vogels, G.D., 1997. The role of the cellulolytic high molecular mass (HMM) complex of the anaerobic fungus *Piromyces* sp. strain E2 in the hydrolysis of microcrystalline cellulose. Arch. Microbiol. 167 (2-3), 137–142.

Domingues, F.C., Queiroz, J.A., Cabral, J.M.S., Fonseca, L.P., 2000. The influence of culture conditions on mycelial structure and cellulase production by *Trichoderma reesei* Rut C-30. Enzym. Microb. Technol. 26 (5-6), 394–401.

Duenas, R., Tengerdy, R.P., Gutierrez-Correa, M., 1995. Cellulase production by mixed fungi in solid-substrate fermentation of bagasse. World J. Microbiol. Biotechnol. 11 (3), 333–337.

Duff, S.J., Murray, W.D., 1996. Bioconversion of forest products industry waste cellulosics to fuel ethanol: a review. Bioresour. Technol. 55 (1), 1–33.

Dunford, N.T., Edwards, J., 2010. Nutritional bioactive components of wheat straw as affected by genotype and environment. Bioresour. Technol. 101 (1), 422–425.

Eggeman, T., Elander, R.T., 2005. Process and economic analysis of pretreatment technologies. Bioresour. Technol. 96 (18), 2019–2025.

El Hattab, M., Culioli, G., Piovetti, L., Chitour, S.E., Valls, R., 2007. Comparison of various extraction methods for identification and determination of volatile metabolites from the brown alga Dictyopteris membranace. J. Chromatogr. A. 1143 (1–2), 1–7.

Enari, T.M., Markkanen, P., 1977. Production of cellulolytic enzymes by fungi. In: Advances in Biochemical Engineering. Vol. 5. Springer, Berlin, Heidelberg, pp. 1–24.

Eriksson, T., Börjesson, J., Tjerneld, F., 2002. Mechanism of surfactant effect in enzymatic hydrolysis of lignocellulose. Enzym. Microb. Technol. 31 (3), 353–364.

Esterbauer, H., Steinier, W., Labudova, I., Hermann, A., Hayn, M., 1991. Production of *Trichoderma* cellulase in laboratory and pilot scale. Bioresour. Technol. 36 (1), 51–65.

Freed, L.E., Vunjak-Novakovic, G., 2000. Tissue engineering bioreactors. In: Lanza, R.P., Langer, R., Vacanti, J. (Eds.), Principles of Tissue Engineering. second ed. Academic Press, San Diego, CA, pp. 143–156.

Fukumori, F., Kudo, T., Horikoshi, K., 1985. Purification and properties of a cellulase from alkalophilic *Bacillus* sp. no. 1139. Microbiology 131 (12), 3339–3345.

Gao, J., Weng, H., Zhu, D., Yuan, M., Guan, F., Xi, Y., 2008. Production and characterization of cellulolytic enzymes from the thermoacidophilic fungal *Aspergillus terreus* M11 under solid-state cultivation of corn stover. Bioresour. Technol. 99 (16), 7623–7629.

Garg, S.K., Neelakantan, S., 1981. Effect of cultural factors on cellulase activity and protein production by *Aspergillus terreus*. Biotechnol. Bioeng. 23 (7), 1653–1659.

Gaur, R., Tiwari, S., 2015. Isolation, production, purification and characterization of an organic-solvent-thermostable alkalophilic cellulase from *Bacillus* vallismortis RG-07. BMC Biotechnol. 15 (1), 19.

Gautam, S.P., Bundela, P.S., Pandey, A.K., Awasthi, M.K., Sarsaiya, S., 2010. Optimization of the medium for the production of cellulase by the *Trichoderma viride* using submerged fermentation. Int. J. Environ. Sci. 1 (4), 656.

Gautam, S.P., Bundela, P.S., Pandey, A.K., Khan, J., Awasthi, M.K., Sarsaiya, S., 2011. Optimization for the production of cellulase enzyme from municipal solid waste residue by two novel cellulolytic fungi. Biotechnol. Res. Int. 2011, 1–8.

Godden, B., Legon, T., Helvenstein, P., Penninckx, M., 1989. Regulation of the production of hemicellulolytic and cellulolytic enzymes by a *Streptomyces* sp. growing on lignocellulose. Microbiology 135 (2), 285–292.

Golovchenko, N.P., Singh, R.N., Velikodvorskaya, G.A., Akimenko, V.K., 1993. Isolation and characteriation of lichenan degrading hydrophobic endoglucanase of *Clostridium thermocellum*. Appl. Microbiol. Biotechnol. 39 (1), 74–79.

Gomathi, D., Muthulakshmi, C., Kumar, D.G., Ravikumar, G., Kalaiselvi, M., Uma, C., 2012. Submerged fermentation of wheat bran by *Aspergillus flavus* for production and characterization of carboxy methyl cellulase. Asian Pac. J. Trop. Biomed. 2 (1), S67–S73.

Gregg, D.J., Saddler, J.N., 1996. Factors affecting cellulose hydrolysis and the potential of enzyme recycle to enhance the efficiency of an integrated wood to ethanol process. Biotechnol. Bioeng. 51 (4), 375–383.

Haq, I., Iqbal, S.H., Qadeen, M.A., 1993. Production of xylanase and CMC cellulase by mold culture. Pak. J. Biotechnol. 4, 403–409.

Heck, J.X., Hertz, P.F., Ayub, M.A., 2002. Cellulase and xylanase productions by isolated Amazon *Bacillus* strains using soybean industrial residue based solid-state cultivation. Braz. J. Microbiol. 33 (3), 213–218.

Hegde, S., Bhadri, G., Narsapur, K., Koppal, S., Oswal, P., Turmuri, N., Hungund, B., 2013. Statistical optimization of medium components by response surface methodology for enhanced production of bacterial cellulose by *Gluconacetobacter persimmonis*. J. Bioprocess Biotechnol. 4 (1), 1–5.

Heinzelman, P., Snow, C.D., Wu, I., Nguyen, C., Villalobos, A., Govindarajan, S., Arnold, F.H., 2009. A family of thermostable fungal cellulases created by structure-guided recombination. Proc. Natl. Acad. Sci. U. S. A. pnas-0901417106.

Helle, S.S., Duff, S.J., Cooper, D.G., 1993. Effect of surfactants on cellulose hydrolysis. Biotechnol. Bioeng. 42 (5), 611–617.

Hendy, N., Wilke, C., Blanch, H., 1984. Enhanced cellulase production using solka floc in a fed batch fermentation. Biotechnol. Lett. 6, 667–672.

Hurst, P.L., Nielsen, J.A.N., Sullivan, P.A., Shepherd, M.G., 1977. Purification and properties of a cellulase from *Aspergillus niger*. Biochem. J. 165 (1), 33–41.

Ibrahim, A.S.S., El-diwany, A.I., 2007. Isolation and identification of new cellulases producing thermophilic bacteria from an egyptian hot spring and some properties of the crude enzyme. Aust. J. Basic & Appl. Sci. 1 (4), 473–478.

Ikram-ul-Haq, M.M.J., Khan, S., Siddiq, Z., 2005a. Cotton saccharifying activity of cellulases produced by co-culture of *Aspergillus niger* and *Trichoderma viride*. Res. J. Agric. Biol. Sci. 1 (3), 241–245.

Ikram-Ul-Haq, Uzma Hameed, Kiran Shahzadi, M., Mohin Javed, Sikander Ali, Qadeer, M.A., 2005b. Cotton saccharifying activity of cellulases by Trichoderma harzianum UM-11 in shake flask. Intl. J. Bot. 1 (1), 19–22.

Ilmen, M., Saloheimo, A.N.U., Onnela, M.L., Penttilä, M.E., 1997. Regulation of cellulase gene expression in the filamentous fungus *Trichoderma reesei*. Appl. Environ. Microbiol. 63 (4), 1298–1306.

Immanuel, G., Dhanusha, R., Prema, P., Palavesam, A., 2006. Effect of different growth parameters on endoglucanase enzyme activity by bacteria isolated from coir retting effluents of estuarine environment. Int. J. Environ. Sci. Technol. 3 (1), 25–34.

Ingram, L.O., Aldrich, H.C., Borges, A.C.C., Causey, T.B., Martinez, A., Morales, F., Zaldivar, J., 1999. Enteric bacterial catalysts for fuel ethanol production. Biotechnol. Prog. 15 (5), 855–866.

Iqbal, H.M.N., Asgher, M., Ahmed, I., Hussain, S., 2010. Media optimization for hyper-production of carboxymethyl cellulase using proximally analyzed agroindustrial residue with *Trichoderma harzianum* under SSF. IJAVMS 4, 47–55.

Irene, K., Mulaa, F., Wamalwa, B., Muge, E., 2016. Isolation and characterization of a thermostable cellulase from *Bacillus licheniformis* strain vic isolated from geothermal wells in the kenyan rift valley. Open Biotechnol. J. 10 (1), 198–207.

Jang, H.D., Chen, K.S., 2003. Production and characterization of thermostablecellulases from Streptomyces transformant T3 -1. World J. Microbiol. Biotechnol. 19, 263–268.

Jecu, L., 2000. Solid state fermentation of agricultural wastes for endoglucanase production. Ind. Crops Prod. 11, 1–5.

Jha, K., Khare, S.K., Gandhi, A.P., 1995. Solid-state fermentation of soyhull for the production of cellulase. Bioresour. Technol. 54 (3), 321–322.

Jo, K.I., Lee, Y.J., Kim, B.K., Lee, B.H., Chung, C.H., Nam, S.W., Lee, J.W., 2008. Pilot-scale production of carboxymethylcellulase from rice hull by *Bacillus amyloliquefaciens* DL-3. Biotechnol. Bioprocess Eng. 13 (2), 182.

Kang, S.W., Park, Y.S., Lee, J.S., Hong, S.I., Kim, S.W., 2004. Production of cellulases and hemicellulases by *Aspergillus niger* KK2 from lignocellulosic biomass. Bioresour. Technol. 91 (2), 153–156.

Kaseke, Z.M., Okaiyeto, K., Nwodo, U.U., Mabinya, L.V., Okoh, A.I., 2016. Optimization of cellulase and xylanase production by micrococcus species under submerged fermentation. Sustainability 8, 11681183.

Kawai, S., Okoshi, H., Ozaki, K., Shikata, S., Ara, K., Ito, S., 1988. Neutrophilic *Bacillus* strain, KSM-522, that produces an alkaline carboxymethyl cellulase. Agric. Biol. Chem. 52 (6), 1425–1431.

Khan, J.A., Singh, S.K., 2011. Production of cellulose using cheap substrates by solid state fermentation. Int. J. Pl. An. Env.Sci. 1 (3), 179–187.

Kim, M.H., Lee, S.B., Ryu, D.D., Reese, E.T., 1982. Surface deactivation of cellulase and its prevention. Enzym. Microb. Technol. 4 (2), 99–103.

Kim, S.W., Kang, S.W., Lee, J.S., 1997. Cellulase and xylanase production by *Aspergillus niger* KKS in various bioreactors. Bioresour. Technol. 59 (1), 63–67.

Kirk, O., Borchert, T.V., Fuglsang, C.C., 2002. Industrial enzyme applications. Curr. Opin. Biotechnol. 13 (4), 345–351.

Kodali, B., Ravindra, P., 2006. Pretreatment studies of rice bran for the effective production of cellulose. Elect. J. Environ. Agric. Food. Chem. 5, 1253–1264.

Korpole, S., Sharma, R., Verma, D., 2011. Characterization and phylogenetic diversity of carboxymethyl cellulase producing bacillus species from a landfill ecosystem. Indian J. Microbiol. 51, 531–535.

Kotchoni, S.O., Shonukan, O.O., 2002. Regulatory mutations affecting the synthesis of cellulase in *Bacillus* pumilus. World J. Microbiol. Biotechnol. 18 (5), 487–491.

Krishna, C., 1999. Production of bacterial cellulases by solid state bioprocessing of banana wastes. Bioresour. Technol. 69 (3), 231–239.

Kubicek, C.P., Mikus, M., Schuster, A., Schmoll, M., Seiboth, B., 2009. Metabolic engineering strategies for the improvement of cellulase production by *Hypocrea jecorina*. Biotechnol. Biofuels 2 (1), 19.

Kumar, S., Sharma, H.K., Sarkar, B.C., 2011. Effect of substrate and fermentation conditions on pectinase and cellulase production by *Aspergillus niger* NCIM 548 in submerged (SmF) and solid state fermentation (SSF). Food Sci. Biotechnol. 20 (5), 1289–1298.

Latifian, M., Hamidi-Esfahani, Z., Barzegar, M., 2007. Evaluation of culture conditions for cellulase production by two *Trichoderma reesei* mutants under solid-state fermentation conditions. Bioresour. Technol. 98 (18), 3634–3637.

Lee, Y.J., Kim, B.K., Lee, B.H., Jo, K.I., Lee, N.K., Chung, C.H., Lee, Y.C., Lee, J.W., 2008. Purification and characterization of cellulose produced by *Bacillus amyloliquefaciens* DL-3 utilizing rice hull. Bioresour. Technol. 99, 378–386.

Lee, B.-H., Kim, B.-K., Lee, Y.-J., Chung, C.-H., Lee, J.-W., 2010. Industrial scale of optimization for the production of carboxymethylcellulase from rice bran by a marine bacterium, *Bacillus subtilis* subsp. subtilis A-53. Enzyme Microb. Technol. 46, 38–42.

Lee, C.K., Darah, I., Ibrahim, C.O., 2011. Production and optimization of cellulase enzyme using *Aspergillus niger* USM AI 1 and comparison with *Trichoderma reesei* via solid state fermentation system. Biotechnol. Res. Int. 2011, 6.

Lejeune, R., Baron, G.V., 1995. Effect of agitation on growth and enzyme production of *Trichoderma reesei* in batch fermentation. Appl. Microbiol. Biotechnol. 43 (2), 249–258.

Levin, L., Forchiassin, F., 1995. Effect of carbon and nitrogen sources on the cellulolytic activity of Trametes trogii. Rev. Argent. Microbiol. 27 (1), 11–20.

Li, P.-J., Jing, D.-B., Hou, Q., Zhang, C.G., 2004. Optimization of solid fermentation of cellulase from *Trichoderma koningii*. J. Environ. Sci. 16 (5), 816–820.

Li, L., Tian, H., Cheng, Y., Jiang, Z., Yang, S., 2006a. Purification and characterization of a thermostable cellulase-free xylanase from the newly isolated *Paecilomyces themophila*. Enzym. Microb. Technol. 38 (6), 780–787.

Li, Y.H., Ding, M., Wang, J., Xu, G.J., Zhao, F., 2006b. A novel thermoacidophilic endoglucanase, Ba-EGA, from a new cellulose-degrading bacterium, *Bacillus* sp. AC-1. Appl. Microbiol. Biotechnol. 70 (4), 430–436.

Lin, S.B., Lin, Y.C., Chen, H.H., 2009. Low molecular weight chitosan prepared with the aid of cellulase, lysozyme and chitinase: characterisation and antibacterial activity. Food Chem. 116 (1), 47–53.

Linko, M., 1977. An evaluation of enzymatic hydrolysis of cellulosic materials. In: Advances in Biochemical Engineering. Vol. 5. Springer, Berlin, Heidelberg, pp. 25–48.

Lonsane, B.K., Ghildyal, N.P., Budiatman, S., Ramakrishna, S.V., 1985. Engineering aspects of solid state fermentation. Enzym. Microb. Technol. 7 (6), 258–265.

Lugani, Y., Singla, R., Sooch, B.S., 2015. Optimization of cellulase production from newly isolated *Bacillus* sp. Y3. J. Bioprocess. Biotechnol. 5, 264.

Lundstedt, T., Seifert, E., Abramo, L., Thelin, B., Nyström, Å., Pettersen, J., Bergman, R., 1998. Experimental design and optimization. Chemom. Intell. Lab. Syst. 42 (3), 3–40.

Lynd, L.R., Weimer, P.J., Van Zyl, W.H., Pretorius, I.S., 2002. Microbial cellulose utilization: fundamentals and biotechnology. Microbiol. Mol. Biol. Rev. 66 (3), 506–577.

Maheshwari, R., Bharadwaj, G., Bhat, M.K., 2000. Thermophilic fungi: their physiology and enzymes. Microbiol. Mol. Biol. Rev. 64 (3), 461–488.

Makoi, J.H., Ndakidemi, P.A., 2008. Selected soil enzymes: examples of their potential roles in the ecosystem. Afr. J. Biotechnol. 7 (3), 181–191.

Mandels, M., 1975. Microbial source of cellulose. Biotechnol. Bioeng. 5, 81–105.

Mandels, M., Reese, E.T., 1963. Inhibition of cellulases and β-glucosidases. In: Reese, E.T. (Ed.), Advances in Enzymic Hydrolysis of Cellulose and Related Materials. Pergamon Press, London, pp. 115–157.

Massadeh, M.I., Wan Yusoff, W.M., Omar, O., Kader, J., 2001. Synergism of cellulase enzymes in mixed culture solid substrate fermentation. Biotechnol. Lett. 23, 1771–1774.

Mawadza, C., Hatti-Kaul, R., Zvauya, R., Mattiasson, B., 2000. Purification and characterization of cellulases produced by two *Bacillus* strains. J. Biotechnol. 83 (3), 177–187.

Michelin, M., de Moraes, M.D.L.T., da Silva, D.P., Ruzene, D.S., Vicente, A.A., Jorge, J.A., Teixeira, J.A., 2011. Production of xylanolytic enzymes by *Aspergillus terricola* in stirred tank and airlift tower loop bioreactors. J. Ind. Microbiol. Biotechnol. 38 (12), 1979–1984.

Michelin, M., Polizeli, M.L.T.M., Ruzene, D.S., Silva, D.P., Teixeira, J.A., 2013. Application of lignocelulosic residues in the production of cellulase and hemicellulases from fungi. Fungal Enzym, 32–59.

Mountfort, D.O., Asher, R.A., 1985. Production and regulation of cellulase by two strains of the rumen anaerobic fungus *Neocallimastix frontalis*. Appl. Environ. Microbiol. 49 (5), 1314–1322.

Muniswaran, P.K.A., Charyulu, N.C.L.N., 1994. Solid substrate fermentation of coconut coir pith for cellulase production. Enzym. Microb. Technol. 16 (5), 436–440.

Muthuvelayudham, R., Viruthagiri, T., 2006. Fermentative production and kinetics of cellulase protein on *Trichoderma reesei* using sugarcane bagasse and rice straw. Afr. J. Biotechnol. 5 (20), 1873–1881.

Nakamura, A., Fukumori, F., Horinouchi, S., Masaki, H., Kudo, T., Uozumi, T., Beppu, T., 1991. Construction and characterization of the chimeric enzymes between the *Bacillus subtilis* cellulase and an alkalophilic *Bacillus* cellulase. J. Biol. Chem. 266 (3), 1579–1583.

Narahara, H., Koyama, Y., Yoshida, T., Pichangkura, S., Ueda, R., Taguchi, H., 1982. Growth and enzyme production in a solid-state culture of Aspergillus oryzae. J. Ferment. Technol. 69, 311–319.

Narasimha, G., Sridevi, A., Buddolla, V., Subhosh, C.M., Rajasekhar, R.B., 2006. Nutrient effects on production of cellulolytic enzymes by *Aspergillus niger*. Afr. J. Biotechnol. 5 (5), 472–476.

Ogundero, V.W., 1982. Cellulose hydrolysis by thermophillic fungi from composting plant materials. Nigerian J. Sci. 16 (2), 83–93.

Ojumu, T.V., Solomon, B.O., Betiku, E., Layokun, S.K., Amigun, B., 2003. Cellulase production by *Aspergillus flavus* Linn isolate NSPR 101 fermented in saw dust, bagasse and corn cob. Afr. J. Biotechnol. 2 (6), 150–152.

Omojasola, P.F., Jilani, O.P., 2008. Cellulase production by *Trichoderma longiAspergillus niger* and *Saccharomyces cerevisiae* cultured on waste materials from orange. Pak. J. Biol. Sci. 11 (20), 2382–2388.

Omajasola, P.F., Jilani, O.P., Ibiyemi, S.A., 2008. Cellulase production by some fungi cultured on pine apple waste. Nat. Sci. 6 (2), 64–79.

Ooshima, H., Sakata, M., Harano, Y., 1986. Enhancement of enzymatic hydrolysis of cellulose by surfactant. Biotechnol. Bioeng. 28 (11), 1727–1734.

Padmavathi, T., Nandy, V., Agarwal, P., 2012. Optimization of the medium for the production of cellulases by *Aspergillus terreus* and *Mucor plumbeus*. Euro. J. Exp. Biol. 2 (4), 1161–1170.

Panda, B.P., Ali, M., Javed, S., 2007. Fermentation process optimization. Res. J. Microbiol. 2, 201–208.

Pandey, A., 1992. Recent developments in solid state fermentation. Process Biochem. 27 (2), 109–117.

Pandey, A., Selvakumar, P., Soccol, C.R., Nigam, P., 1999. Solid state fermentation for the production of industrial enzymes. Curr. Sci., 149–162.

Pandey, A., Soccol, C.R., Nigam, P., Soccol, V.T., 2000. Biotechnological potential of agro-industrial residues. I: sugarcane bagasse. Bioresour. Technol. 74 (1), 69–80.

Park, S., Johnson, D.K., Ishizawa, C.I., Parilla, P.A., Davis, M.F., 2009. Measuring the crystallinity index of cellulose by solid state 13 C nuclear magnetic resonance. Cellulose 16 (4), 641–647.

Paroda, S., Mishra, M.M., 1984. Growth and enzyme production by *Aspergillus terreus* in holocellulose. Ann. Inst. Pasteur Microbiol. 135 (2), 297–302.

Pérez-Avalos, O., Sánchez-Herrera, L.M., Salgado, L.M., Ponce-Noyola, T., 2008. A bifunctional endoglucanase/endoxylanase from *Cellulomonas flavigena* with potential use in industrial processes at different pH. Curr. Microbiol. 57 (1), 39–44.

Pirota, R.D.P.B., Tonelotto, M., da Silva Delabona, P., Fonseca, R.F., Paixão, D.A.A., Baleeiro, F.C.F., Farinas, C.S., 2013. Enhancing xylanases production by a new Amazon Forest strain of *Aspergillus oryzae* using solid-state fermentation under controlled operation conditions. Ind. Crop. Prod. 45, 465–471.

Plaza, M., Amigo-Benavent, M., Del Castillo, M.D., Ibáñez, E., Herrero, M., 2010. Facts about the formation of new antioxidants in natural samples after subcritical water extraction. Food Res. Int. 43, 2341.

Poorna, C.A., Prema, P., 2007. Production of cellulase-free endoxylanase from novel alkalophilic thermotolerent *Bacillus* pumilus by solid-state fermentation and its application in wastepaper recycling. Bioresour. Technol. 98 (3), 485–490.

Pothiraj, C., Balaji, P., Eyini, M., 2006. Enhanced production of cellulases by various fungal cultures in solid state fermentation of cassava waste. Afr. J. Biotechnol. 5 (20), 1882–1885.

Pushalkar, S., Rao, K.K., Menon, K., 1995. Production of β-glucosidase by Aspergillus terreus. Curr. Microbiol. 30 (5), 255–258.

Qi, B., Yaoa, R., Yua, Y., Chena, Y., 2007. Influence of different ratios of rice straw to wheat bran on production of cellulolytic enzymes by *Trichoderma viride* ZY-01 in solid state fermentation. EJEAF Che. 6 (9), 2341–2349.

Rajoka, M.I., 2004. Influence of various fermentation variables on exo-glucanase production in *Cellulomonas flavigena*. Electron. J. Biotechnol. 7 (3), 07–08.

Raju, E.V.N., Divakar, G., 2013. Production of pectinase by using *Bacillus* circulans isolated from dump yards of vegetable wastes. Int. J. Pharm. Sci. Res. 4 (7), 2615.

Rastogi, G., Muppidi, G.L., Gurram, R.N., Adhikari, A., Bischoff, K.M., Hughes, S.R., Sani, R.K., 2009. Isolation and characterization of cellulose-degrading bacteria from the deep subsurface of the Homestake gold mine, Lead, South Dakota, USA. J. Ind. Microbiol. Biotechnol. 36 (4), 585.

Rastogi, G., Bhalla, A., Adhikari, A., Bischoff, K.M., Hughes, S.R., Christopher, L.P., Sani, R.K., 2010. Characterization of thermostable cellulases produced by *Bacillus* and Geobacillus strains. Bioresour. Technol. 101 (22), 8798–8806.

Ray, A.K., Bairagi, A., Ghosh, K.S., Sen, S.K., 2007. Optimization of fermentation conditions for cellulase production by *Bacillus subtilis* CY5 and *Bacillus* circulans TP3 isolated from fish gut. Acta Ichthyol. Piscat. 37 (1), 47–53.

Reczey, K., Szengyel, Z.S., Eklund, R., Zacchi, G., 1996. Cellulase production by *T. reesei*. Bioresour. Technol. 57 (1), 25–30.

Rezaei, F., Joh, L.D., Kashima, H., Reddy, A.P., VanderGheynst, J.S., 2011. Selection of conditions for cellulase and xylanase extraction from switchgrass colonized by *Acidothermus cellulolyticus*. Appl. Biochem. Biotechnol. 164 (6), 793–803.

Romero, M.D., Aguado, J., González, L., Ladero, M., 1999. Cellulase production by *Neurospora crassa* on wheat straw. Enzym. Microb. Technol. 25 (3-5), 244–250.

Ruijter, G.J., Visser, J., 1997. Carbon repression in *Aspergilli*. FEMS Microbiol. Lett. 151 (2), 103–114.

Saha, B.C., 2004. Production, purification and properties of endoglucanase from a newly isolated strain of *Mucor circinelloides*. Process Biochem. 39 (12), 1871–1876.

Sandhu, S.K., Oberoi, H.S., Babbar, N., Miglani, K., Chadha, B.S., Nanda, D.K., 2013. Two-stage statistical medium optimization for augmented cellulase production via solid-state fermentation by newly isolated *Aspergillus niger* HN-1 and application of crude cellulase consortium in hydrolysis of rice straw. J. Agric. Food Chem. 61 (51), 12653–12661.

Sandhya, C., Sumantha, A., Szakacs, G.A., Pandey, A., 2005. Comparative evaluation of neutral protease production by *Aspergillus oryzae* in submerged and solid-state fermentation. Process Biochem. 40 (8), 2689–2694.

Saravanan, P., Muthuvelayudham, R., Kannan, R.R., Viruthagiri, T., 2012. Optimization of cellulase production using *Trichoderma reesei* by RSM and comparison with genetic algorithm. Front. Chem. Sci. Eng. 6 (4), 443–452.

Schrempt, H., Walter, S., 1995. The cellulolytic system of *Streptomyces reticuli*. Int. J. Biol. Macromol. 17 (6), 353–355.

Schülein, M., 2000. Protein engineering of cellulases. Biochim. Biophys. Acta Protein Struct. Mol. Enzymol. 1543 (2), 239–252.

Sethi, S., Datta, A., Gupta, B.L., Gupta, S., 2013. Optimization of cellulase production from bacteria isolated from soil. ISRN Biotechnol. 2013, 985685.

Shah, A.R., Madamwar, D., 2005. Xylanase production under solid-state fermentation and its characterization by an isolated strain of *Aspergillus foetidus* in India. World J. Microbiol. Biotechnol. 21 (3), 233–243.

Shankar, T., Isaiarasu, L., 2011. Cellulase production by *Bacillus* pumilus EWBCM1 under varying cultural conditions. Middle-East J. Sci. Res. 8 (1), 40–45.

Shankar, T., Isaiarasu, L., 2012. Statistical optimization for cellulase production by *Bacillus pumilus* EWBCM1 using response surface methodology. Global J. Biotechnol. Biochem. 7 (1), 1–6.

Sharada, R., Venkateswarlu, G., Venkateshwar, S., Anand, R.M., 2013. Production of cellulase—a review. Int. J. Pharmaceut. Chem. Biol. Sci. 3 (4), 1070–1090.

Sharma, D.K., Tiwari, M., Behera, B.K., 1995. Solid state fermentation of new substrates for production of cellulase and other biopolymer hydrolysing enzymes—Scientific note. Appl. Biochem. Biotechnol. 51, 495–500.

Sharmila, D., 2014. Comparative study on production of cellulase in fresh and sea water by *Fursarium subglutinans MTCC 11891* (Doctoral dissertation). National Institute of Technology Rourkela.

Siddiqui, K.S., Azhar, M.J., Rashid, M.H., Rajoka, M.I., 1997. Stability and identification of active-site residues of carboxymethylcellulases fromAspergillus niger* and *Cellulomonas biazotea*. Folia Microbiol. 42 (4), 312–318.

Siedenberg, D., Gerlach, S.R., Czwalinna, A., Schügerl, K., Giuseppin, M.L.F., Hunik, J., 1997. Production of xylanase by *Aspergillus awamori* on complex medium in stirred tank and airlift tower loop reactors. J. Biotechnol. 56 (3), 205–216.

Singh, A., Hayashi, K., 1995. Microbial cellulases: protein architecture, molecular properties, and biosynthesis. In: Advances in Applied Microbiology. 40. Academic Press, pp. 1–44.

Singh, S., Du Preez, J.C., Pillay, B., Prior, B.A., 2000. The production of hemicellulases by Thermomyces lanuginosus strain SSBP: influence of agitation and dissolved oxygen tension. Appl. Microbiol. Biotechnol. 54 (5), 698–704.

Singh, S., Simmons, B.A., Vogel, K.P., 2009. Visualization of biomass solubilization and cellulose regeneration during ionic liquid pretreatment of switchgrass. Biotechnol. Bioeng. 104 (1), 68–75.

Singh, K., Richa, K., Bose, H., Karthik, L., Kumar, G., Rao, K.V.B., 2014a. Statistical media optimization and cellulase production from marine *Bacillus* VITRKHB. Biotech 4 (6), 591–598.

Singh, S., Moholkar, V.S., Goyal, A., 2014b. Optimization of carboxymethylcellulase production from *Bacillus* amyloliquefaciens SS35. Biotech 4 (4), 411–424.

Singhania, R.R., Sukumaran, R.K., Pandey, A., 2007. Improved cellulase production by *Trichoderma reesei* RUT C30 under SSF through process optimization. Appl. Biochem. Biotechnol. 142 (1), 60–70.

Singhania, R.R., Patel, A.K., Sukumaran, R.K., Larroche, C., Pandey, A., 2013. Role and significance of beta-glucosidases in the hydrolysis of cellulose for bioethanol production. Bioresour. Technol. 127, 500–507.

Sohail, M., Siddiqi, R., Ahmad, A., Khan, S.A., 2009. Cellulase production from *Aspergillus niger* MS82: effect of temperature and pH. N Biotechnol. 25 (6), 437–441.

Sohail, M., Ahmad, A., Khan, S.A., 2011. Production of cellulases from *Alternaria* sp. MS28 and their partial characterization. Pak. J. Bot. 43 (6), 3001–3006.

Soumita, M., Karmakar, M., Ray, R.R., 2011. Production of extra cellular exoglucanase by *Rhizopus oryzae* from submerged fermentation of agro wastes. Recent Res. Sci. Technol. 3 (3), 69–75.

Sreedevi, S., Sajith, S., Benjamin, S., 2013. Cellulase producing bacteria from the wood-yards on Kallai river bank. Adv. Microbiol. 3 (04), 326–332.

Srilakshmi, A., Sai Gopal, D.V.R., Narasimha, G., 2017. Impact of bioprocess parameters on cellulase production by purpureocillium lilacinum isolated from forest soil. Int J Pharm. Bio. Sci 8 (1), 157–165.

Starmans, A.J., Nijhuis, H., 1996. Extraction of secondary metabolites from plant material: a review. Trends Food Sci. Technol. 7, 191–197.

Steiner, J., Socha, C., Eyzaguirre, J., 1994. Culture conditions for enhanced cellulase production by a native strain of *Penicillium purpurogenum*. World J. Microbiol. Biotechnol. 10 (3), 280–284.

Stewart, J.C., Parry, J.B., 1981. Factors influencing the production of cellulase by *Aspergillus fumigatus* (Fresenius). Microbiology 125 (1), 33–39.

Sukumaran, R.K., Singhania, R.R., Mathew, G.M., Pandey, A., 2009. Cellulase production using biomass feed stock and its application in lignocellulose saccharification for bio-ethanol production. Renew. Energy 34 (2), 421–424.

Sukumaran, R.K., Surender, V.J., Sindhu, R., Binod, P., Janu, K.U., Sajna, K.V., Pandey, A., 2010. Lignocellulosic ethanol in India: prospects, challenges and feedstock availability. Bioresour. Technol. 101 (13), 4826–4833.

Sun, Y., Cheng, J., 2002. Hydrolysis of lignocellulosic materials for ethanol production: a review. Bioresour. Technol. 83 (1), 1–11.

Suto, M., Tomita, F., 2001. Induction and catabolite repression mechanisms of cellulase in fungi. J. Biosci. Bioeng. 92 (4), 305–311.

Szabó, I.J., Johansson, G., Pettersson, G., 1996. Optimized cellulase production by *Phanerochaete chrysosporium*: control of catabolite repression by fed-batch cultivation. J. Biotechnol. 48 (3), 221–230.

Szentmihalyi, K., Vinkler, P., Lakatos, B., Illes, V., Then, M., 2002. Rose hip (*Rosa canina* L.) oil obtained from waste hip seeds by different extraction methods. Bioresour. Technol. 82, 195–201.

Tanyildizi, M.S., Ozer, D., Elibol, M., 2005. Optimization of α-amylase production by *Bacillus* sp. using response surface methodology. Process Biochem. 40 (7), 2291–2296.

Tengerdy, R.P., Szakacs, G., 2003. Bioconversion of lignocellulose in solid substrate fermentation. Biochem. Eng. J. 13, 169–179.

Thygesen, A., Anne, B., Schmidt, A.S., Jørgensen, H., Ahring, B., Olsson, L., 2003. Production of cellulose and hemicellulose-degrading enzymes by filamentous fungi cultivated on wet-oxidised wheat straw. Enzyme Microb. Technol. 32, 606–615.

Tokuoka, M., Sawamura, N., Kobayashi, K., Mizuno, A., 2010. Simple metabolite extraction method for metabolic profiling of the solid-state fermentation of *Aspergillusoryzae*. J. Biosci. Bioeng. 110 (6), 665–669.

Umar Dahot, M., Harif Noomrio, M., 1996. Microbial production of cellulases by *Aspergillus fumigatus* using wheat straw as a carbon source. J. Islamic World Acad. Sci. 9 (4), 119–124.

Umikalsom, M.S., Ariff, A.B., Hassan, M.A., Karim, M.I.A., 1998. Kinetics of cellulase production by *Chaetomium globosum* at different levels of dissolved oxygen tensionusing oilpalm empty fruit bunchfiber as substrate. World J. Microbiol. Biotechnol. 14 (4), 491–498.

Vaheri, M.P., Vaheri, M.E., Kauppinen, V.S., 1979. Formation and release of cellulolytic enzymes during growth of *Trichoderma reesei* on cellobiose and glycerol. Eur. J. Appl. Microbiol. Biotechnol. 8 (1-2), 73–80.

Van Wyk, J.P., 2001. Biotechnology and the utilization of biowaste as a resource for bioproduct development. Trends Biotechnol. 19 (5), 172–177.

Várnai, A., Siika-aho, M., Viikari, L., 2010. Restriction of the enzymatic hydrolysis of steam-pretreated spruce by lignin and hemicellulose. Enzym. Microb. Technol. 46 (3-4), 185–193.

Vatai, T., Skerget, M., Knez, Z., 2009. Extraction of phenolic compounds from elder berry and different grape marc varieties using organic solvents and/or supercritical carbon dioxide. J. Food Eng. 90, 246–254.

Wase, D.J., McManamey, W.J., Raymahasay, S., Vaid, A.K., 1985. Comparisons between cellulase production by *Aspergillus fumigatus* in agitated vessels and in an air-lift fermentor. Biotechnol. Bioeng. 27 (8), 1166–1172.

Weimer, P.J., 1996. Why don't ruminal bacteria digest cellulose faster? J. Dairy Sci. 79 (8), 1496–1502.

Wen, Z., Liao, W., Chen, S., 2005. Production of cellulase by *Trichoderma reesei* from dairy manure. Bioresour. Technol. 96 (4), 491–499.

Woodward, J., Zachry, G.S., 1982. Immobilization of cellulase through its carbohydrate side chains—a rationale for its recovery and reuse. Enzym. Microb. Technol. 4 (4), 245–248.

Xin, Z., Yinbo, Q., Peiji, G., 1993. Acceleration of ethanol production from paper mill waste fiber by supplementation with β-glucosidase. Enzym. Microb. Technol. 15 (1), 62–65.

Xiros, C., Katapodis, P., Christakopoulos, P., 2009. Evaluation of Fusarium oxysporum cellulolytic system for an efficient hydrolysis of hydrothermally treated wheat straw. Bioresour. Technol. 100 (21), 5362–5365.

Ye, M., Liu, X., Zhao, L., 2013. Production of a novel salt-tolerant l-glutaminase from *Bacillus amyloliquefaciens* using agro-industrial residues and its application in chinese soy sauce fermentation. Biotechnology 12, 25–35.

Ye, M., Sun, L., Yang, R., Wang, Z., Qi, K., 2017. The optimization of fermentation conditions for producing cellulase of *Bacillus* amyloliquefaciens and its application to goose feed. R. Soc. Open Sci. 4 (10), 171012.

Zambare, V., Christopher, L., 2011. Statistical analysis of cellulase production in *Bacillus* amyloliquefaciens UNPDV-22. Extreme Life Biospeol. Astrobiol. 3 (1), 38–45.

Zhang, Y.H.P., Lynd, L.R., 2004. Toward an aggregated understanding of enzymatic hydrolysis of cellulose: noncomplexed cellulase systems. Biotechnol. Bioeng. 88 (7), 797–824.

Zhang, Y.H.P., Himmel, M.E., Mielenz, J.R., 2006. Outlook for cellulase improvement: screening and selection strategies. Biotechnol. Adv. 24 (5), 452–481.

Zhao, S.-H., Liang, X.-H., Hua, D.-L., Ma, T.-S., Zhang, H.-B., 2011. High yield cellulose production in solid state fermentation by *Trichoderma reesei* SEMCC – 3.217 using water hyacinth (Eichhornia Crassipes). African J. Biotech. 10 (50), 10178–10187.

CHAPTER

7

Comparative Study of Cellulase Production Using Submerged and Solid-State Fermentation

Hemansi[*,a], *Subhojit Chakraborty*[†,a], *Garima Yadav*[*], *Jitendra Kumar Saini*[*], *Ramesh Chander Kuhad*[†,‡]

*Department of Microbiology, School of Interdisciplinary & Applied Life Sciences, Central University of Haryana, Mahendergarh, India †Department of Microbiology, University of Delhi South Campus, New Delhi, India ‡Central University of Haryana, Mahendergarh, India

7.1 INTRODUCTION

Currently, there is a marked increase in global demands of various industrial products and hence an urgent need to explore more sustainable ways for their economical and greener production via microbial fermentations. Considering the limited supply of natural resources, newer raw materials in combination with better technologies and zero or minimum possible waste production must be exploited. Lignocellulosic feedstocks are good alternatives as cheaper and renewable feedstock for the synthesis of many products (food, pharmaceuticals, and chemicals) in an eco-friendly and sustainable manner. Production of bioethanol from lignocellulosic biomass seems very promising due to its ubiquitous nature, abundance, and contribution to reducing greenhouse gas emissions. E10 bioethanol (i.e., a mix of 10% ethanol in petrol) is estimated to contribute up to 6% reduced CO_2 emissions. Lignocellulosic biomass is a natural resource available in plentiful amounts on the globe and can be potentially utilized via chemical or microbial/biochemical conversion platforms (Kuhad et al., 2016). Among these routes, the latter one seems more promising due to its sustainable and renewable nature and noncompetitiveness of the feedstock with food/feed crops. It is important to note that lignocellulose biomass comprises half of the total world biomass that can be utilized for bioethanol production. As shown in Fig. 7.1, lignocellulosic biomass is generally comprised of three components, i.e., cellulose, hemicellulose, and lignin, along with lesser proportions of pectin, proteins, etc. Among all the components, cellulose is present in major proportions (Saini et al., 2015a,b).

Microbial conversion of the cellulosic and hemicellulosic portions of the carbohydrates present in lignocellulose requires the action of enzymes such as cellulases and hemicellulases, respectively. The enzyme cellulase belongs to glycoside hydrolase and its mode of action involves the breakdown of the β-1,4-glycosidic bonds between monomeric units of cellulose (Behera et al., 2017). Cellulase is a multienzyme complex and is comprised of three related but distinct enzymes that act sequentially during the hydrolysis of cellulose: *endo*-β-1,4-glucanases (EGs), cellobiohydrolases (CBHs), and β-glucosidases (BGLs), bringing about the formation of cellobiose, glucose, and various cellulose-derived short oligomers of D-glucose. EG is responsible for making internal incisions in cellulosic chains to expose reducing ends, as well as nonreducing ends. CBH acts on the newly formed reducing as well as nonreducing ends and releases oligomers of glucose along with the formation of cellobiose. The enzyme BGL hydrolyzes the β-1,4-glycosidic links between glucose units of cellobiose and results in the ultimate release of glucose (Saini et al., 2015a,b). The mechanism of cellulase action on its substrate cellulose is shown in Fig. 7.2.

[a] Equal contribution by authors.

From Cellulose to Cellulase: Strategies to Improve Biofuel Production
https://doi.org/10.1016/B978-0-444-64223-3.00007-2

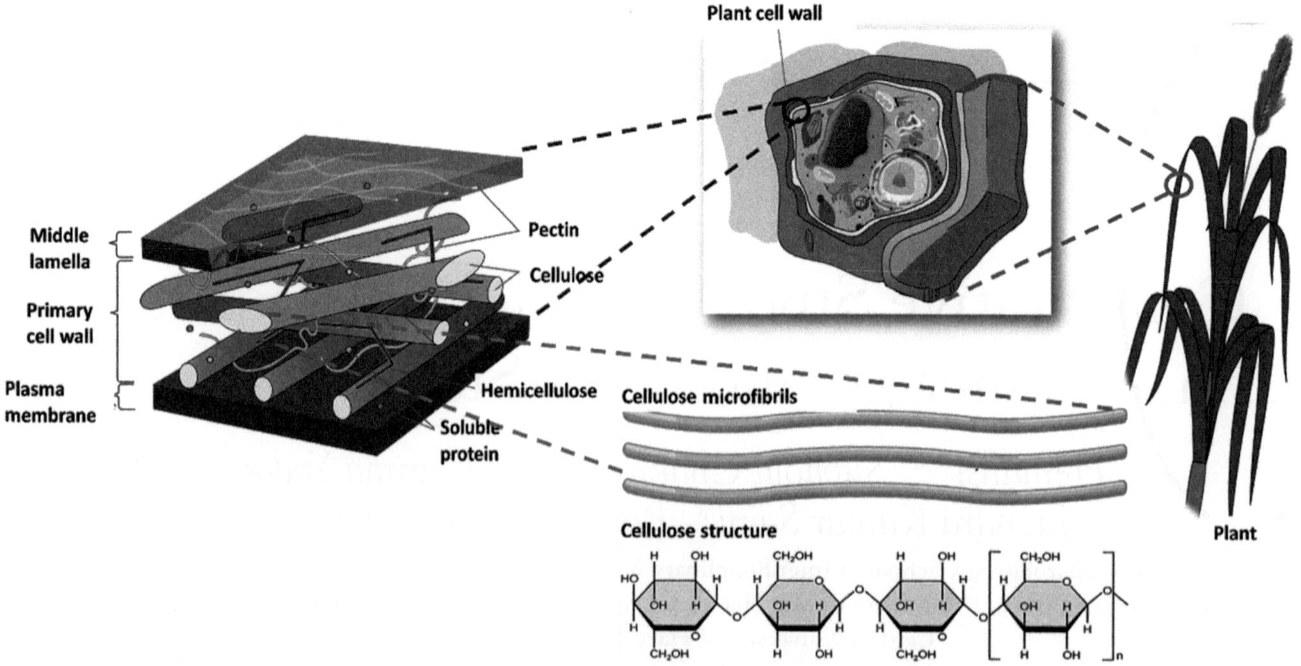

FIG. 7.1 Structure of lignocellulose present in the cell wall of plants. *From https://commons.wikimedia.org/wiki/File:Plant_cell_wall_diagram-en.svg; https://commons.wikimedia.org/wiki/File:Plant_cell_structure_svg_vacuole-ta.svg.*

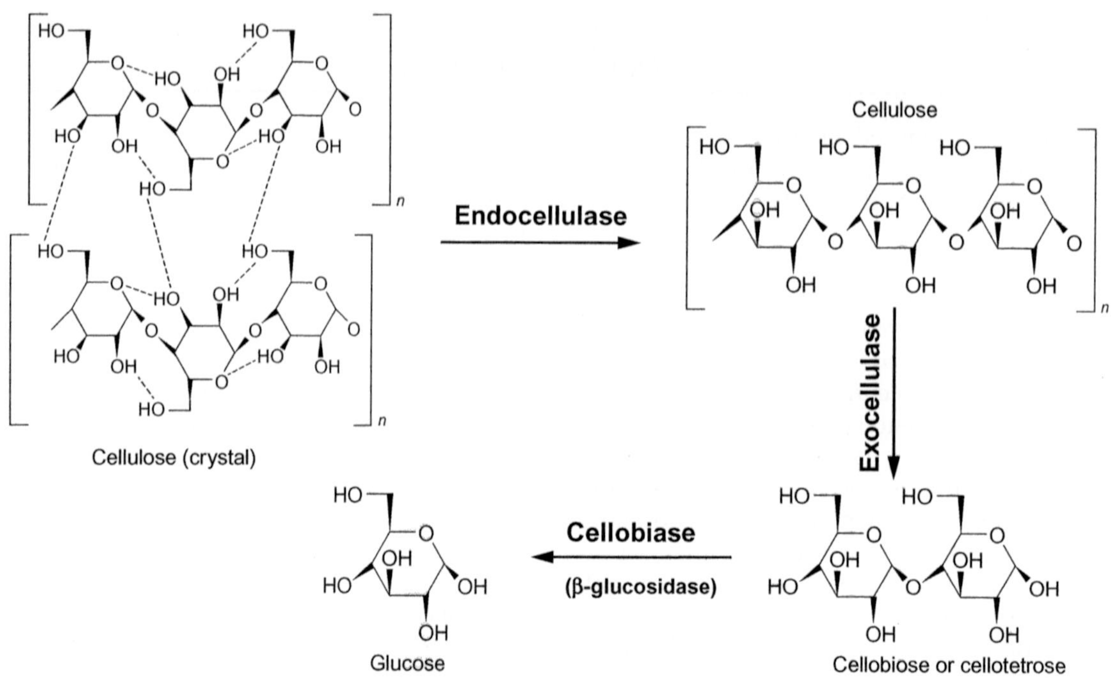

FIG. 7.2 The mechanism of cellulase action on its substrate cellulose. *From https://en.wikipedia.org/wiki/File:Types_of_Cellulase2.png.*

Synergy of all these enzymes is critical for the complete hydrolysis of the cellulosic biomass into its constituent reducing sugar, i.e., glucose.

The application of cellulases is not only limited to biofuel production; in fact this enzyme also finds its uses in the detergent industry, fermentation industry, food industry, paper industry, textile industry, as well as many more industrial sectors as shown in Fig. 7.3 (Kuhad et al., 2011). The demand of cellulases in various industrial sectors is constantly increasing and is expected to grow further. At present, this enzyme is the most thoroughly and widely employed enzyme for lignocellulosic biofuel/bioenergy generation and synthesis of many other biorefinery-based chemicals and products. Lignocellulose-to-bioethanol conversion utilizing the sugar released on hydrolysis by cellulases is foreseen as its largest application and the demands of the cellulase enzyme are expected to increase many

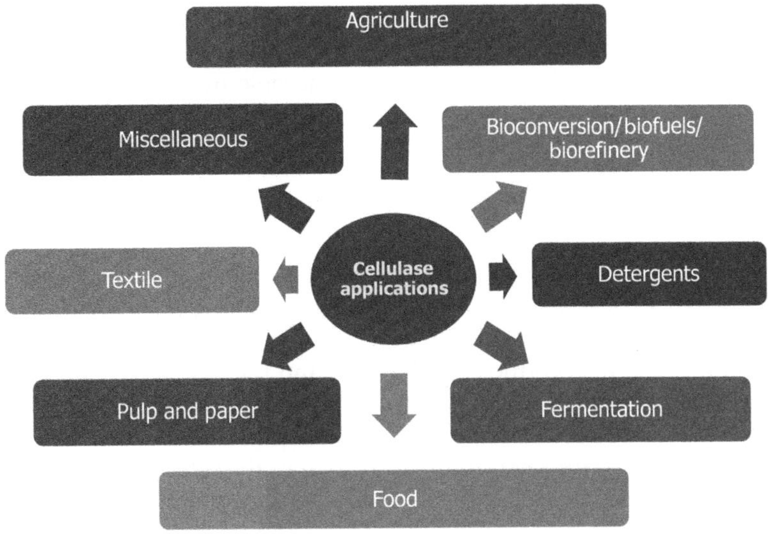

FIG. 7.3 Applications of cellulases in different industrial sectors (Kuhad et al., 2011).

fold as more biofuel plants are set up globally. As compared to previous decades, the worldwide supply of cellulases has significantly improved and its cost has also been comparatively lowered. However, there are still cost-, enzyme titer-, as well as efficiency-related challenges that are major hindrances in the development of sustainable technology for lignocellulosic ethanol production and these need to be suitably addressed. Moreover, until now, only a few industries dominate the cellulase market and there seems much more scope for cost reduction of cellulase production (Kuhad et al., 2011). Microbial cellulases are produced by either of the two possible fermentation configurations, i.e., submerged fermentation (SmF) and solid-state fermentation (SSF) processes. Both these technologies find use in cellulase production and have their own advantages and disadvantages. Both technologies differ significantly in the choice of microorganisms, suitable combination of production parameters, and bioreactor design and operation.

Many different types of microorganisms are employed for the production of cellulase, including both aerobes and anaerobes among fungi, bacteria, as well as actinomycetes. Enhancements in cellulase production require the search for new robust cellulase-producing microorganisms or improvement of the available cellulase-producing microorganisms. Over the past few years, biochemical engineering and fermentation technology aspects of cellulase production have been significantly improved, with improvements in the design of fermentation vessels for SmF and SSF configurations.

Keeping the importance of cellulases in the generation of biofuel in view, this chapter focuses on some of the aforementioned aspects of cellulase production, especially cellulase-producing microorganisms, comparative accounts of SSF and SmF cellulase fermentation technology for their large-scale synthesis, and strategies for the improvement of cellulase-producing microorganisms.

7.2 CELLULASE-PRODUCING MICROORGANISMS

Currently, many diverse kinds of microorganisms are applied by means of fermentation for commercial cellulase production. Various aerobic and anaerobic, mesophilic and thermophilic fungi and bacteria as well as extremophiles produce all three cellulases in high or low titers. Bacteria produce cellulase in single or complex multienzyme form known as cellulosomes, consisting of a number of cellulolytic enzymes bound to a scaffold protein (Behera et al., 2017), whereas fungi mostly secrete cellulases as free molecules (Singhania et al., 2010).

7.2.1 Fungal Cellulases

Cellulases are mainly produced by saprophytic microbes, which grow well on decaying matter. Fungi are the major saprophytes present in nature and hence are widely renowned cellulase producers when grown on cellulosic substrates. The commercial cellulases currently available worldwide are the product of common cellulase producers: the fungi *Trichoderma* and *Aspergillus* (Behera et al., 2017). The reason behind the increased popularity of fungal cellulases is the presence of multiple cellulases in supernatant, which ultimately facilitates depolymerization of cellulosic substrate (Juturu and Wu, 2014). Moreover, the fungal cellulases can easily be harvested and extracted as secreted in surrounding medium, are less complex, and therefore can be easily cloned (Kuhad et al., 2016). Filament-forming fungi like *Aspergillus niger*, *Trichoderma reesei*, *Penicillium* sp., etc. have been utilized in the production of cellulolytic enzymes under SSF conditions. Besides aerobic fungi, anaerobic fungi like *Neocallimastix frontalis*, *Piromonas comunis*, etc. also produce an array of cellulolytic enzymes. Moreover, thermophilic fungi like *Myceliophthora* and *Scytalidium* produce cellulases at high temperature (45°C). Various cellulase producers from different genera are listed in Table 7.1.

7.2.2 Bacterial Cellulases

Several cellulase producers such as *Acidothermus* sp., *Bacillus* sp., *Clostridium* sp., *Cellulomonas* sp., etc. are well known and are being utilized currently for producing enzymes (Table 7.2). However, fungi are better producers of cellulases because bacterial cellulases are produced in low titer and the system is complex and hence less exploited commercially. Nevertheless, researchers are shifting to bacterial cellulases due to several advantages such as high growth rate, easy production of recombinant enzymes, etc. Extremophiles (thermophilic/psychrophilic, alkaliphilic/acidophilic or halophilic) resist various stress conditions and so does the cellulase produced by them, and hence it is found to be more stable under extreme conditions of the process of bioconversion (Sadhu and Maiti, 2013). Some of the extremophile cellulase producers are listed in Table 7.3.

Aerobic as well as anaerobic bacteria are both reported to be cellulase producers and have profound differences in their enzymes. Anaerobic bacteria from genera *Clostridium*, *Acetivibrio*, *Bacteroides*, and *Ruminococcus* break down cellulose and hemicellulose by a multienzyme, extracellular complex known as cellulosomes. Cellulosomes are large aggregated structured enzymes that are adhered to the bacterial surface (Doi and Kosugi, 2004).

TABLE 7.1 Major Fungi Reported for the Production of Cellulase

Microorganism	References	Microorganism	References
Sclerotium rolfsii	Sternberg (1976)	*Formitopsis* sp.	Deswal et al. (2011)
Penicillium chrysosporium	Fan et al. (1987)	*Humicola grisea*	Mello-de-Sousa et al. (2011)
Trichoderma reesei	Fang and Xia (2015)	*Fusarium oxysporum*	Kuhad et al. (1994)
Aspergillus niger	Hemansi et al. (2018)	*Orpinomyces* sp.	Hodrová et al. (1998)
Schizophyllum	Duff and Murray (1996)	*Myceliophthora* sp.	Badhan et al. (2007)
Penicillium oxalicum	Saini et al. (2015a,b)	*Agaricus arvensis*	Jeya et al. (2010)
Thermoascus aurantiacus	Jain et al. (2015)	*Scytalidium thermophilum*	Kaur et al. (2006)
Aspergillus terreus	Gunny et al. (2015)	*Phlebia gigantea*	Niranjane (2006)
Trichoderma asperellum	Raghuwanshi et al. (2014)	*Neurospora crassa*	Romero et al. (1999)
Penicillium brasilianum	Jung et al. (2015)	*Aspergillus fumigatus*	Das et al. (2013)
Penicillium funiculosum	Maeda et al. (2013)	*Acremonium cellulolyticus*	Hideno et al. (2011)

TABLE 7.2 Major Bacteria Reported for Cellulase Production

Microorganism	References	Microorganism	References
Clostridium thermocellum	Chinn et al. (2008)	*Microbispora*	Eida et al. (2012)
Cellulomonas uda	Poulsen et al. (2016)	*Streptomyces drozdowiczii*	de Lima et al. (2005)
Bacillus subtilis	Sreena and Sebastian (2018)	*Paenibacillus*	Eida et al. (2012)
Thermomonospora fusca	Wilson (1988)	*Cohnella*	Eida et al. (2012)
Ruminococcus albus	Leatherwood (1965)	*Acinetobacter amiratus*	Ekperigin (2007)
Bacteroides cellulosilyticus	Robert et al. (2007)	*Fibrobacter succinogenes*	Béra-Maillet et al. (2009)
Erwinia chrysanthemi	Chambost et al. (1987)	*Neocallimastix frontalis*	Srinivasan et al. (2001)
Acetivibrio cellulolyticus	Saddler and Khan (1981)	*Eubacterium cellulosolvens*	Moon and Anderson (2001)

TABLE 7.3 Major Extremophiles Involved in the Production of Cellulase

Microorganism	References
Anoxybacillus sp.	Liang et al. (2010)
Geobacillus sp.	Rastogi et al. (2010)
Anoxybacillus flavithermus	Ibrahim and El-diwany (2007)
Sulfolobus solfataricus	Giuliano et al. (2004)
Alicyclobacillus acidocaldarius	Eckert and Schneider (2003)
Fervidobacterium	Gomes and Steiner (2004)
Rhodothermus	Gomes and Steiner (2004)
Thermotoga	Gomes and Steiner (2004)
Pyrococcus	Gomes and Steiner (2004)
Desulforococcus	Gomes and Steiner (2004)

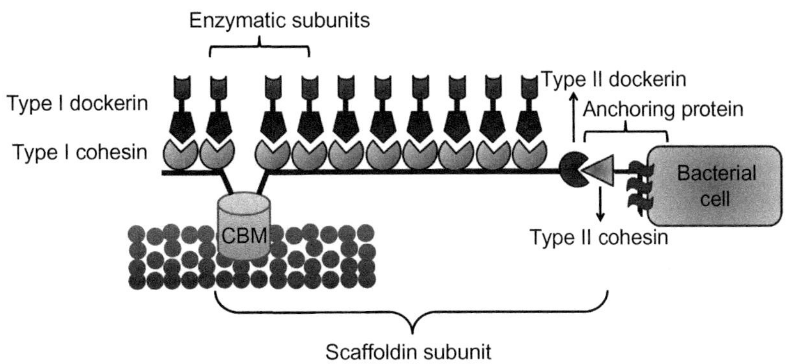

FIG. 7.4 Schematic representation of the bacterial cellulosome.

The composition of a cellulosome is not the same in all bacterial species. The varied nature of cellulosomes is due to disparity in scaffoldin properties in different species. The scaffolding protein called CbpA allows the formation of a complex with numerous different enzymes. Scaffoldins contain approximately nine each of dockerin and cohesins. Dockerin proteins can bind to different cellulosomal enzymes, whereas cohesins bind to another side of the scaffoldin (Juturu and Wu, 2014). Another component of cellulosomes is cellulose-binding domain, also known as cellulose-binding module, which serves as a binding subunit to the enzyme. The most examined cellulosome is that of *Clostridium thermocellum*, a thermophilic bacterium (Sadhu and Maiti, 2013). The schematic structure of a cellulosome is shown in Fig. 7.4.

7.3 ADVANCEMENTS IN THE MICROBIAL PRODUCTION OF CELLULASES

7.3.1 Solid-State Fermentation Technology

SSF is the technique for producing high end value products. Enzymes are the major products that are produced under SSF by utilizing agro-industrial waste (Behera and Ray, 2016). It is the technique in which microbes grow by attaching to solid substrates. This imitates the natural habitat of the microorganisms, in turn giving a fruitful outcome. The SSF technique has emerged as an attractive technology, especially for producing many enzymes. Some of the benefits of SSF are efficient fermentation, highly stable product, which can be easily recovered, negligible catabolite inhibition, and economic aspects (Behera and Ray, 2016; Yoon et al., 2014).

7.3.1.1 *Substrates Used*

SSF cultivations are renowned for the solid substrates they use, which include natural substances for cultivation with the support of liquid medium to moisten them (Ooijkaas et al., 2000). Substrates used for SSF are heterogeneous in nature. They could be procured either from agro-industrial by-products or from agriculture. The molecular structure of the substrate provides information about the properties of the matrix that adsorbs carbon and energy sources. The main factors that influence the selection of appropriate substrates are their basic feature, availability, and price (Krishna, 2005). Due to the heterogeneous nature of the substrate utilized, a pretreatment step of solid biomass is necessary. This is done mainly by three processes: mechanical (size reduction), physical (using high temperature/pressure/radiations), and chemical (using diluted acid/alkali) (Raimbault, 1998). The advantage of adding an extra step of pretreatment is that it increases the availability of substrate along with nutrient supplementation, and pH and moisture content are maintained through mineral solution.

Agricultural waste residues are considered to be one of the best substrates for enzyme production under SSF. The required nutrients for the growth of microbes are easily provided by agrowaste substrates and hence good enzyme yields can be obtained. Some of the commonly used substrates for enzyme production are wheat straw, wheat bran, rice straw, maize, grains, brans, coconut coir and pith, tea, coffee and cassava waste, oil mill waste, sugarcane bagasse and sorghum pulp waste, oil cakes of coconut, mustard, soybean, peanut, and cassava (Krishna, 2005). Natural substrates, however, have a major drawback: the main constituent of the substrate is carbon and during microbial growth this solid structure is disintegrated, changing its characteristics and leading to reduced heat and mass transfer. These limitations lead to the subsequent use of inert support such as polyurethane foam, vermiculite, for their specific properties of minimum interaction with the biological process (Chen, 2013).

7.3.1.2 *SSF Bioreactors for Cellulase Production*

Cellulase is one of the major groups of hydrolytic enzymes for catalyzing cellulose-to-glucose reaction by acting on β-1,4-linkages (Kuhad et al., 2016).

Commonly, fungi like *Trichoderma reesei* and *Aspergillus niger* have been employed as a major source of cellulase production at an industrial scale. Both of the fungi are mesophilic and grow well at 30–35°C. Enzymes produced by thermophiles work well up to 90°C and are more beneficial for industrial processes under harsh conditions due to the fast rate of reaction, and chances of contamination have been reduced at high temperatures (Biswas et al., 2014). At the industrial scale, fermentations have been performed in different types of reactors as per the need. Tray bioreactors and drum-type reactors in batch culture or in continuous modes are some of the processes that are commonly followed at large scale (Behera and Ray, 2016). As reported by Vitcosque et al. (2012) an instrumented lab-scale bioreactor produced 0.55 IU/g FPase, 35.1 IU/g endoglucanase, and 47.7 IU/g xylanase using *Aspergillus niger* as inoculum with a fully automated online monitoring and controlling system by utilizing soybean meal substrate.

7.3.1.3 *Types of SSF Bioreactors*

7.3.1.3.1 TRAY BIOREACTOR

A bioreactor is the system that provides an optimum environment for the growth of microorganisms, which in turn provides maximum yield. Tray bioreactors are used traditionally in making fermented foods mainly in some Asian countries (Arora et al., 2018). Interestingly, many of the processes at commercial scale utilized tray bioreactors in various industries for diverse applications (Thomas et al., 2013). A tray bioreactor is made up of flat trays of wood, metal, or plastic, which can be provided with perforations and packed with substrate support that forms a layer of about 1.5 or 2 cm thickness. These reactors are stacked on each other and kept in chambers with maintained temperature and humidity (Arora et al., 2018; Behera and Ray, 2016). In the case of scaling up with tray bioreactors, either the surface area or number of trays is increased for achieving large quantities of the product (Arora et al., 2018). Brijwani et al. (2010) reported use of a tray bioreactor for producing cellulases under a mixed culture fermentation technique by utilizing *Trichoderma reesei* and *Aspergillus oryzae* grown on soybean meal and wheat bran (4:1) having a 1 cm bed height and optimum operating conditions. An appropriate C:N ratio in the substrate was maintained, which aided improved cellulase titers and brought about the production of all three enzymes in a balanced ratio, which is required for complete conversion of biomass in biofuel production. Higher levels of BGL were reported by Dhillon et al. (2011) while fermenting apple pomace by employing *Aspergillus niger* and *Trichoderma reesei* in a tray bioreactor. Although the tray bioreactor is generally exploited commercially for SSF processes in fermentation industries there are a number of shortcomings such as proper heat transfer due to low thermal conductivity and bed height (Arora et al., 2018).

7.3.1.3.2 PACKED BED REACTOR

A packed bed reactor (PBR) has a characteristic feature of forced aeration through a static bed helping to replenish O_2 and moisture, mitigating the accumulation of heat and CO_2. A PBR is generally employed where mixing is undesirable or deleterious for microbial growth and offers better control. Reactor construction generally consists of a cylindrical glass or metal tube/drum, which houses the substrate, and the walls of the cylinder or drum may be jacketed, which can be provided with cooling plates within the bed to facilitate efficient heat transfer (Arora et al., 2018). *Aspergillus niger* FTCC 5003 in a laboratory using a PBR was carried out, which used palm kernel cake as substrate in the production of cellulase for 7 days and reported cellulase yield as high as 244.53 U/gds, using 100 g of palm kernel cake (Abdeshahian et al., 2011). PBR was used for the production of hydrolases using babassu cake as substrate, which is rich in carbohydrate and known for induction and secretion of an enzyme pool by *Aspergillus awamori*. Higher enzyme yields were reported for cellulases, proteases, endoamylases, and exoamylases except for xylanases (Castro et al., 2015). The main reason for heat generation in a PBR is the deficient removal of the metabolically generated heat, due to the low effective thermal conductivity of the substrate and the low air flow rates employed.

7.3.1.3.3 ROTARY DRUM BIOREACTOR

Rotary drum bioreactors (RDBs) comprise a horizontal cylinder, where a tumbling motion of the solid medium is created due to mixing, facilitated by baffles on the inside wall of the rotating drum (may be perforated or not) (Soccol et al., 2017). Because there is no agitator, uniform mixing is provided gently by improved baffle design (Behera and Ray, 2016). Cellulase production was evaluated in terms of FPase activity by *Trichoderma harzianum* using empty fruit bunches, and in the case of *Penicillium verruculosum* COKE4E alkali-treated empty palm fruit bunch fiber was used as the sole carbon source in an RDB (Alam et al., 2009; Kim and Kim, 2012). The engineering principles used in an RDB reactor, specifically its air circulation and continuous mixing, make it demandable for biofuel production in pilot or lab scale using cellulosic materials.

7.3.1.3.4 FLUIDIZED BED REACTOR

Another type of SSF bioreactor is the fluidized bed bioreactor (FBB) in which substrate and microorganism are maintained in a fluidized state by providing upward flow of air (Arora et al., 2018). The substrate availability increases due to its fluidized state and results in a very effective microbial growth. The greater degree of turbulence aroused superficially by air velocities helps in efficient heat transfer throughout the reactor. An enhanced productivity of ethanol (11.7 g/L/h) was seen when cassava starch was hydrolyzed using coimmobilized glucoamylase in an SSF process with simultaneous fermentation using *Streptomyces cerevisiae* in pectin gel in an FBR (Trovati et al., 2009). FBR was used to develop a process for the production of high bioethanol titer through fed-batch as well as simultaneous saccharification and fermentation of wild, nonedible cassava *Manihot glaziovii*. This allowed fermentation of up to 390 g/L of starch-derived glucose achieving a high bioethanol concentration of up to 190 g/L (24% *v/v*) with yields of around 94% of the theoretical value (Moshi et al., 2014).

7.3.1.3.5 LAB-SCALE BIOREACTOR

Lab-scale column-type bioreactors have been used for SSF operations (Behera and Ray, 2016). *Aspergillus niger* was employed for endoglucanase production and cultivated under SSF using a lab-scale bioreactor equipped with an online automated monitoring and control system. It was evaluated to check its effect of operational conditions. Optimized conditions of the bioreactor system resulted in maximum production of endoglucanase of 56.1 U/gds. The optimized parameters included substrate initial moisture (72%), flow rate (20 mL/min), and inlet air humidity (70%), which resulted in enhanced endoglucanase activity (Farinas et al., 2011).

7.3.2 Submerged Fermentation Technology

The technique called fermentation is the process of converting complex compounds into simpler ones. SmF is another technique apart from SSF that has been used at an industrial scale for producing enzymes using fungi. SmF is favorable in many cases because it is easy to handle and can be monitored effortlessly (dos Reis et al., 2013). In this type of fermentation, the nutrients and oxygen are easily dissolved in liquid and dispersed evenly all over in the vessel, and because of this, transfer of heat and mass takes place efficiently (Biswas et al., 2014).

7.3.2.1 *SmF Bioreactors for Cellulase Production*

As discussed previously, SmF is the technique in which excess water has been present and hence proven to be advantageous at an industrial scale for producing many useful enzymes. Studies on producing cellulases under SmF by utilizing microbes have been widely distributed and generally the organisms present are *Trichoderma reesei* and *Aspergillus niger* (Singhania et al., 2010).

Stirred-tank reactors (STRs) are the reactors commonly utilized for cellulase production. In these types of reactors the major disadvantage is controlling the shear stress, which often proves to be detrimental for the growth of filamentous fungi like *Trichoderma reesei* by destroying the filaments of the fungi. This filament damage and disturbed growth results in lower production of enzymes.

Other bioreactors, i.e., airlift reactors, are better versions of the SmF technique in the case of enhanced rate of oxygen transfer and lesser lethal effects of shear stresses on filamentous fungi (microbes). These types of reactors also minimize power consumption during the process of production because these lack impellers and other moving parts (Biswas et al., 2014).

The rotating fibrous-bed reactor (RFBB) is a type of reactor that is different from the STR when compared for cellulase production by *Trichoderma viride* (Lan et al., 2013). It was found that when *Trichoderma viride* was immobilized on solid support made up of polypropylene for enzyme production, several disadvantages associated with STR use were overcome. Disadvantages such as inhibition in growth of fungi by disintegration of mycelia, fouling of the pH and temperature probes, and clogged sampling vent could be eliminated by the use of RFBB. A 35.5% higher FPase activity was achieved under the operating conditions employed using immobilized mycelia in RFBB in comparison to STR (Biswas et al., 2014; Lan et al., 2013). It has also been observed that *Trichoderma viride* mycelia forms biofilm in RFBB during the process under low-shear conditions, which facilitates the production of enzyme by enhancing it.

7.3.2.2 *SmF Methodologies: Batch SmF, Fed-Batch SmF, and Continuous SmF*

Bioreactor operation mainly has three modes, namely batch, continuous, and fed batch. Batch fermentation reactors are those in which all of the required components are present in the culture medium inside a closed vessel before inoculation. Only buffers and gases (oxygen for aerobes and carbon dioxide or nitrogen for anaerobes) are supplied from outside separately. Another methodology under SmF is the continuous fermentation process in which nutrients and other components in culture medium have been maintained by the inlet and outlet present in the reactor at a constant rate of flow so that the volume inside the reactor remains constant. A fed-batch process lies in between both of the aforementioned processes. In this process, nutrients are supplied to the reactor through an inlet but there is no outlet present for removal of the spent medium during the process (dos Reis et al., 2013).

In the most of the studies of cellulase and xylanase production by fungi, batch fermentation is the most exploited technique. Moreover, the fed-batch process is also reported to enhance enzyme yields during the production process (Ximenes et al., 2007).

Out of the three processes, the fed-batch fermentation method has an extra advantage of utilizing the highly concentrated substrate by adding substrate and other medium components sequentially at intermittent times. This reduces the effects of catabolite repression and viscosity, which consequently will not hinder the production process. Although, fewer studies have taken place that utilized this process, as usual the substrate used for enzyme production is highly complex in nature (more often lignocellulosic biomass) (dos Reis et al., 2013). A higher FPase activity, cellulase productivity, and yield were reported using *Acremonium cellulolyticus* mutant CF-2612 employing batch culture with 5% Solka Floc in a 2L jar fermenter at 30°C, which was 18.0U/mL, 150.0 FPU/L/h, and 360.0 FPU/g carbohydrate, respectively, and when fed-batch culture was used with Solka Floc, these values reached 34.6U/mL, 240.3 FPU/L/h, and 346.0 FPU/g carbohydrate, respectively (Fang et al., 2009). Hydrophobin II is a low-foaming agent, which has been used for cellulase production under SmF by *Trichoderma reesei* strain Rut-C30 on lactose medium (Bailey and Tähtiharju, 2003). dos Reis et al. (2013) reported maximum activities of FPase (8.3U/mL), endoglucanases (37.3U/mL), and xylanases (177U/mL) using *Penicillium echinulatum* mutant strain S1 M29 under fed-batch cultivation with 40g/L of cellulose.

7.3.3 Comparison of SSF and SmF Technologies for Cellulase Production

To enhance the building of biorefinery amenities for supplying sustainable chemical products and energy, interest lies in optimizing the process of producing cellulose-digesting enzymes. The profiles of these enzymes are compatible with the use of submerged and solid-state culturing (Hansen et al., 2015). Commercially, SmF is the mostly utilized method for producing enzymes because it provides advantages such as a well-developed bioreactor equipped with techniques for monitoring and control; however, the SmF method has certain drawbacks, for example, the product yield achieved could be moderate, high cost, and generate large amounts of water as waste (Abd-Elhalem et al., 2015). In Table 7.4 both these strategies are compared.

TABLE 7.4 Advantages and Disadvantages of Submerged and Solid-State Fermentation

		Reason
SSF	Advantages	
	Higher enzyme production	Improved interaction between microorganism and substrate used.
	Requires low operating cost	Lignocellulosic waste used as crude enzyme preparations for degrading the substrate.
	Pragmatic	Waste generated during fermentation utilized in other processes
	Disadvantages	
	Problem faced during mixing	Process of mixing and sensitivity of the microbes to withstand mixing
	Complication of process monitoring and control	Difficulty in parametric control of temperature, pH, and nutrient gradients in the bioreactors
	Incapable of being employed in large-scale production	Complicated process of scale-up
SmF	Advantages	
	Easy process monitoring and control scaled up to industrial production capacity	Better parametric control, i.e., controlling temperature, pH, oxygen, and nutrient availability in the reactor
	Disadvantages	
	Generation of diluted products	Use of higher liquid proportion
	High operating cost	Rigorous monitoring and control of several fermentation parameters
	Not pragmatic	Production of significant amount of wastewater effluent

These drawbacks can be overcome using SSF culturing that provides economic advantages over SmF. SSF requires low-cost substrate such as lignocellulosic waste and provides a platform for better interaction of microorganisms and a substrate that results in higher enzyme concentration. SSF can be used as a substitute for SmF based on its economic analysis in nearly every feature, including equipment, raw material, and final cost of the product (Ashok et al., 2017). Cellulase production by various fungi under SSF and SmF technology is presented in Table 7.5.

TABLE 7.5 Cellulase Production Under Submerged and Solid-State Fermentation

Method	Strain	Substrate	Temp (°C)	Time (Days)	Enzyme Activity (U/g) BGL	CMCase	FPase	References
SmF	*Acremonium cellulolyticus* CF-2612	Wet disk-milled rice straw	30	7	12.5	197.3	5.8	Hideno et al. (2011)
SSF	*Aspergillus niger* (KK2)	Rice straw	28	4–6	100	129	19.5	Kang et al. (2004)
SSF	*Aspergillus niger* NS2	Wheat bran	30	4	33	310	17	Bansal et al. (2012)
SmF[a]	*Aspergillus terreus*	Oil palm fruit (empty) fiber	30	12	10.4	14.1	0.97	Shahriarinour et al. (2011)
SSF	*Fusarium chlamydosporum*HML 0278	Sugarcane bagasse plus wheat bran	30	4	135.2	281.8	95.2	Qin et al. (2010)
SmF[a]	*Penicillium brasilianum* IBT 20888	Solka-Floc cellulose	30	9.6	1.09	98	0.68	Krough et al. (2004)
SSF	*Penicillium chrysogenum* QML-2	Corn stover powder and wheat bran	30	5	321.6	370.15	101.8	Zhang and Sang (2012)
SmF[a]	*Trichoderma harzianum* IOC-4038	Partially delignified cellulignin	30	4	0.7	0.5	0.09	de Castro et al. (2010)
SSF	*Trichoderma reesei* Rut-C30	Wheat bran	30	5	38.1	103	5	Prévot et al. (2013)
SmF[a]	*Trichoderma viride* NCIM 1051	Sugarcane bagasse treated with $NaClO_2$	30	10	0.3	21.8	0.9	Adsul et al. (2004)

[a] *Enzyme activity expressed in U/mL.*
BGL, β-glucosidases; *CMCase*, carboxymethyl cellulase; *FPase*, filter paper activity; *SmF*, submerged fermentation; *SSF*, solid-state fermentation.

7.3.4 Process Scale-Up

Due to the structural difference in the systems, SSF methods could not be applied for scaling up the process, whereas for SmF it is well developed (Singhania et al., 2009). There are many challenges related to process scale-up for SSF such as the generation of heat and heterogeneity of the complete system. Aerobic microbial growth generates much less heat, which suddenly enhances the temperature inside the reactor, which is unwanted especially in the case of the production of heat-labile products such as enzymes, which can be denatured at high temperatures (Soccol et al., 2017). Moreover, loss of water by evaporation is also seen due to raised temperature, which has to be replenished and hence causes adverse local rises in water activity during the process (Behera and Ray, 2016; Soccol et al., 2017). There are several challenges associated with the SSF process such as transfer of biomass and its handling, washing of the pretreated solid portion for removing inhibitors and neutralizing the pH, high solid loading of biomass, and mismatch in optimum temperatures for enzymatic saccharification and fermentation. High solid loading of biomass resulted in higher ethanol yield as shown by many studies during fermentation; nevertheless, high solid loading leads to ineffective mass and heat transfer because of the viscosity of the medium components inside the reactor. Another process known as nonisothermal simultaneous solid-state saccharification, fermentation and separation overcomes the problems faced during traditional SSF processes during bioconversion of lignocellulosic biomass (Ali and Zulkali, 2011). Scale-up of a bioreactor or setting of operating conditions is essential for optimization of production. It can be evaluated either by designing a novel solid-state bioreactor for cellulase production, or by a scale-up approach using simultaneous saccharification and fermentation methodology for the conversion of waste paper sludge to ethanol in continuous stirred-tank reactor configuration (Behera and Ray, 2016).

7.4 SOME RECENT DEVELOPMENTS IN CELLULASE PRODUCTION TECHNOLOGY

In the recent past a number of newer developments in cellulase production technology have been reported, taking advantages of both SSF and SmF technologies, which are discussed briefly in this section.

7.4.1 Sequential SSF and SmF Strategy

In one of the reports (Cunha et al., 2012) sequential SSF and SmF of bagasse was carried out to produce cellulases, employing a fungal strain of *Aspergillus niger*. The reported strategy has the dual advantages of both methodologies, because cultivation was in part under solid state for an initial growth period and in part submerged for synthesis and secretion of cellulases. The results of this study are encouraging because the cellulase produced by the developed methodology was approximately threefold more and the same was also translated to a larger scale, i.e., bubble column fermenter of 5 L capacity. This sequential solid and submerged fermentation strategy seems a promising alternative for current cellulase production.

7.4.2 Use of Mixed Cultures

Besides adopting genetic engineering for improving cellulase production, the co-culture technique can be an alternative for enhancing production under SSF and SmF. For co-culturing two microbes, synchronization is required. Both organisms should have similar growing temperatures and nutritional requirements (carbon source). In the case of fungi, co-culturing is quite easy as compared to that of bacteria. Fungi normally coexist under SSF because it resembles their natural habitat where fungi grow symbiotically on solid substrates (Hölker et al., 2004), whereas developing bacterial co-cultures is problematic because each strain has different optimum conditions for growth and carbon source. Co-culturing fungi for cellulase has several advantages such as adaptability to the surroundings, higher yield, and substrate utilization compared to that of monocultures (Dashtban et al., 2009). Moreover, cellulases produced from co-culture are balanced in respect of producing a complete cellulase system in an appropriate amount. *Trichoderma reesei* and *Aspergillus niger* had been grown in a co-culture under SSF as well as SmF and the enzyme produced exploited at the commercial scale (Kolasa et al., 2014). In one of the reported co-culture approach, the lack of BGL in a *Trichoderma reesei*-derived cellulase system was overcome by co-culturing *Trichoderma reesei* with better BGL producing fungi, such as *Aspergillus niger* or *Aspergillus phoenicis* (Wen et al., 2005). Brijwani et al. (2010) co-cultured *Trichoderma reesei* (ATCC 26921) and *Aspergillus oryzae* (ATCC 12892), and optimized and scaled up the production of cellulase at tray bioreactor level. The cellulases produced under this co-culture were further utilized for hydrolyzing acid- and alkali-treated wheat straw. The results displayed enhanced release of monomeric sugar up to 48 h during hydrolysis of biomass.

7.4.3 One-Pot Cellulase Production, Hydrolysis, and Fermentation

Another important and recently reported innovation associated with SSF for cellulase production included the in-situ hydrolysis of the solid-state fermented biomass containing the produced cellulases (without extraction). Wheat straw or the microcrystalline cellulose was used for cellulase production under SSF by *Penicillium janthinellum* EMS-UV-8. Once the fermentation period was over, whole contents of the flask containing extracellular un-extracted cellulases, fungal mycelium/spores and the remaining substrate were mixed with the buffer in the same flask and maintained under conditions suitable for hydrolysis and later on the produced reducing sugars were also investigated for ethanol fermentation (Singhania et al., 2015). After hydrolyzed sugars were released, the same vessel was used for bioethanol fermentation using a thermotolerant yeast. Although similar ethanol resulted from modified and conventional processes (an SSF with extraction and separate hydrolysis and fermentation), the authors reported the advantages of reducing the enzyme extraction step and thereby reducing economic constraints as well as technical difficulties during large-scale SSF enzyme extraction.

7.5 INDUSTRIAL PRODUCTION OF CELLULASES

To meet the global demands of cellulases, many industries are currently involved in cellulase production for diverse applications, as mentioned earlier. These industries produce cellulases for different commercial applications such as textiles, paper and pulp, and other industries. Two globally renowned companies in the field of producing cellulases for biomass conversion are Genencor and Novozymes. Both of these are significantly good and facilitate the biomass conversion industries by cutting down the cost of the process several fold. These two companies have introduced novel technologies in the market through their dynamic approach toward research. Currently, Accelerase 1500 is a complete cellulase system launched by Genencor (Genencor, n.d.). The enzyme worked specifically on lignocellulosic biomass and proved to be cost effective and proficient for several bioethanol industries. The enzyme was produced by using genetically modified *Trichoderma reesei* having higher expression of BGL and hence higher levels of BGL activity, which is essential for complete conversion of biomass into glucose. Accelerase 1000 was the predecessor to Accelerase 1500 launched by Genencor, which was less efficient in complete conversion of cellobiose to glucose (Genencor, n.d.). Not only cellulases, but the xylanase enzyme Accelerase XY was also launched by Genencor. The blend of cellulase and xylanase from these commercial enzymes ensures the conversion of both C5 (xylan) and C6 (glucan) sugars and enhances the yield of products like bioethanol. Another accessory enzyme Accelerase XC functions similar to the aforementioned xylanase enzyme when blended with other Accelerase enzymes. Accelerase BG is a BGL supplement with other cellulase products, which facilitate biomass hydrolysis (Genencor, n.d.). Another industry, Novozymes, also has a varied range of commercial cellulases. For instance, in textile industries Cellusoft AP and Cellusoft CR have been exploited for bioblasting. Another enzyme called Carezyme is used as a detergent in laundries, and Denimax 601 has been used in stonewash industries (Novozymes, n.d.). Apart from these enzymes, Novozymes has also publicized a range of commercial cellulases that could specifically be exploited for hydrolyzing biomass, but no information is available in the market yet (Novozymes, n.d.). In India, MAP and in Japan Amano Enzyme Inc. are the industries involved in cellulase production. Though many industries are producing cellulases worldwide for many applications, the number is still less for developing cellulases specifically for lignocellulosic biomass conversion. Genencor and Novozymes are the two leading companies that are contributing to bioethanol-producing industries at their level. Table 7.6 shows the commercial cellulases available under various trademarks and their source of origin.

7.6 CONCLUSIONS AND FUTURE PROSPECTS

Though demand for cellulases is on the rise in all sectors, fast-growing biofuel and biorefinery industries are now emerging as the biggest consumers of this particular enzyme. To meet these demands, it is necessary to match cellulase production on a much larger scale, which requires the development and use of more cost-effective, sustainable, and improved production technologies. Moreover, improvements in the quality of the produced cellulases are also needed for better development and growth of lignocellulosic biofuels and product formation. Such developments include improved technologies for production; application of less costly and better feedstocks; applications of improved microbial strains achieved via genetic engineering and recombinant DNA technology; as well as artificially designed cellulases and improved cellulose degrading enzyme mixtures.

Development of cost-effective large-scale commercial lignocellulosic biofuel technology requires constant research and development activities and demands high cellulase availability. Despite many significant developments made

TABLE 7.6 Commercially Available Cellulases, Their Source, and Suppliers

Organism	Supplier	Name of Enzyme
Trichoderma longibrachiatum	Rhodia-Danisco (France)	Cellulase 2000 L
	Rohm-AB Enzymes (Finland)	Rohament CL
	Dyadic (United States)	Viscostar 150 L
	Novozymes	Viscozyme L
Trichoderma reesei	Genencor International	Multifact CL
	Genencor	Accelerase 1500
	Iogen (Canada)	Ultralow microbial
	Alko-EDC (New York)	Econase CE
	Quest International	Cellulase TRI
Trichoderma viride	Amano Enzyme (Troy, VA)	Cellulase TAP106
Trichoderma longibrachiatum and Aspergillus niger	Novozymes, Denmark	Cellubrix (Celluclast)
Aspergillus niger	Novozymes	Novozymes 188
	Amano Enzyme	Cellulase AP30K
	Quest International	Biocellulase A

over the past few years, and corresponding reductions achieved in enzyme costs, still only a few enzyme industries dominate the cellulase market and therefore cellulase has limited accessibility for research and development of biofuels. These challenges prompt researchers to further explore the cellulase production technologies, which are mainly categorized into solid-state and submerged-state fermentation technologies. The main features of the solid-state production process are achievement of better cellulase titers, low cost per unit enzyme, and many others in comparison to more technical processes with a lower production cost compared to SmF, which is currently more favorable for large-scale industrial production due to better monitoring and control, and easier product recovery. Though SSF seems a more promising technology, it is still not able to fully replace SmF for the commercial production of cellulase. Thus much more effort is still needed in solid-state cellulase production technologies to make it a better choice at the industrial level. Several bacteria and actinomycetes have been reported for microbial cellulase production. Currently, industrial cellulase production mainly employs fungal strains due to their abilities to secrete complete cellulase systems, which is desired for effective hydrolysis of lignocellulosic substrate. However, these fungal strains need to be significantly improved genetically or metabolically for optimized secretion of cellulases. Moreover, mixed/co-cultures of fungal strains could be employed for enhanced cellulase production. Apart from these strategies, a sequential SSF/SmF strategy as well as one-pot cellulase production, hydrolysis, and fermentation could be possible innovations that can be time, cost, and labor saving, especially when compared to conventional SSF alone for cellulase production.

Finally, the question "whether SSF or SmF for cellulase production is better" is still open. Of course, there are key criteria such as suitable microorganisms, biomass feedstock/substrate, and design of fermenter that must be carefully considered when comparing both solid-state and submerged cultivations for sustainable and cost-effective cellulase production and not merely enzyme units or titers alone.

Acknowledgment

The authors highly acknowledge financial and infrastructural support from the Central University of Haryana, Mahendergarh, Haryana, India. The authors would also like to thank the Department of Science and Technology (SERB-DST), Government of India, for providing the financial support (SERB/LS/2016/00929). Author Hemansi acknowledges UGC-CSIR for her JRF.

References

Abd-Elhalem, B.T., El-Sawy, M., Gamal, R.F., Abou-Taleb, K.A., 2015. Production of amylases from *Bacillus amyloliquefaciens* under submerged fermentation using some agro-industrial by-products. Ann. Agric. Sci. 60, 193–202. https://doi.org/10.1016/j.aoas.2015.06.001.

Abdeshahian, P., Samat, N., Hamid, A.A., Yusoff, W.M.W., 2011. Solid substrate fermentation for cellulase production using palm kernel cake as a renewable lignocellulosic source in packed-bed bioreactor. Biotechnol. Bioprocess Eng. 16, 238–244. https://doi.org/10.1007/s12257-010-0320-8.

Adsul, M., Ghule, J., Singh, R., Shaikh, H., Bastawde, K., Gokhale, D., Varma, A., 2004. Polysaccharides from bagasse: applications in cellulase and xylanase production. Carbohydr. Polym. 57, 67–72. https://doi.org/10.1016/j.carbpol.2004.04.001.

Alam, M.Z., Mamun, A.A., Qudsieh, I.Y., Muyibi, S.A., Salleh, H.M., Omar, N.M., 2009. Solid state bioconversion of oil palm empty fruit bunches for cellulase enzyme production using a rotary drum bioreactor. Biochem. Eng. J. 46, 61–64. https://doi.org/10.1016/j.bej.2009.03.010.

Ali, H.K.Q., Zulkali, M.M.D., 2011. Design aspects of bioreactors for solid-state fermentation: a review. Chem. Biochem. Eng. Q. 25, 255–266.

Arora, S., Rani, R., Ghosh, S., 2018. Bioreactors in solid state fermentation technology: design, applications and engineering aspects. J. Biotechnol. 269, 16–34. https://doi.org/10.1016/j.jbiotec.2018.01.010.

Ashok, A., Doriya, K., Rao, D.R.M., Kumar, D.S., 2017. Design of solid state bioreactor for industrial applications: an overview to conventional bioreactors. Biocatal. Agric. Biotechnol. 9, 11–18. https://doi.org/10.1016/j.bcab.2016.10.014.

Badhan, A.K., Chadha, B.S., Kaur, J., Saini, H.S., Bhat, M.K., 2007. Production of multiple xylanolytic and cellulolytic enzymes by thermophilic fungus Myceliophthora sp. IMI 387099. Bioresour. Technol. 98, 504–510. https://doi.org/10.1016/j.biortech.2006.02.009.

Bailey, M.J., Tähtiharju, J., 2003. Efficient cellulase production by *Trichoderma reesei* in continuous cultivation on lactose medium with a computer-controlled feeding strategy. Appl. Microbiol. Biotechnol. 62, 156–162. https://doi.org/10.1007/s00253-003-1276-9.

Bansal, N., Tewari, R., Soni, R., Soni, S.K., 2012. Production of cellulases from *Aspergillus niger* NS-2 in solid state fermentation on agricultural and kitchen waste residues. Waste Manag. 32, 1341–1346. https://doi.org/10.1016/j.wasman.2012.03.006.

Behera, S.S., Ray, R.C., 2016. Solid state fermentation for production of microbial cellulases: recent advances and improvement strategies. Int. J. Biol. Macromol. 86, 656–669. https://doi.org/10.1016/j.ijbiomac.2015.10.090.

Behera, B.C., Sethi, B.K., Mishra, R.R., Dutta, S.K., Thatoi, H.N., 2017. Microbial cellulases—diversity & biotechnology with reference to mangrove environment: a review. J. Genet. Eng. Biotechnol. 15, 197–210. https://doi.org/10.1016/j.jgeb.2016.12.001.

Béra-Maillet, C., Mosoni, P., Kwasiborski, A., Suau, F., Ribot, Y., Forano, E., 2009. Development of a RT-qPCR method for the quantification of *Fibrobacter succinogenes* S85 glycoside hydrolase transcripts in the rumen content of gnotobiotic and conventional sheep. J. Microbiol. Methods 77, 8–16. https://doi.org/10.1016/j.mimet.2008.11.009.

Biswas, R., Persad, A., Bisaria, V.S., 2014. Production of cellulolytic enzymes. Bioprocess. Renew. Resour. Commod. Bioprod. 1, 105–132.

Brijwani, K., Oberoi, H.S., Vadlani, P.V., 2010. Production of a cellulolytic enzyme system in mixed-culture solid-state fermentation of soybean hulls supplemented with wheat bran. Process Biochem. 45, 120–128. https://doi.org/10.1016/j.procbio.2009.08.015.

Castro, A.M., Castilho, L.R., Freire, D.M.G., 2015. Performance of a fixed-bed solid-state fermentation bioreactor with forced aeration for the production of hydrolases by *Aspergillus awamori*. Biochem. Eng. J. 93, 303–308. https://doi.org/10.1016/j.bej.2014.10.016.

Chambost, J.P., Boyer, M.H., Cami, B., Barras, E., Cattaneo, J., 1987. Erwinia cellulases. In: Plant Pathogenic Bacteria, Current Plant Science and Biotechnology in Agriculture. Springer, Dordrecht, pp. 150–159.

Chen, H., 2013. Principles and application of solid-state fermentation carried out on inert support materials (adsorbed carrier solid-state fermentation). In: Modern Solid State Fermentation. Springer, Dordrecht, pp. 243–305.

Chinn, M.S., Nokes, S.E., Strobel, H.J., 2008. Influence of moisture content and cultivation duration on *Clostridium thermocellum* 27405 end-product formation in solid substrate cultivation on Avicel. Bioresour. Technol. 99, 2664–2671. https://doi.org/10.1016/j.biortech.2007.04.052.

Cunha, F.M., Esperança, M.N., Zangirolami, T.C., Badino, A.C., Farinas, C.S., 2012. Sequential solid-state and submerged cultivation of *Aspergillus niger* on sugarcane bagasse for the production of cellulase. Bioresour. Technol. 112, 270–274. https://doi.org/10.1016/j.biortech.2012.02.082.

Das, A., Paul, T., Halder, S.K., Jana, A., Maity, C., Das Mohapatra, P.K., Pati, B.R., Mondal, K.C., 2013. Production of cellulolytic enzymes by *Aspergillus fumigatus* ABK9 in wheat bran-rice straw mixed substrate and use of cocktail enzymes for deinking of waste office paper pulp. Bioresour. Technol. 128, 290–296. https://doi.org/10.1016/j.biortech.2012.10.080.

Dashtban, M., Schraft, H., Qin, W., 2009. Fungal bioconversion of lignocellulosic residues; opportunities & perspectives. Int. J. Biol. Sci. 5, 578–595.

de Castro, A.M., Pedro, K.C.N.R., da Cruz, J.C., Ferreira, M.C., Leite, S.G.F., Pereira, N., 2010. *Trichoderma harzianum* IOC-4038: a promising strain for the production of a cellulolytic complex with significant β-glucosidase activity from sugarcane bagasse cellulignin. Appl. Biochem. Biotechnol. 162, 2111–2122. https://doi.org/10.1007/s12010-010-8986-0.

de Lima, G., Luiz, A., Lima, D., Grigorevski, A.L., Nascimento, D., Pires, R., da Bon, S., Pinto, E., Coelho, R.R.R., 2005. *Streptomyces drozdowiczii* cellulase production using agro-industrial by-products and its potential use in the detergent and textile industries. Enzym. Microb. Technol. 37, 272–277.

Deswal, D., Khasa, Y.P., Kuhad, R.C., 2011. Optimization of cellulase production by a brown rot fungus *Fomitopsis* sp. RCK2010 under solid state fermentation. Bioresour. Technol. 102, 6065–6072. https://doi.org/10.1016/j.biortech.2011.03.032.

Dhillon, G.S., Oberoi, H.S., Kaur, S., Bansal, S., Brar, S.K., 2011. Value-addition of agricultural wastes for augmented cellulase and xylanase production through solid-state tray fermentation employing mixed-culture of fungi. Ind. Crop. Prod. 34, 1160–1167. https://doi.org/10.1016/j.indcrop.2011.04.001.

Doi, R.H., Kosugi, A., 2004. Cellulosomes: plant-cell-wall-degrading enzyme complexes. Nat. Rev. Microbiol. 2, 541–551. https://doi.org/10.1038/nrmicro925.

dos Reis, L., Fontana, R.C., da Silva Delabona, P., da Silva Lima, D.J., Camassola, M., da Cruz Pradella, J.G., Dillon, A.J.P., 2013. Increased production of cellulases and xylanases by *Penicillium echinulatum* S1M29 in batch and fed-batch culture. Bioresour. Technol. 146, 597–603. https://doi.org/10.1016/j.biortech.2013.07.124.

Duff, S.J.B., Murray, W.D., 1996. Bioconversion of forest products industry waste cellulosics to fuel ethanol: a review. Bioresour. Technol. 55, 1–33. https://doi.org/10.1016/0960-8524(95)00122-0.

Eckert, K., Schneider, E., 2003. A thermoacidophilic endoglucanase (CelB) from *Alicyclobacillus acidocaldarius* displays high sequence similarity to arabinofuranosidases belonging to family 51 of glycoside hydrolases. Eur. J. Biochem. 270, 3593–3602.

Eida, M.F., Nagaoka, T., Wasaki, J., Kouno, K., 2012. Isolation and characterization of cellulose-decomposing bacteria inhabiting sawdust and coffee residue composts. Microbes Environ. 27, 226–233. https://doi.org/10.1264/jsme2.ME11299.

Ekperigin, M.M., 2007. Preliminary studies of cellulase production by *Acinetobacter anitratus* and *Branhamella* sp. Afr. J. Biotechnol. 6, 28–33.

Fan, L.T., Gharpuray, M.M., Lee, Y.H., 1987. Cellulose Hydrolysis. Springer.

Fang, H., Xia, L., 2015. Cellulase production by recombinant *Trichoderma reesei* and its application in enzymatic hydrolysis of agricultural residues. Fuel 143, 211–216. https://doi.org/10.1016/j.fuel.2014.11.056.

Fang, X., Yano, S., Inoue, H., Sawayama, S., 2009. Strain improvement of *Acremonium cellulolyticus* for cellulase production by mutation. J. Biosci. Bioeng. 107, 256–261. https://doi.org/10.1016/j.jbiosc.2008.11.022.

Farinas, C.S., Vitcosque, G.L., Fonseca, R.F., Neto, V.B., Couri, S., 2011. Modeling the effects of solid state fermentation operating conditions on endoglucanase production using an instrumented bioreactor. Ind. Crop. Prod. 34, 1186–1192. https://doi.org/10.1016/j.indcrop.2011.04.006.

Genencor, http://www.genencor.com/wps/wcm/connect/genencor/genencor/productsandservices/business development/biorefineries/products/accelleraseproductlineen.htm.

Giuliano, M., Schiraldi, C., Marotta, M.R., Hugenholtz, J., De Rosa, M., 2004. Expression of *Sulfolobus solfataricus* α-glucosidase in *Lactococcus lactis*. Appl. Microbiol. Biotechnol. 64, 829–832. https://doi.org/10.1007/s00253-003-1493-2.

Gomes, J., Steiner, W., 2004. The biocatalytic potential of extremophiles and extremozymes. Food Technol. Biotechnol. 42, 223–235.

Gunny, A.A.N., Arbain, D., Jamal, P., Gumba, R.E., 2015. Improvement of halophilic cellulase production from locally isolated fungal strain. Saudi J. Biol. Sci. 22, 476–483. https://doi.org/10.1016/j.sjbs.2014.11.021.

Hansen, G.H., Lübeck, M., Frisvad, J., Lübeck, P.S., Andersen, B., 2015. Production of cellulolytic enzymes from ascomycetes: comparison of solid state and submerged fermentation. Process Biochem. 50, 1327–1341. https://doi.org/10.1016/j.procbio.2015.05.017.

Hemansi, Gupta, R., Kuhad, R.C., Saini, J.K., 2018. Cost effective production of complete cellulase system by newly isolated *Aspergillus niger* RCKH-3 for efficient enzymatic saccharification: medium engineering by overall evaluation criteria approach (OEC). Biochem. Eng. J. 132, 182–190. https://doi.org/10.1016/j.bej.2018.01.019.

Hideno, A., Inoue, H., Tsukahara, K., Yano, S., Fang, X., Endo, T., Sawayama, S., 2011. Production and characterization of cellulases and hemicellulases by *Acremonium cellulolyticus* using rice straw subjected to various pretreatments as the carbon source. Enzym. Microb. Technol. 48, 162–168. https://doi.org/10.1016/j.enzmictec.2010.10.005.

Hodrová, B., Kopečný, J., Káš, J., 1998. Cellulolytic enzymes of rumen anaerobic fungi *Orpinomyces joyonii* and *Caecomyces communis*. Res. Microbiol. 149, 417–427. https://doi.org/10.1016/S0923-2508(98)80324-0.

Hölker, U., Höfer, M., Lenz, J., 2004. Biotechnological advantages of laboratory-scale solid-state fermentation with fungi. Appl. Microbiol. Biotechnol. 64, 175–186. https://doi.org/10.1007/s00253-003-1504-3.

Ibrahim, A.S.S., El-diwany, A.I., 2007. Isolation and identification of new cellulases producing thermophilic bacteria from an Egyptian hot spring and some properties of the crude enzyme. Aust. J. Basic Appl. Sci. 1, 473–478.

Jain, K.K., Dey, T.B., Kumar, S., Kuhad, R.C., 2015. Production of thermostable hydrolases (cellulases and xylanase) from *Thermoascus aurantiacus* RCKK: a potential fungus. Bioprocess Biosyst. Eng. 38, 787–796.

Jeya, M., Nguyen, N.-P.-T., Moon, H.-J., Kim, S.-H., Lee, J.-K., 2010. Conversion of woody biomass into fermentable sugars by cellulase from *Agaricus arvensis*. Bioresour. Technol. 101, 8742–8749. https://doi.org/10.1016/j.biortech.2010.06.055.

Jung, U., Young Yoo, H., Bong Kim, S., Wook Kim, S., 2015. Optimization of medium composition for enhanced cellulase production by mutant *Penicillium brasilianum* KUEB15 using statistical method. J. Ind. Eng. Chem. 25, 145–150.

Juturu, V., Wu, J.C., 2014. Microbial cellulases: engineering, production and applications. Renew. Sust. Energ. Rev. 33, 188–203. https://doi.org/10.1016/j.rser.2014.01.077.

Kang, S.W., Park, Y.S., Lee, J.S., Hong, S.I., Kim, S.W., 2004. Production of cellulases and hemicellulases by *Aspergillus niger* KK2 from lignocellulosic biomass. Bioresour. Technol. 91, 153–156. https://doi.org/10.1016/S0960-8524(03)00172-X.

Kaur, J., Chadha, B.S., Saini, H.S., 2006. Regulation of cellulase production in two thermophilic fungi *Melanocarpus* sp. MTCC 3922 and *Scytalidium thermophilum* MTCC 4520. Enzym. Microb. Technol. 38, 931–936. https://doi.org/10.1016/j.enzmictec.2005.08.036.

Kim, S., Kim, C.H., 2012. Production of cellulase enzymes during the solid-state fermentation of empty palm fruit bunch fiber. Bioprocess Biosyst. Eng. 35, 61–67. https://doi.org/10.1007/s00449-011-0595-y.

Kolasa, M., Ahring, B.K., Lübeck, P.S., Lübeck, M., 2014. Co-cultivation of *Trichoderma reesei* RutC30 with three black *Aspergillus strains* facilitates efficient hydrolysis of pretreated wheat straw and shows promises for on-site enzyme production. Bioresour. Technol. 169, 143–148. https://doi.org/10.1016/j.biortech.2014.06.082.

Krishna, C., 2005. Solid-state fermentation systems—an overview. Crit. Rev. Biotechnol. 25, 1–30. https://doi.org/10.1080/07388550590925383.

Krough, K.B.R., Morkeberg, A., Jorgensen, H., Frisvad, J.C., Olsson, L., 2004. Screening genus *Penicillium* for producers of cellulolytic and xylanolytic enzymes. Appl. Biochem. Biotechnol. 113–116, 389–401.

Kuhad, R.C., Kumar, M., Singh, A., 1994. A hypercellulolytic mutant of *Fusarium oxysporum*. Lett. Appl. Microbiol. 19, 397–400. https://doi.org/10.1111/j.1472-765X.1994.

Kuhad, R.C., Gupta, R., Singh, A., 2011. Microbial cellulases and their industrial applications. Enzyme Res. 2011, 1–10. https://doi.org/10.4061/2011/280696.

Kuhad, R.C., Deswal, D., Sharma, S., Bhattacharya, A., Jain, K.K., Kaur, A., Pletschke, B.I., Singh, A., Karp, M., 2016. Revisiting cellulase production and redefining current strategies based on major challenges. Renew. Sust. Energ. Rev. 55, 249–272. https://doi.org/10.1016/j.rser.2015.10.132.

Lan, T.-Q., Wei, D., Yang, S.-T., Liu, X., 2013. Enhanced cellulase production by *Trichoderma viride* in a rotating fibrous bed bioreactor. Bioresour. Technol. 133, 175–182. https://doi.org/10.1016/j.biortech.2013.01.088.

Leatherwood, J.M., 1965. Cellulase from *Ruminococcus albus* and mixed rumen microorganisms. Appl. Microbiol. 13, 771–775.

Liang, Y., Feng, Z., Yesuf, J., Blackburn, J.W., 2010. Optimization of growth medium and enzyme assay conditions for crude cellulases produced by a novel thermophilic and cellulolytic bacterium, *Anoxybacillus* sp. 527. Appl. Biochem. Biotechnol. 160, 1841–1852. https://doi.org/10.1007/s12010-009-8677-x.

Maeda, R.N., Barcelos, C.A., Anna, L.M.M.S., Pereira, N., 2013. Cellulase production by *Penicillium funiculosum* and its application in the hydrolysis of sugar cane bagasse for second generation ethanol production by fed batch operation. J. Biotechnol. 163, 38–44. https://doi.org/10.1016/j.jbiotec.2012.10.014.

Mello-de-Sousa, T.M., Silva-Pereira, I., Poças-Fonseca, M.J., 2011. Carbon source and pH-dependent transcriptional regulation of cellulase genes of *Humicola grisea* var. thermoidea grown on sugarcane bagasse. Enzym. Microb. Technol. 48, 19–26. https://doi.org/10.1016/j.enzmictec.2010.08.007.

Moon, M., Anderson, K.L., 2001. *Eubacterium cellulosolvens* alters its membrane protein, lipoprotein, and fatty acid composition in response to growth on cellulose. Anaerobe 7 (4), 227–236. https://doi.org/10.1006/anae.2001.0389.

Moshi, A.P., Crespo, C.F., Badshah, M., Hosea, K.M.M., Mshandete, A.M., Mattiasson, B., 2014. High bioethanol titre from *Manihot glaziovii* through fed-batch simultaneous saccharification and fermentation in Automatic Gas Potential Test System. Bioresour. Technol. 156, 348–356. https://doi.org/10.1016/j.biortech.2013.12.082.

Niranjane, A., 2006. Screening diverse cellulase enzymes from the white rot fungus *Phlebia gigantea* for high activity and large scale applications (Doctor of Philosophy, Ph.D., Applied Sciences). RMIT University.

Novozymes, http://www.bioenergy.novozymes.com/cellulosic-biofuel/01/14/2010.

Ooijkaas, L.P., Weber, F.J., Buitelaar, R.M., Tramper, J., Rinzema, A., 2000. Defined media and inert supports: their potential as solid-state fermentation production systems. Trends Biotechnol. 18, 356–360.

Prévot, V., Lopez, M., Copinet, E., Duchiron, F., 2013. Comparative performance of commercial and laboratory enzymatic complexes from submerged or solid-state fermentation in lignocellulosic biomass hydrolysis. Bioresour. Technol. 129, 690–693. https://doi.org/10.1016/j.biortech.2012.11.135.

Poulsen, H.V., Willink, F.W., Ingvorsen, K., 2016. Aerobic and anaerobic cellulase production by *Cellulomonas uda*. Arch. Microbiol. 198, 725–735. https://doi.org/10.1007/s00203-016-1230-8.

Qin, Y., He, H., Li, N., Ling, M., Liang, Z., 2010. Isolation and characterization of a thermostable cellulase-producing *Fusarium chlamydosporum*. World J. Microbiol. Biotechnol. 26, 1991–1997. https://doi.org/10.1007/s11274-010-0383-x.

Raghuwanshi, S., Deswal, D., Karp, M., Kuhad, R.C., 2014. Bioprocessing of enhanced cellulase production from a mutant of *Trichoderma asperellum* RCK2011 and its application in hydrolysis of cellulose. Fuel 124, 183–189. https://doi.org/10.1016/j.fuel.2014.01.107.

Raimbault, M., 1998. General and microbiological aspects of solid substrate fermentation. Electron. J. Biotechnol 1 (3), 174–188.

Rastogi, G., Bhalla, A., Adhikari, A., Bischoff, K.M., Hughes, S.R., Christopher, L.P., Sani, R.K., 2010. Characterization of thermostable cellulases produced by *Bacillus* and *Geobacillus strains*. Bioresour. Technol. 101, 8798–8806. https://doi.org/10.1016/j.biortech.2010.06.001.

Robert, C., Chassard, C., Lawson, P.A., Bernalier-Donadille, A., 2007. *Bacteroides cellulosilyticus* sp. nov., a cellulolytic bacterium from the human gut microbial community. Int. J. Syst. Evol. Microbiol. 57, 1516–1520. https://doi.org/10.1099/ijs.0.64998-0.

Romero, M.D., Aguado, J., González, L., Ladero, M., 1999. Cellulase production by *Neurospora crassa* on wheat straw. Enzym. Microb. Technol. 25, 244–250. https://doi.org/10.1016/S0141-0229(99)00035-6.

Saddler, J.N., Khan, A.W., 1981. Cellulolytic enzyme system of *Acetivibrio cellulolyticus*. Can. J. Microbiol. 27, 288–294.

Sadhu, S., Maiti, T.K., 2013. Cellulase production by bacteria: a review. Br. Microbiol. Res. J. 3 (3), 235–258.

Saini, J.K., Saini, R., Tewari, L., 2015a. Lignocellulosic agriculture wastes as biomass feedstocks for second-generation bioethanol production: concepts and recent developments. 3 Biotech 5, 337–353.

Saini, R., Saini, J.K., Adsul, M., Patel, A.K., Mathur, A., Tuli, D., Singhania, R.R., 2015b. Enhanced cellulase production by *Penicillium oxalicum* for bio-ethanol application. Bioresour. Technol. 188, 240–246. https://doi.org/10.1016/j.biortech.2015.01.048.

Shahriarinour, M., Noor Abdul Wahab, M., Mohamad, R., Shuhaimi, M., Ariff, A., 2011. Cyanobacterial biomass as n-supplement to oil palm empty fruit bunch (OPEFB) fibre for improvement of cellulase production by *Aspergillus terreus* in submerged fermentation. Bioresources 6, 1696–1706. https://doi.org/10.15376/biores.6.2.1696-1706.

Singhania, R.R., Patel, A.K., Soccol, C.R., Pandey, A., 2009. Recent advances in solid-state fermentation. Biochem. Eng. J. 44, 13–18. Invited Review Issue https://doi.org/10.1016/j.bej.2008.10.019.

Singhania, R.R., Sukumaran, R.K., Patel, A.K., Larroche, C., Pandey, A., 2010. Advancement and comparative profiles in the production technologies using solid-state and submerged fermentation for microbial cellulases. Enzym. Microb. Technol. 46, 541–549. https://doi.org/10.1016/j.enzmictec.2010.03.010.

Singhania, R.R., Saini, R., Adsul, M., Saini, J.K., Mathur, A., Tuli, D., 2015. An integrative process for bio-ethanol production employing SSF produced cellulase without extraction. Biochem. Eng. J. 102, 45–48. https://doi.org/10.1016/j.bej.2015.01.002.

Soccol, C.R., da Costa, E.S.F., Letti, L.A.J., Karp, S.G., Woiciechowski, A.L., de Vandenberghe, L.P.S., 2017. Recent developments and innovations in solid state fermentation. Biotechnol. Res. Innov. 1, 52–71. https://doi.org/10.1016/j.biori.2017.01.002.

Sreena, C.P., Sebastian, D., 2018. Augmented cellulase production by *Bacillus subtilis* strain MU S1 using different statistical experimental designs. J. Genet. Eng. Biotechnol. 16, 9–16. https://doi.org/10.1016/j.jgeb.2017.12.005.

Srinivasan, K., Murakami, M., Nakashimada, Y., Nishio, N., 2001. Efficient production of cellulolytic and xylanolytic enzymes by the rumen anaerobic fungus, *Neocallimastix frontalis*, in a repeated batch culture. J. Biosci. Bioeng. 91, 153–158.

Sternberg, D., 1976. Production of cellulase by Trichoderma. Biotechnol. Bioeng. Symp. 6, 35–53.

Thomas, L., Larroche, C., Pandey, A., 2013. Current developments in solid-state fermentation. Biochem. Eng. J. 81, 146–161. https://doi.org/10.1016/j.bej.2013.10.013.

Trovati, J., Giordano, R.C., Giordano, R.L.C., 2009. Improving the performance of a continuous process for the production of ethanol from starch. Appl. Biochem. Biotechnol. 156, 76–90. https://doi.org/10.1007/s12010-009-8562-7.

Vitcosque, G.L., Fonseca, R.F., Rodríguez-Zúñiga, U.F., Bertucci Neto, V., Couri, S., Farinas, C.S., 2012. Production of biomass-degrading multienzyme complexes under solid-state fermentation of soybean meal using a bioreactor. Enzyme Res. 2012, https://doi.org/10.1155/2012/248983.

Wen, Z., Liao, W., Chen, S., 2005. Production of cellulase by *Trichoderma reesei* from dairy manure. Bioresour. Technol. 96, 491–499. https://doi.org/10.1016/j.biortech.2004.05.021.

Wilson, D.B., 1988. Cellulases of *Thermomonospora fusca*. In: Methods in Enzymology, Biomass Part A: Cellulose and Hemicellulose. Academic Press, pp. 314–323.

Ximenes, E.A., Dien, B.S., Ladisch, M.R., Mosier, N., Cotta, M.A., Li, X.-L., 2007. Enzyme production by industrially relevant fungi cultured on coproduct from corn dry grind ethanol plants. Appl. Biochem. Biotechnol. 137–140, 171–183. https://doi.org/10.1007/s12010-007-9049-z.

Yoon, L.W., Ang, T.N., Ngoh, G.C., Chua, A.S.M., 2014. Fungal solid-state fermentation and various methods of enhancement in cellulase production. Biomass Bioenergy 67, 319–338. https://doi.org/10.1016/j.biombioe.2014.05.013.

Zhang, H., Sang, Q., 2012. Statistical optimization of cellulases production by *Penicillium chrysogenum* QML-2 under solid-state fermentation and primary application to chitosan hydrolysis. World J. Microbiol. Biotechnol. 28, 1163–1174. https://doi.org/10.1007/s11274-011-0919-8.

C H A P T E R

8

Advancements in Bioprocess Technology for Cellulase Production

*Roohi**, Kulsoom Bano**, Sadaf Parveen**, Farha Khan**, Mohammed Rehan Zaheer*[†], Mohammed Kuddus*[‡]

*Protein Research Laboratory, Department of Bioengineering, Integral University, Lucknow, India [†]Department of Chemistry, Gagan College of Management and Technology, Aligarh, India [‡]Department of Biochemistry, University of Hail, Hail, Saudi Arabia

8.1 INTRODUCTION

Approximately 70% of plant carbon is fixed into 5–6 carbon sugars and about 200 Gt of CO_2 are converted into organic compounds every year on Earth. Also, almost the same amount of carbon materials has been degraded, 70% of which is accomplished by microorganisms and the rest by plants and animals. Fifty percent of plant dry weight is typically cellulosic material. This plant-based material is a major possible never-ending reservoir of fuels and resources available to humankind. Agricultural leftover materials are the immense reservoir of renewable and economical cellulosic biomass. These renewable resources account for the leaves, stems, and stalks of various plants, and also include corn coir, bagasse, cereal straw, trees, and forest residues (Bhat and Bhat, 1997). Large amounts of agrowastes are also available from industries in the form of citrus peel extract, coconut husk and shells, wood dust, paper mush, municipal cellulosic solid waste, paper mill sludge, etc. Moreover, energy-producing crops such as perennial grasses, switch grasses, and other herbage feedstuffs such as *Miscanthus*, elephant grass, and Bermuda grass are continuously being used for biofuel production. Cellulose comprises major parts of plant cell walls for protection, thereby opposing biodegradation naturally. Cellulase enzymes hydrolyze cellulose, which is a linear polysaccharide molecule having β(1-4)-linked D-glucopyranosyl units. Various types of cellulase have been identified, such as two exo-β-glucanases or cellobiohydrolases (CBHs; EC 3.2.1.91) (CBH I and CBH II), four endo-β-glucanases (EGs; EC 3.2.1.4) (EG I, EG II, EG III, and EG V), and one β-glucosidase (EC 3.2.1.21) (Bhat, 2000).

Cellulase enzymes can be isolated from plants, animals, and microbes but microbial enzymes are preferred since huge quantities of enzyme can be produced cheaply in a single step using a fermenter. Nutrient factors and growth conditions such as pH, temperature, incubation period, carbon-nitrogen sources, metal ions, inhibitors, agitation requirement, etc., which are required for the production of cellulase enzymes, show a cumulative effect (Gusakov, 2011). To achieve the maximum production of enzymes, optimization is the most important step in a successful fermentation process. A detailed research study therefore requires microbial growth conditions for enhanced enzyme production. Several literature reports suggest that the cost of cellulase enzymes depends on the productivity of enzyme-producing microbial strains. Enzymatic degradation of cellulose into glucose can be used for the production of other metabolites like ethanol, organic acids, and maltose syrups (Bhat, 2000). Those processes, which involve selective utilization of cellulose waste into various metabolites and other uses like biofuel production, will definitely reduce the shortage of food and animal feeds; a bold step toward solid waste management and to make our environment clean and green.

From Cellulose to Cellulase: Strategies to Improve Biofuel Production
https://doi.org/10.1016/B978-0-444-64223-3.00008-4

TABLE 8.1 Classification of Cellulase Enzymes

Name of enzyme	EC number	Substrate	Site of cleavage	Product
Endo-1,4-β-D-glucan glucanohydrolase	EC 3.2.1.4	Cut at random at internal amorphous sites of cellulose, lichenin, and cereal β-D-glucans	Endo-1,4-β-D-glucosidic linkages	Oligosaccharides of various lengths
Exoglucanase or 1,4-β-D-glucan cellobiohydrolases (cellobiohydrolases)	EC 3.2.1.91	Nonreducing ends of cellulose and cellotetraose	Exo-1,4-β-D-glucosidic linkages	Cellobiose
Exoglucanases or 1,4-β-D-oligoglucancellobiohydrolases	EC 3.2.1.74	Cellooligosaccharide, *p*-nitrophenyl-β-D-cellobioside	Exo-1,4-β-D-glucosidic linkages	Cellobiose
β-Glucosidases or β-D-glucoside glucohydrolases	EC 3.2.1.21	Terminal nonreducing ends of cellulose	Exo-1,4-β-D-glucosidic linkages	β-D-Glucose
Cellobiose: orthophosphate α-D-glucosyl transferase	EC 2.4.1.49	Cellobiose	Phosphorolytic cleavage	β-D-Glucose
1,4-β-D-oligoglucan: orthophosphate α-D-glucosyl transferase	EC 2.4.1.20	Cellodextrins ranging from cellotriose to cellohexoses	Phosphorolytic cleavage	β-D-Glucose
Cellobiose 2-epimerase	EC 5.1.3.11	Cellobiose	Exo-1,4-β-D-glucosidic linkages	4-*O*-β-D-Glucosylmannose

8.2 CLASSIFICATION OF CELLULASE ENZYMES

Cellulase enzymes are categorized into several classes where differences exist between the enzymes for their substrates, products, and the bond that they hydrolyze, and thereby receive a different enzyme number by the Enzyme Commission. These are listed in Table 8.1.

8.3 MECHANISM OF CELLULOSE BIODEGRADATION

Utilization of synthetic cellulase enzyme for the conversion of cellulosic biomass at a commercial level increases the cost of the process applied to solid waste management. Therefore the idea is to produce enzymes from less costly natural resources, i.e., microorganisms, so a wide range of microorganisms has been studied extensively. Keeping all these points in mind, searching for new and novel isolates is continuous (Shahriarinour et al., 2011). Many organisms produce cellulases for degradation of cellulose, which are necessary for growth and product formation under appropriate conditions. Therefore research should be focused on fermentation conditions so that maximum yields of cellulase enzymes can be obtained. Among the various parameters of fermentation, temperature and pH hold the maximum share for the growth of microbes followed by inducers. Cellulase enzymes hydrolyze β-1,4-linkages in cellulosic constituents. The catalytic units of cellulolytic enzymes have been categorized into various families depending on the structural unit of protein and crystal structures (Table 8.1). Endoglucanase enzymes hydrolyze glycosidic bonds in the middle of the cellulose chain generating short chain oligosaccharides that are further digested by exoglucanases. There are two types of exoglucanases, which act on the long chain oligomers from both ends, i.e., the reducing end (EC 3.2.1.74) or the nonreducing end (EC 3.2.1.91) generating cellobiose (Fig. 8.1), which consequently produces glucose by β-glucosidases (EC 3.2.1.21) through hydrolysis (Klein-Marcuschamer et al., 2012).

8.4 SOURCES OF CELLULASE-PRODUCING MICROBES

A number of scientific reports are available where novel, potential microbes (bacteria and fungi) are mentioned secreting respectable amounts of cellulase enzymes. They undergo an optimization process for the production of enzymes in sole media or in a specifically designed bioreactor (Bhat, 2000). A few of them are mentioned here: *Anaerocellum thermophilum, Bacillus subtilis, Clostridium phytofermentans, Clostridium thermocellum, Clostridium cellulolyticum, Cellulomonas fimi, Trichoderma reesei, Thermomonospora, Aspergillus niger, Ruminococcus, Trichoderma, Erwinia, Acetivibrio, Microbispora, Streptomyces, Aspergillus, Schizophyllum, Penicillium decumbens, Penicillium funiculosum, Penicillium simplicissimum, Penicillium citrinum, Penicillium janthinellum, Penicillium melinii, Penicillium chrysogenum, Penicillium canescens, Penicillium purpurogenum, Penicillium brasilianum, Penicillium verruculosum, Penicillium pinophilum, Penicillium echinulatum, Penicillium occitanis, Penicillium minioluteum*, etc. A detailed description of the bioprocess parameters that affect the production of cellulase enzymes can be found in Table 8.2.

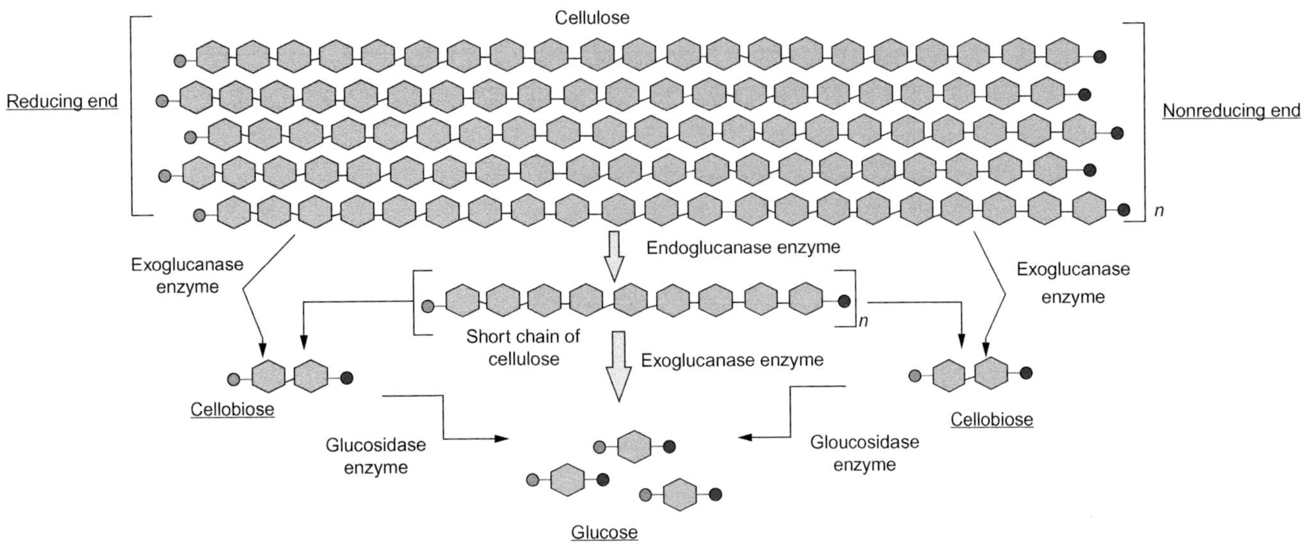

FIG. 8.1 Enzymatic degradation of cellulose.

TABLE 8.2 Optimized Fermentation Conditions for Cellulase-Producing Microbes

Name of microbe	Fungi/bacteria	Substrate	Temp. (°C)	pH	Growth period	References
Purpureocillium lilacinum	Fungi	Xylose	30	5.5	7 days	Srilakshmi et al. (2017)
Bacillus subtilis	Bacteria	Banana fruit stalk	35	7.0	3 days	Krishna (1999)
Aspergillus tubingensis	Fungi	Sorghum straw	40	3.0–9.0	5 days	Adhyaru et al. (2015)
Trichoderma viride	Fungi	Solka Floc BW200	50	4.8	NM	Bélafi-Bakó et al. (2006)
Aspergillus niger LMA	Fungi	Cellulose	30	5.0	72 h	Ahamed and Vermette (2008)
Trichoderma harzianum T2008	Fungi	Empty fruit bunches	32	4.2–6.5	NM	Alam et al. (2009)
Trichoderma sp.	Fungi	Agro-industrial by-products	60	3.0	2 days	Andrade et al. (2011)
Pleurotus sajor-caju	Fungi	Sugarcane	25	5.0	10 days	Pandit and Maheshwari (2012)
Bacillus sp.	Bacteria	NM	40	6.5	5 days	Singh and Pawandeep (2012)
Pseudomonas fluorescens	Bacteria	Agro based	40	10.0	48 h	Sethi et al. (2013)
A. niger	Fungi	Municipal solid waste	40	6.0–7.0	NM	Gautam et al. (2011)
Trichoderma sp.	Fungi	Municipal solid waste	45	6.5	NM	Gautam et al. (2011)
Aspergillus oryzae	Fungi	Soybean hulls, wheat bran	30	5.0	96 h	Brijwani et al. (2010)
Trichoderma reesei	Fungi	Cellulose acetate	40	4.7	NM	Gan et al. (2002)
A. niger	Fungi	Banana peel powder and coir powder	20–50	4.0–9.0	NM	Kiranmayi et al. (2011)

Continued

TABLE 8.2 Optimized Fermentation Conditions for Cellulase-Producing Microbes—cont'd

Name of microbe	Fungi/bacteria	Substrate	Temp. (°C)	pH	Growth period	References
Myceliophthora heterothallica	Fungi	Sugarcane bagasse	35–85	5.0	NM	da Silva et al. (2016)
Cellulomonas	Bacteria	Coconut fiber powder	20–50	5.0–9.0	5 days	Immanuel et al. (2006)
Lysinibacillus sphaericus	Bacteria	Earthworm	30	NM	NM	Jyotsna et al. (2015)
Micrococcus sp.	Bacteria	Cellobiose	30–32	4.5	NM	Andrić et al. (2010)
Aspergillus terreus	Fungi	Rice straw	50	4.8	24 h	Narra et al. (2012)
Streptomyces albaduncus	Fungi	NM	NM	NM	NM	Jyotsna et al. (2015)
Anoxybacillus flavithermus EHP2	Bacteria	Carboxymethyl-cellulose (CMC)	70	7.5	NM	Ibrahim and El-diwany (2007)
Anoxybacillus sp. 527	Bacteria	Crystalline cellulose	70	6.0	NM	Liang et al. (2010)
Bacillus sp. AC-1	Bacteria	CMC	70	4.5–6.5	NM	Li et al. (2006)
Bacillus thuringiensis	Bacteria	Soluble cellulose, CMC, insoluble crystalline cellulose	40	4.0	NM	Lin et al. (2012)
Bacillus sp. NZ	Bacteria	Agricultural residues	50	9.0–10	NM	Nizamudeen and Bajaj (2009)
Bacillus cereus	Bacteria	Palm kernel cake	NM	NM	NM	Lah et al. 2012
Bacillus licheniformis	Bacteria	CMC, filter paper	50–55	6.5–7.0	NM	Acharya and Chaudhary (2012)
Clostridium thermocellum	Bacteria	Cellulose and paper pulp	NM	NM	NM	Zhuang et al. (2007)
Streptomyces sp. BRC1	Bacteria	CMC	26	7.0–7.5	NM	Chellapandi and Jani (2008)
Microbacterium sp. MTCC 10047	Bacteria	CMC	37	7.0	NM	Sadhu et al. (2011)
Bosea sp. MTCC 10045	Bacteria	CMC	37	7.0	NM	Sadhu et al. (2011)
T. viride	Fungi	Birch-to-furfural process waste	45–55	5.0	12 h	Markkanen et al. (1978)
Aspergillus aculeatus	Fungi	Rice straw	45	4.5	NM	Hayashi et al. (1983)
Aspergillus wentii	Fungi	Baggase	50	4.8	NM	Bisaria and Ghose (1977)

8.5 CELLULASE PRODUCTION USING SOLID-STATE FERMENTATION AND SUBMERGED FERMENTATION

In simple terms, fermentation is a technique where complex substrates are biologically converted into simple compounds by various microorganisms. Cellulase is also produced at an industrial level using the same fermentation technology. Over the last few years, fermentation techniques have gained enormous significance because of their economical and ecological reward. Fermentation techniques are classified broadly into two classes because of their brisk processing: solid-state fermentation (SSF) and submerged fermentation (SmF). SSF utilizes solid, nutrient-rich, agrowastes such as bran, bagasse, paddy straw, and paper pulp during fermentation along with microbes that support reduced moisture levels (Sethi and Gupta, 2014). This technique involves the use of industrial and agricultural wastes as cheaper substrates that can be recycled as well (El-Zaher and Fadel, 2010), whereas the SmF technique utilizes liquid substrates, such as molasses and broth, as growth media for the microorganisms that require high moisture content. One of the benefits of this technique is that purifying a product is easier. SSF is predominantly beneficial in producing cellulase for enzyme production by filiform fungi, because it replicates the usual locale of such microorganisms (Shahriarinour et al., 2011). Several litreature reports suggest the significance of estimating the effect of functional parameters on cellulolytic enzyme production from *A.niger* in SSF (Pandey et al., 2000). Downstream processing from SSF is practically simpler and less expensive because it demands comparatively less liquid than SmF. Even genetically modified organisms produce their products more effectively using SSF. The production of cellulases in the SmF process is not favored due to catabolic end product reactions (Berger et al., 2007). SSF is usually preferred over SmF because it gives 2–3 times more enzyme production, properly applies moisture-free fermented solids as an enzyme resource, reduces the cost of downstream processing, produces higher yields in the medium, incorporates natural cellulosic wastes as substrate in contrast to the purified cellulose in SmF, and has the option of carrying out fermentation in nonaseptic conditions too (Dashtban et al., 2009). Besides the numerous advantages, one main limitation of SSF that has been documented is that ~20%–30% of substrate is consumed as compared to almost 5% in the SmF process (Jabasingh, 2011). A few prominent examples of cellulase-producing bacteria in SSF are *Bacillus cereus*, *Bacillus licheniformis*, and *C. thermocellum*.

8.6 BIOPROCESS PARAMETERS FOR THE OPTIMIZATION OF CELLULASE ENZYMES

Enzyme production by microorganisms in a controlled environment can be designed or modified to improve its productivity or give better yields. The majority of bacteria and fungi are the most prominent sources of cellulase enzymes, which hydrolyze the cellulose into glucose in a stepwise manner. Among them, fungi have been studied in more detail due to the filamentous hyphae that can generate automatic pressure on the cellulose composition, resulting in the production of cellulase enzymes in excessive amounts (Sethi and Gupta, 2014). Optimization of both environmental and physiological parameters considerably improves cellulase production by microbes, which is the most important phase to meet the industrial demand for enzymes for various biotechnological applications (Sethi and Gupta, 2014; Bhattacharya et al., 2014). An ideal practice to optimize such parameters is response surface methodology through statistical design. This method employs more validated predictions to ease the optimization process. Cellulase yields mainly depend on the various bioprocess parameters that actually decide the cost of production. These factors are really important and highly influential, and are considered as the major bottleneck of the biotechnological processes. Vast amounts of agricultural, industrial, and municipal cellulosic wastes are consecutively used to minimize the high cost of utilization processes (Alam et al., 2009). Large numbers of the cellulase-producing microbes are cited in the literature reports and the majority of them are summarized here with their optimized fermentation parameters. In Table 8.2, cellulase-producing microbes, pH, temperature, growth period, and the best substrates are listed, while in Table 8.3, the source of cellulase-producing microbes, suitable carbon-nitrogen-metal ions, activators/inhibitors, and the requirement of agitation and its rate are stated.

For fermentation reactions, a controlled environment is necessary for the growth of microorganisms that have to carry out biological reactions, which are available by using bioreactors (Dashtban et al., 2009). The selection of an appropriate type of bioreactor for the maximum production of cellulase is the decisive factor since the bioreactor is the "main controlling point" for the fermentation process. There are many types of bioreactors available and all of them show different performances (Gan et al., 2002). A few of them are mentioned in detail here. A number of fermenters/bioreactors have been designed for cellulase production for the treatment of different microorganisms such as stirred-tank bioreactor, tray-type bioreactor, packed-bed, and rotary drum bioreactors. It was found that the biomass content was higher to a greater extent in the bioreactor as compared to the flask, which indicates that enzyme activity is directly proportional to the increased amount of cellular weight content and disintegration of substrate.

Maximum production of cellulase enzyme on a medium scale was achieved with a 50 L locally developed engineered rotary drum bioreactor. High cellulolytic activity was observed in the bioreactor as compared to the flask. The maximum reactivity was 10.1 ± 1.5 FPA/gds on the next subsequent day of reaction in the bioreactor (Alam et al., 2009). Enhanced production of cellulase in a rotary drum bioreactor is perhaps because of the effect of a few associated factors such as rate of agitation and assimilation, which are not available with an Erlenmeyer flask (Fig. 8.2).

Another bioreactor is the stirred-tank bioreactor (New Brunswick, BioFlo 110) that has a tank volume of 3 L with an operating volume of 1 L under fed-batch growth conditions with lactose. It is equipped with the usual observation

TABLE 8.3 Optimized Fermentation Conditions for Cellulase-Producing Microbes

Source	Name of microbes	Carbon source	Nitrogen source	Metal ions	Inhibitors/ activator	Inoculum size (CFU/ mL)	Agitation rate (rpm)	References
Fungi	*Purpureocillium lilacinum*	Xylose	Ammonium sulfate	NM	Tween 80	2×10^6	150 (Aerobic)	Srilakshmi et al. (2017)
Fungi	*Aspergillus tubingensis*	Xylose	urea	NM	Tween 80	1×10^6	140	Adhyaru et al. (2015)
Fungi	*Aspergillus niger* LMA	Lactose	Peptone	NM	Cellobiose	1.8×10^8	250	Ahamed and Vermette (2008)
Fungi	*Trichoderma* sp.	Wheat bran	Peptone	Na^+, Ca^{2+}, K^+	Co^+, Cu^{2+}, Mn^{2+}	6.22×10^9	NM	Andrade et al. (2011)
Bacteria	*Bacillus* sp.	Lactose	NM	EDTA	NM	1×10^7	200	Singh and Pawandeep (2012)
Bacteria	*Pseudomonas fluorescens*	Glucose	Ammonium sulfate	NM	NM	NM	110	Sethi et al. (2013)
Fungi	*A. niger*, *Trichoderma* sp.	Cellulose and carboxymethyl-cellulose	Peptone and yeast extract	NM	NM	NM	NM	Gautam et al. (2011)
Fungi	*Tricoderma reesei*, *Aspergillus oryzae*	Cellulose	Wheat bran	NM	NM	NM	100	Brijwani et al. (2010)
Fungi	*A. niger*	Agricultural biowaste	NM	NM	Tween 80	NM	NM (Anaerobic)	Kiranmayi et al. (2011)
Fungi	*Myceliophthora heterothallica*	Cardboard	NM	Chloride salts	Tween 80	NM	100	da Silva et al. (2016)
Bacteria	*Streptomyces albaduncus*	Cellulose	Urea	Copper sulfate	EDTA/ Tween 80	NM	NM	Jyotsna et al. (2015)
Fungi	*Aspergillus terreus*	NM	NM	NM	Tween 80, Tween 40	5.04274 μL	120	Narra et al. (2012)
Fungi	*A. terreus*	D-Glucose	Soy peptone	Magnesium sulfate	Calcium chloride	NM	NM	Garg and Neelakantan (1981)
Fungi	*S. albaduncus*	Cellulose	Sodium nitrate	Copper sulfate	Tween 80/ EDTA	NM	NM	Jyotsna et al. (2015)
Bacteria	*Bacillus* sp.	Cellulose material	Ammonium sulfate	NM	NM	NM	NM	Singh and Pawandeep (2012)

system along with controlled factors of production such as proper mixing (250 rpm), pH, temperature (30°C), and aeration for cellulase production. An axial flow impeller (Lightning A100) was used (diameter of impeller: 2.7 in. and pitch ratio: 1.5). The dissolved oxygen value was marked as a time factor and airing was consistently supplied at 2 L/min in a tank without reduction in the value of dissolved oxygen below 20% of air infiltration for all cultures. The pH of the tank was adjusted to 4.8 ± 0.2 and was subsequently restricted to that value by repeated addition of 2 M HCl/NaOH mechanically. Repeated addition of antifoaming agent (silicon) in the tank is required to avoid foam generation for an incubation of 6 days (Ahamed and Vermette, 2008). The tubular membrane reactor is also a modified type of fermenter, which allows high-quality production of cellulase enzymes using microbial populations. A permeable stainless-steel pipe (inner diameter 34 mm, extended up to 148 mm, thickness 0.1 m, developed by Autotrib Ltd., Budapest, Hungary) was layered by fabric coating (Fitevig 500N, made of polyethylene/polyethylene, thickness 1.2 mm, supplier Vignona Ltd., Czech Republic), which are not interlinked. The fine and thin coating imparts assimilation of both the particles and enzymes and further enhances discrimination of the tubular membrane having an exterior of 180 cm^2.

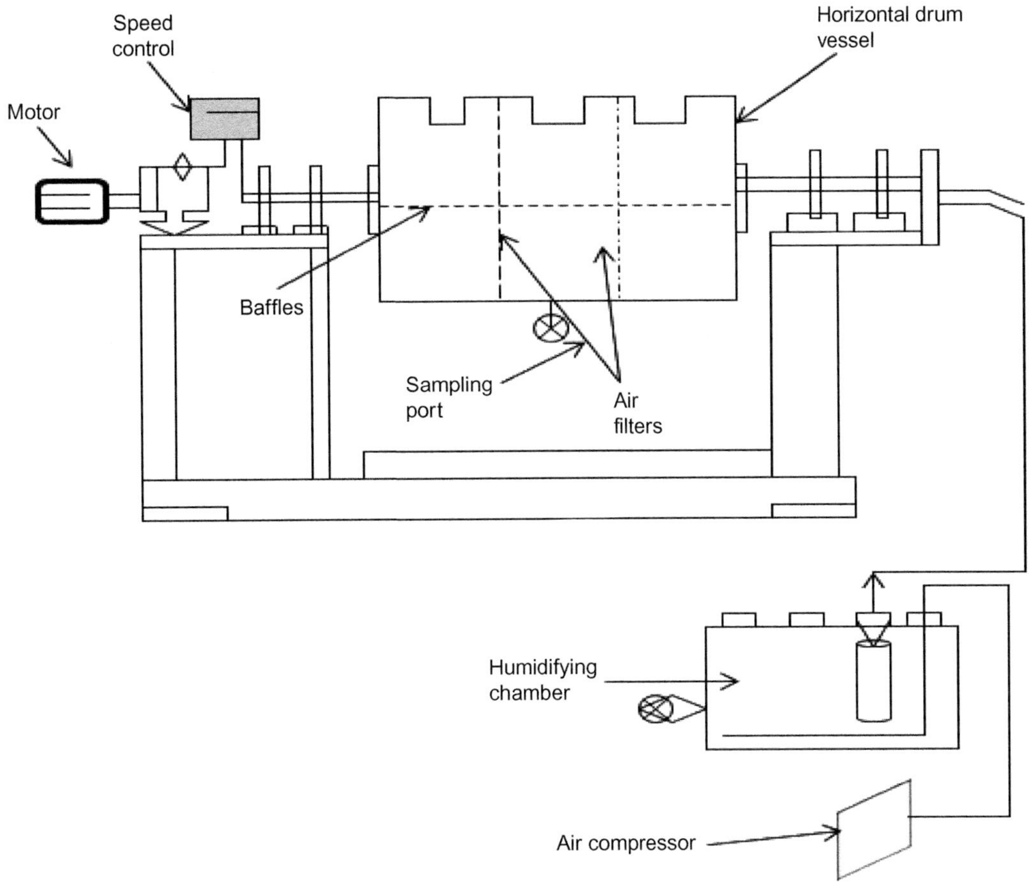

FIG. 8.2 Rotary drum bioreactor used for cellulase production from microbial culture.

At the initial plane of the covering, substrate-bound molecules can be dispersed through the *inlet* and *outlet* ports, while *permeate* can be compiled through the inner side of the tube (Fig. 8.3). Stress in the initial plane can be regulated through the "air inlet" if required. Manometers and stubs are arranged with the glass-coated tubing while the component is insulated to regulate heat. The entire module can be sterilized; consequently, purify biocatalyst provision, although intact living or non-living cells can be applied (Bélafi-Bakó et al., 2006).

In a continuous stirred-tank reactor (CSTR), the reaction and separation zones are intermixed, which is accompanied by a constant long substrate supplement and fitted with electronic cleaning of the covering to regulate parting competence and to facilitate enzyme particle and substrate material entangled within the sluggish film coating on the upper layer of the membrane's exterior. Adequate integration and stirring of the reactant are needed and are accomplished by the inbuilt agitator, which can function at up to 750 rpm. The inbuilt agitator is a distinctive feature of the reactor with reference to other reported systems having continuous recycling in the CSTR where separation of the product partition area is actually situated within cross-flow ultrafiltration module (Fig. 8.4). With the advantage of having a separate zone of agitation and mixing, the recycling CSTR is easy to access and more elastic, where inactivation of enzymatic reactions could be done in large tube and channel like structure. The bench-scale reactor supplied for this purpose was made up from a tailored Amicon dead-end filtration cell (Amicon PM10) with the highest uptake capacity of 2.5 L. The filtration compartment is equipped with a polysulfone ultrafiltration membrane having the following features: molecular weight cutoff (10,000) and disc diameter 150 mm, fitted at the bottom of the cell. The ultrafiltration membrane is unique in its total separation of cellulase and reducing sugar based on total rejection of the enzyme via a mutual effect and ultrafiltration process. The customized bench-scale reactor, placed in a water bath to manage temperature, is operated with nonstop online computation of several physical parameters such as temperature, pH, flow of feed, and filtered drift. A stainless-steel network of diameter 125 mm in the form of a cathode was fixed instantly beneath the membrane, which also performs as a membrane supporter. Similar to the cathode, an anode is placed with 1.0 mm enclosed wires over the surface of the exterior face. Electric cathodes are fitted in a manner that permits the reactor to function through a proscribed electrical back pulse to avoid permanent buildup of enzyme molecules. Usually, a functional power supply ranges between 50 and 400 V, with 0.5–2.0

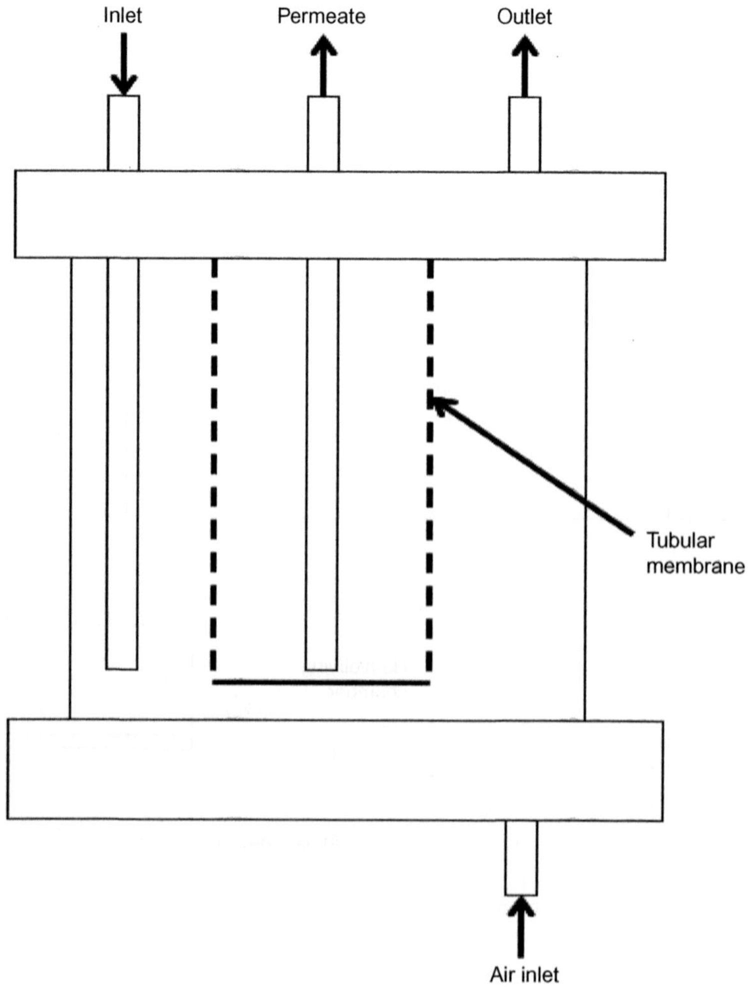

FIG. 8.3 Structure of tubular membrane bioreactor.

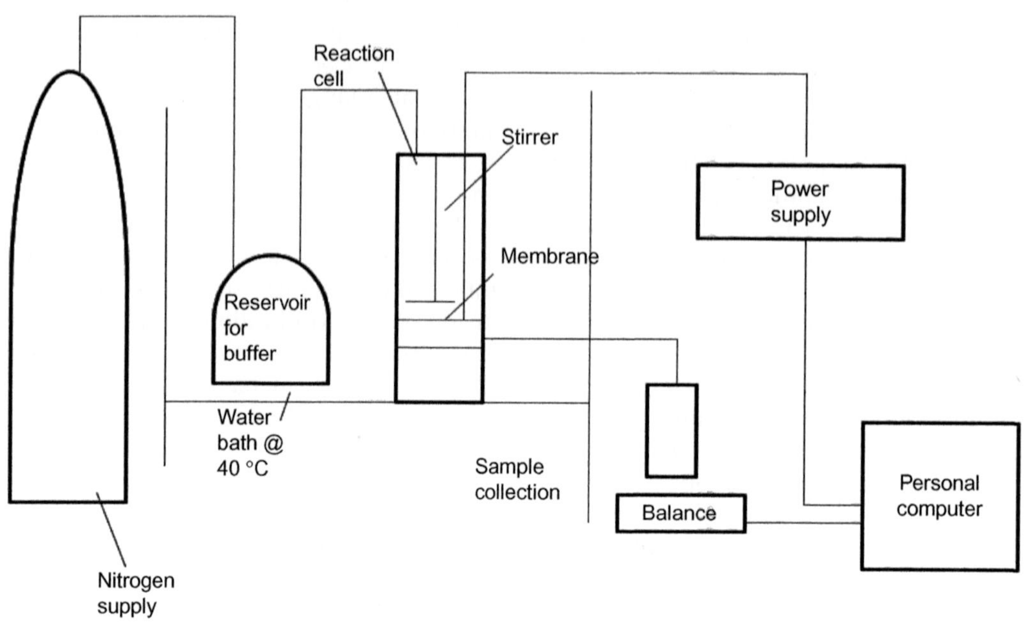

FIG. 8.4 Diagrammatic representation of continuous stirred-tank reactor for enzyme production.

A of electric current. The electrical supply is applied continuously through the computer method within 10–20 s of the impulse period (Gan et al., 2002). It is possible to eject molecules and particles under the controlled condition of electrical back pulsing since they are negatively charged due to the pH of reaction mixture.

8.7 EXPLOITATION OF CELLULASE ENZYMES FOR BIOTECHNOLOGICAL APPLICATIONS

8.7.1 Manufacturing of Paper

Manufacturing of paper and pulp has increased significantly with the use of cellulolytic enzymes over the last 10 years. A large amount of pulp with elevated proportions of thin spikes, mass, and firmness is produced by an automatic mashing process that cleanses and rubs woody substances. Comparing biomechanical pulping with cellulase proceeds with an efficient preservation of energy (20%–40%) in the refining process and enhancement of tensile strength. Mixing of enzymes like cellulases and hemicellulases has been applied to the bioengineering of filament properties for the purpose of increasing evacuation in paper mills during transformation from initial to other stages of pulping. The removal of ink in the recycling of paper utilizes NaOH, coagulates, plasticizers, and emulsifiers. When treated with basic salts, the pulp turns yellow and subsequently the decolorization of alkali-treated pulp is accomplished with hydrogen peroxide. Cellulase enzymes are used here for deinking, as a substitute for basic compounds. This leads to the employment of recycled paper as a feasible option to decrease the amount of trees required for paper synthesis (Martins et al., 2008).

8.7.2 Textile Industry

Textile soaking is flourishing to a great degree by using cellulases, particularly the coating of cellulose-based materials with the aim of enhancing their appearance. The conventional stonewashing of trousers includes amylase-based extraction of starch (desizing) and scraping of trousers with pumice stone in large washing machines. In the same way, cellulases have been effectively applied to the biostoning of trousers and biocoating of fibers and other cellulosic materials. In the course of biostoning, cellulases act on cotton material and snap off the tiny filaments leading to the weft exterior, hence removing the dye that is frequently detached by mechanical scraping in the washing process. However, biopolishing is generally accomplished at the soaking stage, and this involves desizing, scrubbing, cleaning, coloring, and finishing. Cellulases enhance the smoothness and water retention capacity of fibers in acidic conditions, sturdily decrease the affinity for pill synthesis, and give an immaculate layer with less blur (Berger et al., 2007). The exploitation of fabrics with cellulases results in the removal of small filaments, face opaqueness, development of a flat and sleek appearance, and raises color brightness, water retention and dampness, and eco-friendly development. In the manufacturing of cotton material, the projected tiny materials can be successfully detached by using cellulase enzymes to produce a smoother surface of cotton fabrics.

8.7.3 Food-Processing Industry

One of the major applied industries for cellulases is in food biotechnology. Development of the processes of extraction and stabilization is required for fruit and vegetable juices, hence cellulases at this point perform a significant role as a constituent of lignocellulolytic enzyme composites (cellulases, xylanases, and pectinases). The use of such enzyme complexes broadens the yield and course development without extra investment. Such enzymes are required to enhance firmness and consistency and reduce the thickness of tropical mango and papaya fruit purees. The use of pectinases and β-glucosidases results in the dropping of unnecessary bitterness by improving the texture and flavor of citrus fruits (Bhattacharya et al., 2014). Therefore the macerating enzymes (cellulases and pectinases) play a significant role in food biotechnology and their requirement will probably be a boost for the extraction of juices.

8.7.4 Agriculture Industry

When controlling vectors that cause plant diseases, the cellulase category of enzymes, such as β-glucanases from fungal sources, is found to be effective because it has the capability of degrading the cell walls of certain plant parasites. Various soft-rot fungi along with *Trichoderma* sp., *Geocladium* sp., *Chaetomium* sp., and *Penicillium* sp. are significantly important in agriculture by helping the enhancement of plant growth, rapid flowering, seed sprouting, better root structure, and amplified crop yields. Soil fertility could also be improved by using cellulases. Usually, the mixing of dry stalks of plants is marked as a significant strategy to enhance the soil properties and decreases the use of organic fertilizers. Various research has proposed that straw decomposition can be accelerated by using microbes and their enzymes. Cellulolytic fungi such as *Aspergillus*, *Chaetomium*, *Trichoderma*, and *Actinomycetes* are showing capable results. So, by using exogenous cellulases, it is possible to speed up straw decomposition and the fertility of soil increases considerably (Bhat and Bhat, 1997).

8.7.5 Brewing Industry

Enzymes like glucanases, pectinases, and hemicellulases play a significant role in the yield and enhancement of wine in alcoholic beverage-making industries. These enzymes are necessary for clearing up, color removal, fur softening, and for maintaining the superiority and firmness of wine. Adding β-glucosidases to alcoholic drinks like wine results in better aroma improvement. Soft enzyme complexes also have better sensitivity, subsidence, and juice production of grapes applied for fermenting wine. Various marketable enzyme provisions are currently accessible to the wine industry. Beer infusion depends on the enzyme activity generated by the process of steeping and fermentation. Conversion of food grain relies on bean sprouting, which originates from the biogenesis and creation of α- and β-amylases, carboxypeptidase, and β-glucanase that change the seed assets. Earlier, endoglucanase II and exoglucanase II of the *Trichoderma* cellulase system were largely responsible for the decline in the level of polymerization and value thickness (Shahriarinour et al., 2011).

8.7.6 Dye Extraction From Plants

Carotenoids are original dyes extracted from plants. There is an enormous desire for carotenoids as food dyes because of their required assets, such as their common source, high adaptability, nontoxic behavior, and availability in both forms: lipophilic and hydrophilic dyes with colors between red and yellow. The mixture of cellulases with pectinases destroys the cell walls of orange peel, sweet potato, and carrot, and releases carotenoids in the chloroplasts and cell fluids. Such dyes persist in their original form until they react with proteins (Sethi and Gupta, 2014).

8.7.7 Animal Feeding

Using cellulases and hemicellulases in animal feed has established significant consideration due to their nutritional value and economical feed rate. Pretreatment of cultivated fodder such as grasses and grain fodder through cellulases or xylanases can recover its nutritious worth. Use of thermophilic cellulolytic enzymes ensures the edibility and nourishment of the feedstuff, thus enabling a mixture of handling and forage renovation at a subsequent stage. The cellulases and hemicellulases are liable to fractional breaking of cellulosic and hemicellulosic constituents, decortication of seeds, hydrolysis of β-glucans, and improvement in turbidity and elasticity of feedstuff constituents, causing enhancement in the nutritious excellence of feedstock. Cellulolytic enzymes have a progressive outcome on fecal fermentation procedures by increasing the manufacture of propionic acid, which acts as a bacteriostatic agent and therefore can reduce the establishment of disease-causative microbes (Pandey et al., 2000).

8.7.8 Olive Oil Extraction

In common, oils are obtained through mashing and crunching of olives in a pebble and mallet or mortar, which rapidly distributes crushed olive cream over a sequence of straight containers for the improvement of oil. However, throughout this punitive automated method, natural existing enzymes become inactive and corrosion occurs. Therefore, Olivex, which is made up of a pectinase through cellulase and hemicellulase, is added, which is a major profitable enzyme fusion that helps in the recovery of oils from olives. Moreover, the consumption of enzymes amplifies the inhibitor in virgin olive oil and decreases the initiation of rancidity (Martins et al., 2008).

8.7.9 Detergent Industry

Technically, cloth cleansers use protease, lipase, and currently cellulases, and is the latest invention in product manufacturing. Cellulolytic enzymes help to improve radiance, light, sensation, and elimination of dust from the fibers (Sethi and Gupta, 2014).

8.7.10 Bioethanol Production

The utmost common usage of cellulolytic enzyme is presently the biological transition of lignocellulosic resources into beneficial and advanced products such as biofuels. Currently, mostly sugarcane bagasse, corncob, rice straw, *Lantana camara*, switch grass, sawdust, and forestry deposits are enzymatically hydrolyzed in different procedures. These procedures comprise automated, natural, or biotic pretreatment, chemical reaction of synthetic compounds for the yield of digestible particles (e.g., hexose and pentose sugars), biological transition of these minor particles toward the provision of bacterial development and/or yield of organic merchandise, and the partition and distillation of preferred yields (El-Zaher and Fadel, 2010; Gusakov, 2011). Usage of bacterial enzymes considerably decreases the price related to organic reaction in warm surroundings such as pH 4.0–6.0 and temperatures of 45–50°C, which is permitted from decomposition disputes. Protein manufacturing and focused development are dominant implements that can assist the growth of additional resourceful thermophilic cellulolytic enzymes. The retrieval of enzymes is mainly predisposed by surface assimilation of the enzymes against the substrate, particularly to lignin and enzyme inactivation (Gusakov, 2011).

8.8 CONCLUSION

The consumption of cellulosic biomass, which is supposed to be a waste material, into novel valuable products by microorganisms is a revolutionary step toward bioremediation. Microbial cellulases are the enzymes used for cellulose degradation and formation of value-added products, including food, fuel, paper, and textiles. So far, most of the research has been done by using fungal cellulase, but now it is diverted toward bacteria due to their higher growth rate and adaptability to various extreme conditions. The development of fast screening methods for novel cellulase production from microorganisms will be useful for industrial purposes. However, present knowledge of these enzymes' production and their producing organisms is quite substantial. In spite of the developments achieved so far for the production and applications of microbial cellulases, additional efforts are required for the development of novel and specific cellulase enzymes by using enzyme engineering and other molecular techniques.

References

Acharya, S., Chaudhary, A., 2012. Optimization of fermentation conditions for cellulases production by *Bacillus licheniformis* MVS1 and *Bacillus* sp. MVS3 isolated from Indian hot spring. Braz. Arch. Biol. Technol. 55 (4), 497–503.

Adhyaru, D.N., Bhatt, N.S., Modi, H.A., 2015. Optimization of upstream and downstream process parameters for cellulase-poor-thermo-solvent-stable xylanase production and extraction by *Aspergillus tubingensis* FDHN1. Bioresour. Bioprocess. 2 (1), 3.

Ahamed, A., Vermette, P., 2008. Enhanced enzyme production from mixed cultures of *Trichoderma reesei* RUT-C30 and *Aspergillus niger* LMA grown as fed batch in a stirred tank bioreactor. Biochem. Eng. J. 42 (1), 41–46.

Alam, M.Z., Mamun, A.A., Qudsieh, I.Y., Muyibi, S.A., Salleh, H.M., Omar, N.M., 2009. Solid state bioconversion of oil palm empty fruit bunches for cellulase enzyme production using a rotary drum bioreactor. Biochem. Eng. J. 46 (1), 61–64.

Andrade, J.P., Bispo, A.S.D.R., Marbach, P.A.S., do-Nascimento, R.P., 2011. Production and partial characterization of cellulases from *Trichoderma* sp. IS-05 isolated from sandy coastal plains of northeast Brazil. Enzyme Res. 2011.

Andrić, P., Meyer, A.S., Jensen, P.A., Dam-Johansen, K., 2010. Reactor design for minimizing product inhibition during enzymatic lignocellulose hydrolysis. I. Significance and mechanism of cellobiose and glucose inhibition on cellulolytic enzymes. Biotechnol. Adv. 28 (3), 308–324.

Bélafi-Bakó, K., Koutinas, A., Nemestóthy, N., Gubicza, L., Webb, C., 2006. Continuous enzymatic cellulose hydrolysis in a tubular membrane bioreactor. Enzym. Microb. Technol. 38 (1-2), 155–161.

Berger, E., Zhang, D., Zverlov, V.V., Schwarz, W.H., 2007. Two non-cellulosomal cellulases of *Clostridium thermocellum*, Cel9I and Cel48Y, hydrolyse crystalline cellulose synergistically. FEMS Microbiol. Lett. 268 (2), 194–201.

Bhat, M.K., 2000. Cellulases and related enzymes in biotechnology. Biotechnol. Adv. 18 (5), 355–383.

Bhat, M.K., Bhat, S., 1997. Cellulose degrading enzymes and their potential industrial applications. Biotechnol. Adv. 15 (3-4), 583–620.

Bhattacharya, S., Das, A., Patnaik, A., Bokade, P., Rajan, S.S., 2014. Submerged fermentation and characterization of carboxymethyl cellulase from a rhizospheric isolate of *Trichoderma viride* associated with *Azadirachta indica*. J. Sci. Ind. Res. 73 (4), 225–230.

Bisaria, V.S., Ghose, T.K., 1977. Sorption characteristics of cellulases on cellulosic substances. In: Ghose, T.K. (Ed.), Proceedings of the Bioconversion Symposium, IIT Delhi. Indian Institute of Technology, New Delhi, India, pp. 155–164.

Brijwani, K., Oberoi, H.S., Vadlani, P.V., 2010. Production of a cellulolytic enzyme system in mixed-culture solid-state fermentation of soybean hulls supplemented with wheat bran. Process Biochem. 45 (1), 120–128.

Chellapandi, P., Jani, H.M., 2008. Production of endoglucanase by the native strains of *Streptomyces* isolates in submerged fermentation. Braz. J. Microbiol. 39 (1), 122–127.

da Silva, V.D.C.T., de Souza Coto, A.L., de Carvalho Souza, R., Neves, M.B.S., Gomes, E., Bonilla-Rodriguez, G.O., 2016. Effect of pH, temperature, and chemicals on the endoglucanases and β-glucosidases from the thermophilic fungus *Myceliophthora heterothallica* F. 2.1.4. obtained by solid-state and submerged cultivation. Biochem. Res. Int. 2016.

Dashtban, M., Schraft, H., Qin, W., 2009. Fungal bioconversion of lignocellulosic residues; opportunities & perspectives. Int. J. Biol. Sci. 5 (6), 578.

El-Zaher, A., Fadel, M., 2010. Production of bioethanol via enzymatic saccharification of rice straw by cellulase produced by *Trichoderma reesei* under solid state fermentation. New York Sci. J. 3 (4), 72–78.

Gan, Q., Allen, S.J., Taylor, G., 2002. Design and operation of an integrated membrane reactor for enzymatic cellulose hydrolysis. Biochem. Eng. J. 12 (3), 223–229.

Garg, S.K., Neelakantan, S., 1981. Effect of cultural factors on cellulase activity and protein production by *Aspergillus terreus*. Biotechnol. Bioeng. 23 (7), 1653–1659.

Gautam, S.P., Bundela, P.S., Pandey, A.K., Khan, J., Awasthi, M.K., Sarsaiya, S., 2011. Optimization for the production of cellulase enzyme from municipal solid waste residue by two novel cellulolytic fungi. Biotechnol. Res. Int. 2011, .

Gusakov, A.V., 2011. Alternatives to *Trichoderma reesei* in biofuel production. Trends Biotechnol. 29 (9), 419–425.

Hayashi, H., Arai, M., Sakamoto, R., Murao, S., 1983. Enzymatic hydrolysis of cellulosic substances by cellulase mixture of *Aspergillus aculeatus* and *Trichoderma viride*, and ethanol production from the hydrolyzate. Hakko Kogaku Kai shi.

Ibrahim, A.S.S., El-diwany, A.I., 2007. Isolation and identification of new cellulases producing thermophilic bacteria from an Egyptian hot spring and some properties of the crude enzyme. Aust. J. Basic Appl. Sci. 1 (4), 473–478.

Immanuel, G., Dhanusha, R., Prema, P., Palavesam, A., 2006. Effect of different growth parameters on endoglucanase enzyme activity by bacteria isolated from coir retting effluents of estuarine environment. Int. J. Environ. Sci. Technol. 3 (1), 25–34.

Jabasingh, S.A., 2011. Utilization of pretreated coir pith for the optimized bioproduction of cellulase by *Aspergillus nidulans*. Int. Biodeterior. Biodegrad. 65 (8), 1150–1160.

Jyotsna, K.P., Rao, A.R., Devaki, K., 2015. Effect of nutritional factors on cellulase production by *Lyscinibacillus sphaericus* from the gut of earthworm, *Eisenia foetida*. Curr. Biot. 9 (2), 145–152.

Kiranmayi, M.U., Poda, S., Charyulu, P.B.B.N., Vijayalakshmi, M., Krishna, P.V., 2011. Studies on influence of natural biowastes on cellulase production by *Aspergillus niger*. J. Environ. Biol. 32 (6), 695.

Klein-Marcuschamer, D., Oleskowicz-Popiel, P., Simmons, B.A., Blanch, H.W., 2012. The challenge of enzyme cost in the production of lignocellulosic biofuels. Biotechnol. Bioeng. 109 (4), 1083–1087.

Krishna, C., 1999. Production of bacterial cellulases by solid state bioprocessing of banana wastes. Bioresour. Technol. 69 (3), 231–239.

Lah, T.N.T., Rahman, N.N.N., Nama, M.B., 2012. Cellulase activity and glucose production by *Bacillus cereus* monoculture and co-culture utilizing palm kernel cake (PKC) under solid state fermentation. In: International Conference on Environment, Energy and Biotechnology, Singapore, vol. 33, pp. 172–177.

Li, Y.H., Ding, M., Wang, J., Xu, G.J., Zhao, F., 2006. A novel thermoacidophilic endoglucanase, Ba-EGA, from a new cellulose-degrading bacterium, *Bacillus* sp. AC-1. Appl. Microbiol. Biotechnol. 70 (4), 430–436.

Liang, Y., Feng, Z., Yesuf, J., Blackburn, J.W., 2010. Optimization of growth medium and enzyme assay conditions for crude cellulases produced by a novel thermophilic and cellulolytic bacterium, *Anoxybacillus* sp. 527. Appl. Biochem. Biotechnol. 160 (6), 1841–1852.

Lin, L., Kan, X., Yan, H., Wang, D., 2012. Characterization of extracellular cellulose-degrading enzymes from *Bacillus thuringiensis* strains. Electron. J. Biotechnol. 15 (3), 2.

Markkanen, P., Linko, M., Nybergh, P., 1978. Enzymatic hydrolysis of cellulosic waste materials. AIChESymp 74 (172), 89–92.

Martins, L.F., Kolling, D., Camassola, M., Dillon, A.J.P., Ramos, L.P., 2008. Comparison of *Penicillium echinulatum* and *Trichoderma reesei* cellulases in relation to their activity against various cellulosic substrates. Bioresour. Technol. 99 (5), 1417–1424.

Narra, M., Dixit, G., Divecha, J., Madamwar, D., Shah, A.R., 2012. Production of cellulases by solid state fermentation with *Aspergillus terreus* and enzymatic hydrolysis of mild alkali-treated rice straw. Bioresour. Technol. 121, 355–361.

Nizamudeen, S., Bajaj, B.K., 2009. A novel thermo-alkali tolerant endoglucanase production using cost-effective agricultural residues as substrates by a newly isolated *Bacillus* sp. NZ. Food Technol. Biotechnol. 47 (4), 435–440.

Pandey, A., Soccol, C.R., Mitchell, D., 2000. New developments in solid state fermentation. I. Bioprocesses and products. Process Biochem. 35 (10), 1153–1169.

Pandit, N.P., Maheshwari, S.K., 2012. Optimization of cellulase enzyme production from sugarcane pressmud using oyster mushroom *Pleurotussajor caju* by solid state fermentation. J. Bioremed. Biodegrad. 3 (3).

Sadhu, S., Saha, P., Mayilraj, S., Maiti, T.K., 2011. Lactose-enhanced cellulase production by *Microbacterium* sp. isolated from fecal matter of zebra (*Equus zebra*). Curr. Microbiol. 62 (3), 1050–1055.

Sethi, S., Gupta, S., 2014. Optimization of cultural parameters for cellulase enzyme production from fungi. Biolife 2 (3), 989–999.

Sethi, S., Datta, A., Gupta, B.L., Gupta, S., 2013. Optimization of cellulase production from bacteria isolated from soil. ISRN Biotechnol. 2013.

Shahriarinour, M., Wahab, M.N.A., Mohamad, R., Mustafa, S., andAriff, A.B., 2011. Effect of medium composition and cultural condition on cellulase production by *Aspergillus terreus*. Afr. J. Biotechnol. 10 (38), 7459–7467.

Singh, J., Pawandeep, K., 2012. Optimization of process parameters for cellulase production from *Bacillus* sp. JS14 in solid substrate fermentation using response surface methodology. Braz. Arch. Biol. Technol. 55 (4), 505–512.

Srilakshmi, A., SaiGopal, D.V.R., Narasimha, G., 2017. Impact of bioprocess parameters on cellulase production by *Purpureocillium lilacinum* isolated from forest soil. Int. J. Pharm. Biol. Sci. 8 (1), 157–165.

Zhuang, J., Marchant, M.A., Nokes, S.E., Strobel, H.J., 2007. Economic analysis of cellulase production methods for bio-ethanol. Appl. Eng. Agric. 23 (5), 679–687.

Further Reading

Alam, M.Z., Muhammad, N., Mahmat, M.E., 2005. Solid state bioconversion. Am. J. Appl. Sci. 2 (2), 569–572.

CHAPTER

9

Role of Solid-State Fermentation to Enhance Cellulase Production

Navodita Maurice

Laboratory of Immunology, Institute of Genetics, Biological Research Centre, Hungarian Academy of Sciences, Szeged, Hungary

Abbreviations

aa	acid alkaline
HPa	alkaline hydrogen peroxide
BBD	Box-Behnken design
CMCase	carboxymethyl cellulase
FPase	filter paper activity
g	grams
h	hours
H_2O_2	hydrogen peroxide
LPMO	lytic polysaccharide monooxygenases
MnP	manganese peroxidase
NaOH	sodium hydroxide
RSM	response surface methodology
SSC	solid-state culture
SSF	solid-state fermentation
SmF	submerged fermentation

9.1 SOLID-STATE FERMENTATION

Solid-state fermentation (SSF) engages the agitation of solid materials without water but the used substance must hold sufficient moisture to provoke microbial growth and metabolism (Pandey et al., 2001). Naturally, SSF promotes microbial growth on moist solids and initiates fermentation as known from earlier times (Mitchell and Lonsane, 1990). SSF presents multitudinal junctures for the residues of agricultural industries to use agro-industrial wastes as substrates because they require very low energy levels with less waste water production and are also environmentally friendly because they can help in combating the problem of solid waste throwoff (Ogbonna et al., 2001). The development of penicillin by submerged fermentation (SmF) between the two world wars advertised SmF as a standard tool for the production of compounds utilizing the fermentation process. However, research on SSF systems continued during 1950–60. Mycotoxin and protein production as cattle feed by the employment of agro-industrial wastes promoted the SSF arena (Yang et al., 2001), emphasizing it as a strong technology for growing fungi, yeast,

From Cellulose to Cellulase: Strategies to Improve Biofuel Production
https://doi.org/10.1016/B978-0-444-64223-3.00009-6

and bacteria on moist solid substrates over the last few decades (Medeiros et al., 2006). Currently, SSF is universally used for enzyme production, organic compounds, and metabolite manufacturing because it offers numerous assets over SmF in product yield, cost, and ease of use. SSF devices fall within the categories of static and agitated systems. Static systems prefer the use of stationary solid media throughout the fermentation process, while agitated systems follow the mixing of the solid media during the fermentation process (Larroche and Gros, 1997).

Currently, SSF finds its wide applications in hazardous compound bioremediation and biodegradation, agro-industrial residue detoxification by biological means, crop and crop residue by biotransformation for the enrichment of nutritional values, biopulping, and value-added biologically active secondary metabolite production taking command over a number of value-added products ranging from antibiotics to biofuel production. SSF systems have emerged as an encouraging weapon for biopharmaceutical production by utilizing cost-effective substrates with higher yields (Brand et al., 2001).

9.2 GENERAL CONSIDERATIONS OF SSF

Numerous important factors affect SSF processes and the most crucial ones include selection of suitable strains and substrates, along with process parameters by taking their physical, chemical, and biochemical properties into consideration (Pandey et al., 1999). Enzyme production via SSF has achieved greater success (Selvakumar and Pandey, 1999), for example, *Staphylococcus* sp. produce inulinase (Pandey et al., 1998) and *Brevibacterium* sp. produce L-glutamic acid with the help of sugarcane bagasse when solid-state conditions are provided (Nampoothiri and Pandey, 1996). Several classes of filamentous fungi, for example, *Rhizopus* and *Mucor* (Phycomycetes), *Penicillium* and *Aspergillus* (Ascomycetes), and white-rot fungi (Basidiomycetes), are these days extensively used for cellulase production by SSF.

The solid substrate not only provides mandatory elements supporting growth of microorganisms in the SSF processes but also acts as a source of anchorage for the microbial bodies. This concrete material supplies the crucial elements for the survival of microbes and therefore can be considered as the "ideal substrate." It is also possible that some nutrients can occur in minimal proportions, or may be totally deficient, so in such circumstances the nutrients must be provided externally. Some substrates are pretreated (chemically or mechanically) before being used in SSF processes (for example, in the case of lignocellulosic residues) because pretreatment increases the accessibility of the nutrients for the onslaught of microbes. The size of the particles in the substrate and the moisture content or water percentage are the key players for microbial growth on a particular substrate (Liu and Tzeng, 1999). If the substrate particles are small, then they offer a larger surface area for the growth of microbes and are of first preference, but if the substrate particles are too small, then they can cause aggomulation by choking microbial respiration or aeration bringing a declination in microbial growth and thereby giving a lower yield. If bigger particles are selected, they furnish enhanced respiration or aeration possibilities (because of the larger space between the neighboring particles); however, in turn they offer a restricted surface area for the growth of microorganisms (Pandey, 1995).

According to reports from the research conducted so far regarding the selection of suitable substrates, generally tropical agro-industrial crops and residues gathered from a large bulk of plants, for example, soybean, potato, sweet potato, sugar beet, sugarcane bagasse, etc., have been processed under the SSF processes. Wheat bran has been the champion among all the substrates selected so far (Pandey et al., 2000). A number of processes prefer the use of agro-industrial residues as starters for the gross production of bioethanol, enzymatic extracts, organic compounds, etc. Numerous agricultural residues like cassava and sugarcane wastes, pulp, and husk of coffee are considered as the best substrates because they are good organic sources and support efficient microbial growth, thereby enhancing commercial production of industrially useful metabolites. Bioprocesses utilizing the agro-industrial residues, however, act as a means of secondary substrates but also aid in solving the problem of pollution all over the globe (Soccol and Krieger, 1998).

In the SSF processes, however, microbes grow faster under conditions of low water content but if a supplementary inoculum is also added to the substrate, then this added microbe can encounter the existing microorganisms, overruling the rigid restrained operational system of the classical biological reactor units, which is not a mandate for the SSF functioning units. Flexible bioreactor design and lower costs for the SSF processes strengthen its economic advantage over the SmF process. For the cost-effective SSF processes, usually the solid substrate is impregnated with a medium of any mineral salt because it needs a carbon source, which is very cheap and easily available (generally the leftovers of agricultural wastes), demands a low-priced framework, can be used by inexperienced people, and produces a highly concentrated final product (Singhania et al., 2009). Cellulase production via filamentous fungi by SSF is very promising in its cost effectiveness when compared to SmF (Singhania et al., 2010) (Fig. 9.1).

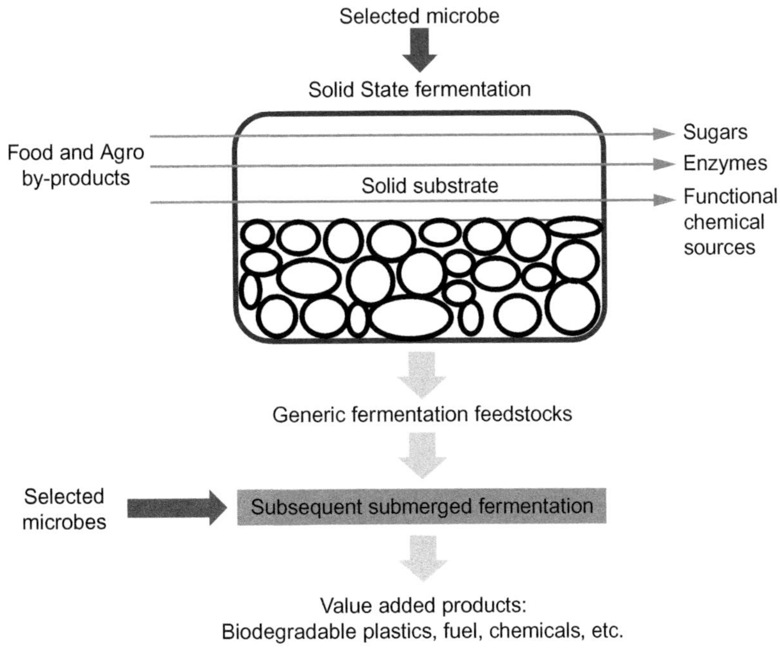

FIG. 9.1 Schematic representation of the solid-state fermentation process. *Adapted from Manan, M.A., Webb, C., 2017. Modern microbial solid state fermentation technology for future biorefineries for the production of added-value products. Biofuel Res. J. 16, 730–740.*

9.3 FUNGAL CELLULASES

Cellulase is the key component participating in the conversion of lignocellulose into efficiently fermentable sugars, thereby producing bioethanol, and the filamentous fungi are key players in cellulases produced for lignocellulosic biomass degradation. This most remarkable of enzymes engaged for biomass degradation is cellulase, which is made up of three distinct enzymes: endocellulase, exocellulase, and β-glucosidase. These three components of cellulase act synergistically and bring down the degradation of cellulose into monosaccharide glucose units (Singh et al., 2017) (Fig. 9.2).

According to research conducted until recently the accessory proteins designated as auxiliary proteins are responsible for the phenomenon of synergism existing between the three distinct cellulase enzymes (Haque et al., 2015). These auxiliary proteins serve as intermediates in the amorphogenesis of the cellulose fibers by (1) breakage of hydrogen bonds among the cellulose fibrils and (2) using their oxidative mechanisms to attack the fiber, resulting in the cleavage of glycosidic bonds. This is due to the presence of high impedance of the lignocellulosic biomass that makes the approach of cellulases to cellulose troublesome (Ezeilo et al., 2017). Crystallinity and recalcitrance reduction, so that the lignocellulosic fiber can be made accessible for attack by cellulases, is the prime target of these accessory proteins. Expansins (mostly present in the plant cell walls) are the well-defined accessory proteins participating in the loosening of the lignocellulosic matrix (Kim et al., 2014). It has been discovered that hydrogen bonds are responsible for keeping the cellulose and hemicellulose fibers intact; expansins break these bonds and therefore enhance the action of cellulases on cellulose, and this potential has shown significant enhancements in lignocellulose biomass degradation (Cosgrove, 2017). Just like expansins, swollenins elicit the swelling of fiber along with changes in the structure of the cellulose fiber, for example, decreasing the particle size and reducing crystallinity, which facilitate access of cellulases (Haque et al., 2015). This positive synergism between expansins and swollenins of the cellulase complex brings enhanced hydrolysis of the lignocellulosic biomass (Rocha et al., 2016).

The lytic polysaccharide monooxygenases (LPMO-EC) are other accessory proteins that catalyze cellulose or chitin, which are highly crystalline substrates. LPMOs, which act on cellulose or chitin, cleave the β-(1-4)-glycosidic bonds and oxidize either the C1 or C4 carbons (Agger et al., 2014). The oxidation of C1 or C4 carbon atoms causes the fibers forming the cellulose to distangle and this activity opens the way for cellulases to act on these cellulosic fibrils by generating areas for enzymatic recognition by the cellulases, indicating that LPMOs work synergistically with cellulases for efficient cellulose degradation (Ezeilo et al., 2017). In the fungal degradation machinery, these LPMOs are a crucial component and their dominion has been demonstrated in the fungal genomes (Berlemont, 2017).

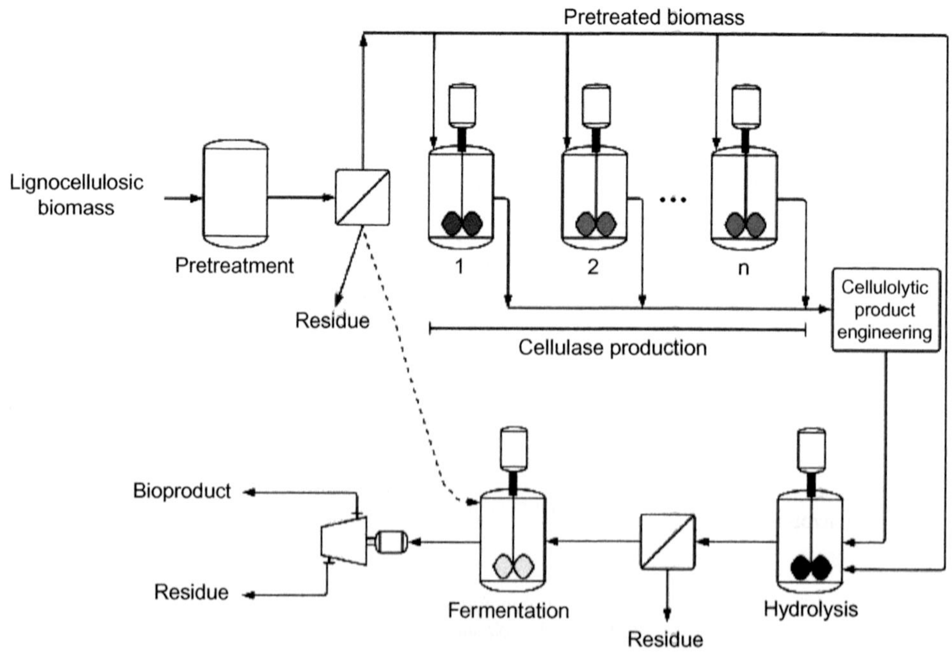

FIG. 9.2 Lignocellulosic biomass conversion. *Adapted from Passos, D.F., Pereira, N., de Castro, A.M., 2018. A comparative review of recent advances in cellulases production by Aspergillus, Penicillium and Trichoderma strains and their use for lignocellulose deconstruction. Curr. Opin. Green Sust. Chem. 14:60–66.*

It has been known that bacteria, actinomycetes, and some fungal species are capable of producing cellulase, but among them the filamentous fungi are primarily considered to be the best sources of cellulase production because they have higher enzyme-secreting capacities. According to the literature, *Trichoderma*, *Humicola*, *Aspergillus*, *Penicillium*, etc. are the commonly known cellulolytic fungi capable of producing all three cellulase components (Dutta et al., 2008). Among the known filamentous fungi, *Trichoderma reesei* is most extensively studied because it can produce all the cellulase components needed for complete cellulose degradation. The availability of physiological information concerning the hypersecreting mutants of *Trichoderma* make it more industrially preferred, but the cellulase produced by *T. reesei* is deficient in the β-glucosidase required for efficient lignocellulosic biomass hydrolysis in sufficient proportions. This is why *T. reesei* cellulase is supplemented with the heterologous β-glucosidase from *Aspergillus niger* for efficient biomass hydrolysis (Sukumaran et al., 2009). *Penicillium* is also emphasized as a potential source of cellulase because it is capable of producing endoglucanase, exoglucanase, and β-glucosidase: the essential cellulase ingredients, and sufficient β-glucosidase production renders this fungal species superior to *Trichoderma* (Gusakov and Sinitsyn, 2012).

Cellulases hydrolyze the β-1,4-D-glucan links of cellulose, which cause the production of glucose, cellobiose, and cellooligosaccharides. Different enzyme classifications can be found in the literature for the cellulases produced by microorganisms, but basically cellulase enzymes fall under three categories, namely cellobiohydrolases, endo-β-1,4-glucanases, and β-glucosidases, which actively hydrolyze the cellulose molecule. Several types of multiple enzymes usually occur within these three classified enzymes (Schülein, 1988). Among the studied fungi, *T. reesei* is the most popularly used fungal strain for the production of cellulase because it secretes two cellobiohydrolases, eight endo-β-1,4-glucanases, and seven β-glucosidases. The endo-β-1,4-glucanases depolymerize the cellulose chains by displaying their condensing and noncondensing edges where cellobiohydrolases work to make space for the cellooligosaccharides and cellobiose units, while cellobiose cleavage is carried out by β-glucosidase, which produces glucose by complete hydrolysis. Therefore cellobiohydrolases, endo-β-1,4-glucanases, and β-glucosidases all work together synergistically to bring the conversion of crystalline cellulose to glucose (Beguin and Aubert, 1994). Generally, the structure of all the cellulases known so far is comprised of two domains, namely a catalyzing domain also known as catalytic domain, and a cellulose domain for binding, or a module for carbohydrate binding, which is joined by a linked peptide. The catalytic domain bears a catalytic site, while the carbohydrate-binding domain participates in the enzyme binding to cellulose (Sakka et al., 2000). Native cellulose degradation demands cooperation of different cellulase levels and the synergism between the endo- and exoglucanases and between exoglucanases is perfect for this action. The action of EC generates free ends for exoglucanases to act in the first category, while exoglucanases work together on the reducing and nonreducing points causing complete cellulose degradation in the

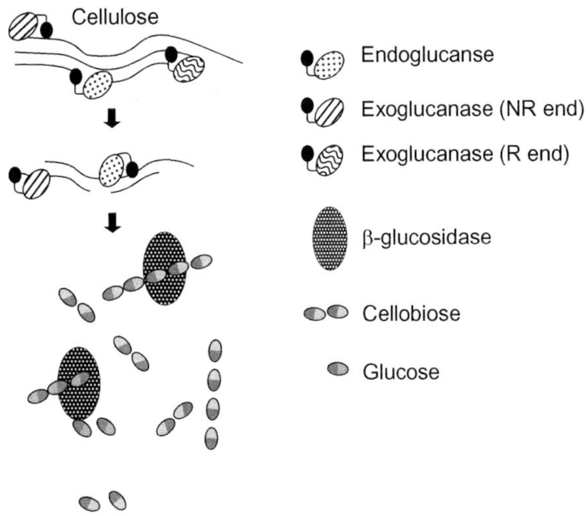

FIG. 9.3 Mechanism of cellulase action (Mathew et al., 2008).

second category (Bhat and Bhat, 1997). Complete cellulose degradation is beyond the scope of an individual enzyme; however, a mixture of enzyme escalates saccharification efficiency (Mathew et al., 2008) (Fig. 9.3).

9.3.1 How are Enzymes of the Cellulase Complex Produced?

It has been well documented that a wide array of microorganisms ranging from bacterial to fungi species along with animals and plants are capable of producing cellulases (Kuhad et al., 2011). Among all these organisms, filamentous fungi are the champions and the genera *Trichoderma*, *Aspergillus*, and *Penicillium* serve as the best used model organisms from small to large industrial-scale cellulase production (Srivastava et al., 2017) (Fig. 9.4).

9.4 CELLULASE PRODUCTION BY SOLID-STATE FERMENTATION

Cellulases are employed in the processing of starch, food production for animals, alcohol fermentation from grains, malting and brewing and textile and paper industries, fruit and vegetable pulp, along with juice extraction (Adsul et al., 2007). Considerable reduction in the cost of efficiently higher enzyme production is conceivable by taking account of the methods for cellulose conversion with the help of microbes capable of producing cellulolytic enzymes. This in turn demands an imperative search for microorganisms that can produce cellulase at a sufficiently higher rate (Kotchoni and Shonukan, 2002). SSF is already known as an attractive strategy for the cost-effective

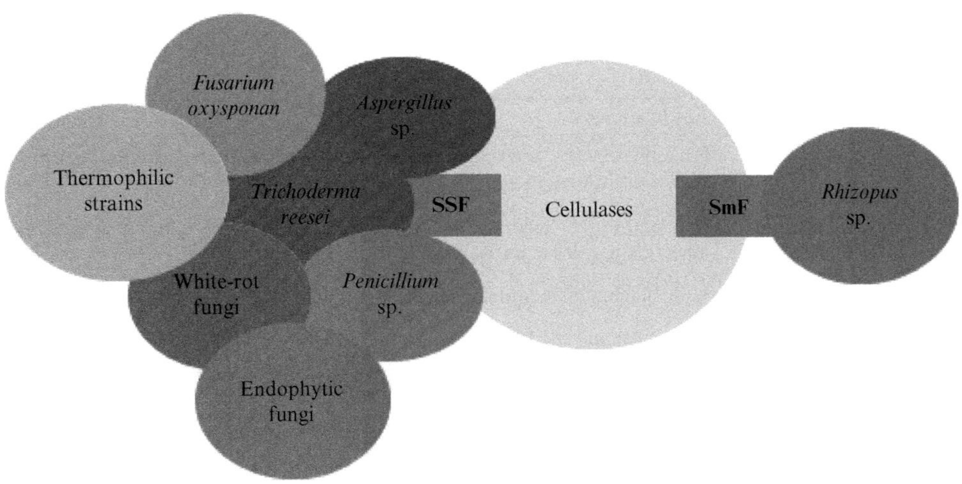

FIG. 9.4 Different fungal species employed in cellulase production by SSF and SmF.

production of cellulase because it requires very low basic expenditure as well as lower operating demands (Xia and Cen, 1999). Cellulase production can be actively affected by the utilization of reasonable and conveniently accessible materials like carbon sources (Senthilkumar et al., 2005). Agricultural wastes like the residues of corn and wheat can be found in abundance at very low prices and they can be very comfortably used as lignocellulosic biomass, which are excellent sources of carbon for the production of microbial enzymes. Significantly increased yields of cellulases by SSF have been achieved by a wide variety of agricultural leftovers by growing microbial inoculum over them (Yang et al., 2006). Commercially available cellulases are basically produced either by the use of thermophilic or mesophilic fungal species in an artificial manner so these Cellulases are of extracellular nature because the cellulose degradation takes place at higher temperatures in the industrial operations (Kim et al., 2005) (Table 9.1).

9.4.1 *Aspergillus* Species

Among the decomposition and decay agents, *Aspergillus* species rank first in possessing the potential to produce a wide range of enzymes inclusive of significantly high production of cellulases. A relatively large number of *Aspergillus* strains participate in the production of cellulases as has already been documented (Wang et al., 2006), but the cellulase-producing potential of *Aspergillus terreus* by SSF cannot be overruled (Emtiazi et al., 2001). Literature suggests that carbon sources influence cellulase biosynthesis in *A. terreus* M11. Carboxymethyl cellulase (CMCase) and filter paper activity (FPase) assays for cellulases and β-glucosidase production by using corn stover transpires to be the best carbon source because of the hemicellulose it possesses as well as its favorable degradation ability (Senthilkumar et al., 2005). Gao et al. (2014) counterclaim that the chief composition of corn stover includes 39% cellulose, 26% hemicellulose, 17% Klason lignin, and 5% ash. However, the chemicals present in the material used as starter and receptiveness of numerous agents, including the physical or chemical associations between them, strongly influence enzyme production efficiency (Panagiotou et al., 2003). Kachlishvili et al. (2006) demonstrated that nitrogen sources produce a variable effect on cellulase production, which is directly dependent on the fungi and compound selected. Different concentrations of the nitrogen source significantly affect enzyme production but higher enzyme activities are obtained with organic nitrogen (Panagiotou et al., 2003). Addition of the yeast extract along with a suitable nitrogen source profoundly enhances cellulase activity to its maximum (Kalogeris et al., 2003a).

A moist environment crucially favors fungal growth in SSF: however, excellent moisture percentage is affected by substrate's nature, nutritional growth of the microorganism, and nature of the desired end product (Kalogeris et al., 2003b). Gao et al. (2014) found that increasing moisture percentage from 60% to 80% enhances cellulase activity by stimulating faster fungal growth, but is also responsible for quick initiation of enzyme production in the case of *A. terreus* M11. The lignocellulosic residue's efficient sources of carbon in SSF and higher moisture content are preferred not only because they speed up *A. terreus* M11 growth but also because they lead to enhanced cellulase production (Kalogeris et al., 2003b). Care must be taken because clump formation can occur if there is excessively higher water content in the medium. These clumps can decrease the aeration efficiency as well as lower hyphal growth, which is ultimately connected to decreased production of enzymes (Panagiotou et al., 2003). Usually, cellulase production remains unaltered within the initial range of pH from 2 to 5 and cellulase activity can reach its maximum at an initial pH of 2. *A. terreus* M11 culture favors the optimal pH of 3–6, pinpointing that this pH is ideal for the degradation of cellulosic materials, and cellulase production can be achieved under acidic conditions as well

TABLE 9.1 Application of Cellulases

Industry	Applications
Biofuel	Ethanol, solvents, and organic acid production, energy-rich animal feed production, along with higher nutritional value
Food and animal feed processing	Fruit and vegetable juice extraction with increased yields, fruit juice clarification, better maceration, color extraction of fruit and vegetables
Textiles	Biopolishing, biostoning, biofinishing
Paper and pulp	Coadditive in pulp bleaching, improved draining, deinking
Agriculture	Control of plant pathogens and disease, improved soil quality
Medical	Antibiofilm agent, treatment of phytobezoar
Others	Reduction in biomass waste, extraction of olive oil and carotenoids

Adapted from Singh, R., Kumar, M., Mittal, A., Mehta, P.K., 2016. Microbial cellulases in industrial applications. Ann. Appl. Biosci. 3, R23–R29.

(Jecu, 2000). Endoglucanase reaches its maximum activity at a pH value of 2, while a pH value of 3 is essential for the maximum activity of β-glucosidase; however, both these enzymes usually prefer a much lower optimal pH (Lechner and Papinutti, 2006).

In SSF, among the most critical substantial factors, temperature plays a prime role in the initiation of cellulase production. *A. terreus* M11 usually requires a temperature between 35 and 45°C but cellulase production reaches its peak if the temperature is 45°C (Maheshwari et al., 2000). If the temperature shoots above 45°C, the biosynthesis of enzymes experiences a significant decline. It has been reported that β-glucosidase and endoglucanase activity for industrial processes require an optimal temperature of 70°C. Enzyme production levels are also affected by time course because at 96 h maximum activity has been reported. After 96 h a gradual decline in enzyme production has been noticed, which also depends on the ratio of amorphous to crystalline cellulose (Ögel et al., 2001). *A. terreus* M11 cellulases when incubated at different times in different buffer solutions with pH ranging from 2 to 7 at 70°C showed that at pH 2–5 endoglucanase is stable, while β-glucosidase is stable at pH 3–5; however, the activities of both enzymes gradually declined if the pH increased above 5. This finding suggests that acidic pH values play a role in maintaining the stability of these enzymes compared to neutral pH values. Endoglucanase and β-glucosidase maintain their activities at 65% and 53% for 6 h at 70°C with pH values of 2 and 3, suggesting that cellulases can be commercially utilized (Gao et al., 2014).

Chandra et al. (2007) cultivated *A. niger* on different, cheaply available lignocellulose sources in SSF and monitored FPase activity for 5 days. On the third day of incubation they observed maximum FPase activity on groundnut fodder, rice bran, and wheat bran except for sawdust. Wheat bran supported the highest FPase activity after the groundnut fodder with *A. niger*, indicating that it can be used for the production of the FPase. From the fourth day onward, wheat bran followed by groundnut fodder and also rice bran showed a decline in FPase activity. Similar to FPase activity, wheat bran yielded CMCase maximum titers followed by groundnut and rice bran, but not sawdust, making it well suited for CMCase production. However, contradictory results were obtained in the case of glucosidase secretion compared with CMCase and FPase activities. Glucosidase production remained basically much less or inconspicuous during the initiation of the incubation process, but gradually increased after the second day of incubation on wheat and rice bran along with fodder from groundnut, as well as sawdust selected as substrates, and remained unchanged during the whole incubation period. Sawdust favored maximum glucosidase production in comparison to other solid substrates. The secretion of extracellular proteins was highest in the case of wheat bran when incorporated with *A. niger*, which also enhanced the secretion of the cellulolytic enzymes on all the selected substrates. After 2–4 days of incubation the pH started to decrease for the selected solid substrates but pH remained unchanged for sawdust and a recovery in pH was noticed when the incubation period was over.

A. niger and *Aspergillus flavus* are capable of producing pectinases, cellulases, hemicellulases, and xylanase on nonstarch polysaccharides like cellulose. Adequate addition of sugars to the lignocellulosic substrates, such as supplementing the substrate with a controlled volume of a specialized medium known as Czapek Dox at the first step, favors propagation and intensification of fungal biomass followed by rapid cellulolytic enzyme secretion when cellulose is present in sufficient quantities. The cellulolytic enzymes released initiate the degradation of cellulosic components of lignocelluloses. Wheat bran increases cellulase production because it is rich in essential elements, it is coarse under damp environment, and it offers an expanded area (Smith et al., 1996). High endoglucanase activity was shown by 1-day old fermented bran because it increases the availability of accessible sites in the cellulose component of lignocelluloses, thereby causing its efficient cleavage at the maximum level. Glucosidase production in SSF by sawdust remains in consistent production because the endoglucanase and allied enzymes by their uniform action cause formation of hydrolyzing products on a regular basis. For higher cellulase production, groundnut fodder follows wheat bran because it is very easily available. Groundnut fodder is a very common cattle feed and its protein content can be increased by the addition of *A. niger* making it more nutritious. *Trichoderma harzianum Rifai* CCMF-470 when grown on wheat bran shows higher titers of CMCase and FPase (Roussos et al., 1991) when compared with cellulase production yield by groundnut fodder, but with *A. niger* grown on distinctive lignocellulose substrates, higher fractions have also been obtained (Hanif et al., 2004). The hydrolyzability of the lignocellulosic materials can be limited by the association of cellulose with lignin and hemicelluloses.

Natural substrates generally favor slow fungal growth but this limitation can be overpowered if the raw substrate is suitably pretreated with mechanical and chemical agents (Cen and Xia, 1999). Structural changes in cellulosic substrates can be induced by pretreatments that can alter the physicochemical properties of the substrate (Zhang and Lynd, 2004). However, their influence on cellulolytic enzyme production by fungal strains in SSF needs more attention (Singhania et al., 2010). Pretreatment methods can alter physical and chemical properties, namely the porosity of the platform, surface volume, as well as the crystalline nature of the solid-state substrate (Brijwani and Vadlani, 2011). Both acidic and alkaline pretreatments augment the enhanced fraction of cellulose compared to the hemicellulosic fraction when soybean hulls undergo pretreatment. A considerable boost in the porous content of the pretreated soybean hulls was observed in comparison to the untreated native soybean hulls. Pretreatment modified the internal structure of the soybean hulls radically, resulting in the partial solubilization and redistribution of the hemicellulose causing substrate swelling (Kumar et al., 2009). The volumetric specific surface (cm^{-1}) for the preprocessed and untreated soybean hulls remained the same because only the external surface area is taken into consideration by particle size analysis. The interfibrillar space or the "internal porosity" is held responsible for allowing large enzyme molecules causing thorough digestibility. The dissolution of cell wall capillaries by chemical pretreatment enlarges the intermicrofibrillar spaces (Esteghlalian et al., 2001).

Finding a simple technique for lumen internal surface area determination is tiresome, and utilizing the external particle diameter as a volumetric specific surface variable does not give a clear picture about the total area taken by the microfibrils (Zhu et al., 2009). However, no significant difference between the chemically processed and native soybean hulls was observed in relation to the volumetric specific surface, although this parameter appears essential for identifying or modifying the current techniques because faster as well as routine investigations of the internal surface area can be easily implemented (Brijwani and Vadlani, 2011). The alkali-pretreated soybean hulls of both mono as well as combinations of *Aspergillus oryzae* and *T. reesei* showed substantially low enzyme production when compared with untreated and steam-treated substrates and substrates pretreated with acid. The alkali pretreatment of the biomass absorbs some of the alkali resulting in lignin and hemicellulose solubilization, distribution, and condensation causing modification of the cellulosic structure (Gossett et al., 1982). The alkali-pretreated sugarcane bagasse gave a significantly decreased yield of cellulase over the untreated bagasse by *T. reesei* (QM 9414) in liquid fermentation (Aiello et al., 1996). Acid pretreatment of lignocellulosic materials can produce additional sugars due to the degradation of lignin, which can act as inhibitor of the reaction. These end products acting as inhibitors can detrimentally affect the production of enzymes as well as the growth of microorganisms (Hendriks and Zeeman, 2009). All cellulolytic activities in *T. reesei* culture showed a substantial enhancement after the steam pretreatment of soybean hulls when compared with the native ones. The soybean hulls pretreated with steam also showed a significantly higher β-glucosidase activity than the native soybean hulls emphasizing that fungal mycelium is in direct correlation with the substrate elements in SSF. However, both steam-pretreated and native soybean hulls showed no significant differences in the case of *A. oryzae* apart from the endoglucanase levels. In the case of *A. oryzae* the soybean hulls pretreated with steam produced a considerably increased elevated fraction of endoglucanases than the native substrate.

Hemicellulose and lignin removal, or degradation by NaOH or H_2SO_4 or H_2O_2 pretreatment, alters the structure of the cell wall and therefore enhances the receptivity of cellulases to cellulose (Ortega et al., 2000). Pretreatment not only upgrades the utilization of the substrate by the microorganisms but also favors increased enzyme production (Pan et al., 2006). For instance, actinomycetes, particularly *Streptomyces* segregated from the soil and grown on Mandel peptone, is also capable of producing endoglucanases (Alani et al., 2008), while *Thermomonospora* sp. (George et al., 2001a) grown in the powder of cellulose paper along with extract of yeast and Tween 80 media gave a higher yield as compared to its growth on wheat bran. *Penicillium* strain *Penicillium brasilianum* IBT, along with yeast extract and pine wood in a bioreactor, gave 0.59 U/mL FPase activity (Jorgensen and Olsson, 2006). *Trichoderma* strain *Trichoderma viride* NCIM 1051 grown on sugarcane bagasse (1.0%) combined with sodium hydroxide (NaOH) gave a good yield of FPase (Adsul et al., 2004). However, the highest concentration of FPase was obtained with *A. niger* IZ9 cultivated on NaOH-treated sugarcane bagasse (Aguiar and Menezes, 2000).

Cellulolytic enzyme production with *A. niger* using potato peel, taking account of the response surface methodology (RSM) in SSF, indicates that enzyme production is significantly influenced by the fermentation time (dos Santos et al., 2012). This can be contributed to the fact that nutrient distribution throughout the fermentation process favors microbial growth and these nutrients decay with time, which in turn affects the enzyme activity of the microbes. As a result, the growth of microbes along with enzyme production also declines gradually. Water content is a very crucial factor in the fermentation process. High water activity decreases substrate porosity causing lower gaseous exchange; however, low water activity decreases microbial growth subsequently lowering enzyme production (Mahanta et al., 2008). Fifty percent moisture content is designated as ideal for enzyme production. Sixty percent moisture content decreases fungal activity by inhibiting fungal growth; however, 40% moisture content favors efficient fungal growth. Xylanase production on a pilot scale by *Aspergillus sulphureus* by SSF with the use of koji noodles (made of fermented rice) is strongly influenced by the water activity of the medium as well as by the dry environment (Lu et al., 2003). *Pleurotus ostreatus* showed a xylanase activity of 25 U/mL when grown on corn cob and oat in liquid fermentation under controlled conditions (Qinnghe et al., 2004).

An expression control mechanism also designated as negative feedback is known to exist among the enzymes and can be activated or switched off by products comprising the medium. In general, the enzymes catalyzing the first steps of a particular metabolic pathway are usually inhibited by the end products in that particular pathway

(Santana et al., 2012). Enzyme production with corn malt is very low at the beginning and is then boosted to its peak, and afterward the fraction of the final products is repressed bringing a decline in its activity (Biazus et al., 2006). Cassava starch hydrolysis after 72 h of fermentation shows higher enzymatic activity by *A. niger* (Omemu et al., 2005; Alva et al., 2007). Enzyme activity undergoes a radical decline if the incubation time is increased because of the formation of by-products due to the microbial metabolism because depletion of nutrients from the medium starts to occur, which further inhibits the growth of fungal cells causing decreased enzyme production (Shafique et al., 2009).

The present-day technique of engineering for economically high production with the help of a unique system of cellulases is comprised of exoglucanases, endoglucanases, and β-glucosidases. This new system has been designed with a brand-new fungal strain, *A. niger* RCKH-3, with the help of the SSF process, and overall evaluation criteria are used as an analysis tool (Hemansi et al., 2018). At 30°C, optimum enzyme production has been reported for *A. niger* RCKH-3 (Saini et al., 2017) with a pH of 5.0 as enzyme denaturation occurs at analogously higher temperatures (Li et al., 2013). The proton motive force is generated in the membranes of the cells and its variations usually cause a decline in enzyme production at optimum pH (Pirt, 1975). Increase in moisture ratio increases enzyme production up to a certain level and then declines because of decreased nutrient diffusion (Gervais and Molin, 2003), and the rate of oxygen transfer also decreases with elevated moisture levels (Mrudula and Murugammal, 2011). An inoculum size of 30% v/w promotes maximum enzyme production due to the presence of an optimum balance between the available nutrients and propagating fungal biomass for enhanced enzyme production (Saini et al., 2017). Enzyme production is greatly affected by changing the inoculum size because insufficient nutrients present in the medium can interfere with enzyme secretion by the microbes (Dutt and Kumar, 2014). A lower size of the inoculum can lead to suboptimum cell biomass production (Saravanan et al., 2013). Enzyme production by using beef extract is boosted because it is a rich source of complex nitrogen, which is mandatory for microbial growth causing increased enzyme production (Gautam et al., 2011). Protein synthesis is strongly affected by ammonium salts (El-Hadi et al., 2014), and PEG 8000 (a nonionic surfactant) modifies cell membrane permeability by releasing cell-bound enzymes, which in turn enhance cellulase production (Abdel-Fatah et al., 2012). Higher individual influences on cellulase production by inoculum age, NH_4Cl, and beef extract have been observed, while PEG 8000 shows a minimum or negligible effect on enzyme production.

9.4.2 *Trichoderma reesei*

Bioconversion of lignocellulosic biomass is activated by the lignocellulolytic enzymes but their rates of production for industrial application and bioprocessing have been a disputable topic for some time (Mendes et al., 2017). This demands the search for microbes capable of producing lignocellulolytic enzymes with higher precision by utilizing easily accessible sources of carbon (Fang and Xia, 2015). Da Silva and coworkers have been successful in producing lignocellulolytic enzymes by growing the filamentous fungus *T. reesei* CCT2768 on the straw of carnauba by the SSF method (da Silva et al., 2018). The native Brazilian palm tree *Carnauba* (*Copernicia prunifera* [Miller] and H. Moore) is extensively employed in carnauba wax extraction, which produces a significantly high amount of fiber. This fiber is rich in lignocellulosic residues with higher carbon content, which can be utilized for bioethanol or cellulase production with the help of biorefining processes (Melo et al., 2012). Lignocellulolytic enzyme production for 168 h in an Erlenmeyer flask by SSF gives outstanding FPase activity and after 96 h of fermentation, *T. reesei* CCT2768 shows maximum protein secretion.

The production of cellulolytic enzymes with hydrogen peroxide-alkaline (HPa) pretreatment has been known to be ideal because at the same time it can effectively remove lignin and hemicellulose. Reports indicate that recalcitrance of lignocellulosic materials can be reduced by pretreatments that further modify digestibility and access of microorganisms to cellulose (Gonçalves et al., 2015). Lignin and hemicellulose removal can increase the access of fungus for cellulosic structure alteration and therefore supports higher enzyme secretion. Moreover, fewer recalcitrant-inhibiting substances are generated by HPa pretreatment (Rabelo et al., 2014). The total time spent in the production of enzymes is also an important factor to be taken into account. The peak FPase activity for *T. reesei* extract pretreated with hydrogen peroxide is generally obtained at 72 h of fermentation (Hansen et al., 2015).

Bioprocesses are greatly affected by decreasing the time involved in producing an enzyme that appears to be a prime component because this factor favors elevated concentrations with reduced operational charges. Pretreated alkaline *Carnauba* straw shows higher xylanase production at 48 h because lignin is removed and no significant alterations in the hemicellulose proportion of the fibers occur (Arora et al., 2015). Enzyme induction by use of carbon sources rich in lignocellulose gathered from wastes of agricultural crops minimizes the enzyme production rates and the enzymes produced can be utilized in various bioindustrial applications like cellulosic ethanol production. A twofold increase in cellulosic enzyme production was achieved with HPa-pretreated *Carnauba* straw; however, the increase was sevenfold when HPa-pretreated *Carnauba* straw was supplemented with the agricultural residues of sugarcane and cashew (da Silva et al., 2018). Higher xylanase production was obtained with HPa-treated *Carnauba* straw residue. Total proteins secreted by *T. reesei* CCT2768 also show equal significance in the case of HPa-pretreated *Carnauba* straw residues, which strengthens the fact that the production of cellulolytic enzymes increases with HPa pretreatment (Rajan and Carrier, 2014). Not only this, different performances have been shown by pretreatment, which in turn depends on production of the cellulolytic enzymes or production of ethanol, for example, for ethanol production by alkaline or acid alkaline, pretreatment causes losses of lignin and pentoses in considerable amounts. This limitation can be overcome by the use of engineering strains that can ferment pentoses (Gonçalves et al., 2016). For the production of enzymes, hemicelluose residue and lignin are mandatory for the production of xylanases because they serve as sources of energy after being burnt; they can also be utilized for the synthesis of polymers (Chen and Wan, 2017).

Residues from corn, for example, corncob in its native untreated form, can serve as a platform for growing *T. reesei* ZU-02 by SSF in batch cultures, and cellulases can be produced for commercial purposes (Xia and Cen, 1999). Right after fermentation of one batch (higher peak of cellulase concentration was observed), soaking and slowly stirring of the koji enzyme are very important for the extraction of cellulase. For the next batch fermentation, after the cellulase protein harvesting step, the material used as a platform for the growth of the fungal mycelia along with spores of the used fungi were used without fresh substrate or inoculum. This whole batch fermentation process was repeated a couple of times and maximum fraction of cellulase was obtained in just 5 days of SSF. Deep trough fermenters give a higher cellulase production with a better quality that can be used for commercial purposes.

Tween 80 is a surfactant that deeply influences the structure of the cell membrane of *T. reesei*, thereby improving the enzyme's secretion (Dutt and Kumar, 2014). Enzyme activity can be considerably increased by the addition of Tween in culture media (Liu et al., 2006) and CMCase induction is done by betaine in the culture medium; however, *Acremonium cellulolyticus* enhances β-glucosidase and xylanase production (Yamanobe et al., 1988). Cellulase production by *Trichoderma longibrachiatum* can be improved by the addition of nonionic surfactants too (Kristensen et al., 2007). With the presence of carboxymethyl cellulose in the growth media of *Penicillium funiculosum*, not only is the endogluconase gene switched on but the xylanase gene is stimulated (Rao et al., 1988). Cellulase and xylanase genes are switched on by lactose addition (Singhania et al., 2010) and they also affect the Xyr1 (xylanase regulator 1) in the case of *T. reesei*; in addition, it also affects the hydrolytic genes, namely xyn1, xyn2, cbh1, cbh2, egl1 transcription that encodes Xylanase I, Xylanase II, Cellobiohydrolases I, Cellobiohydrolases II, and Endoglucanase I (Stricker et al., 2007). Cellulases as well as xylanases are activated by the effect of xyn1, xyn2, cbh1, cbh2, and egl1 genes in *T. reesei* and these genes are located differently on the chromosomes and so are switched on and off in a variable fashion (Ramoni et al., 2017).

The production of lignocellulose-degrading enzymes is significantly influenced by the substrates. The genes associated with CMCase and FPase production in general are induced by cellulose. Lowest lignin levels have been observed in the treated wheat bran but comparatively higher cellulose and hemicellulose levels occur at the same time. Microorganisms can easily access cellulose if the lignin content is low and therefore the treated wheat bran serves as a perfect carbon source aiding in CMCase and FPase production. Xylanase-encoding gene expression can be simultaneously recuperated by hemicellulose and cellulose. The highest cellulose percentage along with the lowest lignin content has been observed in the treated rice husk. However, the untreated rice husk is extremely low in cellulose content with excessively higher lignin levels rendering itself excellent for commercial xylanase production. *Thermomonospora* sp. cultured on corn residue (corncob) impregnated with powder of cellulose paper when given a dose of Tween 80 showed excellent elevation in the production of xylanases (George et al., 2001b). Xylanase production by *Trichoderma* sp. significantly decreases when the culture medium is supplemented with lactose and the straw from both wheat and rice (Sanghvi et al., 2010). Xylanase and CMCase activity by *T. viride* is enhanced up to 70% and 30% when Tween 80 is supplemented with rice straw (Liu et al., 2006). Addition of adequate amounts of betaine along with other additives to the wheat straw, wheat bran, and rice husk media brings a significant increase in CMCase and FPase production (Taherzadeh-Ghahfarokhi et al., 2018).

9.4.3 *Penicillium* Species

During the 19th century the antibiotic penicillin produced by the fungus *Penicillium* emerged as a wonderful blessing against bacterial infections caused by staphylococci and streptococci bacteria. A few years ago it was found that the fungus *Penicillium* is also capable of producing cellulase. It can produce all three of the required cellulase components, including sufficient amounts of β-glucosidase, which makes it superior over the classically used fungus *Trichoderma* (Pandey, 1995). Later, it came into light that another species of *Penicillium*, for instance,

Penicillium chrysosporium commonly known as a white fungus, could produce lignocellulolytic enzymes indicating that it can degrade lignocellulose (Broda et al., 1996). *P. chrysosporium* has its own cellulase system of cellobiohydrolases (CBHI) of which cellobiohydrolase 4 is the chief component along with 1 and 6 CBHI analogs (Covert et al., 1992). It turned out later that *P. chrysosporium* has an endoglucase 28 (EG28), which is 28 kDa long and is deficient of carbohydrate-binding module (Henriksson et al., 1999). The synergism occurring between CBHI and EG28 indicates that this endoglucanase 28 is homologously similar to EGIII found in the *Humicola insolens* and *T. reesei* fungal species.

High titers of cellulase are produced by *Penicillium* species (Jørgensen et al., 2005) and currently numerous reports are available that indicate that improvement of *Penicillium* strains by mutation can significantly increase the production of cellulases (Adsul et al., 2007). According to a report, five different cellulases are produced by *P. brasilianum* (Jørgensen et al., 2003). It has been proven that different species of *Penicillium* offer an exceedingly diversified ratio among FPase and β-glucosidase activity in comparison to the *Trichoderma* species (Syed et al., 2013). According to the literature, some *Penicillium* species display significantly higher fractions of cellulase production and among them *Penicillium verruculosum* (Solov'eva et al., 2005), *Penicillium pinophilum* (Brown et al., 1987), *Penicillium occitanis* (Chaabouni et al., 1994), *Penicillium janthinellum* (Singhania et al., 2014), *P. funiculosum* (de Carvalho et al., 2014), *Penicillium echinulatum* (Reis et al., 2013), *Penicillium decumbens* (Liu et al., 2013a, b, c), *Penicillium crustosum*, *Penicillium chrysogenum* (Hou et al., 2007), *Penicillium citrinum* (Dutta et al., 2008), *P. brasilianum* (Jørgensen et al., 2003), and *Penicillium purpurogenum* (Suto et al., 1991) can be taken into account. The technology of bioconversion for producing cellulase enzymes by *Penicillium* depends on the flexibility of the overall process.

The aerobic microorganisms participating in the secretion of cellulases show resemblance in their growth and cellulose utilization pattern to that occurring in SSF, which is very much different when compared to SmF (Zhu et al., 2009). However, submerged cultures offer more pronounced advantages because completely controlled monitoring and easy handling facilities are available (Mathew et al., 2008). *Penicillium* strains can be easily grown by SSF and can be utilized for the production of cellulases because they have a wider stability range toward pH, metal ions, temperature, and other interfering factors. *P. citrini* isolated in SSF for cellulase production shows higher tolerance against alkali (Dutta et al., 2008). Enzyme extraction after the completion of fermentation from the fermented matter demands higher labor cost but the cellulase biomass produced by SSF does not need any further processing and can be used directly because it is easily bioconvertible (Singhania et al., 2015). Therefore SSF lowers the cost of the operational process, which in turn further reduces the overall technology cost (Singhania et al., 2010).

Extensive information can be obtained from the literature regarding different sources that can enhance cellulase production with the help of *Trichoderma* species; however, little information is currently available concerning *Penicillium* sp. There is a difference between the substances that give positive results with *Trichoderma* sp. that are not compatible with species of *Penicillium*. For instance, cellulase production by *T. reesei* increases if lactose and sophorose are used as inducers, but they fail to cause any effect for *Penicillium* sp. (Kurasawa et al., 1992). Inducers such as acetyl cellobiose in the case of *P. purpurogenum* can induce cellulase production (Suto et al., 1991) and gentiobiose is known for other *Penicillium* sp. (Kurasawa et al., 1992). The presence or absence of light can also cast an effect on cellulase induction despite the carbohydrate or disaccharide sources (Agrawal et al., 2016). Currently, although *Penicillium* sp. have emerged as good replacing agents for the *Trichoderma* sp. employed in producing cellulases (Gusakov and Sinitsyn, 2012) there is a greater need of selection of the easily available disaccharide sources as well as allied chemicals that can boost the production of cellulases. *P. janthinellum* mutant EMS-UV-8 is capable of producing all the cellulolytic activities, including β-glucosidases when supplemented with disaccharides, and its derivatives showed higher cellulase production. The efficiency of this mutant indicates that it can be a potential candidate employed for the commercial production of cellulases only if a suitable inducer is added to it (Singhania et al., 2014).

A few new inducers like cellobiose, octaacetate, turanose, melibiose, α-D-cellobiose, crocin, gentiobiose, and trehalose were tested for cellulase induction by Agrawal et al. (2016). They found that octaacetate appears to be a very promising cellulase inducer in lower concentrations. *P. janthinellum* EMSUV-8 mutant with octaacetate shows remarkably high β-glucosidase as well as FPase activities. Cellulase production by this strain is strongly influenced by light intensity because it affects the growth of mycelia. The potential targets partaking in the regulation of upstream transcription have also been identified. For the *Neurospora crassa* and *T. reesei* species, cellobiose plays the role of natural cellulase inducer because it is a good source of carbon (Zhou et al., 2012). The import of β-glucosidase and cellobiose intracellulary keeps a check on the intracellular cellobiose levels, but cellobiose acts as an inducer of cellulase in *Penicillium*. Cellulase expression can be induced by gentiobiose in *Penicillium* (Kurasawa et al., 1992), which suggests that there is a marked difference between the induction mechanisms of *Penicillium* and *T. reesei*. It is impossible to overtake the popularity of *Trichoderma* cellulases despite the fact that *Penicillium* species can also potentially produce significantly better cellulases. Among the biomass-degrading fungi, *T. reesei* mutant strains rank first in cellulase production, although numerous reports support the superiority of *Penicillium* cellulases in comparison to the cellulases obtained from the strains of strains (Maeda et al., 2011).

CBHI obtained from *P. funiculosum* and endoglucanase from *Acidothermus cellulolyticus* when grown together on corn stover for 72 h converted 69% of cellulose. However, the *T. reesei* CBHI along with endoglucanase from *A. cellulolyticus* converted 52% (MRI, 2004). *Aspergillus awamori* combined with the CBHI obtained from *P. funiculosum* gave 65% conversion efficiency. The practice of using CBHI by recombinant technology shows higher conversion efficiency in a half run in comparison to the full runs performed by using *T. reesei* CBHI only (Castro et al., 2010). China is using *Penicillium oxalicum* species for the production of cellulase commercially. *Penicillium* species show a higher ratio of β-glucosidases, which has brought *Penicillium* species in competition with *T. reesei*. Cellulases produced by *Penicillium* species by utilizing different lignocellulolytic substrates have shown 2–3 times higher yields of glucose hydrolysis than the *T. reesei* enzymes (Gusakov, 2011). Sufficient secretion of endoglucanases and β-glucosidases has been reported for different *Penicillium* strains that can contribute to the successful production of cellulases (Jørgensen et al., 2005). *Penicillium* strains are the chief commercial producers of cellulases in China, although *Trichoderma* is also being used for cellulase production; however, Adisseo (France) uses *P. funiculosum* for the production of Rovabio enzymes composed of β-glucanases, cellulases, and xylanase (Adisseo, 2018). Huang and coworkers (Huang et al., 2015) isolated and screened *P. oxalicum* EU2106, which is a hypermutant of *P. oxalicum* HP7-1, a filamentous fungus from Huaping National Natural Reserve in Guilin, China (Zhang et al., 2014), and concluded that it can produce cellulase after three ^{60}Co γ-ray and two ethylmethane sulfonate or ultraviolet mutagenenic treatments.

Cassava is a widely used crop in the tropical and subtropical regions, following rice, maize, and sugarcane, and is considered to be an economical source of carbohydrate as well as energy. Apart from being a source of fiber it also participates in the production of bioethanol and other commercially important by-products (Zhang et al., 2016). During the process of bioethanol production the cassava residue is usually fed to the animals or is used as a manure. Cellulose, hemicellulose, and lignin are the major constituents of cassava and if they are allowed to degrade naturally they cause pollution because lignocellulosic degradation into easily decomposable sugar units is difficult (Zhang et al., 2011). In recent years, biohydrogen and biomethane production by the bioconversion of cassava residue is in great demand (Zhang et al., 2014). The biorefinement of materials rich in cellulose requires the following actions: lignocellulosic materials must be pretreated, enzymes must be hydrolyzed, and biochemical resultants must be fermented and purified (Xue et al., 2016). Basically, cassava residue is usually pretreated either with acidic or alkaline chemicals and these chemicals usually have higher energy demands along with the use of sophisticated techniques requiring more energy and special machineries that are expensive and require maintenance (Zhang et al., 2016). *P. oxalicum* EU2106 grown on pretreated sugarcane bagasse and pulp of sugarcane bagasse showed elevated cellulase levels in SmF (Xue et al., 2016), but production of cellulase by this strain in SSF is still unclear. Su et al. (2017) used the Box-Behnken design method of RSM to analyze cellulase production by *P. oxalicum* EU2106 in SSF where they pretreated the cassava residue with NaOH-H_2O_2. For elevated cellulase secretion they used wheat bran, which is an economical source of carbon for growing *P. oxalicum* EU2106. Several studies have shown that wheat bran is the perfect agricultural waste for growing *Trichoderma*, *Aspergillus*, *Penicillium*, and other filamentous fungal species for achieving greater success in the commercial production of cellulases and xylanases (Farinas, 2015). The cellooligosaccharides in wheat bran have also been demonstrated to be one of the most significant factors for the induction of cellulase production among the filamentous fungi (Sun et al., 2008).

RSM along with cultivation of substrate, moisture level of the substrate, and pH of the culture media along with the initial and final temperature requirements predict that this technique can boost the production of cellulases; however, selection of the microbe as well as growing substrate is always the most important factor (Farinas, 2015). The liquid/solid ratio with zero-level parameter interaction has an effective on the production of cellulases as has been found in single factor experiments performed in SSF. Microbial growth and metabolism are significantly affected by the moisture levels of the substrate in SSF because water participates in diffusion and nutrient absorption by microbes and maintains functioning and stability of biological macromolecules. Several filamentous fungi, such as *T. reesei* (Brijwani et al., 2010) and *A. niger* (Delabona et al., 2013), have demonstrated considerably higher cellulase and hemicellulase production when the moisture level of the substrate was between 40% and 90%. Yeast or any other microorganism capable of carrying out fermentation can be grown on the cassava residue and may be used as a source of produce for value-added chemicals or biofuels. Xu et al. (2014) observed that *Hypocrea orientalis* gave 81% yield of cellulase when grown on cassava residue, while Su et al. (2017) observed that *P. oxalicum* EU2106 grown on cassava residue pretreated with NaOH-H_2O_2 in SSF completely catalyzed cellulose into glucose between 24 and 96 h, which suggests that *P. oxalicum* EU2106 can serve as an efficient tool for the bioconversion of lignocellulosic residue on a commercial basis.

9.4.4 White-Rot Fungi

9.4.4.1 *Pleurotus ostreatus* and *Pleurotus sajor-caju*

White-rot fungi utilize lignocellulose materials for producing lignocellulases that share diverse functions in the commercial manufacturing of value-added products (Pointing, 2001). It has been well documented that white-rot fungi prefer solid substrates for their growth and therefore find their usefulness in different bioprocesses accompanying agricultural solid waste as substrates (Massadeh et al., 2001). Reddy et al. (2003) assessed the potential of agricultural residues of banana such as the leaf and pseudostem wastes that can serve as sources for growing two white-rot fungal species, namely *P. ostreatus* and *P. sajor-caju* by SSF and therefore lignolytic and cellulolytic enzymes can be produced. These fungal species release extracellular proteins while growing on banana waste and secrete different enzymes, namely CMCase, FPase, xylanase, laccase, lignin, and peroxidase. They were surprised that these fungi when grown on the leaf residues showed a twofold increase in extracellular protein content when compared with the pseudostem residues, pinpointing that growth conditions influence microbial growth because leaf biomass offers a large surface area that directly influences enzyme activities. It was also found that the layout of the fibers of cellulose may affect the degradation potential of several products derived from paper (Venelampi et al., 2003).

Laccase is an extracellular enzyme that modifies the lignin structure by its oxidoreductive action but its function is related to fruiting body development and stage of morphogenesis (Wood, 1984). When mycelia are in the growing stage, laccase concentration is higher and the concentration further increases until mycelia reach their maximum stage of development. *Agaricus bisporus* (Wood, 1980), *Lentius edodes* (Leatham and Stahmann, 1981), *Coprinus congregatus* (Ross, 1982), and *Schizophyllum commune* (Leonard and Phillips, 1973) have shown elevated levels of laccase production during their vegetative growth and it continues until they produce fruiting bodies. Changes in the morphology of the fungus can be noticed easily with the help of laccase because its activity starts to decline when the fungus attains its full mycelial growth. *P. sajor-caju* and *P. ostreatus* produced higher concentrations of laccase in comparison to xylanases and cellulases when grown on the leaf waste of banana. Before the completion of 20 days, *P. sajor-caju* and *P. ostreatus* showed a declination in laccase activity levels while growing on the leaf and pseudostems of banana, and this fact indicates that they require less than 20 days for their mycelia to reach their peak growth phase. *P. chrysosporium* when grown in SSF also showed similar activity levels of lignin peroxidase and laccase (Laplante and Chahal, 1993). Therefore, it is clear that for the production of lignocellulolytic enzymes, residues obtained from banana can be used as a substitute for other residues obtained from agricultural wastes (Martinez et al., 2002). *Pleurotus pulmonarius* and *P. ostreatus* when grown on wheat straw showed variations in the production the lignocellulolytic enzymes when provided with different growth conditions (Velazquez-Cedeno et al., 2002). Elevated laccase production was obtained when *P. sajor-caju* was grown on rubber tree sawdust (Eichlerova et al., 2000), while Reddy et al. (2003) achieved maximum laccase production when *P. ostreatus* and *P. sajor-caju* were allowed to grow on the biomass of leaves. For the commercial production of enzymes, higher amounts are needed and this can be achieved by modifying the growth conditions supplemented with methods that improve the strains and use of boosters that elevate the growth of microbes (Eichlerova-Volakova and Homolka, 1997). Laccase obtained by these fungal species can be used for pollutant detoxification as well as in the treatment of industrial wastewater (Hublik and Schinner, 2000).

The toxic chlorinated and other associated can undergo bioremediation with the help of lignin peroxidases and growing *Pleurotus* residues of banana leads can escalate, simplify as well as bring a much-reduced cost of production of these valuable enzymes (Lamar, 1992). The CMCase and FPase activities were comparatively very low when *P. ostreatus* and *P. sajor-caju* were allowed to grow on the pseudostem and leaf residues of banana. Rai and Saxena (1990) also reported that in *Pleurotus* sp. rice straw there was no FPase activity, while CMCase levels were much lower than the expected levels. When crystalline cellulose and sawdust were used as a growth medium for *L. edodes* and *P. sajor-caju* there was a total absence of cellulolytic activity (Buswell et al., 1993). The efficiency of white-rot fungi to degrade cellulose is limited by a low FPase activity, which is in correlation with the CMCase activity that degrades highly crystalline cellulosic molecules (Leatham, 1985). Some white fungi degrade lignin in this way where there is no loss of cellulose molecules from the lignocellulosic biomass (Ander and Eriksson, 1977). When *P. ostreatus* and *P. sajor-caju* are allowed to grow on residues of banana there is greater reduction in the lignin percentage in comparison to the cellulose fraction (Krishna and Chandrasekharan, 1996).

P. ostreatus and *P. sajor-caju*, however, produce very low levels of xylanase when grown on the residues of banana but they can serve as better tools in the pulp industries. The commonly used bleaching agents such as oxygen and hydrogen peroxide can be merged with xylanases to advertise chlorine-free pulp technology. The utility of banana residue to be used as a food for animals can be enhanced by modifying the time duration of the batch culture for effective SSF (Singh et al., 1989), and *Pleurotus* sp. can be used to solve the problem of environmental pollution, pulp production, and other value-added compounds (Cohen et al., 2002). Banana waste appears to be an economical and easily available source for ensuring the growth of *P. ostreatus* and *P. sajor-caju* for effective enzyme production.

9.4.4.2 *Lentinus edodes*

Basidiomycetous fungi have been in great demand for the commercial production of ligninolytic enzymes and studies are still ongoing to identify new species for this purpose (Mikiashvili et al., 2004). White-rot fungi produce a mixture of enzymes, for example, lignin peroxidases, laccases, and manganese peroxidases (MnPs), which degrade the lignin completely. Few species of white-rot fungi secrete all three enzymes; however, some secrete either one or two types. A dietary fungus, *L. edodes* (shiitake), finds potential use in medicine production along with several environmental and biotechnological utilizations. This fungus can degrade the lignocellulosic biomass into simpler components by secreting hydrolytic and oxidative enzymes essential for its growth (Reddy et al., 2003). The lignin-producing efficiency of this fungal species has been scantily studied (Elisashvili et al., 2006). The degradation of lignocellulosis biomass by basidiomycetes demands complete evaluation of the activities of hydrolytic and oxidative enzyme activities because the hemicellulosic, cellulosic, and lignin components are very complexly joined to each other. The machinery forming the basis of enzyme synthesis in the bioconversion of lignocellulose helps to improve the whole technological process of this edible mushroom species. A large bulk of low-cost enzymes is needed in the biotechnological applications because these enzymes can utilize the lignocellulosic wastes rich in soluble carbohydrates and inducers leading to efficient enzyme synthesis and in turn bring down lignocellulosic degradation (Moldes et al., 2004). Appropriate plant residue selection that can aid in the successful growth of fungus along with efficient enzyme synthesis are the two main factors that form the backup of an efficient biotechnological operation. *L. edodes* and *Pleurotus* spp. must be thoroughly investigated for their oxidative and hydrolytic enzymes when the fruit bodies have been harvested because the substrate left in the form of residue may serve as an economical agent for the production of lignocellulolytic enzymes and therefore act as an efficient tool in the bioremediation and enzyme production processes (Elisashvili et al., 2003).

Basidiomycetes can stimulate enzyme production by lignocellulosic wastes (Elisashvili et al., 2006) and Elisashvili and coworkers (Elisashvili et al., 2008) found that *L. edodes* and *Pleurotus* species share very different origins that can produce lignocellulolytic enzymes by SSF and SmF if raw plant materials are used as substrates. Wide differences occur not only between the two species but also among the strains of the same species even if they are cultivated in identical culture conditions. Heproël and colleagues (Heproël et al., 2000) pointed out that there are quantitative variations in the enzyme activities of *L. edodes* and *Pycnoporus cinnabarinus* strains when cultivated under ideal conditions. Laccase and MnP have been known to be produced by *L. edodes* and several *Pleurotus* species (Elisashvili et al., 2006). The appropriate selection of a suitable growth substrate and the method of fungus cultivation are the key players responsible for enhanced laccase and MnP secretion. When eucalyptus and rice bran residues were used as a substrate for growing strains of *L. edodes*, remarkable differences were observed in the MnP and laccase activities (Silva et al., 2005). Laccase and MnP activities have been best reported for the *Pleurotus dryinus* IBB 903 and *P. ostreatus* 2191 strains, respectively. Laccase production by *P. dryinus* IBB 903 significantly accelerates if copper and xylidine are added to the culture medium but MnP accumulation is enhanced if manganese is added to the mandarin peels in SmF (Elisashvili et al., 2006). It has been proved that *P. dryinus* IBB 903 requires a lignocellulosic substrate for producing MnP because complete loss of enzyme production has been observed if the fungus is allowed to grow in an artificial environment impregnated with distinctive sources of carbon (Elisashvili et al., 2006).

L. edodes can produce moderate amounts of cellulolytic enzymes with lower exo-1,4-β-D-glucanase levels (Leatham, 1985). *P. ostreatus* and *P. sajor-caju* strains when grown on residues of banana and *L. edodes* cultivated on wheat straw in SSF show very low CMCase, FPase, and xylanase levels (Reddy et al., 2003; Mata and Savoie, 1998). Similarly, other brown and white-rot basidiomycetes have also shown very low levels of hydrolytic enzyme production (Nazareth and Sampy, 2003). However, *Funalia trogii* and *Polyporus ciliatus* cultivated on grape vine cuttings, sawdust, and mandarin peels by SmF show radically high activities of CMCase and xylanase (Tsiklauri et al., 1999). Higher cellulase and xylanase activities have been reported for *P. dryinus* IBB 903 and *P. ostreatus* IBB 2191 when cultivated on lignocellulosic materials in both SSF and SmF processes (Elisashvili et al., 2008). These strains remarkably show higher CMCase, xylanase, and FPase activities, which are considered to be the best obtained from other strains of oyster mushrooms. This finding indicates that all the lignocellulosic polysaccharides can be degraded at higher rates by some of these edible mushroom strains with sufficient proportions of carbon and energy sources responsible for mycelia formation causing extensive substrate colonization; finally fruit body formation must be provided. Considerably higher cellulase and xylanase activities have been noticed for both commercial oyster mushroom strains when cultivated on wheat straw and tree leaves in SSF. Apart from the fungal distinctiveness, characteristics of the lignocellulosic material also govern the fungal lignocellulolytic expression. Varying the lignocellulosic growth substrate brings alterations not only in the ratios of hydrolytic and oxidizing enzymes but also variations in the individual enzymes. In SmF, excellent fungal growth has been observed, which gives a clear indication that enzyme production is substrate dependent. Basidiomycetes' biosynthetic potential is highly focused on the method adopted for fungal cultivation. *P. dryinus* IBB 903 when cultured on tree leaves shows five- and threefold higher CMCase and xylanase productivity in SmF than in SSF. MnP secretion increases with plant substrates in SSF, while it decreases or completely inhibits MnP production by *L. edodes* and several *Pleurotus* strains in SmF (Fujian et al., 2001).

9.4.5 *Fusarium oxysporum*

Fusarium oxysporum isolated from cumin is a filamentous fungus already known as a good cellulase- (Christakopoulos et al., 1995) and xylanase- (Christakopoulos et al., 1996) producing agent by SmF absorbing carbon for its growth and metabolism from the residues of lignocellulosic materials. *F. oxysporum* can elicit the bioconversion of cellulose along with xylan into ethanol without any hindrance, where it participates in successive steps of polysaccharide hydrolysis followed by the fermentation of the resulting oligosaccharides (Christakopoulos et al., 1989). Growth in the solid-state culture (SSC) is highly governed by the moisture content of the medium. Studies have shown that with high moisture content and use of lignocellulosic substrates as the main carbon source, fungal growth and enzyme production are enhanced (Kalogeris et al., 1998). It has been reported that for multienzyme production, initial culture pH and optimum moisture level above 80% are very important, and xylanolytic and cellulolytic enzymes have shown higher activities at an initial pH of 7 with 27°C temperature but the temperature must be 33°C for endoglucanase production (Panagiotou et al., 2003). Different concentrations of nitrogen sources significantly affect enzyme production. The commercial production of enzymes by a solid-state bioreactor requires 80% water content, alkaline pH (7), 30°C temperature for growth initiation of microbes, followed by availability of nitrogen source (7%), which produces tiny levels of different cellulases within 24 h of continuous operation. A continuous steady increase was noticed for the xylanases and endoglucanases between 50 and 144 h. The heterogeneous environment that crops up during the process of inoculation generates a gap of 24 h, which affects the production of enzymes by the bioreactor as well as the flask trials. A bioreactor produces 10%–15% decreased enzyme fractions in comparison to the Erlenmeyer flask, and similar findings have been obtained for *Thermoascus aurantiacus* in SSC in the context of the secretion of xylanase (Kalogeris et al., 2003a). The cellulase and hemicellulase activities cannot be quantitatively compared because standard assay conditions are lacking and also their yields are usually lower when wheat straw was used to grow *T. aurantiacus* in an SSC bioreactor (Kalogeris et al., 2003b), but the yields are generally higher than other microorganisms grown in SSC (Topakas et al., 2002). At least four extracellular xylanases have been produced by *F. oxysporum* whose separation can be done by isoelectrofocusing.

9.4.6 Endophytic Fungi

Endophytic fungi form a diversified group of microorganisms inhabiting most of the plant species occurring in the natural ecosystem without displaying any external symptoms (Kogel et al., 2006). The symbiotic relationship of these endophytic fungi with plants as well as their ability to control pests and present valuable compounds intensifies the demand of these microbes in the modern era (Chapla et al., 2014). According to a recent report they have been identified for their ability to produce hydrolases (oxidative ligninolytic enzymes, cellulases, xylanases, etc.) (Amirita et al., 2012). Endophytic fungi can invade and colonize within the plant tissues with the help of these enzymes and can obtain nourishment. However, little information is available about different species of endophytic fungi and therefore their usage in biotechnological applications is at a preliminary stage (Corrêa et al., 2014). Filamentous fungi are capable of producing proteins that can be used on a commercial basis, and cellulases and xylanases produced by them find greater roles in biotechnological processes (Novy et al., 2015). Endophytic fungi depend on suitable media for their effective growth, which in turn releases enzymes capable of degrading plant residues by complete hydrolysis (Katoch et al., 2014). Endophytic fungi like *Aspergillus*, *Penicillium*, *Chaetomium*, *Acremonium*, and *Alternaria* species have been studied in great detail for cellulase production by SSF, while information about other species is still lacking (Onofre et al., 2013; Yoon et al., 2014).

Endophytic fungi cultivated on sugarcane bagasse and wheat bran by SSF show a considerably higher production of cellulases and xylanases (Moretti et al., 2014). They also show higher endoglucanase and β-glucosidase activities in SSF (Damaso et al., 2012) but their FPase activities are slightly variable (Dashtban et al., 2010). Cellulase production by *T. viride* PAJ 01, *Saccharicola* sp. EJC 04, *Botryosphaeria* sp. AM 01, *Ustilaginoidea* sp. XYA 04, *Ustilaginoidea* sp. CV 04, *Cladosporium cladosporioides* PAJ 03, *Paecilomyces* sp. SF 021, and *Phomopsis stipata* SC 04, and on a mixture of lignocellulosic substrates by SSF, showed a remarkable increase. *Botryosphaeria* sp. AM 01 cultivated on cotton seed meal:wheat bran showed higher endoglucanase and β-glucosidase production. *Paecilomyces* sp. SF 021 and *Ustilaginoidea* sp. XYA 04 showed higher endoglucanase production, and *Ustilaginoidea* sp. CV 04 produced both endoglucanases and β-glucosidases when grown on sugarcane bagasse:oat. Better xylanase and β-xylosidase production was obtained with *Botryosphaeria* sp. AM 01 and *Paecilomyces* sp. SF 021 than *Saccharicola* sp. EJC 04 (Marques et al., 2018).

Paecilomyces sp. SF 021 and *Botryosphaeria* AM 01 when grown on sugarcane bagasse:soybean meal failed to give expressive xylanase production. Karmakar and Ray (2010) found lower endoglucanase production by *Rhizopus oryzae* PR7 grown on a mixture of sugarcane bagasse with cotton seed. Rodríguez-Zúñiga et al. (2011) observed a significant increase in cellulase activity when the proportion of wheat bran to soybean meal in the substrate mixture was increased, which was higher when sugarcane bagasse was used as a substrate alone. The activities of cellulases are affected by microbial selection and solid media used for growth and culture environment (Yadav et al., 2016), but excellent cellulase-producing species reported so far are *Botryosphaeria* sp. AM 01 and *Saccharicola* sp. EJC 04 (Marques et al., 2018).

When cotton seed meal:wheat bran were used as a substrate for growing *Botryosphaeria* sp. AM 01 and *Saccharicola* sp. EJC 04 under SSF conditions, maximum endoglucanase activities were for *Botryosphaeria* sp. AM 01 at 192 h and for *Saccharicola* sp. EJC 04 at 96 h, respectively. *Saccharicola* sp. EJC 04 shows higher β-glucosidase production at 192 h. Both *Botryosphaeria* sp. AM 01 and *Saccharicola* sp. EJC 04 show higher FPase activities at 192 and 144 h. The xylanase and β-xylosidase activities were also higher for both these species. Mesophilic fungi when grown in SSF also show maximum enzyme production at 192 h (Zhang and Sang, 2012).

Fungal spores have been extensively used as the initial inoculum substances for cellulase production by most fungal species under SSF (Xu et al., 2018). *A. niger* grown on sawdust with a higher inoculum amount leads to decreased endoglucanase production by SmF (Acharya et al., 2008). *P. pulmonarius* grown on wheat bran by SSF gave a reduced yield of β-glucosidase (Zilly et al., 2012). Increasing the substrate moisture content above 65% decreases the endoglucanase activity of *Botryosphaeria* sp. AM 01; however, the activities of other enzymes remained unchanged for *Saccharicola* sp. EJC 04 suggesting that moisture levels do not have any effect on this species (Maurya et al., 2012). Cotton seed meal:wheat bran when used as substrates for growing *Botryosphaeria* sp. AM 01 and *Saccharicola* sp. EJC 04 by SSF with 65% moisture content at 28°C and for 192 h gave boosted levels of cellulases and xylanases; however, *Saccharicola* sp. EJC 04 produced the highest β-glucosidase levels (Almeida et al., 2011; Wipusaree et al., 2011). *Botryosphaeria* sp. AM 01 endoglucanase is resistant against pH variations and is able to maintain its activity to 81%–94% even after 24 h of incubation at pH 3–8. Between pH levels of 3 and 9 and after 24 h, 50% activity of endoglucanase produced by *Saccharicola* sp. EJC 04 is lost. Similar performances have been shown by the β-glucosidases of both strains with approximately 50%–85% activity at pH 3.5–10.5. However, the xylanases produced by *Botryosphaeria* sp. AM 01 and *Saccharicola* sp. EJC 04 show 50%–90% activity between pH 3 and 7 and their β-xylosidases show 90% activity between pH 3 and 7. *Botryosphaeria* sp. AM 01 endoglucanase is highly durable within the temperature range of 30–50°C after a 1 h incubation period and maintains its 97% activity, while for *Saccharicola* sp. EJC 04 67% activity is maintained after incubation at 45°C. *Botryosphaeria* sp. AM 01 β-glucosidase is stable under a wide temperature range (30–60°C) with 100% activity at 40°C incubation, while enzyme stability is drastically lost for *Saccharicola* sp. EJC 04 if temperature increases beyond 70°C. However, increased stability against pH and temperature variations have been noticed for cellulases (Almeida et al., 2011).

From an industrial processing point of view, *Saccharicola* sp. EJC 04 and *Botryosphaeria* sp. AM 01 enzymes are rather more stable at wider ranges of pH and temperature, which is a very important feature (Nigam, 2013). Cellulase production can be inhibited during the hydrolytic catalysis of enzymes while acting on lignocellulosic material due to the accumulation of the hydrolysis by-products in excess in the form of sugars (Kuusk and Väljamäe, 2017), for example, microbial β-glucosidases can inhibit glucose production (Guimarães et al., 2006). Alcohol inhibition can occur during β-glucosidase production because cellulases also have a role to play in fermentation and saccharification operations (Sun and Cheng, 2002). Ethanol concentrations above 15% reduce the β-glucosidase activity of *Saccharicola* sp. EJC 04 because ethanol can alter the polarity of the medium by forming conformational changes on enzyme structure or enzyme denaturation (Barbagallo et al., 2004). However, β-glucosidase activity of *Botryosphaeria* sp. is found to increase in the presence of ethanol (Krisch et al., 2012), which may be due to β-glycosyltransferase activity (Villena et al., 2006). Ethanol in addition to water can act as an intermediary glycosyl group, an acceptor in this case (Pemberton et al., 1980). For simultaneous saccharification and fermentation the glucosyltransferase activity of β-glucosidase is a very interesting feature (Leite et al., 2008).

The enzymatic extracts derived from *Saccharicola* sp. EJC 04 and *Botryosphaeria* sp. AM 01 from their ideal growth conditions and at the lowest level of variable temperature (50°C) show a higher glucose concentration, but sugar concentration decreases if the temperature is increased (Radecka et al., 2015). Unproductive adsorption on cellulose fiber can occur at higher enzyme loads because reduction in the available surface area can become a limiting factor, which reduces the hydrolysis rate (Bommarius et al., 2008). Hydrolysis products can also inhibit enzyme activity (Griggs et al., 2012). According to the studies done so far it has become clear that glucose concentration can be increased by elevating the dose of enzymes (Pandiyan et al., 2014). Novozym 188 and Celluclast 1.5 L, the commercially available enzymes, or the use of *Miceliophthora thermophila* JCP 1–4 enzymatic extract on ozone-treated sugarcane bagasse for initiating the process, elevated glucose and xylose concentrations after 24 h (Qu et al., 2017; Pereira et al., 2016).

9.4.7 Thermophilic Fungi

Thermophilic strains can be inoculated on nonsterile wastes and therefore can compete with the autochthonous microbiota for enhanced protease activity (El-Bakry et al., 2016). Studies have also proved that selection and adoption of compost microorganisms to be used as a cellulosic substrate for effective cellulolytic degradation is also possible (Eichorst et al., 2013a). Shemekite et al. (2014) demonstrated that for SSF operations, coffee husk can serve as an ideal substrate along with variable fungal communities that share diverse enzymatic activities and this all together can be used commercially for producing compounds of great importance (Narita and Inouye, 2014). Cerda et al. (2017a) analyzed cellulase production by coffee husk by SSF and found the predominance of Dipodascaceae followed by Phaffomycetaceae along with two unidentified families of the class Tramellomycetes.

Yeast species (two kinds) showed predominance among the Phaffomycetaceae family but *Cyberlindnera jardinii* appeared as the most dominant species capable of both metabolizing pentoses and tolerating lignin by-products, and therefore proving itself as a suitable species for lignocellulosic waste treatment (López-González et al., 2015). The second most dominant species was *Barnettozyma californica*, which can ferment xylose and is also capable of xylanase production (Morais et al., 2013). The most representative yeasts were *Dipodascus australiensis* and *Galactomyces* sp., although they have mild halotolerant characteristics but are capable of xylanase production along with cellulose degradation (He et al., 2016). *Trichosporonales* sp. LM659, coarse *Trichosporon*, and *Tremellales* were reported for the class Tremellomycetes. Although the bacterial populations are chiefly responsible for cellulose-degrading activity, generally these bacterial species do not give a clear picture of their machinery operating for the degradation of cellulose (Behera and Ray, 2016). Hemicellulose degradation potential is a key feature of fungal or yeast species but despite their abundance as hemicellulose degraders this feature is least studied. These findings suggest that in bioreactors the overall consortia of microbial communities with different synergistic or complementary actions are known to occur that are usually known in isolation conditions (Kanokratana et al., 2015). Therefore for the complete biodegradation of lignocellulose, diverse actions are needed that initiate the unwinding of cellulose components. During this initial phase, enzymes are brought toward the site of action to elicit depolymerization not only in cellulose but also in lignocellulose (Duarte et al., 2012).

The selection of a microbiota depends on the alimental prerequisites associated with the lignocellulosic environments (López-González et al., 2014). Both lignocellulosic activators and lignocellulosic inhibitors coexist and it cannot be neglected that other inhibitory effects generated by the action of other lignocellulosic enzymes and cellulases also act as a limiting factor (Wongwilaiwalin et al., 2013). The use of inoculum made up of fermented solids for a continuous batch culture reduces the costs of inoculum requirements and in turn potentially reduces the generation of solid wastes (Farinas, 2015). Details of the microbial diversity inhabiting the nonsterile SSF processes are not very well known because a specialized biomass usually generates an enzyme pool with different enzymatic activities concerning lignocellulose degradation. Multienzymatic preparation can improve bioethanol production by SSF (Melikoglu et al., 2013) and such enzymatic pools do not require any additional processing steps but provide better results in comparison to commercial preparations (Lever, 2005).

Hydrolases produced by thermophilic fungi have higher thermostability, high rates of substrate hydrolysis, and optimum activity at elevated temperatures (Lee et al., 1992). The thermophilic fungus *T. aurantiacus* is popularly known for its cellulolytic (Gomes et al., 2000) and xylanolytic enzyme activities by both solid-state and submerged operations (Souza et al., 1999). Kalogeris et al. (2003a) optimized *T. aurantiacus* growth by using cheaper materials rich in elemental carbon and obtained higher cellulolytic enzyme activity levels. They found that for the enhanced production of endoglucanases and β-glucosidases, native wheat straw proved to be the best source of elemental carbon. However, rice straw, wheat bran, oat bran, and Arundo cane gave lower yields of endoglucanases and β-glucosidases, respectively. The use of wheat straw combined with wheat bran in SSC shows increased cellulase production (Fujian et al., 2002). Escalated cellulolytic levels were obtained for *T. aurantiacus* when wheat straw was used as a substrate. Inorganic nitrogen shows higher enzyme activities among the different tested nitrogen sources, which was in contrast to the reports where organic sources of nitrogen have shown high hydrolytic activities with *T. aurantiacus* (Roche et al., 1994). Considerably higher cellulolytic levels were obtained with ammonium sulfate. For cellulase production the growth temperature must be between 45°C and 53°C, for example, *T. aurantiacus* prefers

50°C in SSC conditions for enhanced production of endoglucanases and β-glucosidases (Grajek, 1986). Most of the fungal species prefer acidic pH (pH 4) because above this range growth as well as enzyme production by the fungal species radically declines and a pH range of 4–8 is regarded as the best for obtaining maximum yields of endoglucanases and β-glucosidases (Prior et al., 1992). Several workers have reported that the cellulolytic efficiency of thermophilic and mesophilic fungi when grown under SSC conditions for cellulase production cannot be compared because of lack of any standard enzyme assay conditions. *T. aurantiacus* cellulolytic levels when compared with β-glucosidase activities of other microbes were excellent at a pH of 4.5 but endoglucanase levels were maximum at a pH range of 3.5–4. The perfect temperatures are 80°C for maximum β-glucosidase and 75°C for higher yields of endoglucanases. Incubation of *T. aurantiacus* in buffer solutions between pH 3 and 11 for 24 h gave enhanced enzyme production compared to a pH range between 5 and 8. Endoglucanase has a half-life of 42 min, while it is 18 min for β-glucosidase, and their thermostability can be measured after incubation at 80°C and pH 5. The half-lives of these enzymes are, however, 2.5 days and 1 day at 70°C (Jain et al., 2017).

9.4.8 Specialized Inoculum

A multispecies self-heating thermophilic batch composed of different bacterial species, namely *B. californica*, *Sphingobacterium composti*, and *Pseudoxanthonomas taiwanensis* along with *C. jardinii*, a common yeast species (Cerda et al., 2017a), has already been known for its lignocellulose-degrading efficiency by SSF, thereby resulting in the enhanced production of proteases (Abraham et al., 2013), amylases (Cerda et al., 2016), and cellulases. The refuse from coffee known as coffee husk that can be used as a solid medium for growing specialized consortia in a sequential batch operation for cellulase production has already been conducted and this combination has shown continuous and reproducible cellulase production (Cerda et al., 2017b). Reports for continuous enzyme production are scarce (Cheirsilp and Kitcha, 2015) and studies carried out with SSF are almost negligible (Idris et al., 2017). The SSF processes for enzyme production have mostly been conducted under sterile conditions and by taking a specific set of microorganisms (Behera and Ray, 2016; Kuhad et al., 2016). Cerda et al. (2016) and Wang et al. (2016) strongly recommend the use of multispecies SSF because the autochthonous microbiota (lacking inoculation) guarantees the efficacy of the process by following the metagenomics approach. The metagenomics approach allows the specialized mixed inoculum to propagate properly for escalated production of cellulases on a commercial basis. Different combinations of gases emitted by SSF operations for the production of cellulases, proteases, and lipases have been reported at a pilot scale (Maulini-Duran et al., 2015).

The majority of these cellulase-producing units support 1–25 g of materials to be used as substrates, which is practically very small, along with authentic microbial types (*Aspergillus* or *Trichoderma* spp.) and long fermentation times (96–240 h) for enhanced cellulase and xylanase production. The results of Cerda et al. (2017a)) indicate that the production of cellulases and hemicellulases within a fermentation time slot of 24 h can bring about the degradation of the organic matter and therefore can lead to higher enzyme production. Microbial community identification emerges as an attractive alternative showing direct correlation between substrate consumption and enzyme production. A total of 21 different fungal families have been isolated from the by-products of both propagation and pilot bioreactors. Phaffomycetaceae showed a relative abundance of 43.2% in the pilot-scale SSF. However, *C. jardinii* and *B. californica* showed relatively high abundance in the propagation reactors (Cerda et al., 2017b). These two species partake in the degradation of hemicellulose and can survive if lignin hydrolyzate, a component of bioethanol production, is also present (Morais et al., 2013). Another important xylanase producer is *B. californica* that also produces enzymes for lignin degradation (Martorell et al., 2012). *B. californica* is also known to produce tyrosinase that oxidizes phenolic compounds and therefore enhances lignin hydrolyzate breakdown. The vast affluence of *B. californica* predicts higher lignin degradation in the pilot SSF operations within the reactors (Nordberg et al., 2014). These reactors, however, have similar operational conditions but changes in the scale or operational conditions such as temperature mean that microbial communities undergo a radical shift to bring about cellulase and xylanase production. The results showing the reproducibility of pilot-scale SSF are now at the preliminary stage and demand more attention.

Propagation and pilot fermentation reactors emphasize the relative abundance of about 30 bacterial families, apart from the fungal families, when their by-products were analyzed. Three families, namely Paenibacillaceae, Xanthomonadaceae, and Sphingobacteriaceae, were found in greater dominance (Cerda et al., 2017b). *Flavobacterium anatoliense*, a strict aerobic bacterium, showed its relatively higher abundance in the propagation reactors, and cellulose is totally unsuitable for the growth of this bacterial species; it cannot degrade the complexly linked polysaccharides but can grow on coffee husk components or by-products of associated microbial communities (Kacagan et al., 2013). *Pseudoxanthomonas taiwanensis* is highly abundant in propagation reactors and in original specialized inoculum (Cerda et al., 2017a). β-Glucosidases can be produced by *P. taiwanensis* and it can also influence other cellulolytic bacteria by initiating their growth and metabolisms (Eichorst et al., 2013b). Paenibacillaceae, Xanthomonadaceae, and Sphingobacteriaceae also showed their predominance in the pilot SSF except *Sphingobacterium thermophilum* and *Sphingobacterium arenae* (Sphingobacteriaceae), which were the champions. The development of the different microbes is affected by the harsh conditions generated in the pilot SSF, and bacteria capable of thriving under strong thermophilic and alkaline environments also get a chance to maintain their communities. These bacterial species overpowered the population of *P. taiwanensis* to the solid matrix, because they can produce β-glucosidases too (Yabe et al., 2013). Other cellulases produced during these harsh conditions can be halotolerant and therefore can be utilized in the production of bioethanol (Gladden et al., 2014).

9.5 CONCLUSION

Modern-day techniques, such as SSF, are popularly employed in the production of cellulase enzymes. SSF also finds its applicability in the bioconversion of lignocellulosic biomass and also proves its superiority over the traditionally used SmF in a number of ways. SSF has a lower repression rate of catabolism and it produces highly concentrated end products with lesser generation of effluents that helps to solve the problem of pollution around the globe. Production of cellulases from different fungal species, including filamentous, thermophilic, and endophytic fungi along with the use of specialized inoculum by SSF, not only increases enzyme production but also offers utilization of agricultural or industrial residues by helping in biomass degradation and combating the problem of pollution.

Acknowledgments

I would like to express my sincere gratitude to Dr. Pradeep Kumar Shukla, Assistant Professor, Department of Biological Sciences, Faculty of Science, Sam Higginbottom University of Agriculture, Technology and Sciences, Allahabad-211007, UP, India, for valuable suggestions and technical proof reading of the chapter. I would also like to convey a note of thanks to Prof. (Dr.) Pramod W. Ramteke, Dean PG Studies and Head, Department of Biological Sciences, Faculty of Science, Sam Higginbottom University of Agriculture, Technology and Sciences, Allahabad-211007, UP, India, for this wonderful opportunity to contribute this chapter. Last but not the least I am very much thankful to Gábor Draskovits, Scientific Administrator, Synthetic and Systems Biology Unit, Laboratory of Microbial Experimental Evolution, Biological Research Centre, Hungarian Academy of Sciences, H-6726, Szeged, Temesvári krt. 62, Hungary, for continuous encouragement and motivation.

References

Abdel-Fatah, O.M., Hassan, M.M., Elshafei, A.M., Haroun, B.M., Atta, H.M., Othman, A.M., 2012. Physiological studies on carboxymethyl cellulase formation by *Aspergillus terreus* DSM 826. Braz. J. Microbiol. 43, 1–11.

Abraham, J., Gea, T., Sánchez, A., 2013. Potential of the solid-state fermentation of soy fibre residues by native microbial populations for bench-scale alkaline protease production. Biochem. Eng. J. 74, 15–19.

Acharya, P., Acharya, D., Modi, H., 2008. Optimization for cellulase production by *Aspergillus niger* using saw dust as substrate. Afr. J. Biotechnol. 7, 4147–4152.

Adisseo, 2018. Adisseo website. https://www.adisseo.com/en/.

Adsul, M.G., Bastawde, K.B., Varma, A.J., Gokhale, D.V., 2007. Strain improvement of *Penicillium janthinellum* NCIM 1171 for increased cellulase production. Bioresour. Technol. 98, 1467–1473.

Adsul, M.G., Ghule, J.E., Singh, R., Shaikh, H., Bastawde, K.B., Gokhale, D.V., Varma, A.J., 2004. Polysaccharides from bagasse: applications in cellulase and xylanase production. Carbohydr. Polym. 57, 67–72.

Agger, J.W., Isaksen, T., Várnai, A., Vidal-Melgosa, S., Willats, W.G., Ludwig, R., Hom, S.J., Eijsink VGH Westereng, B., 2014. Discovery of LPMO activity on hemicelluloses shows the importance of oxidative processes in plant cell wall degradation. Proc. Natl. Acad. Sci. U. S. A. 111, 6287–6292.

Agrawal, R., Satlewal, A., Sharma, B., Mathur, A., Gupta, R., Tuli, D., Adsul, M., 2016. Induction of cellulases by disaccharides or their derivatives in *Penicillium janthinellum* EMSUV- 8 mutant. Biofuels 8, 615–622.

Aguiar, C.L., Menezes, T.J.B., 2000. Produção de celulases e xilanase por *Aspergillus niger* IZ-9 usando fermentação submersa sobre bagaço de cana de açúcar. Boletim do Ceppa 18, 57–70 (in Portuguese).

Aiello, C., Ferrer, A., Ledesma, A., 1996. Effect of alkaline treatments at various temperatures on cellulase and biomass production using submerged sugarcane bagasse fermentation with *Trichoderma reesei* QM 9414. Bioresour. Technol. 57, 13–18.

Alani, F., Anderson, W.A., Moo-young, M., 2008. New isolate of *Streptomyces* sp. with novel thermo-alkalotolerant cellulases. Biotechnol. Lett. 30, 123–126.

Almeida, M.N., Guimarães, V.M., Bischoff, K.M., Falkoski, D.L., Pereira, O.L., Gonçalves, D.S.P.O., De Rezende, S.T., 2011. Cellulases and hemicellulases from endophytic *Acremonium* species and its application on sugarcane bagasse hydrolysis. Appl. Biochem. Biotechnol. 165, 594–610.

Alva, S., Anupama, J., Savla, J., Chiu, Y.Y., VyshalI, P., Shruti, M., Yogeetha, B.S., Bhavya, D., Purvi, J., Ruchi, K., Kumudini, B.S., Varalakshmi, K.N., 2007. Production and characterization of fungal amylase enzyme isolated from *Aspergillus* sp. JGI 12 in solid state culture. Afr. J. Biotechnol. 6, 576–581.

Amirita, A., Sindhu, P., Swetha, J., Vasanthi, N.S., Kannan, K.P., 2012. Enumeration of endophytic fungi from medicinal plants and screeening of extracellular enzymes. World J. Sci. Technol. 2, 13–19.

Ander, P., Eriksson, K.E., 1977. Selective degradation of wood components by white-rot fungi. Physiol. Plant. 41, 239–248.

Arora, R., Behera, S., Kumar, S., 2015. Bioprospecting thermophilic/thermotolerant microbes for production of lignocellulosic ethanol: a future perspective. Renew. Sust. Energ. Rev. 51, 699–717.

Barbagallo, R.N., Spagna, G., Palmeri, R., Restuccia, C., Giudici, P., 2004. Selection, characterization and comparison of β-glucosidase from mould and yeasts employable for enological applications. Enzym. Microb. Technol. 35, 58–66.

Beguin, P., Aubert, J.P., 1994. The biological degradation of cellulose. FEMS Microb. Rev. 13, 25–58.

Behera, S., Ray, R., 2016. Solid state fermentation for production of microbial cellulases: recent advances and improvement strategies. Int. J. Biol. Macromol. 86, 656–669.

Berlemont, R., 2017. Distribution and diversity of enzymes for polysaccharide degradation in fungi. Sci. Rep. 7, 222.

Bhat, M.K., Bhat, S., 1997. Cellulose degrading enzymes and their potential industrial applications. Biotechnol. Adv. 15, 583–620.

Biazus, J.P.M., Souza, R.R., Santana, J.C.C., Tambourgi, E.B., 2006. Otimização da secagem do malte de *Zea mays*. Cięnc. Tecnol. Alimentos 26, 787–792 (in Portuguese).

Bommarius, A.S., Katona, A., Cheben, S.E., Patel, A.S., Ragauskas, A.J., Knudson, K., Pu, Y., 2008. Cellulase kinetics as a function of cellulose pretreatment. Metab. Eng. 10, 370–381.

Brand, D., Pandey, A., Leon, J.A.R., Roussos, S., Brand, I., Soccol, C.R., 2001. Packed bed column fermenter and kinetic modeling for upgrading the nutritional quality of coffee husk in solid-state fermentation. Biotechnol. Prog. 17, 1065–1070.

Brijwani, K., Oberoi, H.S., Vadlani, P.V., 2010. Production of a cellulolytic enzyme system in mixed-culture solid-state fermentation of soybean hulls supplemented with wheat bran. Process Biochem. 45, 120–128.

Brijwani, K., Vadlani, P.V., 2011. Cellulolytic enzymes production via solid state fermentation: effect of pretreatment methods on physicochemical characteristics of substrate. Enzyme Res., 1–10.

Broda, P., Birch, P.R.J., Brooks, P.R., Sims, P.F.G., 1996. Lignocellulose degradation by *Phanerochaete chrysosporium*: gene families and gene expression for a complex process. Mol. Microbiol. 19, 923–932.

Brown, J.A., Collin, S.A., Wood, T.M., 1987. Development of a medium for high cellulase, xylanase and β-glucosidase production by a mutant strain (NTG III/6) of the cellulolytic fungus *Penicillium pinophilum*. Enzym. Microb. Technol. 9, 355–360.

Buswell, J.A., Cai, Y.J., Chang, S.T., 1993. Fungi and substrate associated factors affecting the ability of individual mushroom species to utilise different lignocellulsic growth substrates. In: Chang, S.T., Buswell, J.A., Chiu, S.W. (Eds.), Mushroom Biology and Mushroom Products. Hong Kong. Chinese University Press, pp. 141–150.

Castro, A.M., Carvalho, M.L.A., Leite, S.G.F., Pereira Jr., N., 2010. Cellulases from *Penicillium funiculosum*: production, properties and application to cellulose hydrolysis. J. Ind. Microbiol. Biotechnol. 37, 151–158.

Cen, P., Xia, L., 1999. Production of cellulase in solid state feremental. Recent progress in bioconversion of lignocellulosics. Adv. Biochem. Eng. Biotechnol. 65, 69–92.

Cerda, A., El-Bakry, M., Gea, T., Sánchez, A., 2016. Long term enhanced solid-state fermentation: inoculation strategies for amylase production from soy and bread wastes by *Thermomyces* sp. in a sequential batch operation. J. Environ. Chem. Eng. 4, 2394–2401.

Cerda, A., Gea, T., Vargas-García, M., Sánchez, A., 2017b. Towards a competitive solid state fermentation: cellulases production from coffee husk by sequential batch operation and role of microbial diversity. Sci. Total Environ. 589, 56–65.

Cerda, A., Mejías, L., Gea, T., Sánchez, A., 2017a. Cellulase and xylanase production at pilot scale by solid-state fermentation from coffee husk using specialized consortia: the consistency of the process and the microbial communities involved. Bioresour. Technol. 243, 1059–1068.

Chaabouni, E.S., Hadj-Talieb, N., Mostrati, R., Ellouz, R., 1994. Preliminary assessment of *Penicillium occitanis* cellulase: a further useful system. Enzym. Microb. Technol. 16, 538–542.

Chandra, M.S., Viswanath, B., Reddy, B.R., 2007. Cellulolytic enzymes on lignocellulosic substrates in solid state fermentation by *Aspergillus niger*. Indian J. Microbiol. 47, 323–328.

Chapla, V.M., Zeraik, M.L., Leptokarydis, I.H., Silva, G.H., Bolzani, V.S., Young, M.C.M., Pfenning, L.H., Araújo, A.R., 2014. Antifungal compounds produced by *Colletotrichum gloeosporioides*, an endophytic fungus from *Michelia champaca*. Molecules 19, 19243–19252.

Cheirsilp, B., Kitcha, S., 2015. Solid state fermentation by cellulolytic oleaginous fungi for direct conversion of lignocellulosic biomass into lipids: fed-batch and repeated-batch fermentations. Ind. Crop. Prod. 66, 73–80.

Chen, Z., Wan, C., 2017. Biological valorization strategies for converting lignin into fuels and chemicals. Renew. Sust. Energ. Rev. 73, 610–621.

Christakopoulos, P., Kekos, D., Kolisis, F., Macris, B.J., 1995. Controlling simultaneous production of endoglucanase and beta-glucosidase by *Fusarium oxysporum* in submerged culture. Biotechnol. Lett. 17, 883–888.

Christakopoulos, P., Macris, B.J., Kekos, D., 1989. Direct fermentation of cellulose to ethanol by *Fusarium oxysporum*. Enzym. Microb. Technol. 11, 236–239.

Christakopoulos, P., Mamma, D., Nerinckx, W., Kekos, D., Macris, B., Claeyssens, M., 1996. Production and partial characterization of xylanase from *Fusarium oxysporum*. Bioresour. Technol. 58, 115–119.

Cohen, R., Persky, L., Hadar, Y., 2002. Biotechnological applications and potential wood-degrading mushrooms of the genus *Pleurotus*. Appl. Microbiol. Biotechnol. 58, 582–594.

Corrêa, R.C.G., Rhoden, S.A., Mota, T.R., Azevedo, J.L., Pamphile, J.A., de Souza, C.G.M., Polizeli, M.D.L.T.D.M., Bracht, A., Peralta, R.M., 2014. Endophytic fungi: expanding the arsenal of industrial enzyme producers. J. Ind. Microbiol. Biotechnol. 41, 1467–1478.

Cosgrove, D.J., 2017. Microbial expansins. Annu. Rev. Microbiol. 71, 479–497.

Covert, S.F., Wymelenberg, A.V., Cullen, D., 1992. Structure, organization and transcription of a cellobiohydrolase gene cluster from *Phanerochaete chrysosporium*. Appl. Environ. Microbiol. 58, 2168–2175.

da Silva, F.L., de Oliveira Campos, A., dos Santos, D.A., de Oliveira Júnior, S.D., de Araújo Padilha, C.E., de Sousa Junior, F.C., de Macedo, G.R., dos Santos, E.S., 2018. Pretreatments of Carnauba (*Copernicia prunifera*) straw residue for production of cellulolytic enzymes by *Trichorderma reesei* CCT-2768 by solid state fermentation. Renew. Energy 116, 299–308.

Damaso, M.C.T., da Terzi, S., Farias, A.X., Oliveira, A.C.P., de Fraga, M.E., Couri, S., 2012. Selection of cellulolytic fungi isolated from diverse substrates. Braz. Arch. Biol. Technol. 55, 513–520.

Dashtban, M., Maki, M., Leung, K.T., Mao, C., Qin, W., 2010. Cellulase activities in biomass conversion: measurement methods and comparison. Crit. Rev. Biotechnol. 30, 302–309.

de Carvalho, M.L.A., Carvalho, D.F., Gomes, E.D., Maeda, R.N., Anna, L.M.M.S., de Castro, A.M., Pereira Jr., N., 2014. Optimization of cellulase production by *Penicillium funiculosum* in stirred tank bioreactor using multivariate response surface analysis. Enzyme Res., 1–8.

Delabona, P.D., Pirota, R., Codima, C.A., Tremacoldi, C.R., Rodrigues, A., Farinas, C.S., 2013. Effect of initial moisture content on two Amazon rainforest *Aspergillus* strains cultivated on agro-industrial residues: biomass-degrading enzymes production and characterization. Ind. Crop. Prod. 42, 236–242.

dos Santos, T.C., Gomes, D.P.P., Bonomo, R.C.F., Franco, M., 2012. Optimisation of solid state fermentation of potato peel for the production of cellulolytic enzymes. Food Chem. 133, 1299–1304.

Duarte, G.C., Moreira, L.R.S., Jaramillo, P.M.D., Filho, E.X.F., 2012. Biomass-derived inhibitors of holocellulases. BioEnergy Res. 5, 768–777.

Dutt, D., Kumar, A., 2014. Optimization of cellulase production under solid state fermentation by *Aspergillus flavus* (AT-2) and *Aspergillus niger* (AT-3) and its impact on stickies and ink particle size of sorted office paper. Cellul. Chem. Technol. 48, 285–298.

Dutta, T., Sahoo, R., Sengupta, R., Ray, S.S., Bhattacharjee, A., Ghosh, S., 2008. Novel cellulases from an extremophilic filamentous fungi *Penicillium citrinum*: production and characterization. J. Ind. Microbiol. Biotechnol. 35, 275–282.

Eichlerova, I., Homolka, L., Nerud, F., Zadrazil, F., Baldrian, P., Gabriel, J., 2000. Screening of *Pleurotus ostreatus* isolates for their ligninolytic properties during cultivation on natural substrates. Biodegradation 11, 279–287.

Eichlerova-Volakova, I., Homolka, L., 1997. Variability of ligninolytic enzyme activities in basidiospore isolates of the fungus *Pleurotus ostreatus* in comparison with that of protoplast-derived isolates. Folia Microbiol. 42, 583–588.

Eichorst, S., Varanasi, P., Stavila, V., Zemla, M., Auer, M., Singh, S., Simmons, B., Singer, S., 2013a. Community dynamics of cellulose-adapted thermophilic bacterial consortia. Environ. Microbiol. 15, 2573–2587.

Eichorst, S., Varanasi, P., Stavila, V., Zemla, M., Auer, M., Singh, S., Simmons, B., Singer, S., 2013b. Community dynamics of cellulose-adapted thermophilic bacterial consortia. Environ. Microbiol. 15, 2573–2587.

El-Bakry, M., Gea, T., Sánchez, A., 2016. Inoculation effect of thermophilic microorganisms on protease production through solid-state fermentation under non-sterile conditions at lab and bench scale (SSF). Bioprocess Biosyst. Eng. 39, 585–592.

El-Hadi, A.A., El-Nour, S.A., Hammad, A., Kamel, Z., Anwar, M., 2014. Optimization of cultural and nutritional conditions for carboxymethylcellulase production by *Aspergillus hortai*. J. Radiat. Res. Appl. Sci. (1)23–28.

Elisashvili, V., Chichua, D., Kachlishvili, E., Tsiklauri, N., Khardziani, T., 2003. Lignocellulolytic enzyme activity during growth and fruiting of the edible and medicinal mushroom *Pleurotus ostreatus* (Jacq:Fr.) Kumm (Agaricomycetideae). Int. J. Med. Mushrooms 4, 159–166.

Elisashvili, V., Penninckx, M., Kachlishvili, E., Asatiani, M., Kvesitadze, G., 2006. Use of *Pleurotus dryinus* for lignocellulolytic enzymes production in submerged fermentation of mandarin peels and tree leaves. Enzym. Microb. Technol. 38, 998–1004.

Elisashvili, V., Penninckx, M., Kachlishvili, E., Tsiklauri, N., Metreveli, E., Kharziani, T., Kvesitadze, G., 2008. *Lentinus edodes* and *Pleurotus* species lignocellulolytic enzymes activity in submerged and solid-state fermentation of lignocellulosic wastes of different composition. Bioresour. Technol. 99, 457–462.

Emtiazi, G., Naghavi, N., Bordbar, A., 2001. Biodegradation of lignocellulosic waste by *Aspergillus terreus*. Biodegradation 12, 259–263.

Esteghlalian, A.R., Bilodeau, M., Mansfield, S.D., Saddler, J.N., 2001. Do enzymatic hydrolyzability and Simons' stain reflect the changes in the accessibility of lignocellulosic substrates to cellulase enzymes? Biotechnol. Prog. 17, 1049–1054.

Ezeilo, U.R., Zakaria, I.I., Huyop, F., Wahab, R.A., 2017. Enzymatic breakdown of lignocellulosic biomass: the role of glycosyl hydrolases and lytic polysaccharide monooxygenases. Biotechnol. Biotechnol. Equip. 31, 647–662.

Fang, H., Xia, L., 2015. Cellulase production by recombinant *Trichoderma reesei* and its application in enzymatic hydrolysis of agricultural residues. Fuel 143, 211–216.

Farinas, C., 2015. Developments in solid-state fermentation for the production of biomass degrading enzymes for the bioenergy sector. Renew. Sust. Energ. Rev. 52, 179–188.

Fujian, X., Hongzhang, C., Zuohu, L., 2001. Solid-state production of lignin peroxidase (LiP) and manganese peroxidase (MnP) by *Phanerochaete chrysosporium* using steam-exploded straw as substrate. Bioresour. Technol. 80, 149–151.

Fujian, X., Hongzhang, C., Zuohu, L., 2002. Effects of periodically dynamic changes of air on cellulase production in solid-state fermentation. Enzym. Microb. Technol. 30, 45–48.

Gao, D., Haarmeyer, C., Balan, V., Whitehead, T., Dale, B., Chundawat, S., 2014. Lignin triggers irreversible cellulase loss during pretreated lignocellulosic biomass saccharification. Biotechnol. Biofuels 7, 175.

Gautam, S.P., Bundela, P.S., Pandey, A.K., Khan, J., Awasthi, M.K., Sarsaiya, S., 2011. Optimization for the production of cellulase enzyme from municipal solid waste residue by two novel cellulolytic fungi, optimization for the productionof cellulase enzyme from municipal solid waste residue by two novel cellulolytic fungi. Biotechnol. Res. Int. 1–8.

George, S.P., Ahmad, A., Rao, M.B., 2001a. A novel thermostable xylanase from *Thermomonospora* sp.: influence of additives on thermostability. Bioresour. Technol. 78, 221–224.

George, S.P., Ahmad, A., Rao, M.B., 2001b. Studies on carboxymethyl cellulose produced by an alkalothermophilic actinomycete. Bioresour. Technol. 77, 171–175.

Gervais, P., Molin, P., 2003. The role of water in solid-state fermentation. Biochem. Eng. J. 13, 85–101.

Gladden, J., Park, J., Bergmann, J., Reyes-Ortiz, V., D'haeseleer, P., Quirino, B., Sale, K., Simmons, B., Singer, S., 2014. Discovery and characterization of ionic liquid tolerant thermophilic cellulases from a switchgrass-adapted microbial community. Biotechnol. Biofuels 7, 15.

Gomes, I., Gomes, J., Gomes, D.J., Steiner, W., 2000. Simultaneous production of high activities of thermostable endoglucanase and β-glucosidase by the wild thermophilic fungus *Thermoascus aurantiacus*. Appl. Microbiol. Biotechnol. 53, 461–468.

Gonçalves, F.A., Ruiz, H.A., dos Santos, E.S., Teixeira, J.A., de Macedo, G.R., 2015. Bioethanol production from coconuts and cactus pretreated by autohydrolysis. Ind. Crop. Prod. 77, 1–12.

Gonçalves, F.A., Ruiz, H.A., Silvino, E.S., Teixeir, J.A., de Macedo, G.R., 2016. Bioethanol production by *Saccharomyces cerevisiae*, *Pichia stipitis* and *Zymomonas mobilis* from delignified coconut fibre mature and lignin extraction according to biorefinery concept. Renew. Energy 94, 353–365.

Gossett, J.M.D., Stuckey, C., Owen, W.F., McCarty, P.L., 1982. Heat-treatment and anaerobic-digestion of refuse. J. Environ. Eng. Div. ASCE 108, 437–454.

Grajek, W., 1986. Temperature and pH optima of enzyme activities produced by cellulolytic fungi in batch and solid state cultures. Biotechnol. Lett. 8, 587–590.

Griggs, A.J., Stickel, J.J., Lischeske, J.J., 2012. A mechanistic model for enzymatic saccharification of cellulose using continuous distribution kinetics. II: cooperative enzyme action, solution kinetics, and product inhibition. Biotechnol. Bioeng. 109, 676–685.

Guimarães, L.H.S., Peixoto-Nogueira, S.C., Michelin, M., Rizzatti, A.C.S., Sandrim, V.C., Zanoelo, F.F., Aquino, A.C.M.M., Junior, A.B., Polizeli, M.D.L.T.M., 2006. Screening of filamentous fungi for production of enzymes of biotechnological interest. Braz. J. Microbiol. 37, 474–480.

Gusakov, A.V., 2011. Alternative to *Trichoderma* reesei in biofuel production. Trends Biotechnol. 29, 419–425.

Gusakov, A.V., Sinitsyn, A.P., 2012. Cellulases from *Penicillium* species for producing fuels from biomass. Biofuels 3, 463–477.

Hanif, A., Yasmeen, A., Rajoka, M.I., 2004. Induction, production, repression and de-repression of exoglucanase synthesis in *Aspergillus niger*. Bioresour. Technol. 94, 311–319.

Hansen, G.H., Lübeck, M., Frisvad, J.C., Lübeck, P.S., Andersen, B., 2015. Production of cellulolytic enzymes from ascomycetes: comparison of solid state and submerged fermentation. Process Biochem. 50, 1327–1341.

Haque, M.A., Cho, K.M., Barman, D.N., Kim, M.K., Yun, H.D., 2015. A potential cellulose microfibril swelling enzyme isolated from *Bacillus* sp. AY8 enhances cellulose hydrolysis. Process Biochem. 50, 807–815.

He, Y., Liu, F., Gong, L., Di, J., Ding, Y., Ma, C., 2016. Enzymatic in situ saccharification of chestnut shell with high ionic liquid-tolerant cellulases from *Galactomyces* sp. CCZU11-1 in a biocompatible ionic liquid-cellulase media. Bioresour. Technol. 201, 133–139.

Hemansi, G.R., Kuhad, R.C., Saini, J.K., 2018. Cost effective production of complete cellulase system by newly isolated *Aspergillus niger* RCKH-3 for efficient enzymatic saccharification: medium engineering by overall evaluation criteria approach (OEC). Biochem. Eng. J. 132, 182–190.

Hendriks, A.T.W.M., Zeeman, G., 2009. Pretreatments to enhance the digestibility of lignocellulosic biomass. Bioresour. Technol. 100, 10–18.

Henriksson, G., Nutt, A., Henriksson, H., Pettersson, B., Stahlberg, J., Johansson, G., Pettersson, G., 1999. Endoglucanase 28 (Cel12A)—a new *Phanerochaete chrysosporium* cellulase. Eur. J. Biochem. 259, 88–95.

Heproël, I., Moukha, S., Lesage-Meessen, L., Sigoillot, J.C., Asther, M., 2000. Selection of *Pycnoporus cinnabarinus* strains for laccase production. FEMS Microbiol. Lett. 183, 301–306.

Hou, Y., Wang, T., Long, H., Zhu, H., 2007. Cloning sequencing and expression analysis of the first cellulase gene encoding cellobiohydrolase 1 from a cold-adaptive *Penicillium chrysogenum* FS010. Acta Biochim. Biophys. Sinica 39, 101–107.

Huang, Y., Qin, X., Luo, X.M., Nong, Q.D., Yang, Q., Zhang, Z., Gao, Y., Lu, F.X., Chen, Y., Yu, Z.W., Liu, J.L., Feng, J.X., 2015. Efficient enzymatic hydrolysis and simultaneous saccharification and fermentation of sugarcane bagasse pulp for ethanol production by cellulase from *Penicillium oxalicum* EU2106 and thermotolerant *Saccharomyces cerevisiae* ZM1-5. Biomass Bioenergy 77, 53–63.

Hublik, G., Schinner, F., 2000. Characterization and immobilization of the laccase from *Pleurotus ostreatus* and its use for the continuous elimination of phenolic pollutants. Enzym. Microb. Technol. 27, 330–336.

Idris, A., Pandey, A., Rao, S., Sukumaran, R., 2017. Cellulase production through solid-state tray fermentation, and its use for bioethanol from sorghum stover. Bioresour. Technol. 242, 265–271.

Jain, K.K., Kumar, S., Deswal, D., Kuhad, R.C., 2017. Improved production of thermostable cellulase from *Thermoascus aurantiacus* RCKK by fermentation bioprocessing and its application in the hydrolysis of officewaste paper, algal pulp, and biologically treated wheat straw. Appl. Biochem. Biotechnol. 181, 784–800.

Jecu, L., 2000. Solid state fermentation of agricultural wastes for endoglucanase production. Ind. Crop. Prod. 11, 1–5.

Jørgensen, H., Eriksson, T., Borjesson, J., Tjerneld, F., Olsson, L., 2003. Purification and characterization of five cellulases and one xylanase from *Penicillium brasilianum* IBT 20888. Enzym. Microb. Technol. 32, 851–861.

Jørgensen, H., Mørkeberg, A., Krogh, K.B.R., Olsson, L., 2005. Production of cellulases and hemicellulases by three *Penicillium* species: effect of substrate and evaluation of cellulose adsorption by capillary electrophoresis. Enzym. Microb. Technol. 36, 42–48.

Jorgensen, H., Olsson, L., 2006. Production of cellulases by *Penicillium brasilianum* IBT 20888: effect of substrate on hydrolytic performance. Enzym. Microb. Technol. 38, 381–390.

Kacagan, M., Inan, K., Belduz, A., Canakci, S., 2013. *Flavobacterium anatoliense* sp. nov., isolated from fresh water, and emended description of *Flavobacterium ceti*. Int. J. Syst. Evol. Microbiol. 6 (Pt 6), 2075–2081.

Kachlishvili, E., Penninckx, M.J., Tsiklauri, N., Elisashvili, V., 2006. Effect of nitrogen source on lignocellulolytic enzyme production by white-rot basidiomycetes under solid-state cultivation. World J. Microbiol. Biotechnol. 22, 391–397.

Kalogeris, E., Christakopoulos, P., Katapodis, P., Alexiou, A., Vlachou, S., Kekos, D., Macris, B.J., 2003a. Production and characterization of cellulolytic enzymes from the thermophilic fungus *Thermoascus aurantiacus* under solid state cultivation of agricultural wastes. Process Biochem. 38, 1099–1104.

Kalogeris, E., Christakopoulos, P., Kekos, D., Macris, B.J., 1998. Studies on the solid-state production of thermostable endoxylanases from *Thermoascus aurantiacus*. Characterization of two isoenzymes. J. Biotechnol. 60, 155–163.

Kalogeris, E., Iniotaki, F., Topakas, E., Christakopoulos, P., Kekos, D., Macris, B.J., 2003b. Performance of an intermittent agitation rotating drum type bioreactor for solid-state fermentation of wheat straw. Bioresour. Technol. 86, 207–213.

Kanokratana, P., Eurwilaichitr, L., Pootanakit, K., Champreda, V., 2015. Identification of glycosyl hydrolases from a metagenomic library of microflora in sugarcane bagasse collection site and their cooperative action on cellulose degradation. J. Biosci. Bioeng. (4)384–391.

Karmakar, M., Ray, R.R., 2010. Extra cellular endoglucanase production by *Rhizopus oryzae* in solid and liquid state fermentation of agro wastes. Asian J. Biotechnol. 2, 27–36.

Katoch, M., Salgotra, A., Singh, G., 2014. Endophytic fungi found in association with *Bacopa monnieri* as potential producers of industrial enzymes and antimicrobial bioactive compounds. Braz. Arch. Biol. Technol. 57, 714–722.

Kim, I.J., Lee, H.J., Choi, I.G., Kim, K.H., 2014. Synergistic proteins for the enhanced enzymatic hydrolysis of cellulose by cellulase. Appl. Microbiol. Biotechnol. 98, 8469–8480.

Kim, J.Y., Hur, S.H., Hong, J.H., 2005. Purification and characterization of an alkaline cellulase from a newly isolated alkalophilic *Bacillus* sp. HSH-810. Biotechnol. Lett. 27, 313–316.

Kogel, K.H., Franken, P., Hückelhoven, R., 2006. Endophyte or parasite-what decides? Curr. Opin. Plant Biol. 9, 358–363.

Kotchoni, S.O., Shonukan, O.O., 2002. Regulatory mutations affecting the synthesis of cellulase in Bacillus pumilus. World J. Microbiol. Biotechnol. 18, 487–491.

Krisch, J., Bencsik, O., Papp, T., Vágvölgyi, C., Takó, M., 2012. Characterization of a β-glucosidase with transgalactosylation capacity from the zygomycete *Rhizomucor miehei*. Bioresour. Technol. 114, 555–560.

Krishna, C., Chandrasekharan, M., 1996. Banana waste as substrate for alpha-amylase production by *Bacillus subtilis* (CBTK 106) under solid state fermentation. Appl. Microbiol. Biotechnol. 46, 106–111.

Kristensen, J.B., Börjesson, J., Bruun, M.H., Tjerneld, F., Jørgensen, H., 2007. Use of surface active additives in enzymatic hydrolysis of wheat straw lignocellulose. Enzym. Microb. Technol. 40, 888–895.

Kuhad, R., Deswal, D., Sharma, S., Bhattacharya, A., Jain, K., Kaur, A., Pletschke, B., Singh, A., Karp, M., 2016. Revisiting cellulase production and redefining current strategies based on major challenges. Renew. Sust. Energ. Rev. 55, 249–272.

Kuhad, R.C., Gupta, R., Singh, A., 2011. Microbial cellulases and their industrial applications. Enzyme Res. 1–10.

Kumar, P., Barrett, D.M., Delwiche, M.J., Stroeve, P., 2009. Methods for pretreatment of lignocellulosic biomass for efficient hydrolysis and biofuel production. Ind. Eng. Chem. Res. 48, 3713–3729.

Kurasawa, T., Yachi, M., Suto, M., Kamagata, Y., Takao, S., Tomita, F., 1992. Induction of cellulase by gentiobiose and its sulfur-containing analog in *Penicillium purpurogenum*. Appl. Environ. Microbiol. 58, 106–110.

Kuusk, S., Väljamäe, P., 2017. When substrate inhibits and inhibitor activates: implications of β- glucosidases. Biotechnol. Biofuels 10, 7.

Lamar, R.J., 1992. The role of fungal lignin degrading enzymes in xenobiotics degradation. Curr. Opin. Biotechnol. 3, 261–266.

Laplante, S., Chahal, D.S., 1993. Abstracts of General Meeting of American Society of Microbiology. 327.

Larroche, C., Gros, J.-B., 1997. Special transformation processes using fungal spores and immobilized cells. Adv. Biochem. Eng. Biotechnol. 55, 179–220.

Leatham, G.F., 1985. Extracellular enzymes produced by the cultivated mushroom *Lentinus edodes* during degradation of a lignocellulosic medium. J. Appl. Environ. Biol. 50, 859–867.

Leatham, G.F., Stahmann, M.A., 1981. Studies on the laccase of *Lentius edodes*: localization and association with the development of fruiting bodies. J. Gen. Microbiol. 125, 147–157.

Lechner, B.E., Papinutti, V.L., 2006. Production of lignocellulosic enzymes during growth and fruiting of the edible fungus *Lentinus tigrinus* on wheat straw. Process Biochem. 41, 594–598.

Lee, Y.E., Lowe, S.E., Zeikus, J.G., 1992. Molecular biology and biochemistry of xylan degradation by thermoanaerobes. In: Visser, J., Beldman, G., Kusters-van Someren, M.A., Voragen, A.G.J. (Eds.), Xylans and Xylanases, . Progress Biotechnology. 7. Elsevier Science, Amsterdam, pp. 275–293.

Leite, R.S.R., Alves Prado, H.F., Cabral, H., Pagnocca, F.C., Gomes, E., Da Silva, R., 2008. Production and characteristics comparison of crude β-glucosidases produced by microorganisms *Thermoascus aurantiacus* and *Aureobasidium pullulans* in agricultural wastes. Enzym. Microb. Technol. 43, 391–395.

Leonard, T.J., Phillips, L.E., 1973. Study of phenoloxidase activity during the reproductive cycle in *Schizophyllum commune*. J. Bacteriol. 114, 7–10.

Lever, M., 2005. Modelling the energy performance of a farm-scale cellulose to ethanol process with on-site cellulase production and anaerobic digestion. Renew. Energy 74, 893–902.

Li, C., Yang, Z., Zhang, R.H., Chen, S., Ma, L., 2013. Effect of pH oncellulase production and morphology of *Trichoderma reesei* and theapplication in cellulosic material hydrolysis. J. Biotechnol. 168, 470–477.

Liu, B.L., Tzeng, Y.M., 1999. Water content and water activity for the production of cyclodepsipeptide in solid state fermentation. Biotechnol. Lett. 21, 657–661.

Liu, G., Qin, Y., Li, Z., Qu, Y., 2013a. Improving lignocellulolytic enzyme production with *Penicillium*: from strain screening to systems biology. Biofuels 4, 523–534.

Liu, G., Zhang, L., Qin, Y., Zou, G., Li, Z., Yan, X., Wei, X., Chen, M., Chen, L., Zheng, K., Zhang, J., Ma, L., Li, J., Liu, R., Xu, H., Bao, X., Fang, X., Wang, L., Zhong, Y., Liu, W., Zheng, H., Wang, S., Wang, C., Xun, L., Zhao, G.P., Wang, T., Zhou, Z., Qu, Y., 2013b. Long-term strain improvements accumulate mutations in regulatory elements responsible for hyper-production of cellulolytic enzymes. Sci. Rep. 3, 1569.

Liu, G., Zhang, L., Wei, X., Zou, G., Qin, Y., Ma, L., Li, J., Zheng, H., Wang, S., Wang, C., Xun, L., Zhao, G.P., Zhou, Z., Qu, Y., 2013c. Genomic and secretomic analyses reveal unique features of the lignocellulolytic enzyme system of *Penicillium decumbens*. PLoS One 8, e55185.

Liu, J., Yuan, X., Zeng, G., Shi, J., Chen, S., 2006. Effect of biosurfactant on cellulase and xylanase production by *Trichoderma viride* in solid substrate fermentation. Process Biochem. 41, 2347–2351.

López-González, J., Vargas-García, M., López, M., Jurado, M., Moreno, J., 2015. Biodiversity and succession of mycobiota associated to agricultural lignocellulosic waste-based composting. Bioresour. Technol. 187, 305–313.

López-González, J., Vargas-García, M., López, M., Suárez-Estrella, F., Jurado, M., Moreno, J., 2014. Enzymatic characterization of microbial isolates from lignocellulose waste composting: chronological evolution. J. Environ. Manag. 145, 137–146.

Lu, W., Li, D., Wu, Y., 2003. Influence of water activity and temperature on xylanase biosynthesis in pilot-scale fermentation by *Aspergillus sulphureus*. Enzym. Microb. Technol. 32, 305–311.

Maeda, R.N., Serp, V.I., Rocha, V.A.L., Mesquita, R.A.L., Anna, L.M.M.S., Castro, A.M., Driemeier, C.E., Pereira Jr., N., Polikarpov, I., 2011. Enzymatic hydrolysis of pre-treated sugar cane bagasse using *Penicilium funiculosum* and *Trichoderma harzianum*. Proc. Chem. 46, 1196–1201.

Mahanta, N., Gupta, A., Khare, S.K., 2008. Production of protease and lipase by solvent tolerant *Pseudomonas aeruginosa* PseA in solid-state fermentation using *Jatropha curcas* seed cake as substrate. Bioresour. Technol. 99, 1729–1735.

Maheshwari, R., Bharadwaj, G., Bhat, M., 2000. Thermophilic fungi: their physiology and enzymes. Microbiol. Mol. Biol. Rev. 64, 461–488.

Marques, N.P., de Cassia Pereira, J., Gomes, E., da Silva, R., Araújo, A.R., Ferreirae, H., Rodriguese, A., Dussána, K.J., Bocchini, D.A., 2018. Cellulases and xylanases production by endophytic fungi by solid state fermentation using lignocellulosic substrates and enzymatic saccharification of pretreated sugarcane bagasse. Ind. Crop. Prod. 122, 66–75.

Martinez, A.T., Camarero, S., Guillen, F., Guitierrez, A., Munoz, C., Varela, E., Martinez, M.J., Barassa, J., Ruel, K., Pelayo, J., 2002. Progress in biopulping of non-woody materials chemical, enzymeatic and ultrastructural aspects of wheat straw delignification with ligninolytic fungi from the genus *Pleurotus*. FEMS Microbiol. Rev. 13, 265–274.

Martorell, M., Pajot, H., de Figueroa, L., 2012. Dye-decolourizing yeasts isolated from Las Yungas rainforest. Dye assimilation and removal used as selection criteria. Int. Biodeterior. Biodegrad. 66, 25–32.

Massadeh, M., Yusoff, W.M.W., Omar, O., Kader, J., 2001. Synergism of cellulase enzymes in mixed culture solid substrate fermentation. Biotechnol. Lett. 23, 1771–1774.

Mata, G., Savoie, J.-M., 1998. Extracellular enzyme activity in six *Lentinula edodes* strains during cultivation in wheat straw. World J. Microbiol. 14, 513–519.

Mathew, G.M., Sukumaran, R.K., Singhania, R.R., Pandey, A., 2008. Progress in research on fungal cellulases for lignocellulose production. J. Sci. Ind. Res. 67, 898–907.

Maulini-Duran, C., Abraham, J., Rodríguez-Pérez, S., Cerda, A., Jiménez-Peñalver, P., Gea, T., Barrena, R., Artola, A., Font, X., Sánchez, A., 2015. Gaseous emissions during the solid state fermentation of different wastes for enzyme production at pilot scale. Bioresour. Technol. 179, 211–218.

Maurya, D.P., Singh, D., Pratap, D., Maurya, J.P., 2012. Optimization of solid state fermentation conditions for the production of cellulase by *Trichoderma reesei*. J. Environ. Biol. 33, 5–8.

Medeiros, A.B.P., Pandey, A., Freitas, R.J.S., Christen, P., Soccol, C.R., 2006. Optimization of production of aroma compounds by *Kluyveromyces marxianus* in solid-state fermentation using factorial design and response surface methodology. Biochem. Eng. J. 33–39.

Melikoglu, M., Sze, C., Webb, C., 2013. Kinetic studies on the multi-enzyme solution produced via solid state fermentation of waste bread by *Aspergillus awamori*. Biochem. Eng. J. 80, 76–82.

Melo, J.D.D., Carvalho, L.F.M., Medeiros, A.M., Souto, C.R.O., Paskocimas, C.A., 2012. A biodegradable composite material based on polyhydroxybutyrate (PHB) and carnauba fibers. Compos. B Eng. 43, 2827–2835.

Mendes, F.B., Atala, D.I.P., Thoméo, J.C., 2017. Is cellulase production by solid-state fermentation economically attractive for the second generation ethanol production? Renew. Energy 114, 525–533. Part B.

Mikiashvili, N., Wasser, S., Nevo, E., Chichua, D., Elisashvili, V., 2004. Lignocellulolytic enzyme activities of medicinally important basidiomycetes from different ecological niches. Int. J. Med. Mushrooms 6, 63–71.

Mitchell, D.A., Lonsane, B.K., 1990. Definition, characterization and economic evaluation. In: Doelle, H.W., Rolz, C. (Eds.), General Principles of Solid Substrate Fermentation. Rapid Publications, Oxford Ltd, UK.

Moldes, D., Lorenzo, M., Sanroman, M.A., 2004. Different proportions of laccase isoenzymes produced by submerged cultures of *Trametes versicolor* grown on lignocellulosic wastes. Biotechnol. Lett. 26, 327–330.

Morais, C., Cadete, R., Uetanabaro, A., Rosa, L., Lachance, M., Rosa, C., 2013. D-Xylosefermenting and xylanase-producing yeast species from rotting wood of two Atlantic rainforest habitats in Brazil. Fungal Genet. Biol. 60, 19–28.

Moretti, M.M.D.S., Bonfa, E.C., Chierotti, M.C.M., Movio, A.P., da Silva, R., Gomes, E., 2014. Fibrolytic enzyme production of *Myceliophthora thermophila* M.7.7. using inexpensive carbon sources and mineral nutrients. Afr. J. Microbiol. Res. 8, 4013–4019.

MRI, 2004. Midwest Research Institute to Initiate New Energy Resource Alliance: WO078919. https://www.nrel.gov/news/press/1997/09mri.html.

Mrudula, S., Murugammal, R., 2011. Production of cellulase by *Aspergillus niger* undersubmerged and solid state fermentation using coir waste as a substrate. Braz. J. Microbiol. 42, 1119–1127.

Nampoothiri, K.M., Pandey, A., 1996. Solid state fermentation for L-glutamic acid production using *Brevibacterium* sp. Biotechnol. Lett. 16, 199–204.

Narita, Y., Inouye, K., 2014. Review on utilization and composition of coffee silverskin. Food Res. Int. 61, 16–22.

Nazareth, S.W., Sampy, J.D., 2003. Production and characterisation of lignocellulases of *Panus tigrinus* and their application. Int. Biodeterior. Biodegrad. 52, 207–214.

Nigam, P.S., 2013. Microbial enzymes with special characteristics for biotechnological applications. Biomol. Ther. 3, 597–611.

Nordberg, H., Cantor, M., Dusheyko, S., Hua, S., Poliakov, A., Shabalov, I., Smirnova, T., Grigoriev, I., Dubchak, I., 2014. The genome portal of the Department of Energy Joint Genome Institute: 2014 updates. Nucleic Acids Res. 42, 26–31.

Novy, V., Longus, K., Nidetzky, B., 2015. From wheat straw to bioethanol: integrative analysis of a separate hydrolysis and co-fermentation process with implemented enzyme production. Biotechnol. Biofuels 8, 1–12.

Ogbonna, D.N., Sokari, T.G., Achinewhu, S.C., 2001. Development of an owoh-type product from African yam beans (*Sphenostylis stenocarpa*) (Hoechst (ex. A. Rich.) Harms.) seeds by solid substrate fermentation. Plant Foods Hum. Nutr. 56, 183–194.

Ögel, Z.B., Yarangümeli, K., Dü, H., Ifrij, I., 2001. Submerged cultivation of *Scytalidium thermophilum* on complex lignocellulosic biomass for endoglucanase production. Enzym. Microb. Technol. 28, 689–695.

Omemu, A.M., Akpan, I., Bankole, M.O., Teniola, O.D., 2005. Hydrolysis of raw tuber starches by amylase of *Aspergillus niger* AM07 isolated from the soil. Afr. J. Biotechnol. 4, 19–25.

Onofre, S.B., Silva, G.C., Mattiello, S.P., Groth, D., Malagi, I., 2013. Production of cellulolytic enzymes by *Aspergillus flavus* using solid state fermentation based on sugarcane bagasse. Am. J. Biochem. 3, 25–28.

Ortega, N., Busto, M.D., Perez-Mateos, M., 2000. Enzymatic saccharification of pretreated wheat straw by *T. reesei* cellulases and *A. niger* glucosidase. Biocat. Biotrans. 18, 311–330.

Pan, X., Gilkes, N., Kadla, J., Pye, K., Saka, S., Gregg, D., Ehara, K., Xie, D., Lam, D., Saddler, J., 2006. Bioconversion of hybrid poplar to ethanol and co-products using an organosolvent fractionation process: optimization of process yields. Biotechnol. Bioeng. 94, 851–861.

Panagiotou, G., Kekos, D., Macris, B.J., Christakopoulos, P., 2003. Production of cellulolytic and xylanolytic enzymes by *Fusarium oxysporum* grown on corn stover in solid state fermentation. Ind. Crop. Prod. 18, 37–45.

Pandey, A., 1995. Glucoamylase research: an overview. Starch/Stärke 47, 439–445.

Pandey, A., Kavita, P., Selvakumar, P., 1998. Culture conditions for production of 2-1-β-D-fructan-fructanohydrolase in solid culturing on chicory (*Cichorium intybus*) roots. Braz. Arch. Biol. Technol. 41, 231–236.

Pandey, A., Selvakumar, P., Soccol, C.R., Nigam, P., 1999. Solid state fermentation for the production of industrial enzymes. Curr. Sci. 77, 149–162.

Pandey, A., Soccol, C.R., Leo, J.A.R., Nigam, P., 2001. Solid-state fermentation in biotechnology. Asia Tech Publishers, Inc, New Delhi, p. 221.

Pandey, A., Soccol, C.R., Nigam, P., Soccol, V.T., 2000. Biotechnological potential of agro-industrial residues. I. Sugarcane bagasse. Bioresour. Technol. 74, 81–87.

Pandiyan, K., Tiwari, R., Singh, S., Nain, P.K.S., Rana, S., Arora, A., Singh, S.B., Nain, L., 2014. Optimization of enzymatic saccharification of alkali pretreated *Parthenium* sp. using response surface methodology. Enzyme Res., 1–8.

Pemberton, M., Brown, R., Emert, G., 1980. The role of β-glucosidase in the bioconversion of cellulose to ethanol. Can. J. Chem. Eng. 58, 723–729.

Pereira, J., de, C., Travaini, R., Marques, N.P., Bolado-Rodríguez, S., Martins, D.A.B., 2016. Saccharification of ozonated sugarcane bagasse using enzymes from *Myceliophthora thermophila* JCP 1-4 for sugars release and ethanol production. Bioresour. Technol. 204, 122–129.

Pirt, S.J., 1975. Principles of Microbe and Cell Cultivation. Wiley, New York, p. 274.

Pointing, S.B., 2001. Feasibility of bioremediation by white-rot fungi. Appl. Microbiol. Biotechnol. 57, 20–23.

Prior, B.A., Du Preez, J.C., Rein, P.W., 1992. Environmental parameters. In: Doelle, H., Mitchell, D., Rolz, C. (Eds.), Solid Substrate Cultivation. Elsevier Applied Science, London and New York, pp. 1–16.

Qinnghe, C., Xiaoyu, Y., Tiangui, N., Cheng, J., Qiugang, M., 2004. The screening of culture condition and properties of xylanase by white-rot fungus *Pleurotus ostreatus*. Process Biochem. 39, 561–566.

Qu, X., Hu, B., Zhu, M., 2017. Enhanced saccharification of cellulose and sugarcane bagasse by *Clostridium thermocellum* cultures with Triton X-100 and β-glucosidase/Cellic®CTec2 supplementation. RSC Adv. 7, 21360–21365.

Rabelo, S.C., Andrade, R.R., Filho, M., Costa, A.C., 2014. Alkaline hydrogen peroxide pretreatment, enzymatic hydrolysis and fermentation of sugarcane bagasse to ethanol. Fuel 136, 349–357.

Radecka, D., Mukherjee, V., Mateo, R.Q., Stojiljkovic, M., Foulquié-Moreno, M.R., Thevelein, J.M., 2015. Looking beyond *Saccharomyces*: the potential of non-conventional yeast species for desirable traits in bioethanol fermentation. FEMS Yeast Res. 15, 1–13.

Rai, R.D., Saxena, S., 1990. Extracellular enzymes and nonstructural compounds during growth of *Pleurotus sajor-caju* on rice straw. Mushroom J. Trop. 10, 69–73.

Rajan, K., Carrier, D.J., 2014. Effect of dilute acid pretreatment conditions and washing on the production of inhibitors and on recovery of sugars during wheat straw enzymatic hydrolysis. Biomass Bioenergy 62, 222–227.

Ramoni, J., Marchetti-Deschmann, V., Seidl-Seiboth, M., Seiboth, B., 2017. *Trichoderma reesei* xylanase 5 is defective in the reference strain QM6a but functional alleles are present in other wild-type strains. Appl. Microbiol. Biotechnol. 101, 4139–4149.

Rao, M., Gaikwad, S., Mishra, C., Deshpande, V., 1988. Induction and catabolite repression of cellulase in *Penicillium funiculosum*. Appl. Biochem. Biotechnol. 19, 129–137.

Reddy, G.V., Babu, P.R., Komaraiah, P., Roy, K.R.R.M., Kothari, I.L., 2003. Utilization of banana waste for the production of lignolytic and cellulolytic enzymes by solid substrate fermentation using two *Pleurotus species* (*P. ostreatus* and *P. sajor-caju*). Process Biochem. 38, 1457–1462.

Reis, D.L., Fontana, R.C., Delabona, P.S., Da Silva Limba, D.J., Cammasola, M., Pradella, J.G., Dillon, A.J., 2013. Increased production of cellulases and xylanases by *Penicillium echinulatum* SIM29 in batch and fed-batch culture. Bioresour. Technol. 146, 597–603.

Rocha, V.A.L., Maeda, R.N., Pereira Jr., N., Kern, M.F., Elias, L., Simister, R., Steele-King, C., Gómez, L.D., McQueen-Mason, S.J., 2016. Characterization of the cellulolytic secretome of *Trichoderma harzianum* during growth on sugarcane bagasse and analysis of the activity boosting effects of swollenin. Biotechnol. Prog. 32, 327–336.

Roche, N., Desgragnes, C., Durand, A., 1994. Study on the solid-state production of a thermostable α-L-arabinofuranosidase of *Thermoascus aurantiacus* on sugar beet pulp. J. Biotechnol. 38, 43–50.

Rodríguez-Zúñiga, U.F., Farinas, C.S., Neto, V.B., Couri, S., Crestana, S., 2011. Produção de celulases por *Aspergillus niger* por fermentação em estado sólido. Pesq. Agropec. Bras. 46, 912–919 (in Portuguese).

Ross, I.K., 1982. The role of laccase in carpophore initiation in *Coprinus congregatus*. J. Gen. Microbiol. 2763–2770.

Roussos, S., Raimbault, M., Saucedo-Castaneda, G., Viniegra-Gonzalez, G., Lonsane, B.K., 1991. Kinetics and ratios of carboxy-methyl cellulase and filter paper activities of the cellulolytic enzymes produced by *Trichoderma harzianum* on different substrates in solid state fermentation. Micol. Neotrop. Apl. 4, 19–40.

Saini, A., Aggarwal, N.K., Yadav, A., 2017. Cost-effective cellulase production using *Parthenium hysterophorus* biomass as an unconventional lignocellulosic substrate. 3 Biotech 7, 12.

Sakka, K., Kimura, T., Karita, S., Ohmiya, K., 2000. Molecular breeding of cellulolytic microbes, plants, and animals for biomass utilization. J. Biosci. Bioeng. 90, 227–233.

Sanghvi, G.V., Koyani, R.D., Rajput, K.S., 2010. Thermostable xylanase production and partial purification by solid-state fermentation using agricultural waste wheat straw. Mycology 1, 106–112.

Santana, R.M., Gonçalves, Z.S., Bonomo, R.C.F., Franco, M., 2012. Produção de amiloglucosidade utilizando como substrato a palma forrageira. Rev. Caatinga 25, 188–193. (in Portuguese).

Saravanan, P., Muthuvelayudham, R., Viruthagiri, T., 2013. Enhanced production of cellulase from pineapple waste by response surface methodology. J. Eng., 1–8.

Schülein, M., 1988. Cellulases of *Trichoderma reesei*. Methods Enzymol. 160, 234–242.

Selvakumar, P., Pandey, A., 1999. Comparative studies on inulinase synthesis by *Staphylococcus* sp. and *Kluyveromyces marxianus* in submerged culture. Bioresour. Technol. 69, 123–127.

Senthilkumar, S.R., Ashokkumar, B., Raj, K.C., Gunasekaran, P., 2005. Optimization of medium composition for alkali-stable xylanase production by *Aspergillus fischeri* Fxn 1 in solid-state fermentation using central composite rotary design. Bioresour. Technol. 96, 1380–1386.

Shafique, S., Bajwa, R., Shafique, S., 2009. Screening of *Aspergillus niger* and *A. flavus* strains for extra cellular alpha-amylase activity. Pak. J. Bot. 41, 897–905.

Shemekite, F., Gómez-Brandón, M., Franke-Whittle, I.H., Praehauser, B., Insam, H., Assefa, F., 2014. Coffee husk composting: an investigation of the process using molecular and non-molecular tools. Waste Manag. 34, 642–652.

Silva, E.M., Machuca, A., Milagres, A.M.F., 2005. Evaluation the growth and enzyme production from *Lentinula edodes* strains. Process Biochem. 40, 161–164.

Singh, A., Adsul, M., Vaishnav, N., Mathur, A., Singhania, R.R., 2017. Improved cellulase production by *Penicillium janthinellum* mutant. Indian J. Exp. Biol. 55, 436–440.

Singh, R.P., Garcha, H.S., Khanna, P.K., 1989. Biodegradation of lignocellulosics in solid state fermentation by *Pleurotus* spp. Indian J. Microbiol. 29, 49–52.

Singhania, R.R., Gusakov, R.K., Patel, A.K., Larroche, C., Pandey, A., 2010. Advancement and comparative profiles in the production technologies using solid-state and submerged fermentation for microbial cellulases. Enzym. Microb. Technol. 46, 541–549.

Singhania, R.R., Patel, A.K., Soccol, C.R., Pandey, A., 2009. Recent advances in solid-state fermentation. Biochem. Eng. J. 44, 13–18.

Singhania, R.R., Saini, J.K., Saini, R., Adsul, M., Mathur, A., Gupta, R., Tuli, D., 2014. Bioethanol production from wheat straw via enzymatic route employing *Penicillium janthinellum* cellulases. Bioresour. Technol. 169, 490–495.

Singhania, R.R., Saini, R., Adsul, M., Saini, J.K., Mathur, A., Tuli, D., 2015. An integrative process for bio-ethanol production employing SSF produced cellulase without extraction. Biochem. Eng. J. 102, 45–48.

Smith, J.P., Rinzema, A., Tramper, J., Van Sonsbeek, H.M., Knol, W., 1996. Solid state fermentation of wheat bran by *Trichoderma reesei* QM 9414: substrate composition changes, C balance enzyme productions growth and kinetics. Appl. Microbiol. Biotechnol. 46, 489–496.

Soccol, C.R., Krieger, N., 1998. Brazilian experiments for the valorization of agro-industrial residues by solid state fermentation. In: Pandey, A. (Ed.), Advances in Biotechnology. Educational Publishers, New Delhi, pp. 25–40.

Solov'eva, I.V., Okunev, O.N., Vel'kov, V.V., 2005. The selection and properties of *Penicillium verruculosum* mutants with enhanced production of cellulses and xylanases. Microbiology 74, 141–146.

Souza, M.C., de, O., Roberto, I.C., Milagres, A.M.F., 1999. Solid state fermentation for xylanase production by *Thermoascus aurantiacus* using response surface methodology. Appl. Microbiol. Biotechnol. 52, 768–772.

Srivastava, N., Srivastava, M., Mishra, P.K., Gupta, V.K., Molina, G., Rodriguez-Couto, S., Manikanta, A., Ramteke, P.W., 2017. Applications of fungal cellulases in biofuel production: advances and limitations. Renew. Sust. Energ. Rev. 82, 2379–2386.

Stricker, A.R., Steiger, M.G., Mach, R.L., 2007. Xyr1 receives the lactose induction signal and regulates lactose metabolism in *Hypocrea jecorina*. FEBS Lett. 581, 3915–3920.

Su, L.-H., Zhao, S., Jiang, S.-X., Liao, X.-Z., Duan, C.-J., Feng, J.-X., 2017. Cellulase with high β-glucosidase activity by *Penicillium oxalicum* under solid state fermentation and its use in hydrolysis of cassava residue. World J. Microbiol. Biotechnol., 33–37.

Sukumaran, R.K., Singhania, R.R., Mathew, G.M., Pandey, A., 2009. Cellulase production using biomass feed stock and its application in lignocellulose saccharification for bioethanol production. Renew. Energy 34, 421–424.

Sun, X.Y., Liu, Z.Y., Qu, Y.B., Li, X.Z., 2008. The effects of wheat bran composition on the production of biomass-hydrolyzing enzymes by *Penicillium decumbens*. Appl. Biochem. Biotechnol. 1, 119–128.

Sun, Y., Cheng, J., 2002. Hydrolysis of lignocellulosic materials for ethanol production: a review. Bioresour. Technol. 83, 1–11.

Suto, M., Yachi, M., Kamagata, Y., Sasaki, H., Takao, S., Tomita, F., 1991. Cellulase induction by soluble acetyl cellobioses in *Penicillium purpurogenum*. J. Ferment. Bioeng. 72, 352–357.

Syed, S., UL-Hassan, S.R., Johri, S., 2013. A novel cellolose from endophtye, *Penicillium* sp. NFCC 286. Am. J. Microbiol. Res. 1, 84–91.

Taherzadeh-Ghahfarokhi, M., Panahi, R., Mokhtarani, B., 2018. Optimizing the combination of conventional carbonaceous additives of culture media to produce lignocellulose-degrading enzymes by *Trichoderma reesei* in solid state fermentation of agricultural residues. Renew. Energy 131, 946–955.

Topakas, E., Katapodis, P., Kekos, D., Macris, B.J., Christakopoulos, P., 2002. Production and partial characterization of xylanase by *Sporotrichum thermophile* under solid-state fermentation. World J. Microbiol. Biotechnol. 19, 195–198.

Tsiklauri, N.D., TSh, K., Kachlishvili, E.T., Elisashvili, V.I., 1999. Cellulase and xylanase activities of higher basidiomycetes during bioconversion of plant raw material depending on the carbon source in the nutrient medium. Appl. Biochem. Microbiol. 35, 291–295.

Velazquez-Cedeno, M.A., Mata, G., Savoie, J.M., 2002. Waste-reducing cultivation of *Pleurotus ostreatus* and *Pleurotus pulmonarius* on coffee pulp: changes in the production of some lignocellulolytic enzymes. World J. Microbiol. Biotechnol. 18, 201–207.

Venelampi, O., Weber, A., Ronkko, T., Itavaara, M., 2003. The biodegradation and disintegration of paper products in the composting environment. Compost. Sci. Util. 11, 200–209.

Villena, M.A., Iranzo, J.F.U., Gundllapalli, S.B., Otero, R.R.C., Briones, P.A.I., 2006. Characterization of an exocellular β-glucosidase from *Debaryomyces pseudopolymorphus*. Enzym. Microb. Technol. 39, 229–234.

Wang, X.J., Bai, J.G., Liang, Y.X., 2006. Optimization of multienzyme production by two mixed strains in solid-state fermentation. Appl. Microbiol. Biotechnol. 73, 533–540.

Wang, Z., Lu, Z., Shi, J., Xu, Z., 2016. Exploring flavour-producing core microbiota in multiespecies solid-state fermentation of traditional Chinese vinegar. Sci. Rep. 6, 26818.

Wipusaree, N., Sihanonth, P., Piapukiew, J., Sangvanich, P., Karnchanatat, A., 2011. Purification and characterization of a xylanase from the endophytic fungus *Alternaria alternata* isolated from the Thai medicinal plant, *Croton oblongifolius* Roxb. Afr. J. Microbiol. Res. 5, 5697–5712.

Wongwilaiwalin, S., Laothanachareon, T., Mhuantong, W., Tangphatsornruang, S., Eurwilaichitr, L., Igarashi, Y., 2013. Comparative metagenomic analysis of microcosm structures and lignocellulolytic enzyme systems of symbiotic biomass-degrading consortia. Appl. Microbiol. Biotechnol. 97, 8941–8954.

Wood, D.A., 1980. Inactivation of extracellular laccase during fruiting of *Agaricus bisporus*. J. Gen. Microbiol. 117, 339–345.

Wood, D.A., 1984. Microbial process in mushroom cultivation: a largescale solid substrate fermentation. J. Chem. Technol. Biotechnol. 34, 232–240.

Xia, L., Cen, P.L., 1999. Cellulase production by solid state fermentation on lignocellulosic waste from the xylose industry. Process Biochem. 34, 909–912.

Xu, X., Lin, M., Zang, Q., Shi, S., 2018. Solid state bioconversion of lignocellulosic residues by *Inonotus obliquus* for production of cellulolytic enzymes and saccharification. Bioresour. Technol. 247, 88–95.

Xu, X.Q., Wu, X.B., Cui, Y., Cai, Y.X., Liu, R.W., Long, M.N., Chen, Q.X., 2014. Enzymatic saccharification of cassava residues and glucose inhibitory kinetics on β-glucosidase from *Hypocrea orientalis*. J. Agric. Food Chem. 62, 11512–11518.

Xue, J.L., Zhao, S., Liang, R.M., Yin, X., Jiang, S.X., Su, L.H., Yang, Q., Duan, C.J., Liu, J.L., Feng, J.X., 2016. A biotechnological process efficiently co-produces two high value-added products, glucose and xylooligosaccharides, from sugarcane bagasse. Bioresour. Technol. 204, 130–138.

Yabe, S., Aiba, Y., Sakai, Y., Hazaka, M., Kawahara, K., Yokota, A., 2013. *Sphingobacterium thermophilum* sp. nov., of the phylum bacteroidetes, isolated from compost. Int. J. Syst. Evol. Microbiol. 63, 1584–1588.

Yadav, P.R., Chauhan, P.B., Gahlout, M., 2016. Isolation, screening and optimization of process parameters for enhanced production of cellulase by solid state fermentation. Int. J. Adv. Res. Biol. Sci. 3, 21–27.

Yamanobe, T., Mitsuishi, Y., Takasaki, Y., 1988. Method for production of cellulolytic enzymes and method for saccharification of cellulosic materials therewith. USP 4, 742,005.

Yang, S.Q., Yan, Q.J., Jiang, Z.Q., Li, L.T., Tian, H.M., Wang, Y.Z., 2006. High-level of xylanase production by the thermophilic *Paecilomyces themophila* J18 on wheat straw in solid-state fermentation. Bioresour. Technol. 97, 1794–1800.

Yang, X.X., Chen, H.Z., Gao, H.L., Li, Z.H., 2001. Bioconversion of corn straw by coupling ensiling and solid-state fermentation. Bioresour. Technol. 78, 277–280.

Yoon, L.W., Ang, T.N., Ngoh, G.C., Seak, A., Chua, M., 2014. Fungal solid-state fermentation and various methods of enhancement in cellulase production. Biomass Bioenergy 67, 319–338.

Zhang, H., Sang, Q., 2012. Statistical optimization of cellulases production by *Penicillium chrysogenum* QML-2 under solid-state fermentation and primary application to chitosan hydrolysis. World J. Microbiol. Biotechnol. 28, 1163–1174.

Zhang, M., Xie, L., Yin, Z.X., Khanal, S.K., Zhou, Q., 2016. Biorefinery approach for cassava-based industrial wastes: current status and opportunities. Bioresour. Technol. 215, 50–62.

Zhang, Q., Tang, L., Zhang, J., Mao, Z., Jiang, L., 2011. Optimization of thermal-dilute sulfuric acid pretreatment for enhancement of methane production from cassava residues. Bioresour. Technol. 102, 3958–3965.

Zhang, Y.H.P., Lynd, L.R., 2004. Toward an aggregated understanding of enzymatic hydrolysis of cellulose: noncomplexed cellulase systems. Biotechnol. Bioeng. 88, 797–824.

Zhang, Z., Liu, J.L., Lan, J.Y., Duan, C.J., Ma, Q.S., Feng, J.X., 2014. Predominance of *Trichoderma* and *Penicillium* in cellulolytic aerobic filamentous fungi from subtropical and tropical forests in China, and their use in finding highly efficient β-glucosidase. Biotechnol. Biofuels 7, 1–14.

Zhou, Q., Xu, J., Kou, Y., Lv, X., Zhang, X., Zhao, G., Zhang, W., Chen, G., Liu, W., 2012. Differential involvement of beta-glucosidases from *Hypocrea jecorina* in rapid induction of cellulase genes by cellulose and cellobiose. Eukaryot. Cell 11, 1371–1381.

Zhu, J.Y., Wang, G.S., Pan, X.J., Gleisner, R., 2009. Specific surface to evaluate the efficiencies of milling and pretreatment of wood for enzymatic saccharification. Chem. Eng. Sci. 64, 474–485.

Zilly, A., dos, S., Bazanella, G.C., Helm, C.V., de, S., Bracht, C.A.V.A.C.G.M., Bracht, A., Peralta, R.M., 2012. Solid-state bioconversion of passion fruit waste by white-rot fungi for production of oxidative and hydrolytic enzymes. Food Bioprocess Technol. 5, 1573–1580.

Further Reading

Manan, M.A., Webb, C., 2017. Modern microbial solid state fermentation technology for future biorefineries for the production of added-value products. Biofuel Res. J. 16, 730–740.

Passos, D.F., Pereira Jr., N., de Castro, A.M., 2018. A comparative review of recent advances in cellulases production by *Aspergillus, Penicillium* and *Trichoderma* strains and their use for lignocellulose deconstruction. Curr. Opin. Green Sust. Chem. 14, 60–66.

Singh, R., Kumar, M., Mittal, A., Mehta, P.K., 2016. Microbial cellulases in industrial applications. Ann. Appl. Biosci. 3, R23–R29.

CHAPTER

10

Strategies to Improve Solid-State Fermentation Technology

*Darshan M. Rudakiya**,†

*Bioconversion Technology Division, Sardar Patel Renewable Energy Research Institute, Anand, Gujarat, India

†Department of Microbiology, N. V. Patel College of Pure and Applied Sciences, Anand, Gujarat, India

10.1 INTRODUCTION

Enzyme industries are currently growing exponentially, wherein enzymes are applied to food manufacturing, textiles, paper and pulp, cosmetics, pharmaceuticals, animal feed, and also to tools for research and development (Li et al., 2012). There are more than 5000 known enzymes and of these only 20–30 enzymes are truly industrialized. Across the globe, enzyme demand is mainly satisfied by 15–20 major enzyme-producing industries and about 650 minor suppliers. Cellulase is one of the truly industrial enzymes, which have potential applications in textile, detergent, paper and pulp, food processing, brewing, animal feed, and recently in petroleum industries (Himmel et al., 2007). Cellulases principally comprise three types of enzyme, which act on the cellulose in a consecutive manner to depolymerize it. These enzymes are endoglucanase or endo-1,4-β-glucanase (EC 3.2.1.4), exoglucanase or exo-1,4-β-D-glucanase (EC 3.2.1.176), and β-glucosidase or β-D-glucoside glucohydrolase (EC 3.2.1.21). Initially, endoglucanase hydrolyzes the glycosidic bonds of cellulose, generating the long chains of oligosaccharides. Subsequently, exoglucanase unidirectionally acts on the long chain oligomers of cellulose that form cellobiose by cleaving either on nonreducing ends (EC 3.2.1.91) or on reducing ends (EC 3.2.1.176). Lastly, β-glucosidase hydrolyzes the cellobiose and cellobiose is converted into glucose subunits (Juturu and Wu, 2014; Doi and Kosugi, 2004).

Bacterial cellulosome encompasses the scaffolding protein, wherein the subunits of cellulases attach intermittently, which are cohesins and dockerins. Both sides of dockerin bind the single subunit of enzyme and cohesin. Interaction between cohesin and dockerin is important for the bacterial cellulosome. On the contrary, fungal cellulases have a simple architecture as compared to bacterial cellulases. They have one catalytic binding site (CD) and a cellulose-binding domain (CBD). A linker peptide chain connects the CBD and CD. CBD is not involved in hydrolysis of the substrate but its presence is mainly involved in enzyme stability and activity. Fig. 10.1 represents the 3D structures of the bacterial and fungal cellulases (endoglucanase, exoglucanase, and β-glucosidase) with their substrate binding sites (Juturu and Wu, 2014; Doi and Kosugi, 2004).

Solid-state fermentation (SSF) is a type of fermentation process, wherein the surface of solid materials is employed for microbial growth and product formation (Ray and Behera, 2017). This process lacks "free water" content; however, fewer amounts of water are added to moisten the solid substrates that assist microbial growth and metabolism (Pirota et al., 2014). These substrates can provide a unique microenvironment that leads to the favorable growth of microbes and their metabolic activities. Most of the microbes tend to grow in natural habitats rather than in liquid conditions, for instance, 98% of marine microbes grow on solid substrates (Singhania et al., 2009).

SSF technology mainly includes various parameters of upstream processing, fermentation processes, and downstream processing. Enzyme production using the SSF process has many advantages, which mainly includes higher productivity and yield, less capital cost, lower catabolite repression, and easy and efficient enzyme recovery. However, enzyme production is challenging for SSF because inadequate knowledge of different parameters that are involved

From Cellulose to Cellulase: Strategies to Improve Biofuel Production
https://doi.org/10.1016/B978-0-444-64223-3.00010-2

in SSF and lack of appropriate strategies. For this reason, the majority of enzymes, including cellulases, are mainly produced by submerged fermentation (SmF) by the majority of suppliers. So, it is crucial to comprehend the numerous factors and to work on different strategies, which make cellulase production possible at a large scale (Kuhad et al., 2016; Pandey, 2003; Farinas, 2015). Thus various strategies used to improve SSF technology are discussed in this chapter. There are a total of seven strategies: (1) organism-based, (2) substrate-based, (3) process optimization-based, (4) bioreactor-based, (5) process monitoring-based, (6) contamination control-based, and (7) downstream processing-based strategies. All strategies are described sequentially in the ensuing sections.

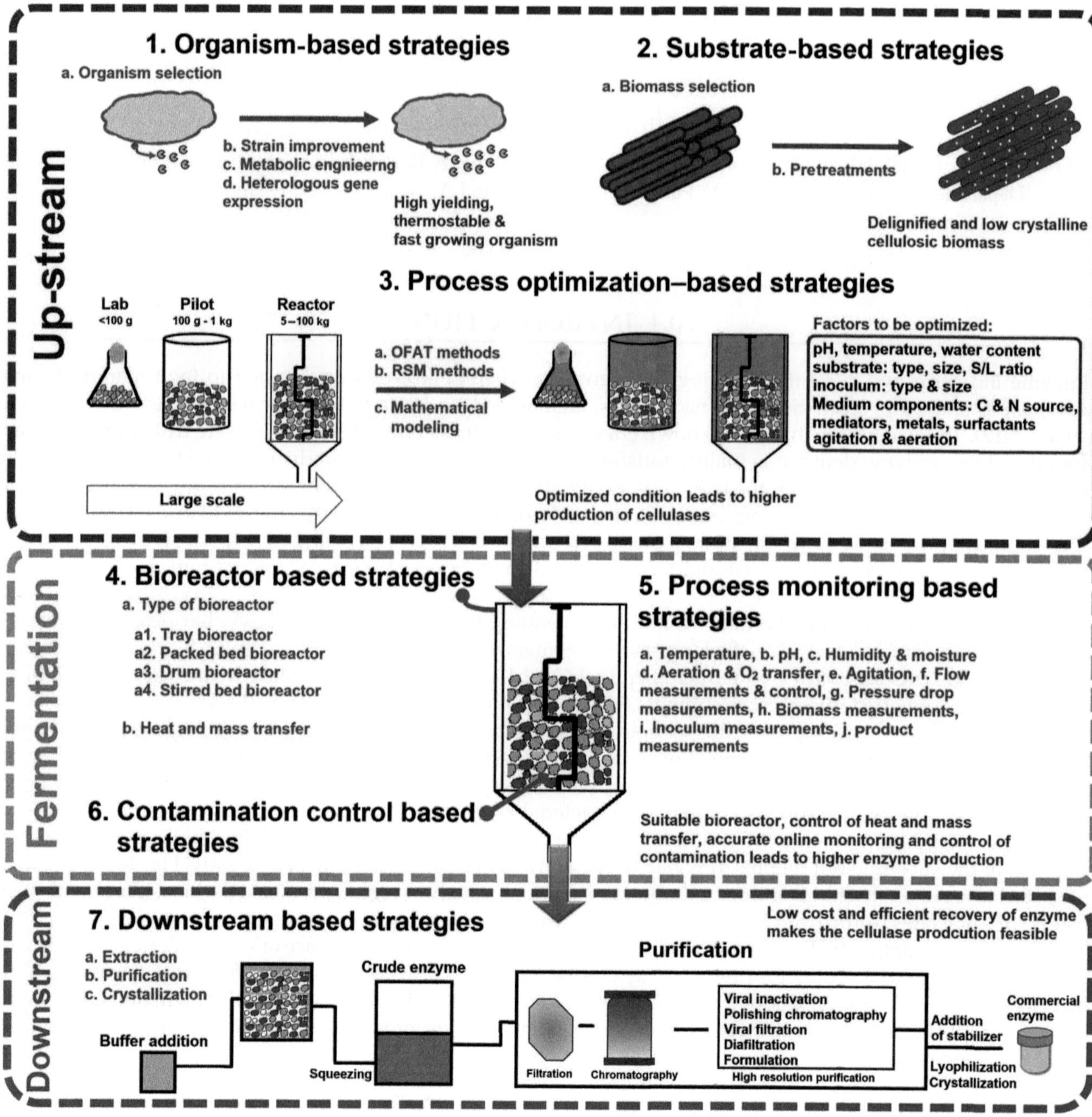

FIG. 10.1 Upstream, fermentation and downstream strategies should be applied to maximize the production of cellulases.

10.2 ORGANISM-BASED STRATEGIES

The most important parameter for cellulase fermentation is the choice of high-yielding cellulase-producing organisms. In the case of SSF, the majority of organisms cannot survive because they lack free water. However, the majority of fungi grows almost without free water. Additionally, lignocelluloses are the natural habitat for fungi, which stimulates fungal growth and the production of cellulase as well. On the contrary, certain bacteria and a few strains of actinomycetes can also produce cellulases under SSF conditions (Ray and Behera, 2017; Behera and Ray, 2016).

The majority of fungal species have been cultivated for cellulase production under SSF for the last few decades. These fungi include genera such as *Aspergillus, Fusarium, Penicillium, Phanerochaete, Malenocarpus, Trametes, Pleurotus, Trichoderma, Lentinus, Pycnoporus, Tricholoma, Irpex, Cepirlopsis, Cerrena,* and *Gloeophyllum*. The majority of cellulase-producing fungi are from terrestrial sources. However, several researchers have isolated the marine fungi *Cladosporium sphaerospermum* and *Chaetomium* sp. for cellulase production (Ray and Behera, 2017; Trivedi et al., 2015; Ravindran et al., 2010). Table 10.1 shows the production of cellulases under SSF by terrestrial and marine fungi

TABLE 10.1 Production of Cellulases and Xylanases by Different Fungi Using Solid-State Fermentation

		Enzymes[d]				
Fungus	**Substrate/supportive materials**	**CMs**	**FPA**	**BGL**	**XYL**	**References**
Trichoderma reesei RUT C-30	Wheat bran	457[a]	–	–	–	(Idris et al., 2017)
Trichoderma veride sp. 3.2942	Polyurethane foam[f]	38.2[a]	10.6[a]	–	–	(Hu et al., 2011)
T. reesei ZU-02	Corncob residue	–	158[c]	–	–	(Xia and Cen, 1999)
Talaromyces stipitatus MTCC 12687	*Parthenium hysterophorus*	53.3[a]	4.51[a]	62.6[a]	934[a]	(Bharti et al., 2018)
Botryosphaeria sp.	Oil palm empty fruit bunch	8.13[a]	3.26[a]	0.11[a]	–	(Ang et al., 2013)
Penicillium citrinum	Wheat bran	1.89[b]	1.72[b]	–	–	(Bahrin et al., 2011)
Penicillium funiculosum	Cellulose (2%) + cellobiose (0.5% mg/L)	7[b]	2.5[b]	8[b]	–	(Dutta et al., 2008)
Penicillium echinulatum SIM29	Elephant grass plus 50% wheat bran	205[a]	32.9[a]	148.9[a]	–	(Rao et al., 1988)
P. echinulatum 9A0251	Cellulose (1%) + glucose (0.15%) + methylxanthines (1–5 μM)	–	3.1[b]	2.9[b]	–	(Camassola and Dillon, 2007)
Penicillium brasilianum IBT-20888	cellulose (20 g/L) + pretreated spruce (35.8 g/L)	3.59[b]	0.35[b]	1.84[b]	–	(Castro et al., 2010)
Pseudolagarobasidium acaciicola AGST3	*Pithecellobium dulce*	–	0.27[b]	231[b]	327[b]	(Rudakiya and Gupte, 2017)
	Tamarindus indica	–	0.41[b]	605[b]	322[b]	
Tricholoma giganteum AGDR1	*P. dulce*	–	0.23[b]	85[b]	336[b]	(Rudakiya and Gupte, 2017)
	T. indica	–	0.26[b]	43[b]	378[b]	
Aspergillus fumigatus SK1	Oil palm trunk	54.2[a]	3.36[a]	4.54[a]	418[a]	(Scholl et al., 2015)
Aspergillus niger ATTC 6275	Palm kernel cake	23.8[a]	–	–	282[a]	(Prasertsan et al., 1992)
A. fumigatus	Wheat straw	16.9[a]	0.98[a]	11.8	56.4[a]	(Sherief et al., 2010)
Cladosporium sphaerospermum[e]	Sea weed *Ulva* sp.	10.2[a]	9.6[a]	–	–	(Trivedi et al., 2015)
Chaetomium sp. (NIOCC 36)[e]	Cotton seeds pH 12	9.8[b]	3.1[b]	6.4[b]	–	(Ravindran et al., 2010)
Thermoascus aurantiacus	Wheat straw	987[a]	4.4[a]	48.8[a]	–	(Kalogeris et al., 1999)

[a] *Enzyme activity: IU/gds.*
[b] *Enzyme activity: IU/mL.*
[c] *Enzyme activity: IU/g koji.*
[d] *Enzymes:* CMs, *CMCase (endogluconase);* FPA, *filter paper activity (exogluconase);* BGL, *β-glucosidase;* XYL, *xylanase.*
[e] *Fungi are from marine habitats.*
[f] *It was used as a supportive material.*

(Idris et al., 2017; Hu et al., 2011; Xia and Cen, 1999; Bharti et al., 2018; Ang et al., 2013; Bahrin et al., 2011; Dutta et al., 2008; Rao et al., 1988; Camassola and Dillon, 2007; Castro et al., 2010; Rudakiya and Gupte, 2017; Scholl et al., 2015; Prasertsan et al., 1992; Sherief et al., 2010; Trivedi et al., 2015; Ravindran et al., 2010; Kalogeris et al., 1999). Most fungi produce all cellulolytic enzymes during SSF, so all enzymes synergistically act on the plant polysaccharides. Because most fungi are mesophilic it is relatively challenging at the large scale where heat and other factors cause the reduction in enzyme productivity. Additionally, fungi grow slowly as compared to bacteria and yeast, so production time can be extended (Ray and Behera, 2017).

In contrast, a few bacterial species also produce cellulolytic enzymes under aerobic and anaerobic conditions, which are mainly *Bacillus*, *Cellulomonas*, *Acidothermus*, *Nocardiopsis*, *Bacteroides*, *Clostridium*, *Alcaligenes*, and *Acetivibrio*. Certain species of actinomycetes, *Streptomyces* and *Thermomonospora*, also produce cellulases. Bacterial cellulases are more complex and are in multienzyme complexes, which provide increased function and synergy. Additionally, production of bacterial cellulases can be useful under SSF because bacteria can survive in a wide variety of industrial and environmental conditions because they are psychrophilic, thermophilic, acidophilic, and alkaliphilic. Additionally, bacterial cultures show a faster growth rate as compared to fungal species, so products can be obtained in a few days (Sadhu and Maiti, 2013).

To overcome the disadvantage of cellulase-producing organisms, certain successful strategies have been used to enhance cellulase activity and stability. Strain improvement using metabolic engineering, recombinant gene expression, and mixed co-culture strategies have been used to improve the productivity of SSF (Ray and Behera, 2017; Behera and Ray, 2016).

10.2.1 Strain Improvement

Selection of higher cellulase-producing organisms can be achieved by screening the organisms using primary and secondary screening techniques. After selecting an organism, cellulase production can be maximized by optimization of process parameters. These parameters and optimization methods will be discussed under process optimization-based strategies (see Section 4). Sometimes, alkaline, acidic, thermostable, and metal-tolerant enzymes are required for specific applications; enzyme screening is also required. In this case, respective tolerant cellulase-producing organisms may not exhibit higher production of enzyme. Simultaneously, metabolic engineering and strain improvement are necessary. Several researchers have studied enhanced cellulase activity by mutagenesis experiments. In such experiments, organisms are incubated with mutagenic agents, e.g., UV radiation, ethyl methane sulfonate, etc., for several seconds to minutes. Subsequently, mutant strains are also tested for cellulase activity. Based on this method, high cellulose-producing mutant organisms can be achieved. Different fungi are used for the mutagenesis experiment to enhance cellulase activity and this has been studied by several researchers (Ray and Behera, 2017; Dhillon et al., 2011a; Dhulgande et al., 2011; El-Ghonemy et al., 2014).

10.2.2 Heterologous Gene Expression

Enzyme production over short periods with higher yield is the major goal for the enzyme industries. Thus expression of cellulases using recombinant technology is the sought-after strategy to produce enzymes in a shorter time. Generally, bacteria, fungi, yeasts, and plants are used to express the cellulase genes (Juturu and Wu, 2012; Delgado et al., 2002).

Bacterial expression system: A wide array of bacteria have been applied to the production of recombinant cellulases. *Escherichia coli*, *Bacillus subtilis*, *Pseudomonas cellulose*, *Zymomonas mobilis*, and *Streptomyces lividans* are frequently used bacteria to produce recombinant cellulases. However, the bacterial expression system mainly faces problems related to the expression and secretion of enzymes due to the lack of posttranslational machineries (Delgado et al., 2002; Herai et al., 2004).

Yeast expression system: The yeast expression system is always preferable for gene expression from fungi because yeasts can perform eukaryotic posttranslational modifications and can also secrete enzymes extracellularly. Additionally, yeast can grow in a short time compared to fungi, be relatively free of toxins, and have high cell densities. Mainly *Pichia pastoris*, *Saccharomyces cerevisiae*, *Hansenula polymorpha*, *Kluyveromyces lactis*, and *Yarrowia lipolytica* have been used for the expression of cellulases (Juturu and Wu, 2014; Ward, 2012).

Fungal expression system: Heterologous expression of cellulases in fungi is an ideal platform because they have posttranslational machineries. Additionally, they produce enzymes in high amounts. Mostly, *Aspergillus nidulans*, *Aspergillus oryzae*, *Aspergillus niger*, *Penicillium chrysosporium*, and *Rhizopus oryzae* are potential fungi, which are used to produce the recombinant cellulases. Recombinant cellulases from fungi are obtained by the following strategies:

(1) gene fusion strategy, (2) overexpression of chaperones, (3) screening the multicopy gene strains, and (4) use of protease-deficient host strains (Ray and Behera, 2017; Garvey et al., 2013).

Plant expression system: The plant expression system is the alternative for animal and microbial cellulase production. Moreover, it is successfully used for the molecular farming of cellulases. Several transgenic plants (e.g., tobacco, maize, wheat, etc.) are grown for the production of cellulases (Pirota et al., 2014; Streatfield, 2007).

10.2.3 Mixed Co-Culture System

The majority of researchers have used the pure culture of a single strain for the production of cellulases using SSF. Substrate utilization is relatively specific in the case of a single strain in a pure culture system, which directly affects the cost and productivity of an SSF bioreactor. Additionally, maintenance and monitoring of single strain sterility throughout production is quite a challenging task. On the contrary, a mixed co-culture system, i.e., the use of more than one culture, leads to a broad substrate range, sterility becomes discretionary, and strain engineering is also not required. These cultures can grow on a wide range of solid substrates with higher cellulase activity (Lin et al., 2011). *A. niger* and *Trichoderma reesei* have been used by Dhillon et al. (Dhillon et al., 2011b) to assess cellulase production using agricultural residues. Results revealed that increased cellulase activities and optimized combinations of β-glucosidase and cellulase activities were achieved using a mixed culture in tray fermentation. A mixed co-culture system may be efficient for the on-site production of cellulases in an SSF bioreactor.

10.3 SUBSTRATE-BASED STRATEGIES

Generally, two types of solid matrices are used for SSF processes: (1) the matrix can be a source of nutrients (e.g., starch, lignocellulose, chitin, etc.) or (2) it is solely supported by infused nutrients (e.g., polyurethane foam and polystyrene). Lignocelluloses are the most abundantly distributed plant biomass, which is useful for potential applications. They are divided into four major categories: hardwood, softwood, grasses, and agriculture wastes. Remarkably, biomass is being produced in huge amounts for food, clothing, biofuel, and pharmaceutical applications. However, the residues of biomass are discarded or burned in large amounts. Utilization of these residues has a great advantage for the production of cellulases. In SSF technology, the majority of these residues can be used for the production of cellulases. Additionally, residues used in SSF production have commercial advantages, wherein sources are easily available, cheap, and a natural food for the organism (Rudakiya and Gupte, 2017; Ravindran and Jaiswal, 2016; Saha et al., 2016; Rabemanolontsoa and Saka, 2016).

Mainly, lignocellulosic substrates have three major components. Cellulose is the main component of lignocellulose (~50% of total composition), which is a linear polymer made up of D-glucose monomers connected by β-1,4-glycosidic bonds. Hemicellulose presents 10%–40% of total composition, and is a branched biopolymer containing pentose, hexose, and/or uronic acids. Lignin is an aromatic polymer, and presents 10%–40% of the total composition of biomass. It is composed from phenylpropanoid precursors. The ratio of cellulose, hemicelluloses, and lignin can vary based on the substrate type, species type, plant age, harvesting season, and culture conditions. A wide array of lignocellulosic substrates has been used for the production of microbial cellulases (Rudakiya and Gupte, 2017) (Table 10.2). Still, cellulase production from native biomass is reasonably lower compared to pretreated biomass because the organisms cannot break the whole lignin structure (except white-rot fungi) and degrade crystalline cellulose. For these reasons, lignocellulosic substrates are pretreated using various methods, which chiefly include physical, chemical, biological, thermal, and combinatorial methods. Certain new technologies are emerging for the pretreatment of biomass, which can be applied to cellulase production. All the pretreatment techniques are discussed in the following sections (Saha et al., 2016; Chen et al., 2017).

10.3.1 Physical Pretreatment

Reduction in the size and crystallinity of lignocellulose and lowering the degree of polymerization are mainly observed in physical pretreatments. During pretreatment, the generation of toxic chemicals is negligible, which is the major advantage of this pretreatment. Physical pretreatments mainly include the different types of methods, i.e., milling, grinding, chipping, freezing, extrusion, and pyrolysis (Rabemanolontsoa and Saka, 2016).

Milling: Milling is the first stage of pretreatment, which reduces the size of lignocellulosic substrates. Afterward, the chipping and grinding processes reduce the size of lignocellulose by up to 10–30 and 0.2–0.5 mm,

TABLE 10.2 Composition (%) of Lignocellulose Components in Lignocellulosic Substrates

Lignocellulosic materials	Type of biomass[a]	Lignin (%)	Hemicellulose (%)	Cellulose (%)	Extractives (%)
Hardwood	–	18–30	12–42	40–45	1–3
Softwood	–	25–35	25–35	45–50	1–4
Pithecellobium dulce	HW	30.4	21.7	41.6	2.6
Tamarindus indica	HW	30.1	22.1	43.6	2.8
Eucalyptus saligna	HW	26.9	12.6	48.1	2.9
Poplar (*Populus* spp.)	HW	25.6	16.6	42.2	3.1
CAFI high-lignin poplar	HW	29.1	20.4	43.8	2.5
Jatropha	EC	20–25	8–7	30–38	2–6
Miscanthus	EC	9–12	43–52	24–33	1–4
Switchgrass	EC	12–16	30–38	42–50	10–15
Sweet sorghum	EC	17–24	25–30	40–45	3–5
Nutshell	AW	30–40	25–30	25–30	1–2
Banana pseudostem	AW	4–8	25–32	35–45	2–7
Corn stover	AW and SW	10–17	25–35	40–45	1–4
Sugarcane bagasse	AW and SW	20–25	20–25	40–45	1–2
Rice straw	AW and SW	10–15	25–30	35–45	1–5
Wheat straw	AW and SW	15–20	45–50	30–35	2–6

[a] *Type of biomass:* HW, *hardwood;* SW, *softwood;* AW, *agriculture waste;* EC, *energy crop.*

respectively. Two-roll milling, ball milling, disk milling, hammer milling, and colloid milling are used for pretreatment. These pretreatments are highly advantageous for the production of enzymes during the SSF process (Hitoshi et al., 2005).

Steam explosion: Steam explosion is an efficient physical pretreatment that weakens the lignocellulose components. Substrates are generally exposed to steam (160–270°C) at 20–50 bar for a few seconds to a few minutes. This increases the surface area of substrate particles. Various substrates are initially employed for steam explosion and then used for the production of cellulases (Rudakiya and Gupte, 2017; Rabemanolontsoa and Saka, 2016).

Pyrolysis: This is an endothermic process wherein lignocellulosic substrates have been treated at higher temperatures (>300°C). Decomposition of the biomass leads to the formation of biogas products, carbon dioxide, and char. However, a higher concentration of lignin leads to the higher onset temperature of the pyrolysis reaction, which is a big disadvantage of using these pretreatments. Thus certain chemical treatments are required to improve the performance of pyrolysis (Rudakiya and Gupte, 2017; Kumari and Singh, 2018).

Freeze: This is a newly developed pretreatment method wherein a freezing-thawing process of bulk biomass is done. In this method, the crystal formation produces the breaking force, which leads to the destruction of lignocellulose (Kumari and Singh, 2018).

Extrusion: This is a type of physical pretreatment wherein lignocellulosic substrate faces shearing, mixing, and heating. This pretreatment leads to a reduction in particle size, and a higher degree of depolymerization is observed after pretreatment (Shirkavand et al., 2016).

10.3.2 Chemical Pretreatment

A chemical agent or solvent is applied to disrupt the components of lignocellulosic biomass. Destruction of the components is highly efficient among all methods; however, it has numerous disadvantages such as the production and usage of toxic chemicals, high cost of the process, and carbohydrate loss. Chemical pretreatment mainly comprises alkaline pretreatment, acid pretreatment, organic solvent and ionic pretreatment, and ozonolysis (Rabemanolontsoa and Saka, 2016).

Alkaline pretreatment: Alkaline pretreatment is efficient for the elimination of lignin and hemicelluloses. Additionally, it causes the swelling and partial deterioration of cellulose, causing it to decrystallize (Shirkavand et al., 2016).

Acid pretreatment: Acid pretreatment is efficient for the hydrolysis of lignin and carbohydrates. Lignin removal leads to a higher concentration of cellulose in the biomass followed by a higher production of cellulases. However, the generation of hazardous compounds during pretreatment and carbohydrate loss are the major disadvantages (Rabemanolontsoa and Saka, 2016).

Organic solvent pretreatment: Organic solvent-based pretreatment can efficiently remove the lignin and hemicellulose, but the cost of the solvent and its recovery is quite challenging. In this pretreatment, acetone, ethylene, ethanol, and methanol are generally applied (Kumari and Singh, 2018).

Ionic liquid pretreatment: Ionic liquids, the third generation of solvents, are liquid organic salts that have gained attention recently. These ionic liquids efficiently remove the lignin of biomass. The major disadvantage of using ionic liquid is cost; the chemicals are too expensive. The most commonly applied ionic liquids are: 1-butyl-3-methylimidazolium hexafluorophosphate, 1-butyl-3-methylimidazoliumacetate, 1-butyl-3-methylimidazoliumm-ethylsulfate, 1-benzyl-3-methylimidazoliumchloride, 1-butyl-3-methylimidazoliumchloride, *N*,*N*-dimethylethanol-ammonium groups, and 1-butyl-1-ethylpyrrolidiniumchloride (Rabemanolontsoa and Saka, 2016).

Ozonolysis: Ozone is a powerful oxidant that can decompose the lignin and hemicellulose structure. It also decreases the crystallinity of cellulose. No toxic by-products are generated during the ozonolysis of lignocellulosic biomass. However, the process is very expensive because of the high amounts of ozone required for the process (Kumari and Singh, 2018; Shirkavand et al., 2016).

10.3.3 Biological Pretreatment

Biological pretreatment is the "green" approach for the destruction of lignocellulosic biomass. In biological pretreatment, mainly fungi are the most suitable and efficient candidates, which degrade the lignin and hemicellulose structure efficiently. Sometimes microbial consortia and crude enzymes are also used to destroy the components. Despite having advantages, this approach takes longer as compared to the other approaches, and also has lower efficiency compared to physical and chemical pretreatments. In fungi, a group from *Basidiomycetes*, called white-rot fungi, has the capacity to mineralize the whole lignin structure along with the degradation of cellulose and hemicellulose. Numerous white-rot fungi have been applied to pretreatment processes, which include genera such as *Phanerochaete*, *Tricholoma*, *Trametes*, *Pleurotus*, *Schizophyllum*, *Cerrena*, *Cerporiopsis*, *Coprinopsis*, etc. (Wan and Li, 2012, 2010; Hatakka, 1983).

10.3.4 Combinatorial Pretreatment

Combinatorial methods are used to overcome the disadvantages of different methods. In this pretreatment, a mixture of physical-chemical, physical-biological, chemical-biological, thermal-biological pretreatments is applied for the production of cellulase. Common examples of physicochemical pretreatments are steam explosion, ammonia fiber explosion, wet oxidation, and carbon dioxide explosion (Hassan et al., 2018). In the case of physical-chemical-thermal-biological pretreatments, lignocellulosic biomass is initially treated by physical, chemical, or thermal methods, followed by biological pretreatments (Rudakiya and Gupte, 2017; Wan and Li, 2010).

10.3.5 Emerging Pretreatments

Several new pretreatments are emerging for the treatment of biomass, wherein their approaches are conducted under extreme and nonclassical conditions. This includes the pulsed-electric field, microwave, ultrasound, high hydrostatic pressure, electron beam, gamma-ray, and homogenization pretreatments. These methods are promising candidates as a sustainable green pretreatment solution for lignocellulosic biomass (Hassan et al., 2018).

Pulsed-electric field (PEF): This process mainly involves a strong electrical field (0.1–80 kV/cm) that is generated for a short time (10^{-2} to 10^{-4} s) between lignocellulosic biomass. Under the effect of PEF, the structure of biomass is disrupted, which facilitates organisms for growth (Hassan et al., 2018; Barba et al., 2015).

Microwave irradiation: Microwaves, an electromagnetic radiation, range from 1 mm to 1 m in wavelength with 300–300,000 MHz of electromagnetic spectrum. Researchers have used microwave radiation for biomass pretreatment under two different conditions, wherein microwave-assisted solvolysis under mild temperatures (<200°C) is used to produce value-added chemicals and microwave-assisted pyrolysis under higher temperature (>400°C) is used to convert biomass into bio-oil or biogas (Lo et al., 2017; Li et al., 2016).

Ultrasound: Ultrasound ranges from 20 kHz to 1 MHz and is generally used for pretreatment purposes. Mainly, it alters the surface structure and produces free radicals, which chemically attack α-*O*-4 and β-*O*-4 linkages of lignin. Subsequently, lignocellulosic components split and small cavitation bubbles form after pretreatment. It represents green technology for the pretreatment of lignocellulosic biomass (Shirkavand et al., 2016; Kumar and Sharma, 2017).

High hydraulic pressure: Lignocellulosic substrates are kept below 100–600 MPa of pressure for 15–60 min. This leads to the breakage of the structure of lignocellulose, which allows for the effective penetration of organisms (Hassan et al., 2018; Oliveira et al., 2012)

Electron beam irradiation (EB): EB ionization is used to pretreat the lignocellulosic biomass, wherein irradiation dosage can be as much as 1000 kGy. This breaks the structure of lignocellulose and decreases cellulose crystallization and the degree of polymerization (Hassan et al., 2018).

Gamma-ray radiation: Gamma rays are generally obtained from the isotopes of Co-60 or Ce-137, which are efficiently tested for the pretreatment of lignocellulosic biomass. Gamma-ray radiation easily breaks the structure, alters the lignin functional groups, and breaks the crystalline region of cellulose (Hyun Hong et al., 2014).

High-pressure homogenization: At a high pressure of 10–100 MPa with homogenization of biomass, the particle size decreases and the surface area increases, which leads to improved accessibility of solid substrate to organisms. This method is simple, green, and highly efficient and can be applied at a large scale for the pretreatment of lignocellulosic biomass (Saelee et al., 2016).

10.4 PROCESS OPTIMIZATION-BASED STRATEGIES

The term "optimization" is generally used in SSF as a means of discovering the optimum conditions for microorganisms, which produce, respectively, metabolites and/or enzymes at the highest level. Optimization is one of the preeminent strategies to improve the production of enzymes (Bezerra et al., 2008; Araujo and Brereton, 1996). Traditionally, enzyme production in SSF has been optimized using single factorial methods; however, their disadvantages are lack of experimental design and a tedious and sluggish process (Lundstedt et al., 1998). To overcome these issues, the optimization of enzyme production has been applied using multivariate statistical techniques (Rüdt et al., 2017; Rudakiya, 2018). These methods provide advantages such as proper experimental design, fewer experiments, and interaction between factors. However, these methods are also not applied at a large scale due to the lack of proper knowledge of microbial growth kinetics, product, and by-product kinetics (Bezerra et al., 2008). Mathematical modeling is the only route where microbial growth and product kinetics can be understood and applied at a large scale (Sangsurasak et al., 1996).

In process optimization, various factors of SSF affect the production of cellulases. Temperature, pH, moistening agent, moisture content, substrate type, substrate particle size, carbon sources, nitrogen sources, inducers, metals, amino acids, vitamins, and surfactants are the most critical parameters for the production of cellulases in SSF (Singhania et al., 2006; Deswal et al., 2011). In SSF, incubation temperature of the bioreactor is a most important factor because most of the fungi grow in the temperature range of 25–32°C, which directly effects the production of enzymes (Singhania et al., 2006). However, cellulase-producing bacteria grow in a higher temperature range (30–40°C) compared to fungi. Similar to temperature, the pH of SSF medium is also a critical parameter for enzyme production, wherein fungi tend to grow in acidic conditions, while bacteria show higher growth toward slight alkaline conditions (Salim et al., 2017). As discussed in previous strategies, the type of lignocellulosic substrate is also important for cellulose production. Generally, higher cellulose-containing substrates enhance the expression of cellulase genes, which leads to the higher production of cellulases (Singhania et al., 2006). Particle size of the lignocellulosic substrates also plays a crucial role in cellulase production because its effect on growth and enzyme production has been studied by several researchers (Wan and Li, 2012; Xin and Geng, 2010). Usually, smaller size particles provide higher surface areas for microbial growth and enzyme production. However, very small particles are not preferable because accumulation of substrate leads to poor aeration in microbes. Similarly, large particles are also not preferable because microbes obtain less surface area for attachment, resulting in poor microbial growth and lower enzyme production (Membrillo et al., 2011; Zhang and Sun, 2014).

Moistening medium and moistening medium-to-substrate (M/S) ratio are the most important parameters for enzyme production. Fungi generally exhibit higher enzyme production from 1:1 to 1.35 of initial M/S ratio (Deswal et al., 2011; Kalogeris et al., 2003). For higher production of enzymes, different nutrients have been added to the moistening medium, by which organisms may obtain them for their initial growth and metabolism. Many researchers provide different moistening mediums for the higher production of cellulases, which comprise nutrients such as carbon sources, nitrogen sources, inducers, metals, amino acids, vitamins, and surfactants

(Membrillo et al., 2011; Zhang and Sun, 2014; Kalogeris et al., 2003). Each of these factors either enhances or diminishes the growth of microbes, which directly affects enzyme production. To obtain a higher production of cellulases, each factor of the moistening medium should be optimized (Deswal et al., 2011). Different optimization methodologies are described in the ensuing sections, which are generally applied for the enhancement of cellulase production.

10.4.1 Single Factorial Methodology

Single factorial or "one-factor-at-a-time (OFAT)" optimization is a classical experimental optimization technique, which optimizes the entire system by varying a single factor at a time. Only one factor changes in each trail; however other factors remain constant. This approach is easy to understand and simple to implement, and assists in the selection of significant parameters and improves production yield (Singh et al., 2011). Additionally, this optimization is preferred when medium, substrate, and/or organism are novel. One can certainly optimize all the parameters by varying the individual factors. However, the OFAT method is usually discouraged in the literature for the higher production of enzymes. Major drawbacks of the OFAT method are: (1) it requires more runs to estimate the effect, (2) it cannot estimate the interaction between two different parameters, (3) it is prone to bias due to the time trends if numerous factors are involved, (4) it lacks a general conclusion, (5) it often fails the optimal settings of factors, (6) it requires more time and is tedious compared to statistical optimization, and (7) its large scale cannot be applied because it requires numerous variables (Frey and Wang, 2006; Frey et al., 2003).

Adaptive OFAT is introduced to overcome the disadvantages of OFAT, so it has been applied for optimization purposes in recent times. In adaptive OFAT, the first factor is optimized by varying the different variables, wherein the variable with higher production yield is selected for the forthcoming experiments (Frey and Wang, 2006). Furthermore, the second factor is optimized by keeping the first factor with that specific variable constant and by varying the different variables of the second factor. In an experiment with the last factor, all preceding factors with the higher yield acquired variables that are selected and only the variables of the last factor are changed. Adaptive OFAT is illustrated by using six different factors (pH, temperature, carbon and nitrogen source, amino acids, and metals) with respective variables in Fig. 10.2 to optimize the cellulase production yield. This approach reduces the errors obtained in the conventional OFAT method (Bezerra et al., 2008; Frey et al., 2003). Despite the many advantages of this method it has numerous disadvantages, e.g., cellulase production has been mainly optimized in recent times using the adaptive OFAT method (Singhania et al., 2006; Deswal et al., 2011; Salim et al., 2017). Brown-rot fungus *Fomitopsis* sp. RCK2010 exhibited a 1.2- to 2.8-fold increment for CMCase, FPase, and β-glucosidase production

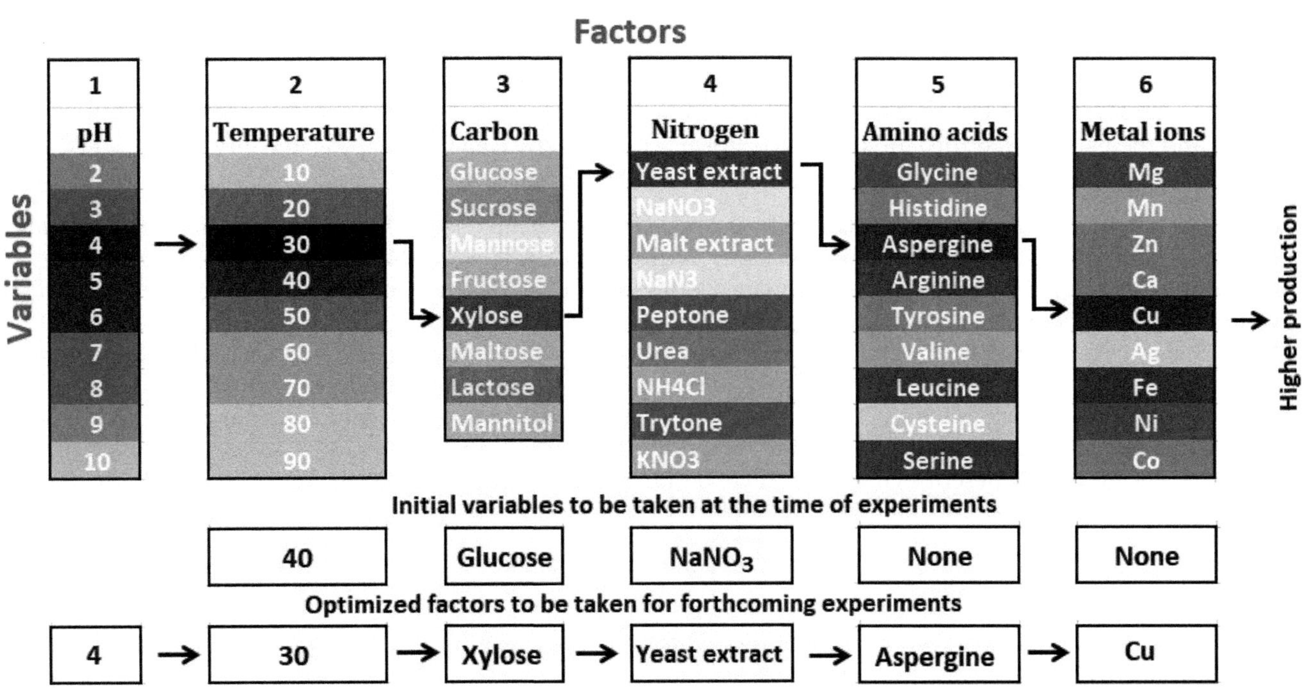

FIG. 10.2 An illustration of the adaptive one-factor-at-a-time method that can be used for the optimization of enzymes and/or other metabolites.

under SSF by using the adaptive OFAT method (Deswal et al., 2011). Likewise, *A. niger* NS-2 showed a 2- to 4-fold increment of CMCase, FPase, and β-glucosidase production when fungi were grown on agricultural and kitchen waste residues (Bansal et al., 2012).

10.4.2 Response Surface Methodology

Design of experiments is a set of specific experiments defined by a matrix and studied by different levels of combination. Additionally, it is an assortment of different tools that are used to study the behavior of a system. This includes the planning of variables and execution of certain experiments to evaluate the effects of variables on particular systems.

10.4.2.1 Different Types of Designs

There are two types of multivariate designs: (1) full or fractional factorial designs and (2) qualitative or quantitative responses. Factorial designs are generally screened initially to obtain knowledge of the system or to select the best factors. Subsequently, responses of the selected factors are attained by performing response designs such as central composite designs (CCD), Box-Behnken designs (BBD), and the Doehlert matrix (DM) (Rüdt et al., 2017; Rudakiya, 2018).

First-order models: First-order models are used in exploratory studies wherein large a number of factors are needed to screen. In this method, the main effects caused by the factors are detected; however, minor effects are discarded. There are two types of first-order model, which include factorial design (full or fractional) and Plackett-Burman design. Factorial designs are the most popular first-order design and it is useful for preliminary studies where all factors are mandatory. In this model, linear surface response is determined and the given linear model is:

$$\hat{y} = b_0 + b_1x_1 + b_2x_2 + b_{12}x_1x_2$$

If one can consider a full factorial design with 11 factors and two-level designs then it would require 2048 (2^{11}) individual experiments with 11 main effects, 55 second-order interactions, and 1981 further interactions. Conversely, the Plackett-Burman design requires fewer numbers of experiments. A large number of factors may have an influence on the response, which directly shows the positive or negative response on the system. This method is extremely confounding. Additionally, it cannot be used to study the individual main effects and interactions between them. Despite these confounding features, it is widely accepted for screening experiments.

Higher-order models: Linear models are not sufficient to characterize the experimental data, so more experiments should be performed after factorial design. The following equation is used to determine the quadratic response:

$$\hat{y} = b_0 + b_1x_1 + b_2x_2 + b_{11}x_1^2 + b_{22}x_2^2 + b_{12}x_1x_2$$

This can be used to predict the factor level, which produces minimum and maximum values. Thereafter, model coefficients and their standard errors are calculated, which is followed by ANOVA analysis to verify the model. Mainly, three types of design have been applied to check the responses (Ferreira et al., 2007):

1. *CCD*: CCD combines two levels of factors, which can be full or partial design. It contains additional data points with one central point of the experimental design. Additionally, CCD is an efficient alternative to the three-level designs, wherein fewer experiments can be conducted in CCD design compared to other factorial designs. Therefore CCD is the mostly accepted design for the response surface analysis (Ferreira et al., 2007).
2. *BBD*: This is a class of rotatable second-order design based on incomplete three-level designs. The experiments are defined as $N = 2k(k - 1) + C_0$, where C_0 is the number of points and k is the number of factors. The overall structure of BBD is shown in Fig. 10.3. The advantage of BBD is that it does not contain the highest and lowest values, wherein results of the extreme condition might be unsatisfactory.
3. *DM*: This design mainly describes the spherical experimental domain with uniform space filling. However, the matrix is not orthogonal or rotatable so it cannot be used significantly for response surface analysis.

10.4.3 Mathematical Modeling

A mathematical model is a collection of equations, which provides insights into the various phenomena of the system. In the case of SSF bioreactors, microbial growth and product formation are the major criteria that are to be evaluated. Mathematical modeling is a most important tool to optimize different bioprocesses at a large scale. Optimized

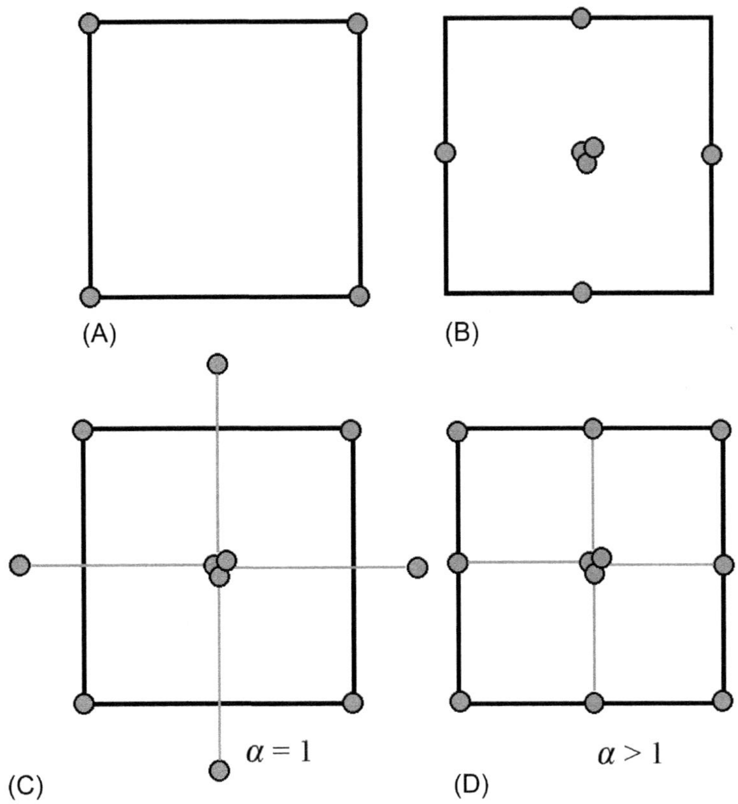

FIG. 10.3 General experimental designs of response surface methodology in two dimensions: (A) a factorial design, (B) a Box-Behnken design, (C and D) central composite designs with star-coded distances $\alpha = 1$ (C) and $\alpha > 1$ (D). *Modified from Bezerra, M.A., Santelli, R.E., Oliveira, E.P., Villar, L.S., Escaleira, L.A., 2008. Response surface methodology (RSM) as a tool for optimization in analytical chemistry. Talanta 76, 965–977; Araujo, P.W., Brereton, R.G., 1996. Experimental design. I. Screening. TrAC Trends Anal. Chem. 15, 26–31; Lundstedt, T., Seifert, E., Abramo, L., Thelin, B., Nyström, Å., Pettersen, J., Bergman, R., 1998. Experimental design and optimization. Chemom. Intell. Lab. Syst. 42, 3–40; Rüdt, M., Briskot, T., Hubbuch, J., 2017. Advances in downstream processing of biologics—spectroscopy: an emerging process analytical technology. J. Chromatogr. A 1490, 2–9.*

and validated models guide the design and operation of SSF bioreactors used for metabolite or enzyme production. Additionally, they also provide information regarding the various phenomena within the fermentation system that interact with each other and control the overall performance of the bioreactor. Mathematical models are mainly divided into two categories: microscale models and macroscale models. Microscale models are based on the phenomena that occur within and on the substrate particles, which do not describe the performance of the bioreactor. On the contrary, macroscale models are based on the phenomena that occur in the bioreactors, which mainly comprise heat and mass transfer across the substrate bed. These models describe the overall performance of the bioreactor. Fig. 10.4 shows the schematic representation of a bioreactor model encountering various phenomena and it is evaluated using mathematical modeling (Gelain et al., 2015).

A mathematical model of a bioreactor contains certain state variables, which are mainly temperature of substrate (T) and amount of biomass (X) in the bioreactor. These variables also describe the state of the bioreactor at a specific instant. The value of these variables varies as the independent variables change. The independent variables are the variables do not depend on the system. In such cases, the system is dependent on these variables, where variables change even at a minute level, productivity of whole system is altered. Lastly, operating variables can be controlled because their values highly affect the performance of the bioreactor. Examples of these variables are mainly temperature of the inlet air (T_{in}), humidity of the inlet air (H_{in}), air flow rate (F), and temperature of the surroundings (T_{surr}) (Montoya et al., 2014).

Different biological and physical parameters are involved in the mathematical modeling of the bioreactor. Four parameters must be included to build the model of the bioreactor: (1) *design of bioreactor*: this includes the height, width, and depth of the bioreactor; (2) *transport parameter*: this is related to the transfer of energy and mass within the bioreactor. It comprises the diffusivity of O_2, nutrients and other products, thermal conductivity, and heat transfer; (3) *thermodynamic parameter*: this is related to the quantity of energy and the equilibrium state of the biomass. It encompasses the saturation humidity of air, heat capacity of air, heat capacity of substrate particles, enthalpy of the

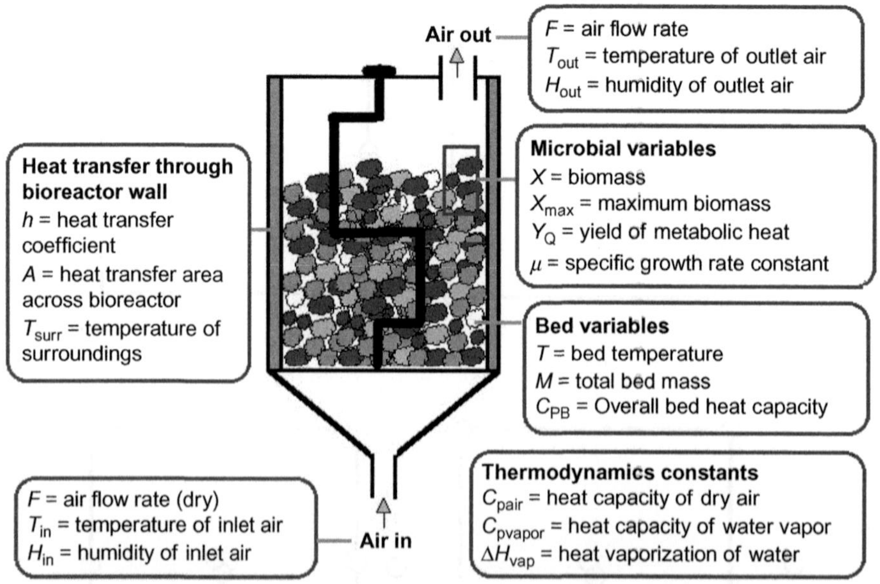

FIG. 10.4 A mathematical model is predicted to evaluate the temperature within the stirred packed-bed bioreactor. There are various variables involved in this model and a list of variables is given in the boxes. *Modified from Mitchell, D.A., Berovič, M., Krieger, N., 2006. Solid-state fermentation bioreactor fundamentals: introduction and overview. In: Solid-State Fermentation Bioreactors. Springer, Berlin, Heidelberg, pp. 1–12.*

TABLE 10.3 Equations are Used to Calculate Microbial Growth Kinetics in Mathematical Modeling

Model	Equation	References
Linear	$\frac{dX}{dt} = K$	Mitchell et al. (2006)
Exponential	$\frac{dX}{dt} = \mu X$	Mitchell et al. (2006)
Logistic	$\frac{dX}{dt} = \mu X\left(1-\frac{X}{Xm}\right)$	Mitchell et al. (2006)
Bovill	$\frac{dX}{dt} = \mu X\left(1-\frac{X}{Xm}\right)\left(\frac{Q}{1+Q}\right)$	Bovill et al. (2000)
Arrhenius	$\mu_{max}(T) = k_g^0 \exp\left(\frac{-E_g}{RT}\right) - K_g \exp\left(\frac{-E_d}{RT}\right)$	–
Fanaei and Vaziri	$\mu = \left(\frac{\mu_s + (T_{max}-T_{opt})}{(T_{max}-T_{opt})}\right)\left(\frac{\mu_{opt}+(T_{max}-T)}{\mu_s+(T_{max}-T)}\right)$	Fanaei and Vaziri (2009)
Weber	$\mu_{max}(T)=c_0(T-T_{min})[1-\exp(c_1(T-T_{max}))]^2$	Ratkowsky et al. (1983)
Gelain	$\left(\frac{dX}{dt}\right)_c = \mu_m\left(\frac{S}{1+S}\right)\left(1-\frac{X}{Xm}\right)^r X$	Gelain et al. (2015)
Montoya	$\frac{dC_b}{dt} = \mu_{max}C_b\left(1-\left(\frac{C_b}{C_{bm}}\right)^n\right)$	Montoya et al. (2014)

dX/dt = biomass production rate; K = constant; X and C_b = biomass concentration; C_{bm} = maximum biomass concentration; μ = specific growth rate; μ_{max} = maximum specific growth rate; μ_{opt} = optimum specific growth rate; X_m = maximum biomass concentration; X_0 = initial biomass concentration; T_{min} = minimum temperature growth; T_{max} = maximum temperature growth; T_{opt} = optimum temperature growth; R = universal gas constant; T = temperature; K_g and k_g^0 = constants; r = exponent for growth repression; S = cellulose concentration; n = reaction order.

vaporization water, and equilibrium concentration of O_2 in the substrate; (4) *biological parameter*: this is related to the behavior of the organism. It includes maximum specific growth rate, maximum biomass, and yield coefficients. In mathematical modeling, different kinetic submodels are used to calculate the microbial growth kinetics. Table 10.3 represents the different equations, which have been used for the kinetic submodels for microbial growth (Gelain et al., 2015; Montoya et al., 2014; Ratkowsky et al., 1983).

10.5 BIOREACTOR-BASED STRATEGIES

Strategies based on the bioreactor are crucial for the overall performance of SSF production. Design of bioreactors greatly affects the productivity of enzymes and performance of the SSF process. Similarly, heat and mass transfer is also a key parameter, which affects the overall productivity of the SSF bioreactor (Ray and Behera, 2017; Behera and Ray, 2016).

10.5.1 Design of Bioreactor

Four different types of bioreactors are classified based on aeration and mixing, which are generally used for the production of cellulases.

10.5.1.1 Tray Bioreactor (Group I)

Tray fermenters or "Group I" bioreactors are the simplest bioreactors in SSF. They do not require higher instrumentation because fermentation takes place in small-, medium-, or large-sized trays. Traditionally, this type of fermentation has been taking place for several centuries and the fermentation room is called the koji room. The produced crude enzymes can directly be used for sale or for downstream processing. Organisms with their culture conditions are shown in Table 10.4 under tray fermentation. Heat transfer is possible through conduction, and in the case of low thermal conductivity, substrate cannot pass heat. A large scale is only possible by incubating multiple trays. Sterility of the product as well as the system cannot be maintained, which leads to the risk of contamination (Mitchell et al., 2006; Manan and Webb, 2016; Dhillon et al., 2011c; Díaz et al., 2013) (Fig. 10.5).

TABLE 10.4 Different Types of Solid-State Fermenter Bioreactors are used for the Production of Cellulases

Bioreactor	Substrate used	Enzyme produced	Microorganism used	Enzyme production	References
Tray	Wheat bran, soybean hulls, and rapeseed mean	Cellulase	*Aspergillus awamori*, *Aspergillus oryzae*	Varies based on the substrate	(Manan and Webb, 2016)
Tray	Apple pomace	β-Glucosidase	*Aspergillus niger*	64.18 IU/gds	(Dhillon et al., 2011c)
Tray	Grape pomace and orange peel	CMCase	*A. awamori*	2.16 IU/gds	(Díaz et al., 2013)
Static bed	Wheat bran	Endoglucanase	*A. niger*	29.8 IU/gds	(Farinas et al., 2011)
Packed bed	Soybean bran	Cellulase	*Trichoderma reesei* NRRL-6156	4.2 FPU/g (FPase), 7.4 U/g (exocellulase), 2.5 U/g endocellulase	(Gasparotto et al., 2015)
Static bed with forced aeration	Wheat bran	Endoglucanase	*A. niger*	50.2 IU/gds	(Farinas et al., 2011)
Packed bed	Babassu cake	Cellulase	*A. awamori*	6.2 U/g	(Castro et al., 2015)
Packed bed	Wheat bran and sugarcane bagasse	Endoglucanase	*Myceliophtora* sp. I-1D3b	878 IU/gds	(Zanelato et al., 2012)
Rotating drum	Empty palm fruit bunch fiber	CMCase	*Penicillium verruculosum*	6.5 U/gds	(Kim and Kim, 2012)
Rotating drum	Empty palm fruit bunch fiber	Endoglucanase Exoglucanase β-Glucosidase	*A. niger*	135 U/L 52 U/L 161 U/L	(Noratiqah et al., 2013)
Rotating drum	Grape pomace and orange peels	CMCase	*A. awamori*	3.9 IU/gds	(Diaz et al., 2009)
Rotating drum	Palm oil lignocellulosic biomass	FPase	*Trichoderma harzianum*	8.2 FPU/gds	(Alam et al., 2009)
Rotating drum	Grape pomace	Cellulase	*A. awamori*	3.69 IU/gds	(Diaz et al., 2009)
Gas double dynamic SSF	–	Cellulase	*Penicillium decumbens* JUA10	18 IU/gds	(Fujian et al., 2002)

CMCase, carboxymethyl cellulase; *SSF*, solid-state fermenter.

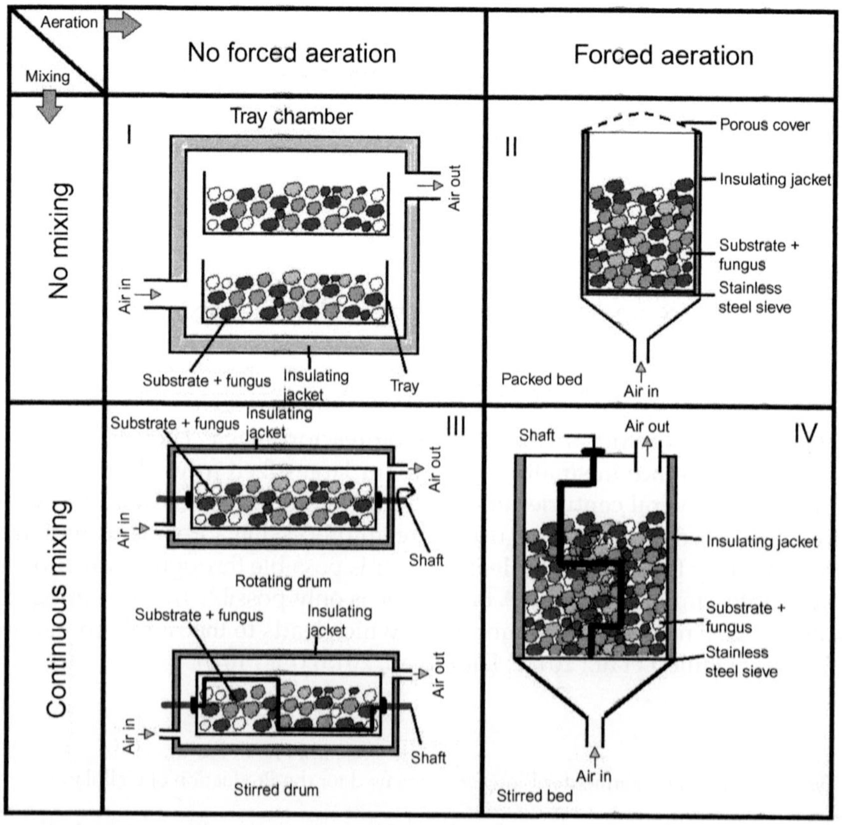

FIG. 10.5 Schematic diagram of solid-state fermentation bioreactors that are used for cellulase production, in which they are classified based on aeration and mixing. *Adapted from Mitchell, D.A., Berovič, M., Krieger, N., 2006. Solid-state fermentation bioreactor fundamentals: introduction and overview. In: Solid-State Fermentation Bioreactors. Springer, Berlin, Heidelberg, pp. 1–12.*

To calculate the maximum depth of the tray where O_2 concentration cannot fall to zero in the tray bioreactor, the following equation is applied:

$$D_c = \sqrt{\frac{2DCY_{XO}}{R_{XM}}}$$

where Y_{XO} is the yield coefficient of biomass from O_2, C is the O_2 concentration in the tray chamber, D is the effective diffusivity of O_2 in the bed, and R_{XM} is the maximum growth rate.

The temperature profile within the tray bioreactor can be calculated by the following equation:

$$T = \frac{(T_s + 273) + \Theta + N_{BI}}{N_{BI} + 1} + \frac{N_{BI}}{N_{BI} + 1}(T_S - T_a + \Theta)z - z^2\Theta$$

where T_S is the temperature at the bed surface, T_a is the temperature of the surrounding air, z is the dimensionless fraction of total bed height, N_{BI} is the Biot number, k is the thermal conductivity of the bed, and Θ is the temperature difference between bottom and top surface of the tray.

Microbial growth can be calculated by using the following equation:

$$Rx = \mu X\left(1 - \frac{X}{X_{\max}}\right)$$

where Rx is the overall growth rate, μ is the specific growth rate constant, X is the biomass density, and $X_{\max}$ is the maximum value for the biomass density.

Advantages: It requires lower energy because mixing of the substrate is not required. Temperature can be controlled by incubating the trays in a temperature-controlled room. Additionally, these rooms are closed so moisture loss can be avoided. There is no requirement for any highly sophisticated operating technology and industrial-scale instruments.

Limitations: The transfer of gas is limited, so the bed size of the fermenter should not be higher (less than 3 cm). Additionally, limited gas transfer leads to problems of heat and mass transfer. It requires a huge space and a large workforce because large-scale production is only achieved by multiple trays.

10.5.1.2 Packed-Bed Bioreactor (Group II)

Packed-bed bioreactors are Group II-type reactors where forced aeration is applied but the substrate bed is not mixed. The column of this type of reactor lies horizontally, vertically, or at any angle, which alters the relative directions of the forces due to gravity and air pressure. Mainly, two strategies have been applied to enhance the production of enzymes: (1) a traditional and known packed bed type can be used with lower bed height, and (2) a Zymotis-type bioreactor can be used with internal heat plates. If the organism can tolerate the infrequent mixing event, then a traditional packed-bed bioreactor can be useful for the production of enzymes. If organisms are sensitive to mixing events, then a Zymotis design should be strongly considered for enzyme production. Table 10.4 represents the key organisms with their culture conditions that are used for cellulase production under a packed-bed bioreactor (Mitchell et al., 2006; Farinas et al., 2011; Gasparotto et al., 2015; Castro et al., 2015; Zanelato et al., 2012) (Fig. 10.5).

Traditional packed-bed reactor: Microbial growth kinetics is calculated by using the following logistic formula:

$$\frac{dX}{dt} = \mu X\left(1 - \frac{X}{X_m}\right)$$

The mathematical model used for the traditional packed-bed reactor is:

$$\rho_b C_{pb}\left(\frac{\partial T}{\partial t}\right) + \rho_a\left(C_{pa} + \frac{dH_{\mathrm{sat}}}{dT}\lambda\right)V_Z\left(\frac{\partial T}{\partial z}\right) = k_b\left(\frac{\partial^2 T}{\partial z^2}\right) + \rho_s(1-\varepsilon)Y_Q\left(\frac{dX}{dt}\right)$$

wherein the calculation of storage of energy due to temperature increases is shown in the following equation: $\rho_b C_{pb}\left(\frac{\partial T}{\partial t}\right)$.

The following equation is for axial convection, wherein the air remains saturated as temperature increases is assumed and calculated by: $\rho_a\left(C_{pa} + \frac{dH_{\mathrm{sat}}}{dT}\lambda\right)V_Z\left(\frac{\partial T}{\partial z}\right)$.

This equation is for axial conduction: $k_b\left(\frac{\partial^2 T}{\partial z^2}\right)$.

This equation is for heat generation from microbial growth: $\rho_s(1-\varepsilon)Y_Q\left(\frac{dX}{dt}\right)$.

Zymotis packed-bed reactor: Microbial growth kinetics is calculated by using the logistic formula: $\frac{dX}{dt} = \mu X\left(1 - \frac{X}{X_m}\right)$.

The mathematical model used for the Zymotis packed-bed reactor is:

$$\rho_b C_{pb}\left(\frac{\partial T}{\partial t}\right) + \rho_a\left(C_{pa} + \frac{dH_{\mathrm{sat}}}{dT}\lambda\right)V_Z\left(\frac{\partial T}{\partial z}\right) = k_b\left(\frac{\partial^2 T}{\partial x^2}\right) + SY_Q\left(\frac{dX}{dt}\right) + Y_Q X\left(\frac{dS}{dt}\right)$$

wherein the calculation of storage of energy due to the temperature changes in the bed is shown in the following equation: $\rho_b C_{pb}\left(\frac{\partial T}{\partial t}\right)$.

The following equation is for axial convection and evaporation:

$$\rho_a\left(C_{pa} + \frac{dH_{\mathrm{sat}}}{dT}\lambda\right)V_Z\left(\frac{\partial T}{\partial z}\right)$$

The following equation is for horizontal conduction: $k_b\left(\frac{\partial^2 T}{\partial x^2}\right)$.

This equation is for heat generation from microbial growth: $SY_Q\left(\frac{dX}{dt}\right) + Y_Q X\left(\frac{dS}{dt}\right)$.

Subscript *a* is for air, *b* is for bed, *pb* is for packed bed, *pa* is for air, sat is for saturation, and *s* is for substrate particles. μ is the specific growth rate constant, X_m is the maximum biomass concentration, *X* is the biomass density, *S* is the dry substrate, Y_Q is the yield of metabolic heat produced by microbial growth, *T* is the temperature, V_z is the superficial velocity of air, λ is the enthalpy of water vaporization, ε is the void fraction from bed, k is the thermal conductivity, C is the heat capacity, and ρ is the density.

Advantages: It is an appropriate bioreactor for fungi, which are sensitive to shearing force. Flexible modifications are possible for specific requirements. The bioreactor is manually agitated even though no mechanical agitation is applied.

Limitations: The bioreactor has a limited bed height of 30 cm. Temperature gradients are possible through the substrate bed. Oxygen supply is also a limiting step for this reactor.

10.5.1.3 Drum Bioreactor (Group III)

Group III bioreactors are drum bioreactors that have aeration through the headspace of the bioreactor. There are two types of bioreactors: the rotary drum bioreactor and stirred drum bioreactor. In the rotary drum bioreactor, the whole drum is rotating from the axis to rotate the bed. However, the drum is stationary in the case of the stirred drum bioreactor but the paddle-containing shaft rotates the whole bed. The bioreactor is generally used for continuous fermentation (Fig. 10.5). Heat transfer is possible through convection, evaporation, cooling, and conduction. For that purpose, the bioreactor possesses 30%–40% of working volume, i.e., a substrate bed to enhance enzyme productivity. The rotating drum bioreactor is not suitable for scale-up. Sterilization is achieved by injecting high-pressure steam. A higher shear force is observed in this type of bioreactor, which can damage fungal mycelia (Mitchell et al., 2006; Zanelato et al., 2012; Kim and Kim, 2012; Noratiqah et al., 2013; Diaz et al., 2009; Alam et al., 2009).

A summary of mathematical modeling for a well-mixed rotary drum bioreactor is given as follows.

Microbial growth kinetics is calculated by using the logistic formula:

$$\frac{dX}{dt} = \mu X\left(1-\frac{X}{X_m}\right)$$

$$\frac{dM}{dt} = \frac{dX}{dt}\left(1-\frac{1}{Y_{XS}}\right) - m_S X \text{ Equation for the overall mass of the dry solid}$$

$$\frac{d[MW]}{dt} = \left(Y_W\frac{dX}{dt} + m_W X\right) - k_w A_{bh}\left(W - W_{sat}\right) \text{Equation for bed water balance}$$

$$\frac{d[GH]}{dt} = F\left(H_{in} - H\right) - k_w A_{bh}\left(W - W_{sat}\right) \text{Equation for headspace water balance}$$

$$\frac{d\left[T_b M\left(C_{pm} + C_{pw}W\right)\right]}{dt} = \left(Y_Q\frac{dX}{dt} + m_Q X\right) + h_{bh}A_{bh}\left(T_h - T_b\right) + h_{bw}A_{bw}\left(T_w - T_b\right) - k_w A_{bh}\left(W - W_{sat}\right)\lambda$$

Equation for bed energy balance

$$\frac{d\left[T_h G\left(C_{pg} + C_{pv}H\right)\right]}{dt} = FC_{pg}\left(T_{in} - T_h\right) + FC_{pv}\left(H_{in}T_{in} - HT_h\right) + h_{hw}A_{hw}\left(T_w - T_h\right) - h_{bh}A_{bh}\left(T_h - T_b\right) \text{Equation for headspace energy balance}$$

$$\frac{d\left[T_w BC_{pw}\right]}{dt} = h_{ws}A_{hs}\left(T_s - T_w\right) - h_{bw}A_{bw}\left(T_w - T_b\right) - h_{hw}A_{hw}\left(T_w - T_h\right) \text{Equation for drum wall energy balance}$$

Subscript *h* is for headspace, *b* is for bed, *w* is for bioreactor wall, *s* is for surroundings, *in* is for inlet, sat is for saturation, *pg* is for dry air, *pv* is for water vapor, *pw* is for liquid water, *pb* is for bioreactor wall, *pm* is for dry matter, *bw* is for bed/wall, *ws* is for wall/surroundings, and *bh* is for bed/headspace. X_m is the maximum biomass concentration, C is the heat capacity, μ is the specific growth rate constant, h is the heat transfer coefficient, X is the biomass density, ρ is the density, Y_Q is the yield of metabolic heat produced by microbial growth, Y_b is the yield of biomass, S is the dry substrate, T is the temperature, V_z is the superficial velocity of air, λ is the enthalpy of water vaporization, k is the mass transfer coefficient, A is the contact area, and P is the pressure.

Advantages: It increases heat transfer because the majority of the particles are exposure to the air. An external water jacket assists in reducing the temperature of the SSF process.

Limitations: Shearing force affects the fungal growth throughout the SSF process. Power consumption is higher if rotational speed is 10% higher. Because higher rotation speed inhibits fungal growth, substrate agglomeration is observed. It is not suitable for small and soft solid substrate particles.

10.5.1.4 Stirred-Bed Bioreactor (Group IV)

A stirred-bed bioreactor is an intensely aerated and continuously mixed type of bioreactor. In this bioreactor, inlet air from the packed bed remains saturated during fermentation, which is a major advantage over a conventional packed-bed reactor. Heat is transferred mostly by convection and conduction. In situ sterilization is achieved by bioreactor design. In this type of bioreactor, bed capacity is comparatively higher than group I to III-type bioreactors. A large number of skilled workers, sophisticated instrumentation, and a control facility are required for this type of fermenter (Mitchell et al., 2006).

A summary of mathematical modeling for a well-mixed rotary drum bioreactor is given as follows:

$$\frac{\partial X}{\partial t} = \mu X\left(1 - \frac{X}{X_m}\right) \text{Equation for microbial growth kinetics}$$

$$\frac{\partial S}{\partial t} = \left(1 - \frac{1}{Y_{BS}}\right)\frac{\partial (XS)}{\partial t} \text{Equation for consumption of dry solid}$$

$$\frac{\partial (SW)}{\partial t} = Y_{WB}\left(S\frac{\partial X}{\partial t} + X\frac{\partial S}{\partial t}\right) - Ka(W - W_{sat})\text{Equation for solid} - \text{phase water balance}$$

$$\varepsilon\rho_g\left(\frac{\partial H}{\partial t}\right) = -G\left(\frac{\partial H}{\partial z}\right) + Ka(W - W_{sat})\text{Equation for gas} - \text{phase water balance}$$

$$S\left(C_{ps} + WC_{Pw}\right)\frac{\partial T_s}{\partial t} = ha\left(T_g - T_s\right) + Y_Q\left(S\frac{\partial X}{\partial t} + X\frac{\partial S}{\partial t}\right) - Ka(W - W_{sat})\lambda \text{ Equation for solid} - \text{phase energy balance}$$

$$\varepsilon\rho_g\left(C_{pg} + HC_{Pw}\right)\frac{\partial T_g}{\partial t} = -\left(C_{pg} + HC_{Pw}\right)G\frac{\partial T_g}{\partial z} - ha\left(T_g - T_s\right)\text{Equation for gas} - \text{phase energy balance}$$

Subscript g is for the gas phase, s is for the solid phase, sat is for saturation, pg is for dry air, pv is for water vapor, pw is for liquid water, and ps is for dry solids. X_m is the maximum biomass concentration, X is the biomass density, ρ is the density, Y_{WB} is the yield of biomass, T is the temperature, Y_Q is the yield of metabolic heat produced by microbial growth, λ is the enthalpy of water vaporization, C is the heat capacity, h is the heat transfer coefficient, P is the pressure, ε is the void fraction from bed, a is the water activity, and μ is the specific growth rate constant.

Advantages: Difficulties of metabolic heat accumulation can be resolved without temperature control. It is quite easy to add water, pH, and nutrients to the substrates. Substrate homogeneity is higher compared to other types of bioreactor.

Limitations: Sticky solid particles prevent effective fluidization because they are agglomerated. Substrate particle size may play an important role in fluidization and if particles are uneven in size; enzyme production can be affected. It has a greater possibility of damaging fungal mycelia.

10.5.2 Heat and Mass Transfer

An SSF bioreactor exchanges energy and mass between the bioreactor and its surroundings, wherein the airstream contains mass (vapor, CO_2, O_2, and N_2) and energy into and outside the SSF bioreactor. Due to fermentation, energy releases between substrate particles are ultimately observed on the bioreactor wall as well as within the bioreactor. Within the bioreactor, the substrate bed is the major component that releases the heat and mass transfer. Metabolic heat production, conduction, diffusion of gas components, convective heat transfer, evaporation, and convective mass transfer are the possible reasons for these phenomena within the bioreactor. On the contrary, temperature of the environment, type of bioreactor, and conduction are the probable reasons for heat and mass transfer across the bioreactor wall. Thus these phenomena play significant roles to improve the performance of the SSF bioreactor (Mitchell et al., 2006). Mathematical models are used to understand these phenomena, which contain mainly energy and mass balance equations. The equation of energy balance is:

$$m_{\text{bed}} C_{P\text{bed}} \frac{dT_{\text{bed}}}{dt} = \pm Q_a \pm Q_b \pm Q_c \pm \ldots \pm R_Q$$

where m_{bed} is the mass of the bed, T_{bed} is the bed temperature, $C_{P\text{bed}}$ is the overall heat capacity of the bed, R_Q is the rate of metabolic heat production, and Q_a, Q_b, and Q_c is the different rates of heat transport phenomena.

The equation of mass balance is:

$$\frac{dM_{\text{water}}}{dt} = \pm R_a \pm R_b \pm R_c \pm \ldots \pm r_w$$

where r_w is the rate of metabolic water production, R_a, R_b, and R_c are the mass transfer phenomena that involve water, and M_{water} is the mass of water in the bed.

Fujian et al. (2002) investigated heat and mass transfer during cellulase production under SSF using *Penicillium decumbens* JUA10. Numerous parameters such as limit of air pressure, air pulsation period, air circulation rate, and dynamic air changes are accounted for on *P. decumbens* JUA10 growth and enzyme activities.

10.6 PROCESS MONITORING-BASED STRATEGIES

Monitoring and controlling the environmental factors of a bioreactor are critical during the SSF process. There are numerous possibilities for disturbances, i.e., variation in solid substrates, changes in bioreactor environment, equipment failure, and human errors during the SSF process. If the possible faults are not detected during this process, they will directly affect the performance of SSF. As a result, accidents, product loss, or products without any specification are observed (Pérez-Conrea and Agosin, 1999; Aidoo et al., 1982). To reduce such accidents, reproduce the process, and monitor the complete process, certain variables should be measured frequently or continuously during the SSF process. The chief factors for process control and monitoring during SSF include pH, temperature, water content, aeration, agitation, flow control, and pressure drop measurements. Direct measurement of moisture content, temperature, air flow, and agitation are considered by employing classical sensors. However, biomass content is monitored indirectly by employing respirometry and pressure drop. Certain parameters such as pH, substrate, and product concentration are measured using offline techniques (Bellon-Maurel et al., 2003; Lonsane et al., 1985).

10.6.1 Temperature

Temperature of the bioreactor as well as within the SSF process is critical for scale-up of the SSF process. A considerable amount of heat is produced throughout SSF that is correlated with the metabolism of organisms and depth of the substrate bed. In 1 kg of assimilated substrate, metabolic heat can reach up to 3000 K at the center of the bioreactor. The produced heat must be dispersed immediately otherwise higher temperature leads to sporulation inhibition, lower microbial growth, and reduced cellulase production (Hesseltine, 1987; Ghildyal et al., 1994). The temperature is regulated near 30°C by applying several temperature-controlling techniques, which include: (1) a submerged bioreactor in a temperature-controlled water bath, (2) water circulation in a jacket around the bioreactor, (3) a bioreactor in the temperature-controlled room, (4) protection of the bioreactor using water-soaked burlap, and (5) forcing large amounts of air through the bioreactor. Among all the methods, temperature regulation along with moisture is the efficient approach. The input moisture content exchanges water and heat with the substrate particles, which helps in regulating the temperature at ±4°C (Pandey, 2004). Generally, temperature sensors are implanted at different distances on the bioreactor, which are directly linked to the control system. These sensors are protected by a tube, which is made up of stainless steel or ceramic depending on the bioreactor type. A thermocouple, resistant thermometer detector, thermistor, and integrated sensor are the major sensing systems used to detect the temperature of the SSF bioreactor (Pandey et al., 2008; Hölker and Lenz, 2005).

10.6.2 pH

A change in metabolic activity is directly associated with variations in pH during the SSF process. This leads to the agglomeration of solid particles, and sometimes organisms change the morphological stages due to pH changes. Overall, enzyme productivity is affected. It is quite difficult to use the potentiometric electrode or standard pH electrode. Therefore, there are no existing electrodes that can record pH during SSF due to the lack of water. In this case, offline pH measurements only are possible, wherein small amounts of the sample are taken, mixed with distilled water, and the pH is measured. Thereafter, pH of the SSF process can be maintained by addition of acid or base. There are also new pH probe models available, which are connected to the interfaces, and certain output signals can be obtained using network protocols. However, it is somewhat challenging to predict accurate pH values of a bioreactor (Lonsane et al., 1985). Alternatively, solid substrates that have good buffering capacity are used for the SSF process. Initially, these types of solid substrates are adjusted within a specific pH range and the desired pH is obtained after moistening the substrate (Pandey, 2004; Pandey et al., 2008).

10.6.3 Water Activity (Humidity and Moisture)

Moisture and humidity are key factors of the SSF process. Moisture or water activity is mainly directed by type of substrate, organism, and production process. A higher moisture level leads to reduced porosity and oxygen diffusion, increased bacterial contamination, decreased gas volume, and decreased gas exchange. On the contrary, a lower moisture level in the SSF process causes reduced swelling of substrate, increased water tension, and microbial growth. Initial humidity or moisture content generally ranges from 30% to 75%, which mainly depends on the substrate and organism type. However, the moisture level is decreased during microbial growth because an increase in metabolic activities leads to higher temperature with lower moisture content. Thus bioreactors have inlet and outlet passages for the moist air, where humidity or moisture level within the bioreactor is maintained. To measure the moisture or water activity of an SSF bioreactor, there are certain online capacitive sensors available, which are designed to regulate the humidity of the bioreactor. Some advanced sensors have transducer-capacitive elements that absorb the water vapor from the surrounding environment and measure the moisture content. Other electric-based sensors are also available that respond according to the water activity of the SSF process. Online sensors have great advantages such as controlling humidity and temperature, estimating enzyme productivity, and studying mass transfer limitations (Hesseltine, 1987; Pandey, 2004).

10.6.4 Aeration

Aeration in a bioreactor can be achieved by forcefully introducing sterile air. Aeration is controlled and introduced based on the organism type, oxygen consumption, and heat produced during the process. To estimate the growth kinetics, oxygen consumption and carbon dioxide production of the organism are calculated. In SSF, oxygen has only paramagnetic properties and a strong affinity toward magnetic fields. On the contrary, other gases produced during SSF do not have magnetic properties. Thus paramagnetic detectors are used to detect oxygen consumption, which are well adapted for the SSF process (Aidoo et al., 1982; Bellon-Maurel et al., 2003).

10.6.5 Agitation

Agitation is beneficial for the SSF process and provides homogeneity, promotion of growth, prevention of particle aggregation, enhanced gas transfer, and exposure of individual particles to a bioreactor's environment. However, agitation is mainly possible for bacterial and yeast fermentation. In the case of fungi, agitation can break the fungal mycelia that lead to the reduced production of enzymes. So, speed of agitation is necessary and should be optimized for efficient enzyme production without mycelia damage (Pérez-Conrea and Agosin, 1999).

10.6.6 Flow Measurements and Control

Sometimes, air flow during the SSF process of enzyme production is also recorded online by certain flow sensors, which are fitted at the exit of the bioreactor. Different types of devices are used to record the air flow, which include thermal mass flow meters, badge meters, anemometers, and rotameters. At times high-pressure regulators are also installed in the bioreactor to monitor outlet gases. Even though many flow meters are available, adequate flow meters should be selected due to their applicability, cost, space, pressure drop, and precision constraints in enzyme SSF (Bellon-Maurel et al., 2003; Pandey, 2004).

10.6.7 Pressure Drop Measurements

Pressure drop is simple for measurements and can be detected online by using several probes. It can also vary because the chemical composition of fermentation changes. Pressure drop can be used to measure the biomass growth, which requires a minimum gas flow within the bioreactor, pressure drop sensor, or differential pressure sensor to measure the gas flow. The ratio of pressure drop and length of bioreactor indicates the changes in organism growth (germination, growth, and sporulation) and quantitative changes in biomass. Therefore it should be applied as an "alarm sensor" to detect abnormal changes during the SSF process (Richard et al., 1990; Auria et al., 1993).

10.6.8 Other Monitoring Factors

In SSF, there is no reliable method for continuously measuring biomass. So, offline methods are convenient to measure the biomass of organisms. In the case of fungi, the ergosterol method is efficient to evaluate the biomass. In the case of bacteria and yeast, several spectrophotometric methods are useful to calculate the biomass. On the contrary, monitoring respiratory gases is also useful to infer the how an organism is growing (Bellon-Maurel et al., 2003).

Monitoring the substrate concentration is a critical aspect throughout the SSF process. Initial concentration of substrate is enough for general batch-type fermentation, but sometimes it also affects enzyme production that generally takes place after the growth phase. At that time, measurement of respective nutrients is necessary so that nutrient concentration can be measured and regular feed can be added to the SSF process. It is quite difficult to measure, so offline and certain online tools can be used in the SSF process. After measurements, required nutrients can be added in the SSF bioreactor (Mitchell et al., 2006; Lonsane et al., 1985).

Enzyme production largely depends on all the process parameters as discussed in the preceding sections. Thus the determination of enzymes is necessary during the SSF process, which confirms enzyme quality and productivity. Mainly, offline monitoring of cellulase production is conducted by taking samples at specific time intervals and assaying the enzyme activity. It is a reliable and efficient technique for monitoring enzyme production (Mitchell et al., 2006).

10.7 STRATEGIES TO CONTROL CONTAMINATION

Contamination in the SSF process has a significant financial impact on enzyme manufacturers. Lack of microbial contamination in the SSF process can lead to an increment in enzyme production of up to 100 fold. The contamination source, type, and scale of the SSF process and complexity of the fermenter are the major considerations for contamination. Thus sterilization is the crucial strategy to overcome the microbial contamination issue. The control of contaminants or exogenous organisms is mandatory during the fermentation process.

Exogenous organisms include bacteria, fungi, yeast, and algae, which can grow commonly with indigenous organism(s) during the fermentation process. Generally, bacterial contamination is infrequent in the SSF process because the moisture level is too low to support bacterial growth (Hesseltine, 1987). Still, there are possibilities for higher bacterial contamination when the moisture level increases, oxygen level and gas volume lowers, and gas exchange decreases. In the case of fungal contamination, there are greater opportunities to grow unwanted fungi during SSF. Specifically, maximum contamination risk of *Aspergillus*, *Penicillium*, *Candida*, *Trichoderma*, etc. is witnessed when different white-rot fungi are grown for cellulase fermentation. Contamination in white-rot fungi is plausible due to the extended cultivation time (generally, it takes 10–15 days); lower pH, solid substrates, nutrients, and lower humidity favor the growth of unwanted fungi. These exogenous organisms affect the fermentation process by obtaining the nutrients growing on the surface of solid matrix and excreting gases like CO_2, NH_3, and SO_2. Additionally, these organisms produce organic acids, polysaccharides, and gummy biofilms (Lonsane et al., 1985). Due to these consequences, it is impossible for the indigenous organism(s) to survive, which reduces the performance of the SSF process by 20%–50%.

Many methods have been used to detect the different types of contaminant. Among them, counting of microbial cells based on plate assay and fluorescent microscopy are the most commonly applied methods, which demonstrate the type of organisms. However, these methods eliminate the nonculturable organisms not grow on selective media and slow-growing contaminants. Moreover, these methods are labor and time consuming and sometimes they are insensitive to specific types of contaminants. On the contrary, some advanced techniques, i.e., DNA-based methods, are used for the identification of culturable and nonculturable organisms. However, these methods are expensive and necessitate certain amounts of contaminant cells for detection. Nevertheless, the metabolic footprint-based techniques do not require higher amounts of cells and they are highly specific for the different types of the microorganism. Thus efficient and time-saving methods should be used to improve the performance of the SSF process (Maukonen et al., 2006; Hobson et al., 1996; Sue et al., 2011).

SSF contains various components such as bioreactor, substrate, and medium that control the overall performance of bioprocesses and enzyme production yield. Sterilization of all components is comparatively easy at laboratory or flask scale. Substrates are generally sterilized in Erlenmeyer flasks by autoclaving, which depends on the type of lignocellulosic substrate. The medium used in SSF is also sterilized either by autoclaving or filtering (sterile nylon filter membranes). Similarly, sterility can be maintained during the moistening process and the organism inoculating process using laminar air flow (horizontal or vertical) (Lee et al., 2016).

Contamination issues are generally experienced during the scale-up of fermentation if proper precautions are not taken. Mostly, contaminant concentration is found in smaller amounts compared to inoculum. However, there are greater probabilities for contaminants to grow within the bioreactor and if the SSF process takes a long time, then performance of SSF is drastically reduced (Durand, 2003). Among all bioreactor types, continuous-type and type IV bioreactors have higher risk of contamination. Designing a perfect SSF bioreactor with a control system is challenging because it includes many components such as air flow, humidity, water activity, substrate, and bioreactor and it is quite difficult to maintain their sterility (Lonsane et al., 1985). However, certain SSF processes like koji fermentation are carried out under nonsterile conditions, so fermentation contains 10^9 contaminant bacteria per gram (Durand, 2003; Yokotsuka, 1985).

To reduce contamination, fast-growing fungi and bacteria should be used for the SSF process. Inoculum of the desired organism should be pure and its purity should be checked because if contaminants grow from the inoculum stages there would be higher risk of contamination. Use of the desired organism (inoculum) in greater amounts would reduce the risk of contamination in the fermentation process. In the case of fungal SSF, lower pH and moisture level reduce the risk of bacterial contamination. Certain researchers have used different antibiotics (penicillin G, tetracycline, ciprofloxacin, etc.) in tray- and laboratory-scale fermentation to minimize bacterial growth. In white-rot fungal SSF, certain fungicides such as carbendazim, mancozeb, etc. have been used to minimize the contamination of pathogenic fungi. Lignocellulosic substrates should be sterilized by autoclaving them properly. Likewise, bioreactor walls and inoculum tank surfaces should be properly sterilized, so that risk of contamination would be reduced. Additionally, air flow and humidity channels should heat and/or filter sterilized properly. Thus the preclusion of contaminates can be accomplished by obtaining and maintaining pure cultures, designing good equipment, following standard operating procedures, and proper sterility (Lee et al., 2016; Ahmed et al., 1987).

10.8 DOWNSTREAM PROCESSING-BASED STRATEGIES

Numerous organisms, substrates, technologies, and techniques have been used to increase cellulase production; however, their recovery is equally imperative in SSF processes. Nevertheless, they include a major cost factor (up to 50%) in bioprocessing, e.g., tools and techniques applied in recovery are very expensive. Thus fast, cost-effective, and efficient techniques should be used to enhance the recovery of enzymes. Enzyme recovery is comparatively easy and efficient in the SSF process compared to the SmF process. Additionally, downstream processing plays an important role in obtaining pure enzymes because certain industries require pure cellulases with higher activities. However, some industries also need crude cellulases with higher activities for various applications (van Beilen and Li, 2002). The separation methods studied for cellulase recovery from the SSF process mainly include extraction, filtration, chromatography, and crystallization.

Typically, industries optimize downstream processing by heuristic-, experimental-, and model-based approaches. These approaches are very useful to recover enzymes in greater amounts and sometimes their combinatorial approaches are also used. The heuristic approach mainly includes expert knowledge and databases, rule of thumb, and a platform process, which generate the initial dataset for the experimental approach. In the experimental approach, high-throughput experiments, design of experiments, and optimal experimental design are involved that create new data for the model-based approach. In the case of the model-based approach, empirical models, mathematical models, and in silico optimization give the model validation of an experimental-based approach and provide a new platform for a heuristic approach. Thus combinatorial approaches maximize the recovery of enzymes (Baumann and Hubbuch, 2017).

In the SSF process, if the produced cellulase is extracellular, then the substrate-containing biomass of desired organisms is mixed with extraction buffer, which is followed by mixing procedures to extract the enzymes. If enzymes are intracellular, then the desired organisms should be removed from the substrate and cells have to be broken using different methods. Thus intracellular cellulase recovery includes further steps such as cell harvesting, cell breakage, removal of unwanted proteins and cell debris, and enzyme recovery (Doble, 2016).

At the initial stages of enzyme recovery, the retention of catalytic activity is important for the success of the extraction process. A vast range of extraction buffers has been used for the extraction of cellulases, which include mainly Na-acetate, Na-phosphate, glycine-HCl/NaOH, Na-citrate, tris-HCl/NaOH, and carbonate-bicarbonate buffer (Zin et al., 2014; Yang et al., 2017; Okereke et al., 2017). The pH and molarity of the buffer is highly affected by enzyme activity, so these parameters should be optimized before proceeding to a larger scale. After extraction, proteins are precipitated by $(NH_4)_2SO_4$, trichloroacetic acid, ethanol, or acetone precipitation methods. Furthermore, enzymes are purified using a set of purification techniques wherein chromatography and filtration are the efficient techniques. In the case of chromatography, ion exchange-, hydrophobic interaction-, gel permeation-, affinity-, and fast-performance liquid chromatography are mainly used for the purification of cellulases (Devi and Kumar, 2017; Dadheech et al., 2018; Prajapati et al., 2018). On the contrary, several filtration methods, e.g., ultrafiltration, nanofiltration, membrane filtration, ceramic microfiltration, etc. methods, are also available to purify the cellulases. Several other methods like the aqueous two-phase system, dye-ligand chromatography, and foam fractionation have been employed at a small scale; however, their commercial exploitation is anticipated (Ho et al., 2017; Cowan and Daniel, 1996; Banerjee et al., 1993; Sinha et al., 1996).

After purification of enzymes, enzyme stability is the crucial parameter for the SSF process. Certain inactivating agents like physical, chemical, and biological agents inhibit the purified enzymes. Mechanical forces, pH, freezing and thawing, and heating are the most promising physical agents that inactivate enzymes. Organic H-bond formers (e.g., urea, guanidium-HCl, formaldehyde, etc.), oxidizing agents, neutral salts (e.g., NaCl, KCl, $(NH_4)_2SO_2$, KSCN, $CaCl_2$, etc.), organic solvents (e.g., ethylene glycol, polyols, monohydric alcohol, etc.), surfactants (triton, sodium dodecyl sulfate, Tween, etc.), and chelating agents (e.g., 1,10-phenanthroline, ethylenediaminetetraacetic acid) are the possible chemical agents that inhibit the enzymes (Bychkov et al., 2011; Krstić et al., 2007). Proteases are the main class of biological inactivating agents that inhibit enzymes. To stabilize the enzymes, protein engineering, addition of stabilizers, and certain operational conditions are the most important strategies to enhance enzyme stability. Lower temperature and moderate pH are the effective methods to enhance enzyme stability. The addition of substrates, inhibitors, cofactors, vitamins, metals, alteration of humidity near the cellulase, and secondary reaction inhibition make the enzymes more stable (Yang et al., 2017). Due to protein engineering, internal hydrophobic interactions are increased, amino acid residues are compactly packed, hydrophobic residues are reduced, and native conformation is rigidified by introducing new disulfide bonds. All three methods enhance the stability of enzymes (Weijers and Van't Riet, 1992; Pachauri et al., 2018).

10.9 CONCLUDING REMARKS

The current market scenario gives the impression of cellulase production by SmF, wherein companies like Novozymes (Denmark), Genencor (USA), Rohm-AB enzymes (Finland), Rhodia-Danisco (France), Lyven (France), Alko-EDC (USA), and Iogen (Canada) cover the entire market of cellulases. The overall cost of the produced cellulase by the SmF process is 20$/kg, while the cost of the produced cellulase by the SSF process is 0.2$/kg.

Even though the SSF process has a high yield and is cost effective, it possesses certain problems like pH during growth, cell metabolism, and operational conditions during scale-up. In the case of operational conditions, heat transfer is the most challenging parameter for the SSF process, wherein heat transfer during interparticles and particles to the bioreactor is lower than that of the SmF process. To eliminate heat transfer, the SmF process has the option of mixing, while fungal mycelia can be damaged by mixing in the case of the SSF process. All these reasons make the SSF process most challenging at a large scale. In SSF, computational fluid dynamics approaches are the appropriate techniques to minimize heat transfer during fermentation and to observe heat transfer among interparticles and particle bioreactors. Yet, these studies are at the development stage for SSF bioreactors, which may improve the performance of SSF. Lastly, thermostable, high-yielding and fast-growing organisms, pretreated lignocellulosic substrates, optimized production parameters, suitable bioreactors, control of heat transfer, accurate online monitoring, cheap and efficient downstream processing, and control of contaminants can definitely make the SSF process promising at the large scale. This will create a cost-effective market of cellulases by using SSF technology.

References

Ahmed, S.Y., Lonsane, B.K., Ghildyal, N.P., Ramakrishna, S.V., 1987. Design of solid state fermentor for production of fungal metabolites on large scale. Biotechnol. Tech. 1, 97–102.

Aidoo, K.E., Hendry, R., Wood, B.J.B., 1982. Solid substrate fermentations. Adv. Appl. Microbiol. 28, 201–237.

Alam, M.Z., Mamun, A.A., Qudsieh, I.Y., Muyibi, S.A., Salleh, H.M., Omar, N.M., 2009. Solid state bioconversion of oil palm empty fruit bunches for cellulase enzyme production using a rotary drum bioreactor. Biochem. Eng. J. 46, 61–64.

Ang, S.K., Shaza, E.M., Adibah, Y., Suraini, A.A., Madihah, M.S., 2013. Production of cellulases and xylanase by *Aspergillus fumigatus* SK1 using untreated oil palm trunk through solid state fermentation. Process Biochem. 48, 1293–1302.

Araujo, P.W., Brereton, R.G., 1996. Experimental design. I. Screening. TrAC Trends Anal. Chem. 15, 26–31.

Auria, R., Morales, M., Villegas, E., Revah, S., 1993. Influence of mold growth on the pressure drop in aerated solid state fermentors. Biotechnol. Bioeng. 41, 1007–1013.

Bahrin, E.K., Seng, P.Y., Abd-Aziz, S., 2011. Effect of oil palm empty fruit bunch particle size on cellulase production by *Botryosphaeria* sp. under solid state fermentation. Aust. J. Basic Appl. Sci. 5, 276–280.

Banerjee, R., Agnihotri, R., Bhattacharyya, B.C., 1993. Purification of alkaline protease of *Rhizopus oryzae* by foam fractionation. Bioprocess Eng. 9, 245–248.

Bansal, N., Tewari, R., Soni, R., Soni, S.K., 2012. Production of cellulases from *Aspergillus niger* NS-2 in solid state fermentation on agricultural and kitchen waste residues. Waste Manag. 32, 1341–1346.

Barba, F.J., Parniakov, O., Pereira, S.A., Wiktor, A., Grimi, N., Boussetta, N., Saraiva, J.A., Raso, J., Martin-Belloso, O., Witrowa-Rajchert, D., Lebovka, N., Vorobiev, E., 2015. Current applications and new opportunities for the use of pulsed electric fields in food science and industry. Food Res. Int. 77, 773–798.

Baumann, P., Hubbuch, J., 2017. Downstream process development strategies for effective bioprocesses: trends, progress, and combinatorial approaches. Eng. Life Sci. 17, 1142–1158.

Bovill, R., Bew, J., Cook, N., D'agostino, M., Wilkinson, N., Baranyi, J., 2000. Predictions of growth for Listeria monocytogenes and *Salmonella* during fluctuating temperature. Int. J. Food Microbiol. 59, 157–165.

Behera, S.S., Ray, R.C., 2016. Solid state fermentation for production of microbial cellulases: recent advances and improvement strategies. Int. J. Biol. Macromol. 86, 656–669.

Bellon-Maurel, V., Orliac, O., Christen, P., 2003. Sensors and measurements in solid state fermentation: a review. Process Biochem. 38, 881–896.

Bezerra, M.A., Santelli, R.E., Oliveira, E.P., Villar, L.S., Escaleira, L.A., 2008. Response surface methodology (RSM) as a tool for optimization in analytical chemistry. Talanta 76, 965–977.

Bharti, A.K., Kumar, A., Kumar, A., Dutt, D., 2018. Exploitation of *Parthenium hysterophorous* biomass as low-cost substrate for cellulase and xylanase production under solid-state fermentation using *Talaromyces stipitatus* MTCC 12687. J. Radiat. Res. Appl. Sci.

Bychkov, A.L., Buchtoyarov, V.A., Lomovsky, O.I., 2011. Thermal and mechanical denaturation of cellulase in solution and in solid phase. J. Int. Sci. Publ. Mater. Methods Technol. 5, 4–10.

Camassola, M., Dillon, A.J.P., 2007. Effect of methylxanthines on production of cellulases by *Penicillium echinulatum*. J. Appl. Microbiol. 102, 478–485.

Castro, A.M., Castilho, L.R., Freire, D.M., 2015. Performance of a fixed-bed solid-state fermentation bioreactor with forced aeration for the production of hydrolases by *Aspergillus awamori*. Biochem. Eng. J. 93, 303–308.

Castro, A.M., de Carvalho, M.L.D.A., Leite, S.G.F., Pereira, N., 2010. Cellulases from *Penicillium funiculosum*: production, properties and application to cellulose hydrolysis. J. Ind. Microbiol. Biotechnol. 37, 151–158.

Chen, H., Liu, J., Chang, X., Chen, D., Xue, Y., Liu, P., Lin, H., Han, S., 2017. A review on the pretreatment of lignocellulose for high-value chemicals. Fuel Process. Technol. 160, 196–206.

Cowan, D.A., Daniel, R.M., 1996. Rapid purification of two thermophilic proteinases using dye-ligand chromatography. J. Biochem. Biophys. Methods 31, 31–37.

Dadheech, T., Shah, R., Pandit, R., Hinsu, A., Chauhan, P.S., Jakhesara, S., et al., 2018. Cloning, molecular modeling and characterization of acidic cellulase from buffalo rumen and its applicability in saccharification of lignocellulosic biomass. Int. J. Biol. Macromol. 113, 73–81.

Delgado, O.D., Martínez, M.A., Abate, C.M., Siñeriz, F., 2002. Chromosomal integration and expression of green fluorescent protein in *Zymomonas mobilis*. Biotechnol. Lett. 24, 1285–1290.

Deswal, D., Khasa, Y.P., Kuhad, R.C., 2011. Optimization of cellulase production by a brown rot fungus *Fomitopsis* sp. RCK2010 under solid state fermentation. Bioresour. Technol. 102, 6065–6072.

Devi, M.C., Kumar, M.S., 2017. Production, optimization and partial purification of cellulase by *Aspergillus niger* fermented with paper and timber sawmill industrial wastes. J. Microbiol. Biotechnol. Res. 2, 120–128.

Dhillon, G.S., Brar, S.K., Valero, J.R., Verma, M., 2011c. Bioproduction of hydrolytic enzymes using apple pomace waste by *A. niger*: applications in biocontrol formulations and hydrolysis of chitin/chitosan. Bioprocess Biosyst. Eng. 34, 1017–1026.

Dhillon, G.S., Oberoi, H.S., Kaur, S., Bansal, S., Brar, S.K., 2011a. Value-addition of agricultural wastes for augmented cellulase and xylanase production through solid-state tray fermentation employing mixed-culture of fungi. Ind. Crop. Prod. 34, 1160–1167.

Dhillon, G.S., Oberoi, H.S., Kaur, S., Bansal, S., Brar, S.K., 2011b. Value-addition of agricultural wastes for augmented cellulase and xylanase production through solid-state tray fermentation employing mixed-culture of fungi. Ind. Crop. Prod. 34, 1160–1167.

Dhulgande, G.S., Dhale, D.A., Pachkore, G.L., Satpute, R.A., 2011. Mutagenic effectiveness and efficiency of gamma rays and ethyl methanesulphonate in pea (*Pisum sativum* L.). J. Exp. Sci. 2, 7–8.

Diaz, A., de Ory, I., Caro, I., Blandino, A., 2009. Solid state fermentation in a rotating drum bioreactor for the production of hydrolytic enzymes. Icheap-9. In: 9th International Conference on Chemical and Process Engineering, pp. 1–3.

Díaz, A.B., Alvarado, O., de Ory, I., Caro, I., Blandino, A., 2013. Valorization of grape pomace and orange peels: improved production of hydrolytic enzymes for the clarification of orange juice. Food Bioprod. Process. 91, 580–586.

Doble, M., 2016. Principles of Downstream Techniques in Biological and Chemical Processes. Apple Academic Press.

Doi, R.H., Kosugi, A., 2004. Cellulosomes: plant-cell-wall-degrading enzyme complexes. Nat. Rev. Microbiol. 2, 541.

Durand, A., 2003. Bioreactor designs for solid state fermentation. Biochem. Eng. J. 13, 113–125.

Dutta, T., Sahoo, R., Sengupta, R., Ray, S.S., Bhattacharjee, A., Ghosh, S., 2008. Novel cellulases from an extremophilic filamentous fungi *Penicillium citrinum*: production and characterization. J. Ind. Microbiol. Biotechnol. 35, 275–282.

El-Ghonemy, D.H., Ali, T.H., El-Bondkly, A.M., Moharam, M.E.S., Talkhan, F.N., 2014. Improvement of *Aspergillus oryzae* NRRL 3484 by mutagenesis and optimization of culture conditions in solid-state fermentation for the hyper-production of extracellular cellulase. Antonie Van Leeuwenhoek 106, 853–864.

Fanaei, M.A., Vaziri, B.M., 2009. Modeling of temperature gradients in packed-bed solid-state bioreactors. Chem. Eng. Process. 48, 446–451.

Farinas, C.S., 2015. Developments in solid-state fermentation for the production of biomass-degrading enzymes for the bioenergy sector. Renew. Sust. Energ. Rev. 52, 179–188.

Farinas, C.S., Vitcosque, G.L., Fonseca, R.F., Neto, V.B., Couri, S., 2011. Modeling the effects of solid state fermentation operating conditions on endoglucanase production using an instrumented bioreactor. Ind. Crop. Prod. 34, 1186–1192.

Ferreira, S.L.C., Korn, M.D.G.A., Ferreira, H.S., da Silva, E.G.P., Araújo, R.G.O., et al., 2007. Application of multivariate techniques in optimization of spectroanalytical methods. Appl. Spectrosc. Rev. 42, 475–491.

Frey, D.D., Engelhardt, F., Greitzer, E.M., 2003. A role for "one-factor-at-a-time" experimentation in parameter design. Res. Eng. Des. 14, 65–74.

Frey, D.D., Wang, H., 2006. Adaptive one-factor-at-a-time experimentation and expected value of improvement. Technometrics 48, 418–431.

Fujian, X., Hongzhang, C., Zuohu, L., 2002. Effect of periodically dynamic changes of air on cellulase production in solid-state fermentation. Enzym. Microb. Technol. 30, 45–48.

Garvey, M., Klose, H., Fischer, R., Lambertz, C., Commandeur, U., 2013. Cellulases for biomass degradation: comparing recombinant cellulase expression platforms. Trends Biotechnol. 31, 581–593.

Gasparotto, J.M., Werle, L.B., Foletto, E.L., Kuhn, R.C., Jahn, S.L., Mazutti, M.A., 2015. Production of cellulolytic enzymes and application of crude enzymatic extract for saccharification of lignocellulosic biomass. Appl. Biochem. Biotechnol. 175, 560–572.

Gelain, L., da Cruz Pradella, J.G., da Costa, A.C., 2015. Mathematical modeling of enzyme production using *Trichoderma harzianum* P49P11 and sugarcane bagasse as carbon source. Bioresour. Technol. 198, 101–107.

Ghildyal, N.P., Gowthaman, M.K., Rao, K.R., Karanth, N.G., 1994. Interaction of transport resistances with biochemical reaction in packed-bed solid-state fermentors: effect of temperature gradients. Enzym. Microb. Technol. 16 (3), 253–257.

Hassan, S.S., Williams, G.A., Jaiswal, A.K., 2018. Emerging technologies for the pretreatment of lignocellulosic biomass. Bioresour. Technol. 262, 310–318.

Hatakka, A., 1983. Pretreatment of wheat straw by white-rot fungi for enzymatic saccharification of cellulose. Eur. J. Appl. Microbiol. Biotechnol. 18, 350–357.

Herai, S., Hashimoto, Y., Higashibata, H., Maseda, H., Ikeda, H., Ōmura, S., Kobayashi, M., 2004. Hyper-inducible expression system for streptomycetes. Proc. Natl. Acad. Sci. 101, 14031–14035.

Hesseltine, C.W., 1987. Solid state fermentation—an overview. Int. Biodeterior. 23, 79–89.

Himmel, M.E., Ding, S.Y., Johnson, D.K., Adney, W.S., Nimlos, M.R., Brady, J.W., Foust, T.D., 2007. Biomass recalcitrance: engineering plants and enzymes for biofuels production. Science 315, 804–807.

Hitoshi, N., Toshirou, H., Toshimitsu, I., Keiji, T., Shinji, M., Makoto, K., et al., 2005. Bio-methanol production and CO_2 emission reduction from forage grasses, trees, and crop residues. In: Originally Presented at the Third USDA Symposium on Greenhouse Gases & Carbon Sequestration in Agriculture and Forestry. Baltimore, USA.

Ho, S.L., Lan, J.C.W., Tan, J.S., Yim, H.S., Ng, H.S., 2017. Aqueous biphasic system for the partial purification of *Bacillus subtilis* carboxymethyl cellulase. Process Biochem. 58, 276–281.

Hobson, N.S., Tothill, I., Turner, A.P.F., 1996. Microbial detection. Biosens. Bioelectron. 11, 455–477.

Hölker, U., Lenz, J., 2005. Solid-state fermentation—are there any biotechnological advantages? Curr. Opin. Microbiol. 8, 301–306.

Hu, T., Zhou, Y., Dai, L., Wang, Y., Liu, D., Zhang, J., Liu, H., 2011. Enhanced cellulase production by solid state fermentation with polyurethane foam as inert supports. Procedia Eng. 18, 335–340.

Hyun Hong, S., Taek Lee, J., Lee, S., Gon Wi, S., Ju Cho, E., Singh, S., et al., 2014. Improved enzymatic hydrolysis of wheat straw by combined use of gamma ray and dilute acid for bioethanol production. Radiat. Phys. Chem. 94, 231–235.

Idris, A.S.O., Pandey, A., Rao, S.S., Sukumaran, R.K., 2017. Cellulase production through solid-state tray fermentation, and its use for bioethanol from sorghum stover. Bioresour. Technol. 242, 265–271.

Juturu, V., Wu, J.C., 2012. Microbial xylanases: engineering, production and industrial applications. Biotechnol. Adv. 30, 1219–1227.

Juturu, V., Wu, J.C., 2014. Microbial cellulases: engineering, production and applications. Renew. Sust. Energ. Rev. 33, 188–203.

Kalogeris, E., Christakopoulos, P., Katapodis, P., Alexiou, A., Vlachou, S., Kekos, D., Macris, B.J., 2003. Production and characterization of cellulolytic enzymes from the thermophilic fungus *Thermoascus aurantiacus* under solid state cultivation of agricultural wastes. Process Biochem. 38, 1099–1104.

Kalogeris, E., Fountoukides, G., Kekos, D., Macris, B.J., 1999. Design of a solid-state bioreactor for thermophilic microorganisms. Bioresour. Technol. 67, 313–315.

Kim, S., Kim, C.H., 2012. Production of cellulase enzymes during the solid-state fermentation of empty palm fruit bunch fiber. Bioprocess Biosyst. Eng. 35, 61–67.

Krstić, D.M., Antov, M.G., Peričin, D.M., Höflinger, W., Tekić, M.N., 2007. The possibility for improvement of ceramic membrane ultrafiltration of an enzyme solution. Biochem. Eng. J. 33, 10–15.

Kuhad, R.C., Deswal, D., Sharma, S., Bhattacharya, A., Jain, K.K., et al., 2016. Revisiting cellulase production and redefining current strategies based on major challenges. Renew. Sust. Energ. Rev. 55, 249–272.

Kumar, A.K., Sharma, S., 2017. Recent updates on different methods of pretreatment of lignocellulosic feedstocks: a review. Bioresour. Bioprocess. 4, 7.

Kumari, D., Singh, R., 2018. Pretreatment of lignocellulosic wastes for biofuel production: a critical review. Renew. Sust. Energ. Rev. 90, 877–891.

Lee, S.Y., Nielsen, J., Stephanopoulos, G., 2016. Industrial Biotechnology: Products and Processes. John Wiley & Sons.

Li, H., Qu, Y., Yang, Y., Chang, S., Xu, J., 2016. Microwave irradiation—a green and efficient way to pretreat biomass. Bioresour. Technol. 199, 34–41.

Li, S., Yang, X., Yang, S., Zhu, M., Wang, X., 2012. Technology prospecting on enzymes: application, marketing and engineering. Comput. Struct. Biotechnol. J. 2, e201209017.

Lin, C.W., Wu, C.H., Tran, D.T., Shih, M.C., Li, W.H., Wu, C.F., 2011. Mixed culture fermentation from lignocellulosic materials using thermophilic lignocellulose-degrading anaerobes. Process Biochem. 46, 489–493.

Lo, S.L., Huang, Y.F., Chiueh, P.T., Kuan, W.H., 2017. Microwave pyrolysis of lignocellulosic biomass. Energy Procedia 105, 41–46.

Lonsane, B.K., Ghildyal, N.P., Budiatman, S., Ramakrishna, S.V., 1985. Engineering aspects of solid state fermentation. Enzym. Microb. Technol. 7, 258–265.

Lundstedt, T., Seifert, E., Abramo, L., Thelin, B., Nyström, Å., Pettersen, J., Bergman, R., 1998. Experimental design and optimization. Chemom. Intell. Lab. Syst. 42, 3–40.

Manan, M.A., Webb, C., 2016. Multi-enzymes production studies in single tray solid state fermentation with opened and closed system. J. Life Sci. 10, 342–356.

Maukonen, J., et al., 2006. Suitability of the fluorescent techniques for the enumeration of probiotic bacteria in commercial non-dairy drinks and in pharmaceutical products. Food Res. Int. 39, 22–32.

Membrillo, I., Sánchez, C., Meneses, M., Favela, E., Loera, O., 2011. Particle geometry affects differentially substrate composition and enzyme profiles by *Pleurotus ostreatus* growing on sugar cane bagasse. Bioresour. Technol. 102, 1581–1586.

Mitchell, D.A., Berovič, M., Krieger, N., 2006. Solid-state fermentation bioreactor fundamentals: introduction and overview. In: Solid-State Fermentation Bioreactors. Springer, Berlin, Heidelberg, pp. 1–12.

Montoya, S., Sánchez, Ó.J., Levin, L., 2014. Mathematical modeling of lignocellulolytic enzyme production from three species of white rot fungi by solid-state fermentation. In: Advances in Computational Biology. Springer, Cham, pp. 371–377.

Noratiqah, K., Madihah, M.S., Aisyah, B.S., Eva, M.S., Suraini, A.A., Kamarulzaman, K., 2013. Statistical optimization of enzymatic degradation process for oil palm empty fruit bunch (OPEFB) in rotary drum bioreactor using crude cellulase produced from *Aspergillus niger* EFB1. Biochem. Eng. J. 75, 8–20.

Okereke, O.E., Akanya, H.O., Egwim, E.C., 2017. Purification and characterization of an acidophilic cellulase from *Pleurotus ostreatus* and its potential for agrowastes valorization. Biocatal. Agric. Biotechnol. 12, 253–259.

Oliveira, S.C.T., Figueiredo, A.B., Evtuguin, D.V., Saraiva, J.A., 2012. High pressure treatment as a tool for engineering of enzymatic reactions in cellulosic fibres. Bioresour. Technol. 107, 530–534.

Pachauri, P., More, S., Aranganathan, V., Sullia, S.B., Deshmukh, S., 2018. Kinetic Study and Characterization of Cellulase Enzyme from Isolated *Aspergillus niger* subsp. *awamori* for Cellulosic Biofuels.

Pandey, A., 2003. Solid-state fermentation. Biochem. Eng. J. 13, 81–84.

Pandey, A., 2004. Concise Encyclopedia of Bioresource Technology. No. C/620.803 C6. Food Products Press, New York.

Pandey, A., Soccol, C.R., Larroche, C. (Eds.), 2008. Current Developments in Solid-State Fermentation. Springer Science & Business Media.

Pérez-Conrea, R., Agosin, E., 1999. Solid substrate fermentation, automation. In: Flickingen, M.C., Drew, S.W. (Eds.), Encyclopedia of bioprocess technology: fermentation, biocatalysis and bioseparation. John Wiley & Sons, USA, pp. 2429–2446.

Pirota, R.D., Delabona, P.S., Farinas, C.S., 2014. Simplification of the biomass to ethanol conversion process by using the whole medium of filamentous fungi cultivated under solid-state fermentation. BioEnergy Res. 7, 744–752.

Prajapati, B.P., Suryawanshi, R.K., Agrawal, S., Ghosh, M., Kango, N., 2018. Characterization of cellulase from *Aspergillus tubingensis* NKBP-55 for generation of fermentable sugars from agricultural residues. Bioresour. Technol. 250, 733–740.

Prasertsan, P., H-kittikul, A., Chitmanee, B., 1992. Isolation and selection of cellulolytic fungi from palm oil mill effluent. World J. Microbiol. Biotechnol. 8, 614–617.

Rabemanolontsoa, H., Saka, S., 2016. Various pretreatments of lignocellulosics. Bioresour. Technol. 199, 83–91.

Rao, M., Gaikwad, S., Mishra, C., Deshpande, V., 1988. Induction and catabolite repression of cellulase in *Penicillium funiculosum*. Appl. Biochem. Biotechnol. 19, 129–137.

Ratkowsky, D.A., Lowry, R.K., McMeekin, T.A., Stokes, A.N., Chandler, R.E., 1983. Model for bacterial culture growth rate throughout the entire biokinetic temperature range. J. Bacteriol. 154, 1222–1226.

Ravindran, C., Naveenan, T., Varatharajan, G.R., 2010. Optimization of alkaline cellulase production by the marine-derived fungus *Chaetomium* sp. using agricultural and industrial wastes as substrates. Bot. Mar. 53, 275–282.

Ravindran, R., Jaiswal, A., 2016. A comprehensive review on pre-treatment strategy for lignocellulosic food industry waste: challenges and opportunities. Bioresour. Technol. 199, 92–102.

Ray, R.C., Behera, S.S., 2017. Solid state fermentation for production of microbial cellulases. In: Biotechnology of Microbial Enzymes, pp. 43–79.

Richard, A., Sergio, H., Maurice, R., Sergio, R., 1990. Ion exchange resin: a model support for solid state growth fermentation of *Aspergillus niger*. Biotechnol. Tech. 4 (6), 391–396.

Rudakiya, D.M., 2018. Metal tolerance assisted antibiotic susceptibility profiling in *Comamonas acidovorans*. BioMetals 31, 1–5.

Rudakiya, D.M., Gupte, A., 2017. Degradation of hardwoods by treatment of white rot fungi and its pyrolysis kinetics studies. Int. Biodeterior. Biodegrad. 120, 21–35.

Rüdt, M., Briskot, T., Hubbuch, J., 2017. Advances in downstream processing of biologics—spectroscopy: an emerging process analytical technology. J. Chromatogr. A 1490, 2–9.

Sadhu, S., Maiti, T.K., 2013. Cellulase production by bacteria: a review. Br. Microbiol. Res. J. 3, 235.

Saelee, K., Yingkamhaeng, N., Nimchua, T., Sukyai, P., 2016. An environmentally friendly xylanase-assisted pretreatment for cellulose nanofibrils isolation from sugarcane bagasse by high-pressure homogenization. Ind. Crop. Prod. 82, 149–160.

Saha, B.C., Qureshi, N., Kennedy, G.J., Cotta, M.A., 2016. Biological pretreatment of corn stover with white-rot fungus for improved enzymatic hydrolysis. Int. Biodeterior. Biodegrad. 109, 29–35.

Salim, A.A., Grbavčić, S., Šekuljica, N., Stefanović, A., Tanasković, S.J., Luković, N., Knežević-Jugović, Z., 2017. Production of enzymes by a newly isolated *Bacillus* sp. TMF-1 in solid state fermentation on agricultural by-products: the evaluation of substrate pretreatment methods. Bioresour. Technol. 228, 193–200.

Sangsurasak, P., Nopharatana, M., Mitchell, D., 1996. Mathematical modeling of the growth of filamentous fungi in solid-state fermentation. J. Sci. Ind. Res. 55, 333–342.

Scholl, A.L., Menegol, D., Pitarelo, A.P., Fontana, R.C., Zandoná Filho, A., et al., 2015. Elephant grass pretreated by steam explosion for inducing secretion of cellulases and xylanases by *Penicillium echinulatum* S1M29 solid-state cultivation. Ind. Crop. Prod. 77, 97–107.

Sherief, A.A., El-Tanash, A.B., Atia, N., 2010. Cellulase production by *Aspergillus fumigatus* grown on mixed substrate of rice straw and wheat bran. Res. J. Microbiol. 5, 199–211.

Shirkavand, E., Baroutian, S., Gapes, D.J., Young, B.R., 2016. Combination of fungal and physicochemical processes for lignocellulosic biomass pretreatment—a review. Renew. Sust. Energ. Rev. 54, 217–234.

Singh, S.K., Singh, S.K., Tripathi, V.R., Khare, S.K., Garg, S.K., 2011. Comparative one-factor-at-a-time, response surface (statistical) and bench-scale bioreactor level optimization of thermoalkaline protease production from a psychrotrophic *Pseudomonas putida* SKG-1 isolate. Microb. Cell Factories 10, 114.

Singhania, R.R., Patel, A.K., Soccol, C.R., Pandey, A., 2009. Recent advances in solid-state fermentation. Biochem. Eng. J. 44, 13–18.

Singhania, R.R., Sukumaran, R.K., Pillai, A., Prema, P., Szakacs, G., Pandey, A., 2006. Solid-state fermentation of lignocellulosic substrates for cellulase production by *Trichoderma reesei* NRRL 11460.

Sinha, R., Singh, S.P., Ahmed, S., Garg, S.K., 1996. Partitioning of a *Bacillus* alkaline protease aqueous two-phase systems. Bioresour. Technol. 55, 163–166.

Streatfield, S.J., 2007. Approaches to achieve high-level heterologous protein production in plants. Plant Biotechnol. J. 5, 2–15.

Sue, T., Obolonkin, V., Griffiths, H., Villas-Bôas, S.G., 2011. An exometabolomics approach to monitoring microbial contamination in microalgal fermentation processes by using metabolic footprint analysis. Appl. Environ. Microbiol. 77, 7605–7610.

Trivedi, N., Reddy, C.R.K., Radulovich, R., Jha, B., 2015. Solid state fermentation (SSF)-derived cellulase for saccharification of the green seaweed Ulva for bioethanol production. Algal Res. 9, 48–54.

van Beilen, J.B., Li, Z., 2002. Enzyme technology: an overview. Curr. Opin. Biotechnol. 13, 338–344.

Wan, C., Li, Y., 2010. Microbial delignification of corn stover by *Ceriporiopsis subvermispora* for improving cellulose digestibility. Enzym. Microb. Technol. 47, 31–36.

Wan, C., Li, Y., 2012. Fungal pretreatment of lignocellulosic biomass. Biotechnol. Adv. 1447–1457.

Ward, O.P., 2012. Production of recombinant proteins by filamentous fungi. Biotechnol. Adv. 30, 1119–1139.

Weijers, S.R., Van't Riet, K., 1992. Enzyme stability in downstream processing part 1: enzyme inactivation, stability and stabilization. Biotechnol. Adv. 10, 237–249.

Xia, L., Cen, P., 1999. Cellulase production by solid state fermentation on lignocellulosic waste from the xylose industry. Process Biochem. 34, 909–912.

Xin, F., Geng, A., 2010. Horticultural waste as the substrate for cellulase and hemicellulase production by *Trichoderma reesei* under solid-state fermentation. Appl. Biochem. Biotechnol. 162, 295–306.

Yang, X., Zhou, S., Li, M., Wang, R., Zhao, Y., 2017. Purification of cellulase fermentation broth via low cost ceramic microfiltration membranes with nanofibers-like attapulgite separation layers. Sep. Purif. Technol. 175, 435–442.

Yokotsuka, T., 1985. Traditional fermented soybean foods. In: Moo-Young, M., Drew, S., Blanch, H.W. (Eds.), Comprehensive Biotechnology. Pergamon Press, Oxford, New York, pp. 395–427.

Zanelato, A.I., Shiota, V.M., Gomes, E., Silva, R.D., Thoméo, J.C., 2012. Endoglucanase production with the newly isolated *Myceliophtora sp.* i-1d3b in a packed bed solid state fermentor. Braz. J. Microbiol. 43, 1536–1544.

Zhang, L., Sun, X., 2014. Effects of rhamnolipid and initial compost particle size on the two-stage composting of green waste. Bioresour. Technol. 163, 112–122.

Zin, H.W., Park, K.H., Choi, T.J., 2014. Purification and characterization of a carboxymethyl cellulase from *Artemia salina*. Biochem. Biophys. Res. Commun. 443, 194–199.

CHAPTER

11

Knowledge Update on Bioreactor Technology for Cellulase Production

*Desikan Ramesh**, *Iniya Kumar Muniraj*†, *Kiruthika Thangavelu**, *Subburamu Karthikeyan**

*Tamil Nadu Agricultural University, Coimbatore, India †Kumaraguru Institute of Agriculture, Erode, India

11.1 INTRODUCTION

In recent years the demand for production of microbial secondary metabolites (enzymes, antibiotics, bacterial toxins, drugs, etc.) has increased due to their applications in medicine and agriculture (Adney et al., 1991). Among the microbial metabolites, microbial enzymes find an attractive place in the market because of their use in widespread industries (Baker et al., 1998). Use of microbial enzymes reduces process time, cost, and energy input, and they are nontoxic. In addition, microbial enzymes are capable of degrading toxic chemicals and converting them into nontoxic by-products. Among the numerous microbial enzymes known (amylase, cellulases, catalase, laccase, pectate lyase, lipase, proteases, etc.), one of the most widely used enzymes with wide applications and adaptability with many benefits are cellulases. They are a group of three enzymes, namely exoglucanases, endoglucanases, and β-glucosidases, that synergistically act on homopolysaccharide cellulose (Bayer et al., 1998).

Cellulose is the most abundant polysaccharide present in the plant cell wall, produced by some microorganisms and tunicates. In the terrestrial environment, 100 billion tons of cellulose are produced by photosynthesis every year. It is one of the predominant renewable biomasses from agricultural wastes and serves as raw material for the production of different value-added chemicals, fuels, and others (Baker et al., 1998). Cellulose is a linear homopolysaccharide consisting of long glucose chains linked by β-1,4-glycosidic linkages. Its degree of polymerization ranges from 200 to 10,000. Cellulose exists in crystalline form in the plant cell wall and is interlaced with lignin and hemicellulose. The percentage of cellulose, hemicellulose, and lignin varies in the range of 40–50, 20–30, and 20, respectively. Cellulose is degraded by a complex of enzymes produced by various microorganisms, including fungi, bacteria, and actinobacteria (Bischof et al., 2016).

Degradation of cellulose takes place initially with the hydroxylation of the amorphous region of cellulose and further by endoglucanase (endo-1,4-β-D-glucanase, EC 3.2.1.4) by random cleavage of β-1,4 linkages of reducing and nonreducing ends forming cellooligosaccharides, cellobiose, and glucose units. Then cellobiohydralase, or exoglucanase, 1,4-β-D-glucan-cellobiohydrolase (EC 3.2.1.91) hydrolyzes the end products of endoglucanase and releases cellobiose (Bayer et al., 2007). Finally, β-1,4-glucosidase hydrolyzes reducing and nonreducing ends and releases glucose. Cellulase finds its way in the textile, food and feed, pulp and paper, and biofuel industries and the demand for the enzyme is increasing in a steady manner; it is imperative to produce the enzyme in higher quantities with economically feasible processes. Traditionally, cellulases have been commercially produced by submerged fermentation (SmF); however, the recent trend suggests that solid-state fermentation (SSF) has many more advantages than SmF (Bhat, 2000). Similarly, new production techniques in bioreactors have emerged, for example, a higher rate of cellulase production by co-culturing fungal strains as fed batch in a stirred-tank bioreactor

From Cellulose to Cellulase: Strategies to Improve Biofuel Production
https://doi.org/10.1016/B978-0-444-64223-3.00011-4

are evolving. There is a lot of scope in large-scale cellulase production using different bioreactors. In light of this, this chapter focuses on updating the knowledge of bioreactors (their design, heat and mass transfer, production process, etc.) for cellulase production.

11.1.1 Bioreactor

The bioreactor is a mechanical device that provides in vitro cultivation conditions for microorganisms to obtain desired products from the substrate. For cellulase production, the operation of the bioreactor may be based on either the SSF or SmF process. The bioreactor design for cellulase production is based on different parameters such as substrate characteristics (moisture content, particle size/surface area, bulk density, surface tension, rheology properties), information of microorganisms used, interaction and behavior of multiphase reactants (solid, liquid, and gas), mode of mixing of reactants, oxygen transfer rate, mass and heat transfer, reaction conditions (fluid velocity, temperature, duration, pH, etc.), and final products (Arora et al., 2018). The bioreactor should have instrumentation and a data-recording system for online monitoring of process and process analysis. The bioreactor for SSF or SmF may operate on batch, semibatch, or continuous mode. The steps that should be adopted for process design and bioreactor design to overcome the issues regarding design, development, and testing stages are shown in Fig. 11.1. These two stages are both time- and cost-consuming processes for bioreactor development. The bioreactor design is completely different from chemical reactor design due to the usage of microorganisms. Before bioreactor development, process design for cellulase production should be planned. Process design will give an idea of the various unit processes involved in cellulase production from given substrate materials (Formenti et al., 2014).

The outcomes from process design of any bioproduct production are the identification of unit operations, process, process flow chart, and machinery required. Bioreactor development is one of the activities of process design.

1. Preliminary studies on raw materials used as a substrate.
2. Basic information on cellulase-producing microorganisms.
3. Information on growth kinetics of selected microorganisms.
4. The process employed for cellulase production.
5. Requirements of a bioreactor.
6. Heat and mass transfer rate and oxygen transfer.

11.1.2 Preliminary Studies on Substrates

Basic data required for reactor design is obtained from preliminary studies of different substrates used for cellulase production. The characteristics of the substrate such as particle size, density, moisture content, solid loading, rheology properties, and flow behavior inside the reactor should be analyzed (Alasepp et al., 2014).

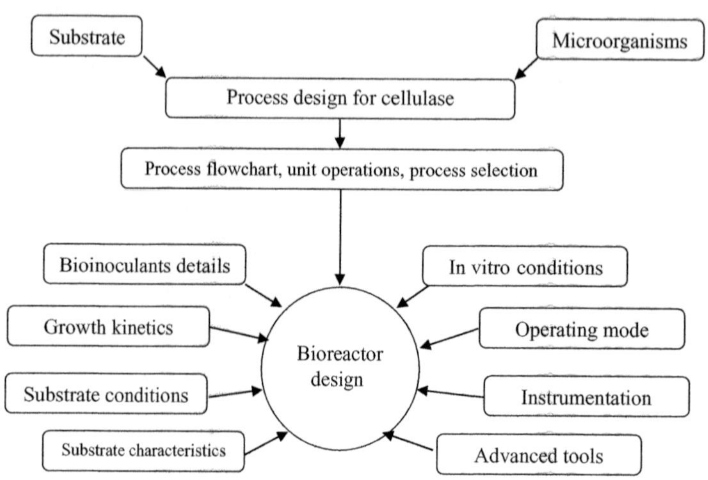

FIG. 11.1 Steps involved in process design and bioreactor design for cellulase production.

11.2 BASIC INFORMATION ON CELLULASE-PRODUCING MICROORGANISMS

Cellulases are produced by several microbial groups, including Fungi, Bacteria, and Actinobacteria. Screening of microorganisms with higher titers from nature on the one hand, and strain improvement using molecular approaches on the other hand, go simultaneously with respect to cellulase production. Table 11.1 shows the different microbial groups used for cellulase production. Cellulase-producing organisms are often found abundantly as saprophytes; other sources include soil samples from forest and nature reserves, hot water springs, compost, sewage, animal manure, bovine rumen, etc. (Sharma et al., 2016).

Cellulase producers generally do not prefer lipids and proteins as their carbon source, rather they can produce more on carbohydrate-rich substrates. Among the microbial group, fungi are the preferred source for cellulase production due to their extracellular secretion of cellulase. Many bacteria accumulate the cellulases inside their cells and it is rare that a single bacterium possesses all three enzymes. Hence commercial production of cellulase is mostly concentrated on the fungal system. In addition, the thermophilic nature of cellulases (isolated from thermopiles) is mostly preferred over others because of their temperature and pH tolerance (Chandrasekaran et al., 2010). Since the microorganisms are collected from different habitats, the growth kinetics of microorganisms should be studied. These data could be used in both process design and reactor development for cellulase production.

11.2.1 The Process Employed for Cellulase Production

Until recently, cellulases were commercially produced in SmF; this technique allows the industry to better monitor the culture conditions and productivity. However, lately SSF has replaced SmF due to the advantages of minimal water requirement, higher titers, use of natural substrates for enzyme production, nonaddition of inducers, easy extraction of enzymes, etc. According to Singhania et al. (2014), the points to note for the perfect design of a reactor for the SSF process are: (1) control systems for environments (temperature, air flow rate, and humidity), (2) protection system for contamination, (3) homogeneous in water activity, (4) provision for removal of CO_2 produced in the process, and (5) simple operation and easy to upscale.

11.2.2 Solid-State Fermentation

SSF involves fermentation in the absence of water or near absence of water, using the natural substrate to provide a hostile environment for the microorganisms. SSF has been in use for fermentation of different bioproducts for a long time; however, its importance has been under review (Dias et al., 2018; Doriya et al., 2016). At present, many bioproducts, especially cellulases, are produced by SSF involving bacteria, fungi, and Actinobacteria. Studies related to SSF suggest that not only can higher titers be achieved but also balanced activities of enzymes can be produced. Cellulase is an inducible enzyme; it undergoes both induction and repression by a natural substrate. At the same time the by-products of hydrolysis of natural substrate inhibit cellulase activity and when designing a fermenter these parameters have to be taken into account (Acharya et al., 2010; Aggarwal et al., 2017; Camassola and Dillon, 2010). A comparison of SSF and SmF in cellulase production is described in Table 11.2.

TABLE 11.1 Different Microorganisms Used for Cellulase Production (Bischof et al., 2016; Bhat, 2000; Kuhad et al., 2011)

S. No.	Microbial group producing cellulases
I.	**Fungi** *Aspergillus niger, Aspergillus nidulans, Aspergillus oryzae, Fusarium oxysporium, Fusarium solani, Humicola insolens, Penicillium brasilaniam, Trichoderma reesei, Trichoderma harzianum*, etc.
II.	**Bacteria** *Acedothermus cellulolyticus, Bacillus* sp., *Bacillus subtilis, Clostridium thermocellum, Pseudomonas cellulosa*
III.	**Actinobacteria** *Celullomonas filmi, Cellulomonas uda, Thermomonospora fusca*, etc.

TABLE 11.2 Comparison of Solid-State Fermentation (SSF) and Submerged Fermentation (SmF) (Viesturs et al., 1987; Zhang et al., 2013, 2015)

Parameters	SSF	SmF
Cellulase yield	Higher in a shorter time	Lower titers
Water requirement	Less consumption	High
Downstream processing	Simple but requires water to extract the enzyme from the substrate	Simpler than SSF with lower water requirements
Inducer requirement	Substrate itself acts as inducer	A high amount of lactose (as inducer) is required to obtain maximum titers
Energy requirement	Less energy intensive	Highly energy intensive
Capital investment	High	Low
Possibility of contamination	Lower since water availability is low	High due to high water requirement

Although SSF seems promising it lacks flexibility in scale-up, sufficient knowledge in types of fermenter to be used and its designing and modeling, heat and mass transfer ratios, a search of newer SSF bioreactors are important.

11.2.3 Requirements of Bioreactor

The bioreactor should fulfill the following requirements and these points should be considered for reactor design:

1. Material selection for bioreactor construction based on metals with higher strength, corrosion resistance, and low price.
2. An inbuilt facility for the sterilization process for the bioreactor.
3. Efficient handling method for inputs and final products.
4. Automatic control system for monitoring and regulating the process parameters.

11.2.4 Heat and Mass Transfer Rate and Oxygen Transfer

Although several advantages can be attained while using SSF, the fermentation process to commercialization is limited due to difficulties in monitoring and controlling the different process parameters. Hence it has become an important task for the understanding of heat and mass transfer effects in SSF (Han et al., 2010). A study was conducted to estimate the impact of heat and mass transfer in a static tray fermenter under SSF conditions and maximum cellulase production obtained under room temperature conditions with higher relative humidity. In another study, moisture content airflow and temperature were the major factors in deciding the maximum enzyme titers (Formenti et al., 2014; Anbar and Bayer, 2012; Nidetzky et al., 1994).

Both SSF and SmF processes are aerobic and exothermic processes. The oxygen is essential for the respiration and *metabolic* activities of microorganisms. To maintain the reaction in the bioreactor, oxygen is continuously supplied via suitable aeration systems. A sparger is employed for distributing oxygen in uniformly sized bubbles in SmF bioreactors, which can help to increase the oxygen transfer rate.

To overcome the issues related to reactor operation, simulation studies on the bioprocess used in the bioreactor should be performed. This step will avoid shortfalls in reactor design. For simulation studies, computational fluid dynamics (CFD) is a good choice because its results can predict the possible problems associated with bioreactor operation. By using CFD analysis, process conditions of the bioreactor can be simulated and analyzed. In addition, we can generate data for a proposed reactor model on the mixing pattern of reactants, turbulent flow, gas holdup, heat and mass transfer coefficients, gas dispersion, hydrodynamics, and shear stresses (Kumar et al., 2012). The commercially available CFD software with packages will suit the conditional requirements of the bioreactor. Simulation software such as Open FOAM, FLUENT, ANSYS, COSMOL, etc. can be used for this kind of study. Details of CFD studies conducted in bioreactors for cellulase production are shown in Table 11.3.

TABLE 11.3 Computational fluid dynamics (CFD) Analysis Used in Different Bioreactors for Cellulase Production

Type of bioreactor used	CFD analysis was done for	Strain	Cellulase yield	References
Stirred-tank bioreactors	Flow pattern effects on cellulase deactivation	*Humicola*	NA	(Berovic, 1999)
Airlift reactors	Hydrodynamics on pellet formation and cellulase production	*Trichoderma reesei* RUT-C30	NA	(Calderon-Cortes et al., 2010)
Stirred-tank reactors (two six-bladed Rushton turbine impellers)	Bubble size distribution	*T. reesei* RUT-C30	3.5 IU/mL	(Niu et al., 2016)

11.3 CLASSIFICATION OF BIOREACTORS FOR CELLULASE PRODUCTION

11.3.1 SSF Process

Recent studies on SSF focus on the development of newer bioreactors with batch, fed-batch, or continuous mode of fermentation. Generally, several types of bioreactor are employed for cellulase production via SSF or SmF processes operating on batch, semicontinuous, or continuous modes (Roberts and Davies, 2012). The design and operation of these bioreactors are different to each other and they also have their own constraints and advantages. The classification of bioreactors used in SSF or SmF processes for cellulase production is shown in Fig. 11.2. The bioreactors mostly used for the SSF process are the tray bioreactor, packed-bed bioreactor, rotary drum bioreactor, and fluidized-bed bioreactor (Cho et al., 2012; Yoon et al., 2007). Details of individual SSF process-based bioreactors are discussed next.

11.3.2 Tray Bioreactors

The main components of this reactor are trays, supporting stands, humidifier, heater, and thermostat-controlled chamber with provisions for air inlet and outlet. The dimensions of the trays may vary with different reactor capacities. The inoculums and substrates are uniformly spread over the trays in thin layers and kept in supporting stands. The thickness of the thin layer used in the tray is 1.5 or 2 cm (Yoon et al., 2007). The thickness of the layer is a limiting factor in this reactor due to poor oxygen distribution in the layer and heat accumulation problems. Based on reactor design, vertical spacing between the trays and also horizontal spacing for supporting stands are maintained. The perforations

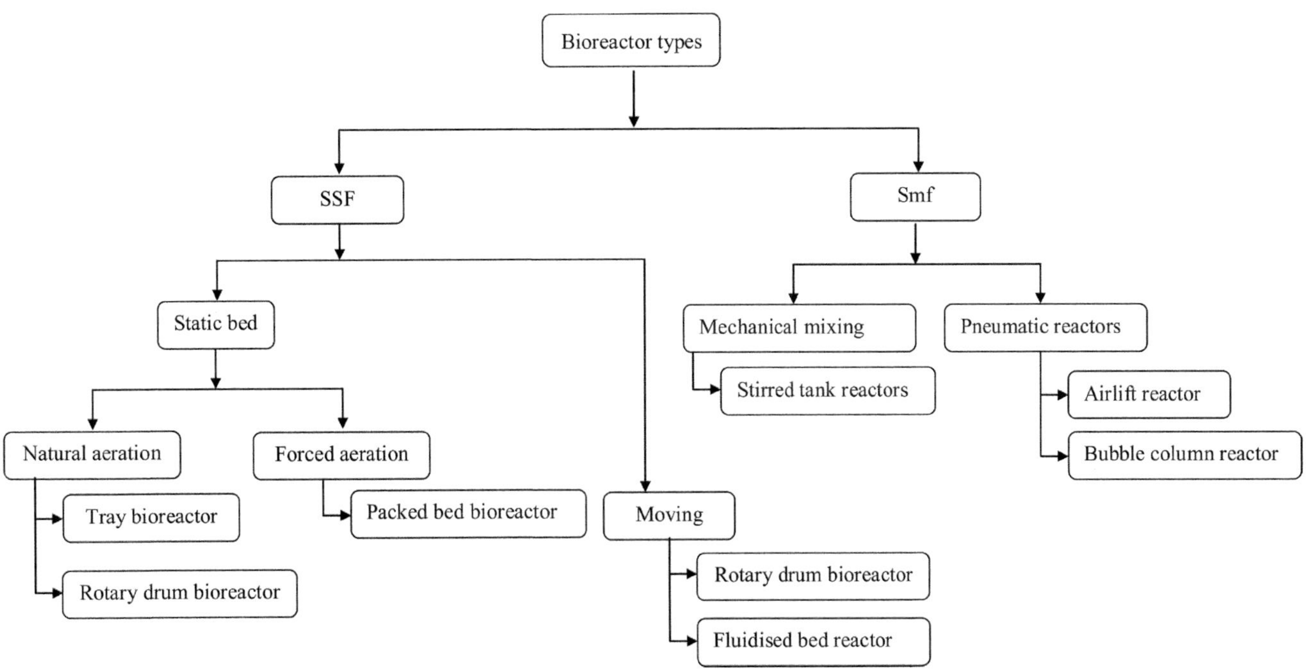

FIG. 11.2 Different types of bioreactor used in solid-state fermentation and submerged fermentation for cellulase production.

in the bottom of the trays would help to enhance the free movement of air through the substrate. This kind of arrangement would serve two purposes: (1) to provide aeration for microorganisms and (2) to remove heat from the substrate. Heat removal from the trays takes place by natural convection (Chandrasekaran et al., 2010; Jorgensen, 2018; Lai et al., 2016; Muller dos Santos et al., 2004). For the oxygen requirement of microorganisms, air can be supplied via the air inlet at the bottom of the chamber and heat removed via an outlet fitted at the top of the chamber. The thermostat chamber is employed for maintaining constant temperature and humidity for the entire reaction period.

11.3.3 Packed-Bed Bioreactor

Generally, packed-bed bioreactors are static-bed reactors used in cellulase production. This reactor does not utilize any mechanical device for continuous or intermittent mixing of substrate and it will prevent damage to the sensitive microorganisms. As compared with a tray bioreactor, it has a forced aeration system for enhancement of better air circulation in the bed and also simultaneously removes heat and CO_2 from the bioreactor (Arora et al., 2018). The major components of this reactor are reactor drum with/without cooling plates, packing bed, medium reservoir, pump, and other accessories. The substrate can be kept in the packed-bed arrangement inside the reactor. The reactor drum is made of glass or metal with perforations at the bottom of the reactor. During operation, a pump is employed to supply air through the perforated reactor bottom. Oxygenated nutrient medium is supplied to the top of the packing bed and collected from the reservoir, and recirculation is done by closed loop operation. Cooling plates placed in the bed enhance heat removal. This kind of arrangement would help to remove heat and CO_2 from the bioreactor. Problems with the existing packed-bed bioreactor are overdrying of the substrate at the bottom (i.e., an inlet for air entry) and uneven distribution of air in the entire packed bed.

11.3.4 Rotating Drum Bioreactor

The rotating drum bioreactor consists of a horizontal drum, electrical motor, gearbox, pump, and arrangements for both inlet/outlet to air and heat removal. In this reactor, mixing can be achieved either by rotating horizontal drum or rotating baffle blade fitted in the central axis of the unit. Gentle mixing of the substrate inside the reactor can be attained at a lower speed of drum/baffle blade and vice versa (Ortiz et al., 2015). Gentle mixing would support microbial growth and also thermal equilibrium, whereas vigorous mixing would result in damage to/shear of the microorganisms. Continuous or intermittent mixing of the substrate can be achieved by varying drum rotational speed. The substrate is filled up to half of its drum capacity and the air is supplied at the bottom of the substrate for enhancing oxygen transfer (Akiyama et al., 1994; Beermann et al., 1991; Boekelheide et al., 1980; Bullock et al., 2000). The heat accumulated in the bed could be removed by air supply and conduction via a drum. Based on the axis of rotation, the rotating drum bioreactor is further divided into two types, namely horizontal and vertical drum rotary reactors (Figs. 11.3 and 11.4). Among these reactors, a horizontal drum rotary reactor is mostly used.

11.3.5 Fluidized-Bed Bioreactor

The major parts of a fluidized-bed bioreactor are distributor plate, reactor column, clump breaker, cyclone, and disengagement zone. Fluidization refers to the suspension of a solid substrate by supplying a working fluid (gas)

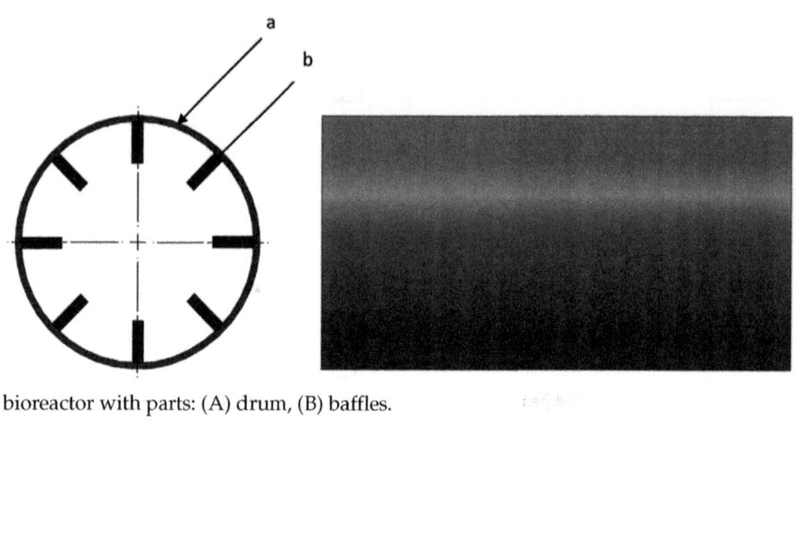

FIG. 11.3 Rotating drum bioreactor with parts: (A) drum, (B) baffles.

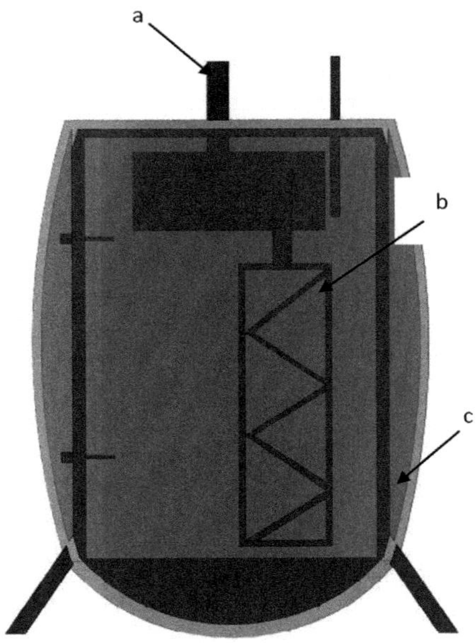

FIG. 11.4 Vertical drum rotary reactor with parts: (A) shaft, (B) agitator, and (C) supporting frame.

at higher velocities. The distributor is placed at the bottom of the reactor and high-velocity fluid is supplied via perforations in the distributor plates. This arrangement maintains the fluidization of the solid substrate particles in the reactor column, which also helps in the mixing of reactants (Thomas et al., 2016). The velocity of working fluid in the fluidization zone should not exceed the terminal velocity of the substrate particles. The main purpose of the disengagement zone in this reactor is to separate the working fluid from solid particles. Cellulase production from various substrates via SSF process-based bioreactors is shown in Table 11.4.

11.3.6 Comparison of Different Types of SSF Bioreactors

Among the SSF-based bioreactors, the tray bioreactor uses the simplest design procedure. Since the thin layer thickness of the substrate placed in the trays is limited, reactor capacity is directly proportional to the number of trays and surface area of the bioreactor. It is very difficult to maintain the reactor temperature in a larger-capacity bioreactor. The merits of the tray bioreactor are uncomplicated reactor design, simple operation, easy upscaling, and minimal operational maintenance. The deterrents of the larger-scale tray-type reactor are labor intensive, difficulty with sterilization, loading and unloading trays, more reactor space required, and costlier process. The merits of the packed-bed bioreactor are larger substrate bed and easy product recovery (Arora et al., 2018). The large-scale packed-bed bioreactor had similar problems to the tray bioreactor, and also overdried substrate, uniform growth, gradients in bed temperature and O_2 distribution, formation of air channels, and difficulties in end product recovery are reported. By comparing static bed to moving bed, capacity of the rotary drum bioreactor is limited to 30% of total reactor volume. Upscaling of the rotary drum bioreactor is restricted due to the constraint in working volume and high energy inputs for plant operation. Maintenance of the rotary drum reactor is higher than that of static-bed reactors due to moving components such as a rotary drum, bearings, etc. In comparison with other bioreactors, skilled persons are required to operate the fluidized-bed bioreactor. Upscaling of this reactor is difficult due to (1) energy intensity, (2) high shear rate, (3) low working volume, and (4) poor reactor performance at a large scale.

11.3.7 SmF Process

SmF bioreactors are different to SSF reactors due to the process, reactor design, and operation. SmF bioreactors use either mechanical or pneumatic mixing. The following bioreactors fall under the SmF technology category and their details are discussed next.

1. Stirred-tank reactors
2. Airlift reactor
3. Bubble column reactor

TABLE 11.4 Cellulase Production Via Solid-State Fermentation Process-Based Bioreactors

	Substrate	Strain	Cellulase yield	References
Tray bioreactor	Wheat bran	*Trichoderma reesei* RUT C-30	959.53 IU/gDS CMCase	(Wood et al., 2005; Narra et al., 2016; Oberoi et al., 2014)
	Grape pomace and orange peels	*Aspergillus awamori*	2.16 IU/gDS CMCase	(Diaz-Godinez et al., 2001)
	Apple pomace	*Aspergillus niger* NRRL567	383.7 ± 17.9 IU/gDS FPase, 425.3 ± 21.3 IU/gDS CMCase, 336.1 ± 16.2 IU/gDS BGL	(Diaz-Godinez et al., 2001)
	Apple pomace	*A. niger* NRRL 2001	401 ± 23.8 IU/gDS FPase, 544.7 ± 24.5 IU/gDS CMCase, 285.4 ± 11.7 IU/gDS BGL	
	Rice straw:wheat bran (3:2)	*A. niger* and *T. reseei*	35.8 IU/gDS FPase, 33.71 IU/gDS BGL, 131.34 IU/gDS CMCase	(Diaz-Godinez et al., 2001)
Packed-bed bioreactor	Babassu cake	*A. awamori*	6.2 U/g	(de Castro et al., 2015, 2010)
	Wheat bran and sugarcane bagasse	*Myceliophthora* sp. I-1D3b	878 U/gDS endoglucanase	(Muller dos Santos et al., 2004)
	Coffee husk	Bacteria (*Pseudoxanthomonas taiwanensis* and *Sphingobacterium composti*), yeasts (*Cyberlindnera jardinii* and *Barnettozyma californica*)	3.1 ± 0.5 FPU/gDM	(Cerda et al., 2017)
	Olive pomace, exhausted grape marc, and vine shoot trimmings	*Aspergillus uvarum* MUM 08.01	38.51 U/g	(Silva et al., 2013)
	Palm kernel cake	*A. niger* FTCC 5003	244.53 U/gDS	(Gardelly et al., 2016)
Rotary drum bioreactor	Wheat bran and glycerol	*A. niger* NRRL 3112	9.3 U/mL BGL	(Kim et al., 2012)
	Empty palm fruit bunch fiber	*Penicillium verruculosum*	6.5 U/gDS CMCase	(Kim et al., 2012, 2016)
Fluidized-bed reactor	Carboxymethyl cellulose	*Cellulomonas flavi-gena*	1.2 U/mg CMCase	(Rojas-Rejon et al., 2011)
	Chitosan	–	13.62 IU/g cellulase	(Li et al., 2018)

BGL, β-glucosidase; *CMCase*, carboxymethyl cellulase; *FPase*, filter paper activity.

11.3.8 Stirred-Tank Reactors

Generally, the stirred-tank reactor consists of an impeller, tank, sparger/air distribution plate, sampling ports, baffles, foam breaker, gearbox, electrical motor, and controller for temperature, foam, and pH. Among the reactors used for SmF, stirred-tank reactors are widely used due to their advantages (Fig. 11.5). Reaction temperature and oxygen transfer can be easily controlled in this reactor as compared to other reactor types. The substrate medium could be sent to the tank and an impeller can be used for efficient mixing of reactants. The air can be uniformly spread in the medium by supplying air via the sparger. The selection of impellers for stirred-tank reactors depends on density and viscosity of the reactants used, quantity to be used, and bioreactor size. The main purpose of baffles fitted in the reactor wall is to reduce vortexing and swirling effects on the medium.

11.3.9 Airlift Reactor

According to Ritter et al. (2013) airlift refers to characteristics of pneumatic contact of the two phases (gas and liquid) or multiphase (gas, liquid, and solid) due to fluid circulation in a closed loop. The airlift reactor consists of channels for upflow and downflow, riser, downcomer, base, and gas separator. As compared with a stirred-tank reactor, there is no mechanical mixing device used for both airlift and bubble column reactors. In this reactor, efficient mixing

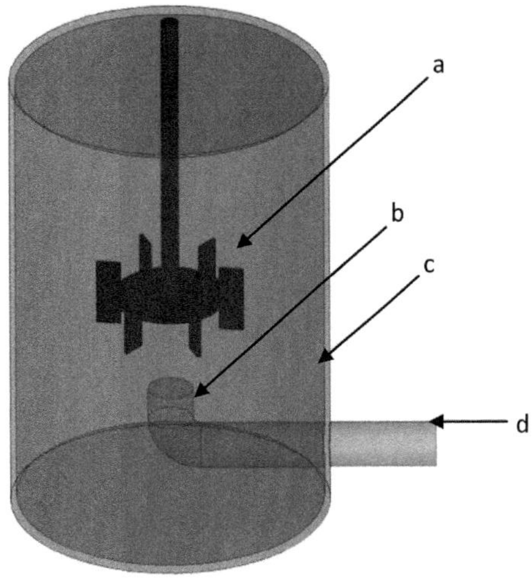

FIG. 11.5 Stirred-tank reactor with parts: (A) impeller, (B) sparger, (C) tank, and (D) air inlet.

is achieved by continuous circulation of the fluid via channels in a closed loop. The fluid continuously circulating in this arrangement will reduce the coalescence of air bubbles (Ritter et al., 2013). Two types of airlift reactors are more popularly available based on the flow of fluid in the reactor: (1) external loop reactors and (2) internal loop reactors.

11.3.10 Bubble Column Reactor

The bubble column reactor is well suited for highly shear-sensitive microorganisms, namely filamentous fungus cells, and mammalian and plant cells used in the SmF process. The bubble column reactor consists of a cylindrically shaped reactor column, sparger, sampling ports, and inlet and outlet for air (Fig. 11.6). The stirred-tank reactor uses a mechanical device for mixing, whereas turbulence caused by upward continuously moving air

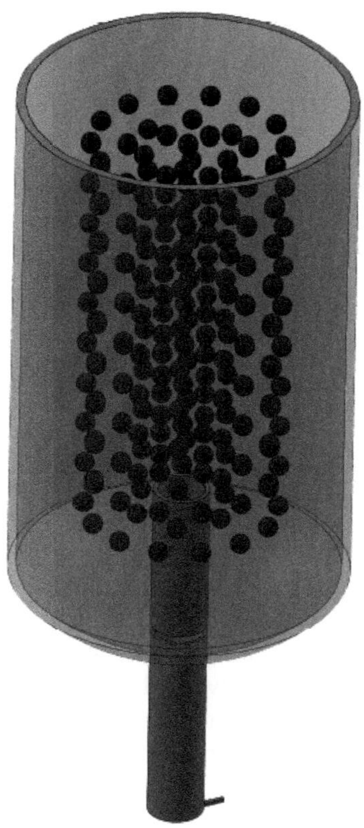

FIG. 11.6 Airlift reactor.

TABLE 11.5 Cellulase Production Via Submerged Fermentation Process-Based Bioreactors

Reactor type	Substrate	Strain	Cellulase yield	References
Stirred-tank reactors	Wheat bran and glycerol	*Aspergillus niger* NRRL 3112	1.78 U/mL/day BGL	(Acharya et al., 2010; Berovic, 1999)
	Sugarcane bagasse	*A. niger* A12	1599 ± 66 IU/L endoglucanase	(Esperanca et al., 2014)
	Wheat bran	*Penicillium oxalicum* RE-10	12.69 U/mL FPase	(Han et al., 2017)
	Glucose sophorose sugar mixture	*Trichoderma reesei*	90.28 IU/mL FPase	(Xia et al., 2016)
Airlift reactor	Cellulose culture medium	*T. reesei* RUT-C30	17 U/mL FPase, 11.8 U/mL CMCase	(Ritter et al., 2013)
	Cellulose and sorbitol	*Penicillium echinulatum*	1.0 U/mL FPase, 6.4 U/mL endoglucanase	(Ritter et al., 2013)
Bubble column reactor	Domestic wastewater	*Trichoderma harzianum* HBA03	10.2 U/L h FPase, 64.6 U/L h CMCase	(Esperanca et al., 2014)
	Avicel	*T. reesei* RUT-C30	1.9 FPU/mL	(Eibl et al., 2014)

BGL, β-glucosidase; *CMCase*, carboxymethyl cellulase; *FPase*, filter paper activity.

bubbles inside the reactor would lead to mixing. In other words, this reactor uses pneumatic air instead of a mechanical device for efficient mixing. The medium is filled in the reactor column and air bubbles are continuously supplied from the sparger plate. Cellulase production from various substrates via SmF process-based bioreactors is shown in Table 11.5.

11.3.11 Comparison of Different Types of SmF Bioreactors

Stirred-tank reactors on a continuous mode of operation perform better than other SmF bioreactors. The deterrents of stirred-tank reactors are energy and labor intensity for batch-scale operation. There is no mechanical device used in pneumatic reactors (both airlift and bubble column reactors).

11.3.12 Comparison of SSF and SmF Bioreactors for Large-Scale Cellulase Production

There is no simple bioreactor design with a fully automatic process control system for the SSF process. The stirred-tank reactor that falls under the SmF process can address large-scale cellulase production and its capacity varies from 1000 to 10,000 L (Patel et al., 2017). Overall comparison of both SSF and SmF reactors used for cellulase production is shown in Table 11.6.

11.4 CONCLUSION

Production of green fuels and high-value chemicals from biomass under the biorefinery concept in the circular bioeconomy will be the future of biomass valorization. It is expected that cellulase production would be doubled to meet future demand. Design and development of energy efficient bioreactors (low energy inputs and capital investment) is a challenge and there is a need for research into cellulase production from multifeedstocks. As compared with the SmF process, SSF bioreactors have several added advantages such as low energy input requirement, easy product recovery, higher product concentration, and low production cost. Even though the SSF process offers several advantages, an SSF bioreactor with a simple design and automatic process control is not yet fully established and commercialized. A stirred-tank reactor with an automatic process control system falling under the SmF category is the preferred unit for large-scale production of cellulase.

TABLE 11.6 Overall Comparison of Both Solid-State Fermentation (SSF) and Submerged Fermentation (SmF) Reactors Used for Cellulase Production

		SSF bioreactors				SmF bioreactors		
S. No.	**Particulars**	**Tray**	**Packed bed**	**Rotary drum**	**Fluidized bed**	**Stirred tank**	**Airlift**	**Bubble column**
1	Reactor design and fabrication	Simple	Simple	Moderate	Difficult	Difficult	Difficult	Difficult
2	Scale-up	Difficult	Difficult	Difficult	Difficult	Moderate	Difficult	Difficult
3	Sterilization	Difficult	Difficult	Difficult	Moderate	Moderate	Moderate	Moderate
4	Temperature control	Difficult	Difficult	Moderate	Moderate	Easy	Easy	Easy
6	Labor	Intensive	Intensive	Intensive	Less intensive	Less intensive	Less intensive	Less intensive
7	Energy input	Less intensive	Intensive	Intensive	Intensive	Intensive	Intensive	Intensive
8	Reactor space required	More	Less	More	Less	Less	Less	Less
9	Operation	Simple	Simple	Moderate	Moderate	Simple	Moderate	Moderate
10	Skilled personnel for plant operation	Not required	Not required	Yes	Yes	Yes	Yes	Yes
11	Commercialization of technology	Yes	No	Yes	No	Yes	No	No
12	Damage to microbial growth	Nil	Nil	Less	Less	Less	Moderate	Moderate
13	Aeration	Natural	Forced	Forced	Forced	Forced	Forced	Forced
14	Operating mode	Batch	Batch	Continuous or batch				

References

Acharya, B.K., Mohana, S., Jog, R., Divecha, J., Madamwar, D., 2010. Utilization of anaerobically treated distillery spent wash for production of cellulases under solid-state fermentation. J. Environ. Manag. 91, 2019–2027.

Adney, W.S., Rivard, C.J., Ming, S.A., Himmel, M.E., 1991. Anaerobic digestion of lignocellulosic biomass and wastes. Cellulases and related enzymes. Appl. Biochem. Biotechnol. 30, 165–183.

Aggarwal, N.K., Goyal, V., Saini, A., Yadav, A., Gupta, R., 2017. Enzymatic saccharification of pretreated rice straw by cellulases from *Aspergillus niger* BK01. 3 Biotech 7, 158.

Akiyama, T., Whitaker, B., Federspiel, M., Hughes, S.H., Yamamoto, H., Takeuchi, T., et al., 1994. Tissue-specific expression of mouse tyrosinase gene in cultured chicken cells. Exp. Cell Res. 214, 154–162.

Alasepp, K., Borch, K., Cruys-Bagger, N., Badino, S., Jensen, K., Sorensen, T.H., et al., 2014. In situ stability of substrate-associated cellulases studied by DSC. Langmuir: ACS J. Surf. Colloids 30, 7134–7142.

Anbar, M., Bayer, E.A., 2012. Approaches for improving thermostability characteristics in cellulases. Methods Enzymol. 510, 261–271.

Arora, S., Rani, R., Ghosh, S., 2018. Bioreactors in solid state fermentation technology: design, applications and engineering aspects. J. Biotechnol. 269, 16–34.

Baker, J.O., Ehrman, C.I., Adney, W.S., Thomas, S.R., Himmel, M.E., 1998. Hydrolysis of cellulose using ternary mixtures of purified cellulases. Appl. Biochem. Biotechnol. 70-72, 395–403.

Bayer, E.A., Chanzy, H., Lamed, R., Shoham, Y., 1998. Cellulose, cellulases and cellulosomes. Curr. Opin. Struct. Biol. 8, 548–557.

Bayer, E.A., Lamed, R., Himmel, M.E., 2007. The potential of cellulases and cellulosomes for cellulosic waste management. Curr. Opin. Biotechnol. 18, 237–245.

Beermann, F., Ruppert, S., Hummler, E., Schutz, G., 1991. Tyrosinase as a marker for transgenic mice. Nucleic Acids Res. 19, 958.

Berovic, M., 1999. Scale-up of citric acid fermentation by redox potential control. Biotechnol. Bioeng. 64, 552–557.

Bhat, M.K., 2000. Cellulases and related enzymes in biotechnology. Biotechnol. Adv. 18, 355–383.

Bischof, R.H., Ramoni, J., Seiboth, B., 2016. Cellulases and beyond: the first 70 years of the enzyme producer *Trichoderma reesei*. Microb. Cell Factories 15, 106.

Boekelheide, K., Graham, D.G., Mize, P.D., Jeffs, P.W., 1980. The metabolic pathway catalyzed by the tyrosinase of *Agaricus bisporus*. J. Biol. Chem. 255, 4766–4771.

Bullock, T.N., Colella, T.A., Engelhard, V.H., 2000. The density of peptides displayed by dendritic cells affects immune responses to human tyrosinase and gp100 in HLA-A2 transgenic mice. J. Immunol. 164, 2354–2361.

Calderon-Cortes, N., Watanabe, H., Cano-Camacho, H., Zavala-Paramo, G., Quesada, M., 2010. cDNA cloning, homology modelling and evolutionary insights into novel endogenous cellulases of the borer beetle Oncideres albomarginata chamela (Cerambycidae). Insect Mol. Biol. 19, 323–336.

Camassola, M., Dillon, A.J., 2010. Cellulases and xylanases production by *Penicillium echinulatum* grown on sugar cane bagasse in solid-state fermentation. Appl. Biochem. Biotechnol. 162, 1889–1900.

Cerda, A., Gea, T., Vargas-Garcia, M.C., Sanchez, A., 2017. Towards a competitive solid state fermentation: cellulases production from coffee husk by sequential batch operation and role of microbial diversity. Sci. Total Environ. 589, 56–65.

Chandrasekaran, A., Bharadwaj, R., Park, J.I., Sapra, R., Adams, P.D., Singh, A.K., 2010. A microscale platform for integrated cell-free expression and activity screening of cellulases. J. Proteome Res. 9, 5677–5683.

Cho, E.J., Jung, S., Kim, H.J., Lee, Y.G., Nam, K.C., Lee, H.J., et al., 2012. Co-immobilization of three cellulases on Au-doped magnetic silica nanoparticles for the degradation of cellulose. Chem. Commun. 48, 886–888.

de Castro, A.M., Castilho, L.R., Freire, D.M., 2015. Multivariate optimization and supplementation strategies for the simultaneous production of amylases, cellulases, xylanases, and proteases by *Aspergillus awamori* under solid-state fermentation conditions. Appl. Biochem. Biotechnol. 175, 1588–1602.

de Castro, A.M., de Albuquerque de Carvalho, M.L., Leite, S.G., Pereira Jr., N., 2010. Cellulases from *Penicillium funiculosum*: production, properties and application to cellulose hydrolysis. J. Ind. Microbiol. Biotechnol. 37, 151–158.

Dias, L.M., Dos Santos, B.V., Albuquerque, C.J.B., Baeta, B.E.L., Pasquini, D., Baffi, M.A., 2018. Biomass sorghum as a novel substrate in solid-state fermentation for the production of hemicellulases and cellulases by *Aspergillus niger* and *A. fumigatus*. J. Appl. Microbiol. 124, 708–718.

Diaz-Godinez, G., Soriano-Santos, J., Augur, C., Viniegra-Gonzalez, G., 2001. Exopectinases produced by *Aspergillus niger* in solid-state and submerged fermentation: a comparative study. J. Ind. Microbiol. Biotechnol. 26, 271–275.

Doriya, K., Jose, N., Gowda, M., Kumar, D.S., 2016. Solid-state fermentation vs submerged fermentation for the production of L-asparaginase. Adv. Food nutr. Res. 78, 115–135.

Eibl, R., Loffelholz, C., Eibl, D., 2014. Disposable bioreactors for inoculum production and protein expression. Methods Mol. Biol. 1104, 265–284.

Esperanca, M.N., Cunha, F.M., Cerri, M.O., Zangirolami, T.C., Farinas, C.S., Badino, A.C., 2014. Gas hold-up and oxygen mass transfer in three pneumatic bioreactors operating with sugarcane bagasse suspensions. Bioprocess Biosyst. Eng. 37, 805–812.

Formenti, L.R., Norregaard, A., Bolic, A., Hernandez, D.Q., Hagemann, T., Heins, A.L., et al., 2014. Challenges in industrial fermentation technology research. Biotechnol. J. 9, 727–738.

Gardelly, M., Trimech, B., Belkacem, M.A., Harbach, M., Abdelwahed, S., Mosbah, A., et al., 2016. Synthesis of novel diazaphosphinanes coumarin derivatives with promoted cytotoxic and anti-tyrosinase activities. Bioorg. Med. Chem. Lett. 26, 2450–2454.

Han, B., Wang, L., Li, S., Wang, E., Zhang, L., Li, T., 2010. Ethanol production from sweet sorghum stalks by advanced solid state fermentation (ASSF) technology. Sheng wu gong cheng xue bao = Chinese J. Biotechnol. 26, 966–973.

Han, X., Wang, Z., Chen, M., Zhang, X., Tang, C.Y., Wu, Z., 2017. Acute responses of microorganisms from membrane bioreactors in the presence of NaOCl: protective mechanisms of extracellular polymeric substances. Environ. Sci. Technol. 51, 3233–3241.

Jorgensen, H., 2018. Test of Efficacy of Cellulases for Biomass Degradation. Methods Mol. Biol. 1796, 283–297.

Kim, A.J., Choi, J.N., Kim, J., Yeo, S.H., Choi, J.H., Lee, C.H., 2012. Metabolomics-based optimal koji fermentation for tyrosinase inhibition supplemented with *Astragalus radix*. Biosci. Biotechnol. Biochem. 76, 863–869.

Kim, B.H., Park, K.C., Park, J.H., Lee, C.G., Ye, S.K., Park, J.Y., 2016. Inhibition of tyrosinase activity and melanin production by the chalcone derivative 1-(2-cyclohexylmethoxy-6-hydroxy-phenyl)-3-(4-hydroxymethyl-phenyl)-propenone. Biochem. Biophys. Res. Commun. 480, 648–654.

Kuhad, R.C., Gupta, R., Singh, A., 2011. Microbial cellulases and their industrial applications. Enzyme Res. 2011: 280696.

Kumar, N.V., Rani, M.E., Gunaseeli, R., Kannan, N.D., Sridhar, J., 2012. Modeling and structural analysis of cellulases using *Clostridium thermocellum* as template. Bioinformation 8, 1105–1110.

Lai, X., Soler-Lopez, M., Wichers, H.J., Dijkstra, B.W., 2016. Large-scale recombinant expression and purification of human tyrosinase suitable for structural studies. PLoS One 11, e0161697.

Li, K., Zhang, J.W., Liu, C.G., Bai, F.W., 2018. Impact of inhibitors on commercial cellulases in lignocellulosic ethanol production. Protein Pept. Lett.

Muller dos Santos, M., Souza da Rosa, A., Dal'Boit, S., Mitchell, D.A., Krieger, N., 2004. Thermal denaturation: is solid-state fermentation really a good technology for the production of enzymes? Bioresour. Technol. 93, 261–268.

Narra, M., Balasubramanian, V., James, J.P., 2016. Enhanced enzymatic hydrolysis of mild alkali pre-treated rice straw at high-solid loadings using in-house cellulases in a bench scale system. Bioprocess Biosyst. Eng. 39, 993–1003.

Nidetzky, B., Steiner, W., Claeyssens, M., 1994. Cellulose hydrolysis by the cellulases from *Trichoderma reesei*: adsorptions of two cellobiohydrolases, two endocellulases and their core proteins on filter paper and their relation to hydrolysis. Biochem. J. 303, Pt 3. 817–823.

Niu, H., Shah, N., Kontoravdi, C., 2016. Modelling of amorphous cellulose depolymerisation by cellulases, parametric studies and optimisation. Biochem. Eng. J. 105, 455–472.

Oberoi, H.S., Rawat, R., Chadha, B.S., 2014. Response surface optimization for enhanced production of cellulases with improved functional characteristics by newly isolated *Aspergillus niger* HN-2. Antonie Van Leeuwenhoek 105, 119–134.

Ortiz, G.E., Guitart, M.E., Cavalitto, S.F., Alberto, E.O., Fernandez-Lahore, M., Blasco, M., 2015. Characterization, optimization, and scale-up of cellulases production by trichoderma reesei cbs 836.91 in solid-state fermentation using agro-industrial products. Bioprocess Biosyst. Eng. 38, 2117–2128.

Patel, A.K., Singhania, R.R., Pandey, A., 2017. Production, purification, and application of microbial enzymes. In: Brahmachari, G., Demain, A.L., Adrio, J.L. (Eds.), Biotechnology of Microbial Enzymes Production, Biocatalysis and Industrial Applications. Academic Press is an imprint of Elsevier, UK, pp. 13–41.

Ritter, C.E., Fontana, R.C., Camassola, M., da Silveira, M.M., Dillon, A.J., 2013. The influence of sorbitol on the production of cellulases and xylanases in an airlift bioreactor. Bioresour. Technol. 148, 86–90.

Roberts, S.M., Davies, G.J., 2012. The crystallization and structural analysis of cellulases (and other glycoside hydrolases): strategies and tactics. Methods Enzymol. 510, 141–168.

Rojas-Rejon, O.A., Poggi-Varaldo, H.M., Ramos-Valdivia, A.C., Martinez-Jimenez, A., Cristiani-Urbina, E., de la Torre Martinez, M., et al., 2011. Production of cellulases and xylanases under catabolic repression conditions from mutant PR-22 of *Cellulomonas flavigena*. J. Ind. Microbiol. Biotechnol. 38, 257–264.

Sharma, A., Tewari, R., Rana, S.S., Soni, R., Soni, S.K., 2016. Cellulases: classification, methods of determination and industrial applications. Appl. Biochem. Biotechnol. 179, 1346–1380.

Silva, J.C., Guimaraes, L.H., Salgado, J.C., Furriel, R.P., Polizeli, M.L., Rosa, J.C., et al., 2013. Purification and biochemical characterization of glucose-cellobiose-tolerant cellulases from *Scytalidium thermophilum*. Folia Microbiol. 58, 561–568.

Singhania, R.R., Saini, J.K., Saini, R., Adsul, M., Mathur, A., Gupta, R., et al., 2014. Bioethanol production from wheat straw via enzymatic route employing *Penicillium janthinellum* cellulases. Bioresour. Technol. 169, 490–495.

Thomas, L., Ram, H., Kumar, A., Singh, V.P., 2016. Production, optimization, and characterization of organic solvent tolerant cellulases from a lignocellulosic waste-degrading actinobacterium, *Promicromonospora* sp. VP111. Appl. Biochem. Biotechnol. 179, 863–879.

Viesturs, U.E., Strikauska, S.V., Leite, M.P., Berzins, A.J., Tengerdy, R.P., 1987. Combined submerged and solid substrate fermentation for the bioconversion of lignocellulose. Biotechnol. Bioeng. 30, 282–288.

Wood, J.M., Chavan, B., Hafeez, I., Schallreuter, K.U., 2005. Regulation of tyrosinase by tetrahydropteridines—what is real? A critical reanalysis of H. Wojtasek's view. Biochem. Biophys. Res. Commun. 331, 891–893.

Xia, C., Ma, D., Gao, B., Hu, X., Yue, Q., Meng, Y., et al., 2016. Characteristics and trihalomethane formation reactivity of dissolved organic matter in effluents from membrane bioreactors with and without filamentous bulking. Bioresour. Technol. 211, 183–189.

Yoon, J.H., Park, J.E., Suh, D.Y., Hong, S.B., Ko, S.J., Kim, S.H., 2007. Comparison of dyes for easy detection of extracellular cellulases in fungi. Mycobiology 35, 21–24.

Zhang, B.B., Lu, L.P., Xu, G.R., 2015. Why solid-state fermentation is more advantageous over submerged fermentation for converting high concentration of glycerol into Monacolin K by Monascus purpureus 9901: a mechanistic study. J. Biotechnol. 206, 60–65.

Zhang, L., Li, Z., Dai, B., Zhang, W., Yuan, Y., 2013. Effect of submerged and solid-state fermentation on pigment and citrinin production by *Monascus purpureus*. Acta Biol. Hung. 64, 385–394.

CHAPTER

12

Downstream Processing Technology for Cellulase Production

S.M. Bhatt*, Shilpa†

*Biotechnology Department, SBBS University, Jalandhar, India †Department of Biotechnology, CGC Landron, Chandigarh, India

12.1 INTRODUCTION

Lignocellulosic biomass is an indispensable biomass source for biofuel production. However, due to its complex structure, conversion of biomass is tedious. Microbial cellulase production is intended to be highly useful in the industrial production of biofuels from cellulosic biomass. Therefore the production of cellulase must be economical. Currently, at the industrial scale, *Trichoderma* is being used to produce cellulase in various modes. The principal enzyme involved in the conversion of lignocellulosic biomass is cellulase, which can easily convert cellulose to fermentable sugars for bioethanol production.

Lignocellulosic biomass composed of various other components (Fig. 12.1) comes from starchy plants, molasses, algae, and cellulosic biomass (which is made up of high-fraction cellulose and hemicellulose covered by lignin) as shown in Table 12.1 (Isikgor and Becer, 2015). Cellulose is one of the most abundantly available biomasses on earth and can be utilized for various value-added products. Various technologies exist for the conversion of lignocellulose into biomass such as thermochemical conversion, biochemical conversion, and enzymatic conversion (Mabee et al., 2011).

In addition, lignocellulosic biomass is considered to be the most promising and sustainable biofuel today, but there are many challenges that need to be overcome for industrial production (Kumar et al., 2016). As compared to 1G biofuel (edible), lignocellulosic biomass-based biofuel (2G biofuel) is economical and highly available as waste (Znameroski et al., 2012). The main challenge is the recalcitrant nature of plant biomass, which needs to be converted into crystalline form useful for fermentation (Xiao and Anderson, 2013).

The presence of high hemicellulose (C5 sugar) in addition to cellulose (C6 sugar) is another challenge; however, removal of lignin is essential in bioprocessing and is accomplished by using pretreatment steps such as enzymatic, chemical, or thermal treatments (Kumar and Sharma, 2017). Pretreatment of lignocellulosic biomass effectively removes lignin, which produces various inhibitors for cellulase production, which is also a challenge (Kumar and Sharma, 2017). Therefore understanding the physicochemical nature of biomass is essential in selecting the correct pretreatment methods.

The selection criteria for suitable pretreatment method depends on: (1) the size of biomass, (2) hemicellulose fraction, (3) degradation of by-products, (4) minimum energy requirements, and (5) economical pretreatment catalyst (Wyman, 2013). To reduce the cost of downstream processing steps, evaluation of operating costs, overall capital costs in biomass handling pretreatment, fraction costs, etc. must be considered (Lynd et al., 2002).

Pretreatment techniques are broadly divided into the following classes: biochemical, enzymatic, and thermochemical (Laser et al., 2009) based on the operating temperature.

From Cellulose to Cellulase: Strategies to Improve Biofuel Production
https://doi.org/10.1016/B978-0-444-64223-3.00012-6

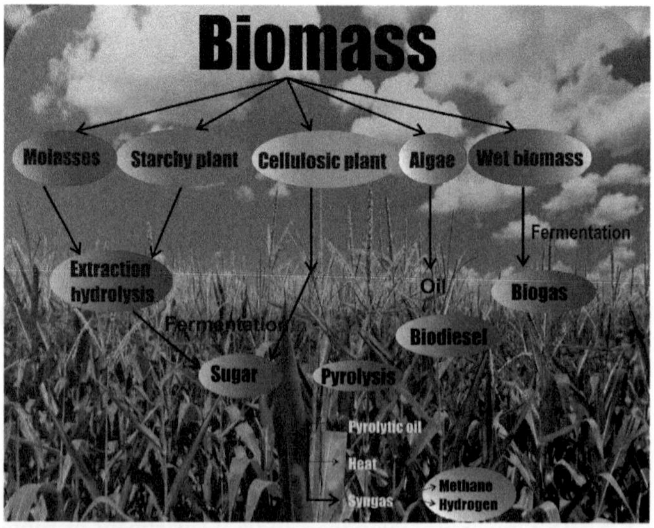

FIG. 12.1 Biomass processing and product formation.

TABLE 12.1 Biomass Fraction of Some Lignocellulosic Feedstocks (Isikgor and Becer, 2015)

Lignocellulosic feedstocks	Cellulose (%)	Hemicellulose (%)	Lignin (%)
Sugarcane bagasse	42	25	20
Softwood	45–50	25–35	25–35
Hardwood	40–55	24–40	18–25
Corn stover	38	26	19
Corn cobs	45	35	15
Nut shells	25–30	25–30	30–40
Rice straw	32	24	18
Grasses	25–40	25–50	10–30
Newspaper	40–55	25–40	18–30
Banana waste	13.2	14.8	14
Wheat straw	29–35	26–32	16–21
Bagasse	54.87	16.52	23.33
Sponge gourd fibers	66.59	17.44	15.46
Agricultural residues	5–15	37–50	25–50
Hardwood	20–25	45–47	25–40
Softwood	30–60	40–45	25–29
Grasses	0	25–40	35–50
Waste papers from chemical pulps	6–10	50–70	12–20
Newspaper	12	40–55	25–40
Sorted refuse	60	20	20
Leaves	15–20	80–85	0
Cotton seed hairs	80–95	5–20	0
Paper	85–99	0	0–15
Switch grass	45	31.4	12
Sweet sorghum	45	27	21

12.2 PRETREATMENT OF BIOMASS

As compared to starch, the lignocellulosic biomass fractional component includes cellulose, hemicellulose, and other materials such as dead lignin covering the entire biomass structure; however, some do not contain lignin, for example, switch grass, which is very common in North America.

The main aim of pretreatment is to get rid of dead lignin and pectin, but this is not required in biomass such as switch grass. Pretreatment is performed mostly by biological, chemical, or physical means, which increases accessibility to microbes and thus increases cellulase production. Therefore improved hydrolysis of cellulose is mandatory by one or a combination of pretreatment steps to enhance cellulase production.

Recent research shows much improvement in downstream operation costs for cellulase production, in addition to the application of better pretreatment methods with enhanced hydrolysis of biomass and less by-product formation (Shirkavand et al., 2016; Bhatt, 2015). Biochemical conversion is slower and thus finds limited application in small-scale reactors, while at a larger scale, the integrative approach is quite useful. For instance, a combination of fungal and biochemical approaches may enhance the hydrolysis of biomass (Rezania et al., 2017).

Lignin-cementing cellulose and hemicellulose are deoxidized into inhibitors during harsh pretreatment and bind to enzymes to deactivate them (Sasmal and Mohanty, 2018). Sindhu et al. (2011) compared the combined effect of pretreatment and their advantages and disadvantages in various chemicals (acid and base). Acid pretreatment is beneficial for short durations, while for long durations it has many disadvantages: it erodes the reactor wall and produces more inhibitors as compared to alkali pretreatment (Bhatt, 2015). The reducing sugar released is also more prevalent in alkali pretreatments as compared to acid pretreatments (Sukumaran et al., 2009). It has been reported that delignification of lignocellulosic biomass with sulfite pretreatment increases the breakdown of lignin into soluble lignin components, which can be useful for various applications (Bu et al., 2012). Alkaline pretreatment further enhances lignin solubility making downstream processing easy.

12.3 CELLULASE COMPOSITION

Cellulase is a multienzyme complex containing endoglucanases, exoglucanases or cellobiohydrolases CBHs), and β-glucosidases (Figs. 12.2 and 12.3A). Most of the cellulase is produced by fungi, some bacteria, and some plants; for example, *Aspergillus*, *Schisophyllum*, *Trichoderma reesei*, and *Penicillium* (Baeyens et al., 2015). Augmented cellulase production has been observed by *Bacillus subtilis*, which produces more cellulase in optimized media (Sreena and Sebastian, 2018).

Cellulase production also depends on species (Li et al., 2017), but mainly three types of cellulase exist in fungal species such as *Trichoderma*, which consists of (1) endoglucanase and xylanase and (2) CBH, which is involved in efficient sacharrification. In *Trichoderma*, CBH is present at around 56%, and is the main enzyme required for cellulose hydrolysis. *Trichoderma* also releases 10%–15% endoglucanase and 1%–10% xylanase (Liu et al., 2017). The action of CBH produces cellobiose from cellulose as shown in the reaction in Fig. 12.3B. CBH is an exoglucanase produced in various bacteria and fungi (Annamalai et al., 2016). There are two types of CBH in *T. reesei*, mainly classified as CBH-I and CBH-II.

CBH-I attacks the reducing ends, while CBH-II attacks the nonreducing ends. CBH-II is highly active against its counterpart CBH-I (Liao et al., 2011). Improvement in CBH-II activity may have enhanced the cellulolytic activity of *T. reesei* (Wang et al., 2012). In some fungi, three types of CBH are reported, namely (EC 3.2.1.91), CBH-IA, CBH-IB, and CBH-II from *Talaromyces emersonii* (Tuohy et al., 2002).

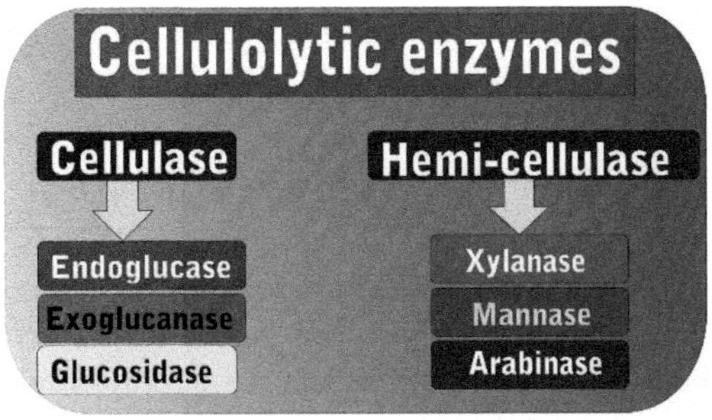

FIG. 12.2 Cellulolytic enzymes composition and types. *Adapted from Sajith, S., Priji, P., Sreedevi, S., Benjamin, S., 2016. An overview on fungal cellulases with an industrial perspective. J. Nutr. Food Sci. 6, 1–13, https://doi.org/10.4172/2155-9600.1000461.*

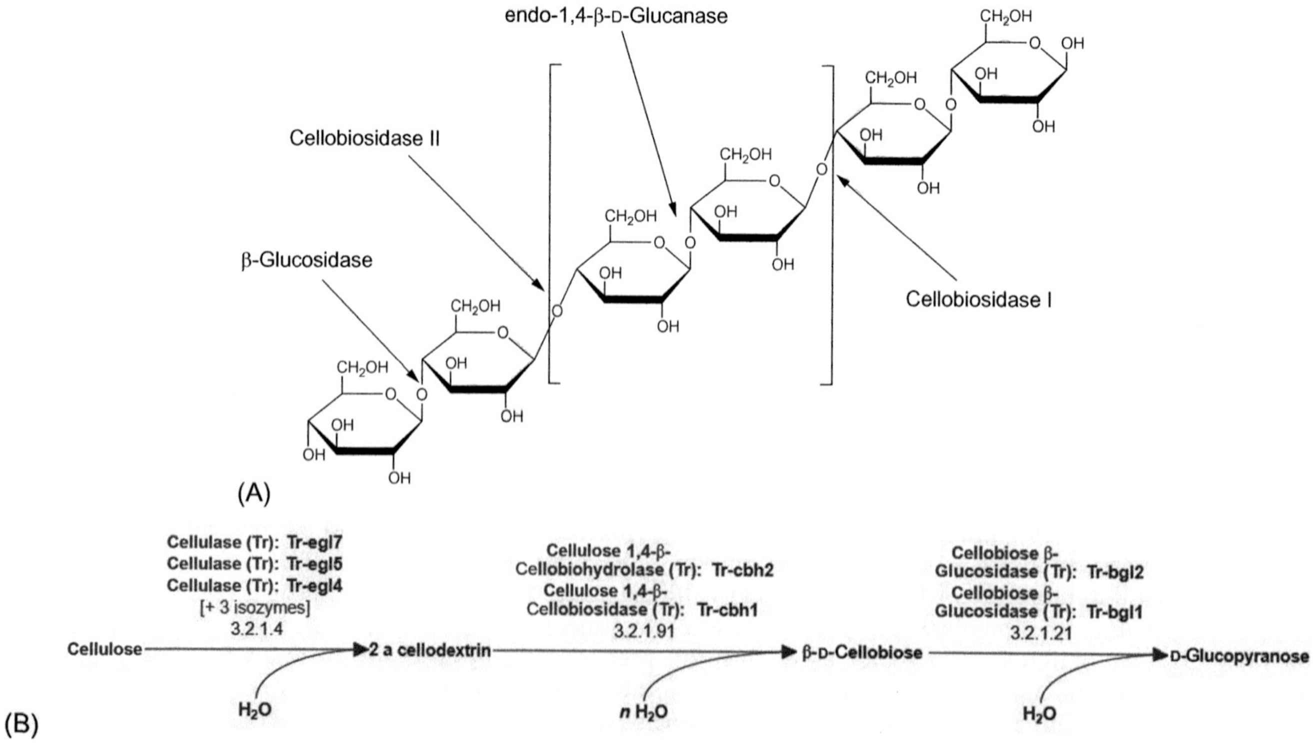

FIG. 12.3 (A) Action of different cellulases over the bond of cellulose. (B) Reaction and products (https://biocyc.org/META/NEW-IMAGE?type=NIL&object=PWY-6788).

Cellulases have been classified into glycoside hydrolases such as Cel6A (CBH-II), Cel7A (CBH-I), Cel12A (EG28), and Cel45A (EG25) (Bukhtojarov et al., 2004). For industrial use, currently *T. reesei* has been used for cellulase production, while the addition of CBH (GH7 CBH) TreCel7A along with *Trichoderma* cells has been reported for a cocktail preparation for the cellulose hydrolysis process (Borisova et al., 2018).

In the recent past, thermophilic fungi have been the center of research in many biotechnological applications, including biomass degradation. Thermophilic fungi mostly grow in the temperature range of 20–50°C and are efficient for biomass degradation. Thermostable enzymes offer hydrolysis of lignocellulosic substrates at higher temperature, which is desirable during the pretreatment of biomass, thus fewer enzymes are required. In harsh conditions, the thermostable enzyme could offer better hydrolysis of pretreated lignocellulosic substrates. Thermostable CBHs have been isolated from several microbes, namely *Fomitopsis pinicola* (Shin et al., 2010), *Humicola grisea* var. *thermoidea* (Momeni et al., 2014), and *Bacillus* and *Geobacillus* strains (Rastogi et al., 2010).

Therefore improved hydrolysis and performance of enzymes are observed when enzymes are more stable and thus the designing of process configurations becomes easy when scaling up the production of enzymes. Thermostable fungal enzymes have now been used in several hydrolyses of biomass. For instance, three most promising enzymes of thermostable cellulase, identified in their categories as CBH, endoglucanase, and P-glucosidase, have been cloned in *T. reesei* and thus a novel mixture of thermostable cellulase was obtained.

Engineered CBHs are shown to have enhanced specific activity with reduced use of CBHs and increased thermostability (Himmel et al., 2018). In some reports, engineered CBH is helpful in cellulose conversion even in single steps (Brady et al., 2015). The optimal temperature of thermostable enzymes is mostly in the range of 45–55°C, at optimum pH 5.0–6.0 (Infinite Enzymes, 2018).

12.4 METABOLIC ENGINEERING

For enhancing enzyme efficiency, sometimes a cocktail of enzymes using two different species (*Aspergillus and Trichoderma*) for on-site production in biorefineries is recommended (Zoglowek et al., 2015). In addition, anaerobic cellulolytic bacteria produce large complexes of cellulolytic enzymes called cellulosomes that can attach to the surface of cellulose, while the fungal enzymes do not form any such complexes and do not attach, although some of the enzymes have been shown to cooperate and act in synergy. Cellulosomes have been isolated from various microbes, e.g., *Clostridium thermocellum* (Hirano et al., 2016).

Metabolic engineering of genes shows enhanced activity of CBH and glycoside hydrolase, especially Family7 (GH7) (Taylor et al., 2018). Particularly, GH7 cellulase has been used in cocktail preparations.

Biorefineries are predominantly using fungal-based cellulase nowadays for enhanced hydrolysis systems. Therefore GH7 CBHs have been useful for cellulase-based engineering.

To date, the structure of GH7 CBH has been studied in *T. reesei* Cel7A, which is a multimodular enzyme (TrCel7A) and is responsible for cellulose binding via the catalytic domain. However, there is still a need for more research in cellulase engineering to establish structure-activity relationships.

Microbial cellulase is a class of inducible enzymes and thus its production depends on several factors in media, namely temperature, pH, agitation, substrate concentration, type of nitrogen, source of sugar, etc. Several optimization tools are used such as Box-Behnken, Plackett-Burman, response surface methodology (RSM), central composite designs, and ANN. Singhania et al. (2007) used both Plackett-Burman and central composite designs (CCDs) to optimize cellulase production by *T. reesei* RUT C30; as a result, cellulase production increased 6.2-fold (37.5% moisture content and 30°C). In another work with carboxymethyl cellulose, the following factors were optimized, namely $CaCl_2$, temperature, Tween-20, and temperature using Plackett–Burman design and RSM for improved cellulase production up to 42 IU/mL (Shajahan et al., 2017). Cellulose production was carried out using the RSM method with a cellulose activity of 8.61 IU/mL (Saravanan et al., 2013). *T. reesei* RUT C30-based cellulase production was optimized using CCDs when rice straw and banana fiber were used as substrate (Muthuvelayudham and Viruthagiri, 2010).

In another work the sugarcane bagasse-based substrate was used for economical cellulase production by *T. reesei* RUT C30. Box-Behnken experimental design was used for optimization of enhanced cellulase yield (25.6 FPU/g dry substrate) (Mekala et al., 2008). Cellulose production was also carried out on corn stover using *Aspergillus tubingensis* IMMIS2 (extracted from rotten tomato) and thus cellulose activity was 81 ± 1.5 μg/mL/min from six substrates used (Imran et al., 2017).

Cellulase production optimization requires various media conditions to be used in the proper way so that cellulase production can be enhanced. There has been various efforts in the recent past to optimize the media condition (Sun et al., 2018; Tabssum et al., 2018; Tariq et al., 2018; Bentil et al., 2018; Maftukhah and Abdullah, 2018).

Production of cellulase enzyme from lignocellulose-based biomass is still low. However, some studies show that in solid-state fermentation (SSF) mode, enzyme production can be enhanced by up to 3.12 U/mL in 4 days (Maftukhah and Abdullah, 2018). The synergistic effect of adding xylanase during cellulase hydrolysis has been reported and found to be effective in improving the performance of cellulase over the hydrolysis of lignocellulosic substrates. As a result, the final ethanol concentration reported was 0.695 g/L after 96 h fermentation and the overall yield was 0.254 g of ethanol produced per gram of carbohydrate consumed, corresponding to 49.8% of theoretical value (Yang et al., 2018).

A further experiment was conducted by mixing three enzymes, 1-endo-1,4-β-glucanase, 2-1,4-β-D-glucan CBHB precursor, and 3-1,4-β-glucosidase (Tang et al., 2018), while experiments with *Fomitopsis* sp. RCK2010 under optimized SSF and activity reported carboxymethyl cellulase (71.526 IU/g), filter-paper activity (FPase) (3.268 IU/g), and β-glucosidase (50.696 IU/g).

Some authors reported enhanced enzyme activity in the presence of surfactant (Alhammad et al., 2018). Another factor that acts as an inducer of enzyme cellulase is Mn^{++} (Chen et al., 2018), while another attempt to produce the enzyme *T. reesei* RUT C30 in limited phosphorus supply resulted in enhanced cellulase production due to prolonged stationary phase (Callow et al., 2015).

However, in a limited nitrogen supply, cellulase production ceases (Callow et al., 2016). Sometimes silencing of *cre1* gene (silenced strain C88) in *Myceliophthora thermophila* ATCC-42464 worked well and enhanced enzyme activity many fold. In addition, a regulator of enzymes for enhanced xylanase and cellulase production was reported in the thermophilic fungus *M. thermophila* strain ATCC 42464 MHR1. In such an attempt the *mhr1*-silenced strain MtR5 was developed and was shown to have enhanced FPase of 1.33-fold, while β,1-4-endoglucanase activity was 1.65-fold, and xylanase activity was 1.48-fold, compared to the parental strain.

As compared to other strains, *Penicillium* proves to be better in terms of cellulase production, especially β-glucosidase. For better biomass hydrolysis, *Penicillium* proves to be a key enzyme in cocktail production with *T. reesei* (Vaishnav et al., 2018). The reason for this was lack of β-glucosidase enzyme in *T. reesei*. Therefore often an *Aspergillus niger* strain is also mixed with the *Trichoderma* strain for better cocktail production to compensate the deficient enzyme (Sukumaran et al., 2009). Not all microbes are fit for cellulase, endoglucanase, or exoglucanase production. Some strains produce better β-glucosidase (Camassola and Dillon, 2007). Therefore in the preparation of cocktail enzymes, fungal cocktail mixing of different enzymes is used in process optimization so that an increase in hydrolysis can be observed.

In a recent work a small-scale stirred-tank reactor was used to enhance cellulase production (Deshavath et al., 2018) with alkaline pretreatment. Better solubilization of lignocellulosic components such as hemicellulose or lignin and thus improved cellulase production were reported (up to 51.38 FPU/g substrate). The major cause of reduced cellulase production reported was the presence of glucose and cellobiose production, which is a well-known repression mechanism. For reducing the repression effect of these substrates, sometimes lactose and sophorose are used as inducers of cellulase production in *T. reesei*; however, the same trick does not work for *Penicillium* sp. (Morikawa et al., 1995). Therefore some cocktail preparations using *Aspergillus*, *Penicillium*, and *Trichoderma* were reported to have enhanced cellulase production (de França Passos et al., n.d.).

12.5 MODE OF CELLULASE PRODUCTION

Cellulase production is done either using SSF mode or submerged fermentation (SmF). For instance, white-rot basidiomycetous fungi produce cellulase in extracellular mode in SSF. SSF is characterized by using a solid support, which is usually a water-free medium of agrowaste biomass. The main disadvantage of SSF is control of mass and heat transfers, but it has the benefit that growth of filamentous fungi canmimic natural conditions (Annamalai et al., 2016). However, SmF creates an aerobic environment during fermentation (Liao et al., 2011). The main disadvantage of SmF is the use of costly electricity during purification, sterilization, and agitation; however, better heat and mass transfers can be achieved.

Nevertheless, limited studies have been undertaken at pilot scale in SSF mode for cellulase production. One such study shows high enzyme production in SSF mode (Marín et al., 2018). In this work in SSF mode, a mixture of coffee husk and wood chips was mixed with inoculating compost for high cellulase production. The steps involved in pilot-scale downstream processing are shown in Fig. 12.4, which basically involve extraction of enzymes either by conventional methods or by static, agitated, or recirculation methods.

CMC, Carboxymethyl cellulose; *SmF*, submerged fermentation; *SSF*, solid-state fermentation.

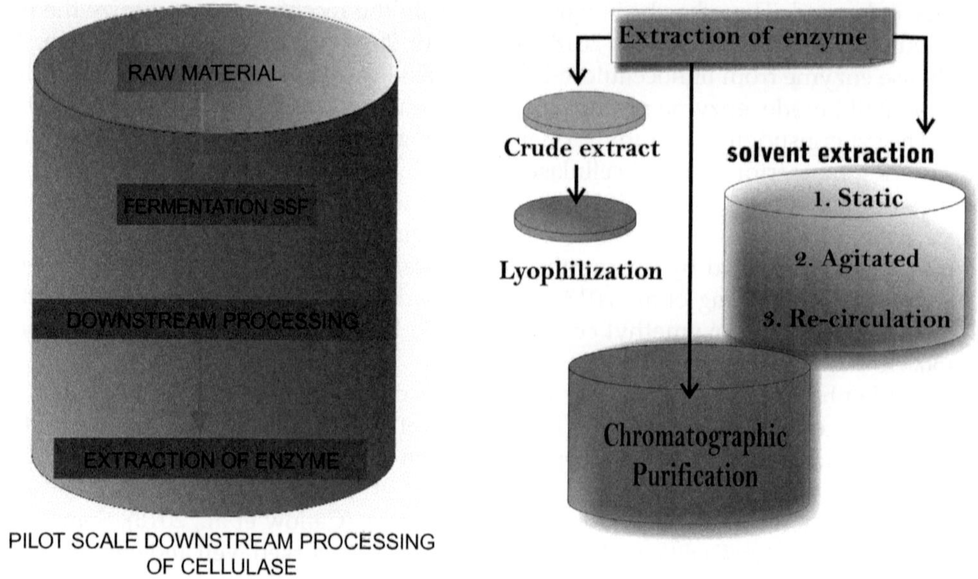

FIG. 12.4 Pilot-scale downstream processing of cellulase sources (Marín et al., 2018).

12.6 CELLULASE EXTRACTION

Agrowaste substrate is used for cellulase production, for instance, willow (Reczey et al., 1996), banana peel in SSF mode (Sun et al., 2011), cassava bagasse in SSF mode (Sugumaran et al., 2013), coconut coir pith (Muniswaran and Charyulu, 1994), paddy straw (Wati et al., 2007), *Jatropha curcas* seed cake (Ncube et al., 2012), fruit waste as substrate (Rathnan, 2011), groundnut shell (Vyas et al., 2005), wheat straw (Ahmed et al., 2010), sugar beet (Moosavi-Nasab and Majdi-Nasab, 2008), and many more as listed in Table 12.2.

Downstream processing mainly involves extraction, purification, and recovery steps. It involves 70% cost of production, therefore it can be divided into three main steps. Solvent extraction improves extraction efficiency often in

TABLE 12.2 Substrate for Cellulase Production

Fungi	Substrate	Mode (SSF/SmF)	Cellulase (U/gds)	References
Aspergillus flavus BS1	Tapioca flour and sawdust	SSF	5408.5	(Sajith et al., 2016)
Aspergillus fumigatus	Rice straw and wheat bran	SSF	14.7	(Sherief et al., 2010)
Aspergillus niger	Wheat bran	SSF	3.2	(Chandra et al., 2007)
A. niger 38	Wheat bran and wheat straw	SSF	14.8	(Jecu, 2000)
A. niger KK2	Rice straw	SSF	130	(Kang et al., 2004)
A. niger MS82	Grass	SSF	100	(Tariq et al., 2018)
Aspergillus terreus M11	Corn stover	SSF	581	(Gao et al., 2008)
A. fumigatus Z5	Corn stover	SSF	526.3	(Liu et al., 2011)
Aspergillus spp. MAM-F35	Wheat straw	SSF	487	(Hammad et al., 2010)
Fomitopsis sp. RCK2010	Wheat bran	SSF	84	(Deswal et al., 2011)
A. niger USM AI	Sugarcane bagasse and palm kernel cake	SSF	3.2	(Lee et al., 2011)
Penicillium citrinum	Rice bran	SSF	2	(Dutta et al., 2008)
Penicillium decumbens	Wheat straw	SSF	52.8	(Mo et al., 2004)
Penicillium echinulatum 9A02S1	Sugarcane bagasse and wheat bran	SSF	282.4	(Camassola and Dillon, 2007)
Trichoderma reesei	Sugarcane bagasse and palm kernel cake	SSF	2.2	(Lee et al., 2011)
T. reesei LW1	Corn straw and wheat bran	SSF	452.5	(Wang et al., 2005)
T. reesei NRRL11460	Sugarcane bagasse	SSF	154.6	(Singhania et al., 2007)
T. reesei LM-UC4	Bagasse	SSF	38.6	(Duenas et al., 1995)
Aspergillus phoenicis QM 329		SSF		
Trichoderma viride	Wheat straw	SSF	555	(Pothiraj et al., 2006)
Thermoascus aurantiacus	Wheat straw	SSF	1572	(Kalogeris et al., 2003)
Aspergillus heteromorphus	Wheat straw	SmF	83	(Singh et al., 2009)
A. flavus BS1	CMC	SmF	2793	(Sajith et al., 2016)
A. fumigatus	Wheat straw	SmF	321	(Saqib et al., 2010)
A. niger	CMC	SmF	1.6	(Narasimha et al., 2006)
A. niger	Maize straw	SmF	102	(Milala et al., 2005)
A. niger	Banana peel	SmF	12.4	(Duenas et al., 1995)
A. niger	Coir waste	SmF	3.3	(Mrudula and Murugammal, 2011)
A. niger	Sawdust	SmF	3.9	(Devi and Kumar, 2012)
A. niger MS82	Grass	SmF	6.8	(Sohail et al., 2009)
Penicillium fellutanum	Lactose	SmF	81	(Kathiresan and Manivannan, 2010)
Rhizopus oryzae	Water hyacinth	SmF	450	(Karmakar and Ray, 2011)
Trichoderma harzianum	CMC	SmF	150	(Rubeena et al., 2013)
T. harzianum MTCC 8230	Rice straw	SmF	1.7	(Gautam et al., 2010)
T. reesei RUT C30	Sugarcane baggase	SmF	121	(da Silva Delabona et al., 2012)
T. reesei ZU-02	Corn cob residue	SmF	5.5	(Liming and Xueliang, 2004)
T. viride	CMC	SmF	173	(Neethu et al., 2012)

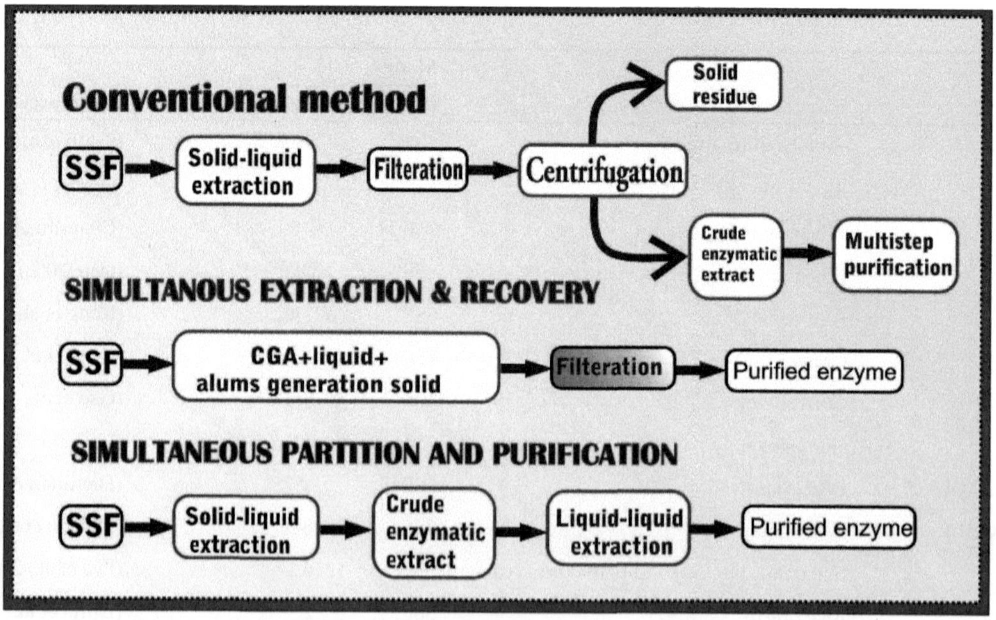

FIG. 12.5 (A) Conventional, (B) solvent extraction, and (C) aqueous two-phase system (Abu Yazid et al., 2017).

SSF mode. For instance, Marín et al. (2018) used a solvent:solid ratio of 1:5 to achieve maximum extraction in the static condition in SSF mode. In many cases, a crude enzyme produced during SSF mode has been used for bio-processing of bioproducts like hydrogen, which is a conventional method as shown in Fig. 12.5A to obtain purified enzyme by the simultaneous extraction and recovery method.

Szabo et al. (2015) found that sonication improved the recovery of enzyme yields by more than twice from SSF mode. Extraction followed by diafiltration removes inhibitory phenolic compounds from extracts, thus further improving recovery (Rezaei et al., 2011). Solid/liquid extraction has been reported to improve the recovery of cellulase by 74% (Poletto et al., 2015).

Another extraction method using aphron flotation in SSF mode is shown in Fig. 12.5. In another single step enzyme recovery process, application of aqueous two-phase extraction (ATPE) after SSF mode fermentation (Fig. 12.5) is better than other existing procedures (Bhavsar et al., 2012).

There have been many literature reviews of the application of cellulase in various extractions. One such attempt was made using cellulase-based extraction of the components cited in Li et al. (2012) where cellulase enzyme-assisted extractions of the naphthodianthrones hypericin and pseudohypericin from *Hypericum perforatum* L. were carried out. In addition, cellulase is finding use in various other applications such as single-cell protein production, ethanol production (Dahnum et al., 2015), which can be utilized in biofuel production, rice bran oil extraction (Sharma et al., 2001), and pigment extraction (Ghosh and Biswas, 2016).

However, the cellulase produced is often low in activity and has low loading capacity, whereas cellulase is required to be highly active and with a high loading capacity of at least 95%.

12.6.1 Cellulase Recovery

Cellulase recovery is quite challenging because of extracellular secretion (Bentil et al., 2018) from the hyphal structure. High growth is linked with cellulase recovery and there is also speculation of mixing with substrate to make recovery difficult. Normally, it is assumed that recovery of endoglucanase is quite difficult but some reports disapprove of this belief and report the recovery of endoglucanase (Valášková and Baldrian, 2006). However, recovery of CBHs was quite difficult. The author also discussed that β-glucosidase is mostly attached to cell walls, thus its isolation is also difficult. In another report it was stated that in liquid medium, recovery of β-glucosidase is not very difficult. Other causes may be lack of optimized media. Media condition mostly activates the enzyme-related gene system. In media conditions, some factors such as surfactant play a vital role. In one article the author worked on the role of surfactant and reported that out of acetone, ethanol, and methanol, acetone was the best recovery solvent (Yuan et al., 2014). In a fed-batch production system it was reported that regular recovery of cellulase produced from *Trichodema reesei* increased its production and activity up to 58% while increase in the total harvested cellulase and β-glucosidase was 24%–62% and 22%–76% respectively (Wang et al., 2017). However, in some cases the aqueous two-phase system has been reported to be suitable for the safe recovery of enzymes. It is economical due to use of the polyethylene glycol (PEG)/salt system (Salmon et al., 2014). For enhanced recovery of up to 99%, use of affinity precipitation of cellulase using pH-response polymer with Cibacron Blue F3GA has been found to be useful. After elution, total cellulase activity was 84.4%, while endoglucanase activity was 99.8% (Ding and Cao, 2013). In some cases, some workers noticed that high activity with high recovery was obtained when pretreatment of lignocellulosic biomass was done in alkaline mode (Damisa and Ameh, 2008). Enzyme recovery with membrane filtration gives around 74% recovery (Mores et al., 2001). Use of microaffinity ligand in three-phase partitioning after ammonium sulfate precipitation led to 96% recovery of enzymes (Sharma et al., 2003). However, in some bioprocessing steps there is no need for recovery of enzymes and direct conversion is performed by using consolidated bioprocessing (CBP) (Fan, 2014). CBP is related to direct conversion of lignocellulosic biomass to glucose with the help of microbes. Sometimes consortia of microbes release all the enzymes required for complete hydrolysis. Therefore there is no point in recovery of enzymes since enzymes are used in hydrolysis. However, for commercial purposes, high recovery with purity is required. Also, it has been observed that CBP enzyme activity never drops too low and also the rate of conversion is quite high. This is the ultimate technique for biofuel production in which only a few steps are involved, and also since this is purly biological there is no pollution generated.

The type of pretreatment decides the actual amount of cellulase released, for example, acidic pretreatments hydrolyze mostly hemicellulose, leaving the cellulose and lignin undigested. The benefit is that pure cellulase is induced because of the presence of cellulose, but in addition the presence of lignin may act as an inhibitor and thus a high yield is expected (Agbor et al., 2014). Organic solvent pretreatments solubilize mostly hemicellulose and lignin, thus pure cellulose is more readily available as compared to cellulose pretreated with acidic or basic pretreatment, thus a better result is expected.

Some rumen bacteria such as *Ruminococcus albus*, *Ruminococcus avefaciens*, and *Fibrobacter succinogens* have cellulosome systems and are reported to be activated during cellulase secretion extracellularly.

12.6.2 e-CBP Versus r-CBP

CBP is modern bioprocessing technique in consolidated mode where microbes, substrate, and media are enclosed in a single reactor. In this regard, e-CBP refers to engineered CBP techniques and r-CBP refers to use of rumen bacteria for CBP. Complete success of this technique is single-step conversion of lignocellulose. As discussed at several place in this article, activation of cellulase depends, on various factors such as type of pretreatment method done, media conditions, pH, temperature, and mode of action such as SSF and SmF. e-CBP is more powerful as compared to a combination of rumen microbes-based CBP (Lin et al., 2015).

12.6.3 Microbial Consortia in Cellulase Production

There is increasing research into the screening of proper microbial consortia. Certain microbial consortia have been explored that secrete huge amounts of cellulase (Tantayotai et al., 2017). Microbial consortia can be divided into either (1) natural, (2) synthetic, or (3) engineered.

Microbial consortia were isolated from natural samples, including horse manure and decomposed wood, and their cellulase activities were characterized as endo-β-glucanase (0.417 and 0.434 U/mg), exo-β-glucanase (0.116 and 0.184 U/mg), and β-glucosidase (1.069 and 3.184 U/mg), which resulted in improved biogas yields of up to 109.60 and 161.49 mL/g-VS.

Engineered microbial consortia (glucose-methanol-choline—GMC) (Brenner et al., 2008) is an innovative way to enhance cellulase production. GMC has the benefit that it can secrete groups of enzymes called cellulosomes, which act on cellulose to hydrolyze it into glucose. The reason behind this may be a supportive growth factor for each other and suitable environment in which they can act faster. One such example is modifying two *Bacillius subtilis* bacteria, one with a scaffold and the other with endoglucanase and xylanase, and thus enzyme activity is enhanced in a co-culture because of quorum sensing in which microbes communicate with each other (Arai et al., 2007; Amore et al., 2013).

12.7 CONCLUSION AND FUTURE PROSPECTS

Liu et al. worked on the price evaluation of cellulase production in two modes: SSF and SmF (Liu et al., 2015). The price of cellulase was \$15.67 kg cellulase^{-1} in SSF mode and the price of cellulase was \$40.36 kg cellulase^{-1} in SmF mode, while the corresponding market price was over \$90.00/kg.

Economical ethanol production using sweet sorghum bagasse was estimated, based on phosphoric acid-catalyzed pretreatment in (L + SScF) mode using an engineered *Escherichia coli* strain (van et al., 2018). The workers concluded that ethanol and gasoline prices could be lowered by decreasing the cost of cellulolytic enzymes or the development of more value-added products from lignin. Further use of the economical substrate for cellulase production can further lower the cost of production; one such example is sugarcane bagasse.

Sugarcane (*Saccharum* spp.) processing produces waste such as sugarcane bagasse, which can be used as a lignocellulosic feedstock for cellulase production.

Finally, using a simple downstream method such as a vacuum drum filter (well suited to fungal biomass separation) may also contribute to reduced fungal enzyme cost.

ATPS may further be effectively used for separation and purification of cellulase enzymes (Brumbauer et al., 1999; Ülger and Salam, 2001; Kim et al., 2004; IKE et al., 2013). Silvério et al. (2013) worked on the ATPS system using a thermoseparating polymer for separation of laccase enzyme and reported loss of activity of around 88% as compared with the classical PEG/salt systems. In addition, to achieve high purity, a PEG/salt system is a better option.

Acknowledgments

The author acknowledges the support of all those who directly or indirectly supported him during article writing, especially Dr. J.S. Bal VC, SBBS University, Jalandhar, Dr. V. Dhir, SBBS University, Dr. Neha (BHU), Dr. Pradeep Mishra (HOD, Chemical Engineering Department, BHU), my wife Dr. Shuchi (Principal, Sarup School, Jalandhar), and my son and daughter (Master Anshman Bhatt and Tejaswini Bhatt).

References

Abu Yazid, N., Barrena, R., Komilis, D., Sánchez, A., 2017. Solid-state fermentation as a novel paradigm for organic waste valorization: a review. Sustainability 9, 224. https://doi.org/10.3390/su9020224.

Agbor, V., Carere, C., Cicek, N., et al., 2014. Biomass pretreatment for consolidated bioprocessing (CBP). In: Advances in Biorefineries: Biomass and Waste Supply Chain Exploitation, pp. 234–258.

Ahmed, I., Zia, M.A., Muhammad, H., Iqbal, N., 2010. Bioprocessing of proximally analyzed wheat straw for enhanced cellulase production through process optimization with *Trichoderma viride* under SSF. World Acad. Sci. Eng. Technol. 37, 1208–1214.

Alhammad, A., Adewale, P., Kuttiraja, M., Christopher, L.P., 2018. Enhancing enzyme-aided production of fermentable sugars from poplar pulp in the presence of non-ionic surfactants. Bioprocess Biosyst. Eng. https://doi.org/10.1007/s00449-018-1942-z.

Amore, A., Giacobbe, S., Faraco, V., 2013. Consolidated bioprocessing for improving cellulosic ethanol production. In: Lignocellulose Conversion: Enzymatic and Microbial Tools for Bioethanol Production. pp. 169–196.

Annamalai, N., Rajeswari, M.V., Sivakumar, N., 2016. Cellobiohydrolases: role, mechanism, and recent developments. In: Microbial Enzymes in Bioconversions of Biomass, pp. 29–36.

Arai, T., Matsuoka, S., Cho, H.-Y., et al., 2007. Synthesis of *Clostridium cellulovorans* minicellulosomes by intercellular complementation. Proc. Natl. Acad. Sci. 104, 1456–1460. https://doi.org/10.1073/pnas.0610740104.

Baeyens, J., Kang, Q., Appels, L., et al., 2015. Challenges and opportunities in improving the production of bio-ethanol. Prog. Energy Combust. Sci. 47, 60–88.

Bentil, J.A., Thygesen, A., Mensah, M., et al., 2018. Cellulase production by white-rot basidiomycetous fungi: solid-state versus submerged cultivation. Appl. Microbiol. Biotechnol. 1–13.

Bhatt, S.M., Shilpa, 2015. Lignocellulosic feedstock conversion, inhibitor detoxification and cellulosic hydrolysis—a review. Biofuels, 1–17.

Bhavsar, K., Kumar, V.R., Khire, J.M., 2012. Downstream processing of extracellular phytase from *Aspergillus niger*: chromatography process vs. aqueous two phase extraction for its simultaneous partitioning and purification. Process Biochem. 47, 1066–1072.

Borisova, A.S., Eneyskaya, E.V., Jana, S., et al., 2018. Correlation of structure, function and protein dynamics in GH7 cellobiohydrolases from *Trichoderma atroviride*, *T. reesei* and *T. harzianum*. Biotechnol. Biofuels 11: 5. https://doi.org/10.1186/s13068-017-1006-7.

Brady, S.K., Sreelatha, S., Feng, Y., et al., 2015. Cellobiohydrolase 1 from *Trichoderma reesei* degrades cellulose in single cellobiose steps. Nat. Commun. 6: 10149. https://doi.org/10.1038/ncomms10149.

Brenner, K., You, L., Arnold, F.H., 2008. Engineering microbial consortia: a new frontier in synthetic biology. Trends Biotechnol. 26, 483–489.

Brumbauer, A., Johansson, G., Réczey, K., 1999. Fractionation of cellulase and β-glucosidase in a *Trichoderma reesei* culture liquid by use of two-phase partitioning. Bioseparation 7, 287–295. https://doi.org/10.1023/A:1008153418019.

Bu, L., Xing, Y., Yu, H., et al., 2012. Comparative study of sulfite pretreatments for robust enzymatic saccharification of corn cob residue. Biotechnol. Biofuels 5, 87. https://doi.org/10.1186/1754-6834-5-87.

Bukhtojarov, F.E., Ustinov, B.B., Salanovich, T.N., et al., 2004. Cellulase complex of the fungus *Chrysosporium lucknowense*: isolation and characterization of endoglucanases and cellobiohydrolases. Biochemist 69, 542–551. https://doi.org/10.1023/B:BIRY.0000029853.34093.13.

Callow, N.V., Ray, C.S., Kelbly, M.A., Ju, L.K., 2016. Nutrient control for stationary phase cellulase production in *Trichoderma reesei* Rut C-30. Enzym. Microb. Technol. 82, 8–14. https://doi.org/10.1016/j.enzmictec.2015.08.012.

Callow, N.V., Ray, C.S., M a, K., Ju, L.-K., 2015. Nutrient control for stationary phase cellulase production in *Trichoderma reesei* Rut C-30 nutrient control for stationary phase cellulase production in *Trichoderma reesei* Rut C-30. Enzym. Microb. Technol. Microb. Technol. 82, 8–14. https://doi.org/10.1016/j.enzmictec.2015.08.012.

Camassola, M., Dillon, A.J.P., 2007. Production of cellulases and hemicellulases by *Penicillium echinulatum* grown on pretreated sugar cane bagasse and wheat bran in solid-state fermentation. J. Appl. Microbiol. 103, 2196–2204.

Chandra, M.S., Viswanath, B., Reddy, B.R., 2007. Cellulolytic enzymes on lignocellulosic substrates in solid state fermentation by *Aspergillus niger*. Indian J. Microbiol. 47, 323–328.

Chen, Y., Shen, Y., Wang, W., Wei, D., 2018. Mn^{2+} modulates the expression of cellulase genes in *Trichoderma reesei* Rut-C30 via calcium signaling. Biotechnol. Biofuels 11, https://doi.org/10.1186/s13068-018-1055-6.

da Silva Delabona, P., Farinas, C.S., da Silva, M.R., et al., 2012. Use of a new *Trichoderma harzianum* strain isolated from the Amazon rainforest with pretreated sugar cane bagasse for on-site cellulase production. Bioresour. Technol. 107, 517–521.

Dahnum, D., Tasum, S.O., Triwahyuni, E., et al., 2015. Comparison of SHF and SSF processes using enzyme and dry yeast for optimization of bioethanol production from empty fruit bunch. In: Energy Procedia, pp. 107–116.

Damisa, D., Ameh, U.V.J., 2008. Effect of chemical pretreatment of some lignocellulosic wastes on the recovery of cellulase from *Aspergillus niger* AH3 mutant. Afr. J. Biotechnol. 7, 2444–2450.

de França Passos D, Pereira N Jr, de Castro AM A comparative review of recent advances in cellulases production by *Aspergillus*, *Penicillium* and *Trichoderma* strains and their use for lignocellulose deconstruction. Curr. Opin. Green Sustain. Chem. doi: https://doi.org/10.1016/j.cogsc.2018.06.003

Deshavath, N.N., Sahoo, S.K., Panda, M.M., et al., 2018. The cost-effective stirred tank reactor for cellulase production from alkaline-pretreated agriculture waste biomass. In: Utilization and Management of Bioresources. Springer, Singapore, pp. 25–35.

Deswal, D., Khasa, Y.P., Kuhad, R.C., 2011. Optimization of cellulase production by a brown rot fungus *Fomitopsis* sp. RCK2010 under solid state fermentation. Bioresour. Technol. 102, 6065–6072. https://doi.org/10.1016/j.biortech.2011.03.032.

Devi, M.C., Kumar, M.S., 2012. Production, optimization and partial purification of cellulase by *Aspergillus niger* fermented with paper and timber sawmill industrial wastes. J. Microbiol. Biotechnol. Res. 2, 120–128.

Ding, Z., Cao, X., 2013. Affinity precipitation of cellulase using pH-response polymer with Cibacron Blue F3GA. Sep. Purif. Technol. 102, 136–141. https://doi.org/10.1016/j.seppur.2012.10.001.

Duenas, R., Tengerdy, R.P., Gutierrez-Correa, M., 1995. Cellulase production by mixed fungi in solid-substrate fermentation of bagasse. World J. Microbiol. Biotechnol. 11, 333–337.

Dutta, T., Sahoo, R., Sengupta, R., et al., 2008. Novel cellulases from an extremophilic filamentous fungi *Penicillium citrinum*: production and characterization. J. Ind. Microbiol. Biotechnol. 35, 275–282.

Fan, Z., 2014. Consolidated bioprocessing for ethanol production. In: Biorefineries: Integrated Biochemical Processes for Liquid Biofuels, pp. 141–160.

Gao, J., Weng, H., Xi, Y., et al., 2008. Purification and characterization of a novel endo-β-1,4-glucanase from the thermoacidophilic *Aspergillus terreus*. Biotechnol. Lett. 30, 323–327.

Gautam, S.P., Bundela, P.S., Pandey, A.K., et al., 2010. Optimization of the medium for the production of cellulase by the *Trichoderma viride* using submerged fermentation. Int. J. Environ. Sci. 1, 656–665.

Ghosh, D., Biswas, P.K., 2016. Enzyme-aided extraction of carotenoids from pumpkin tissues. Indian Chem. Eng. 58, 1–11. https://doi.org/10.1080/00194506.2015.1046697.

Hammad, A.I., Swelim, M., Gannam, R.B., 2010. Enhanced production of cellulase (S) by *Aspergillus* spp. isolated from agriculture wastes by solid state fermentation. Am. J. Agric. Environ. Sci. 8, 402–410.

Himmel, M.E., Adney, W.S., Baker, J.O., et al., 2018. Cellulases: structure, function, and applications. In: Handbook on Bioethanol. Routledge, pp. 143–161.

Hirano, K., Kurosaki, M., Nihei, S., et al., 2016. Enzymatic diversity of the *Clostridium thermocellum* cellulosome is crucial for the degradation of crystalline cellulose and plant biomass. Sci. Rep. 6, https://doi.org/10.1038/srep35709.

IKE, M., PARK, J., TABUSE, M., TOKUYASU, K., 2013. Controlled preparation of cellulases with xylanolytic enzymes from *Trichoderma reesei* (*Hypocrea jecorina*) by continuous-feed cultivation using soluble sugars. Biosci. Biotechnol. Biochem. 77, 161–166. https://doi.org/10.1271/bbb.120696.

Imran, M., Anwar, Z., Irshad, M., et al., 2017. Optimization of cellulase production from a novel strain of *Aspergillus tubingensis* IMMIS2 through response surface methodology. Biocatal. Agric. Biotechnol. 12, 191–198. https://doi.org/10.1016/j.bcab.2017.10.005.

Infinite Enzymes, 2018. Exocellulase CBH II – Purified Enzyme | Cellulase Enzyme.

Isikgor, F.H., Becer, C.R., 2015. Lignocellulosic biomass. Polym. Chem.

Jecu, L., 2000. Solid state fermentation of agricultural wastes for endoglucanase production. Ind. Crop. Prod. 11, 1–5.

Kalogeris, E., Christakopoulos, P., Katapodis, P., et al., 2003. Production and characterization of cellulolytic enzymes from the thermophilic fungus *Thermoascus aurantiacus* under solid state cultivation of agricultural wastes. Process Biochem. 38, 1099–1104.

Kang, S.W., Park, Y.S., Lee, J.S., et al., 2004. Production of cellulases and hemicellulases by *Aspergillus niger* KK2 from lignocellulosic biomass. Bioresour. Technol. 91, 153–156.

Karmakar, M., Ray, R.r., 2011. Current trends in research and application of microbial cellulases. J. Microbiol. 6, 41–53. https://doi.org/10.3923/jm.2011.41.53.

Kathiresan, K., Manivannan, S., 2010. Cellulase production by Penicillium fellutanum isolated from coastal mangrove rhizosphere soil. Res. J. Microbiol. 5, 1160–1164.

Kim, Y., Park, K., Lee, H., et al., 2004. Purification of native and modified enzymes using a reactive aqueous two-phase system. J. Ind. Eng. Chem. 10, 384–388.

Kumar, A.K., Sharma, S., 2017. Recent updates on different methods of pretreatment of lignocellulosic feedstocks: a review. Bioresour. Bioprocess. 4: 7. https://doi.org/10.1186/s40643-017-0137-9.

Kumar, R., Tabatabaei, M., Karimi, K., Sárvári Horváth, I., 2016. Recent updates on lignocellulosic biomass derived ethanol—a review. Biofuel Res. J. 3, 347–356. https://doi.org/10.18331/BRJ2016.3.1.4.

Laser, M., Jin, H., Jayawardhana, K., Lynd, L.R., 2009. Coproduction of ethanol and power from switchgrass. Biofuels Bioprod. Biorefin. 3, 195–218.

Lee, C.K., Darah, I., Ibrahim, C.O., 2011. Production and optimization of cellulase enzyme using *Aspergillus niger* USM AI 1 and comparison with *Trichoderma reesei* via solid state fermentation system. Biotechnol. Res. Int. 2011.

Li, Y., Zhang, X., Xiong, L., et al., 2017. On-site cellulase production and efficient saccharification of corn stover employing cbh2 overexpressing *Trichoderma reesei* with novel induction system. Bioresour. Technol. 238, 643–649.

Li, Z., Wang, X., Shi, G., et al., 2012. Enzyme-assisted extraction of naphthodianthrones from *Hypericum perforatum* L. by 12C6+-ion beam-improved cellulases. Sep. Purif. Technol. 86, 234–241. https://doi.org/10.1016/j.seppur.2011.11.009.

Liao, H., Zhang, X.Z., Rollin, J.A., Zhang, Y.H.P., 2011. A minimal set of bacterial cellulases for consolidated bioprocessing of lignocellulose. Biotechnol. J. 6, 1409–1418. https://doi.org/10.1002/biot.201100157.

Liming, X., Xueliang, S., 2004. High-yield cellulase production by *Trichoderma reesei* ZU-02 on corn cob residue. Bioresour. Technol. 91, 259–262.

Lin, P.P., Mi, L., Morioka, A.H., et al., 2015. Consolidated bioprocessing of cellulose to isobutanol using *Clostridium thermocellum*. Metab. Eng. 31, 44–52. https://doi.org/10.1016/j.ymben.2015.07.001.

Liu, D., Zhang, R., Yang, X., et al., 2011. Thermostable cellulase production of *Aspergillus fumigatus* Z5 under solid-state fermentation and its application in degradation of agricultural wastes. Int. Biodeterior. Biodegrad. 65, 717–725.

Liu, Y., Xu, J., Zhang, Y., et al., 2015. Sequential bioethanol and biogas production from sugarcane bagasse based on high solids fed-batch SSF. Energy 90, 1199–1205. https://doi.org/10.1016/j.energy.2015.06.066.

Liu, Z., Inokuma, K., Ho, S.-H., et al., 2017. Improvement of ethanol production from crystalline cellulose via optimizing cellulase ratios in cellulolytic *Saccharomyces cerevisiae*. Biotechnol. Bioeng. 114, 1201–1207.

Lynd, L.R., Weimer, P.J., van Zyl, W.H., Pretorius, I.S., 2002. Microbial cellulose utilization: fundamentals and biotechnology. Microbiol. Mol. Biol. Rev. 66, 506–577. table of contents, https://doi.org/10.1128/MMBR.66.3.506-577.2002.

Mabee, W.E., McFarlane, P.N., Saddler, J.N., 2011. Biomass availability for lignocellulosic ethanol production. Biomass Bioenergy 35, 4519–4529. https://doi.org/10.1016/j.biombioe.2011.06.026.

Maftukhah, S., Abdullah, A., 2018. Cellulase Enzyme production from rice straw using solid state fermentation and fungi *Aspergillus niger* ITBCC L74. MATEC Web Conf. 156. 01010. https://doi.org/10.1051/matecconf/201815601010.

Marín, M., Artola, A., Sánchez, A., 2018. Optimization of down-stream for cellulases produced under solid-state fermentation of coffee husk. In: Waste and Biomass Valorization, pp. 1–12.

Mekala, N.K., Singhania, R.R., Sukumaran, R.K., Pandey, A., 2008. Cellulase production under solid-state fermentation by *Trichoderma reesei* RUT C30: statistical optimization of process parameters. Appl. Biochem. Biotechnol. 151, 122–131. https://doi.org/10.1007/s12010-008-8156-9.

Milala, M.A., Shugaba, A., Gidado, A., et al., 2005. Studies on the use of agricultural wastes for cellulase enzyme production by *Aspergillus niger*. Res. J. Agric. Biol. Sci. 1, 325–328.

Mo, H., Zhang, X., Li, Z., 2004. Control of gas phase for enhanced cellulase production by *Penicillium decumbens* in solid-state culture. Process Biochem. 39, 1293–1297.

Momeni, M.H., Goedegebuur, F., Hansson, H., et al., 2014. Expression, crystal structure and cellulase activity of the thermostable cellobiohydrolase Cel7A from the fungus *Humicola grisea* var. *thermoidea*. Acta Crystallogr. Sect. D Biol. Crystallogr. 70, 2356–2366. https://doi.org/10.1107/S1399004714013844.

Moosavi-Nasab, M., Majdi-Nasab, M., 2008. Cellulase production by *Trichoderma reesei* using sugar beet pulp. Iran Agric. Res. 25, 107–116. https://doi.org/10.1016/j.jbiotec.2012.01.001.

Mores, W.D., Knutsen, J.S., Davis, R.H., 2001. Cellulase recovery via membrane filtration. In: Applied Biochemistry and Biotechnology—Part A Enzyme Engineering and Biotechnology, pp. 297–309.

Morikawa, Y., Ohashi, T., Mantani, O., Okada, H., 1995. Cellulase induction by lactose in *Trichoderma reesei* PC-3-7. Appl. Microbiol. Biotechnol. 44, 106–111. https://doi.org/10.1007/BF00164488.

Mrudula, S., Murugammal, R., 2011. Production of cellulase by *Aspergillus niger* under submerged and solid state fermentation using coir waste as a substrate. Braz. J. Microbiol. 42, 1119–1127.

Muniswaran, P.K.A., Charyulu, N.C.L.N., 1994. Solid substrate fermentation of coconut coir pith for cellulase production. Enzym. Microb. Technol. 16, 436–440. https://doi.org/10.1016/0141-0229(94)90161-9.

Muthuvelayudham, R., Viruthagiri, T., 2010. Application of central composite design based response surface methodology in parameter optimization and on cellulase production using agricultural waste. Int. J. Chem. Biol. Eng. 97–104.

Narasimha, G., Sridevi, A., Buddolla, V., et al., 2006. Nutrient effects on production of cellulolytic enzymes by *Aspergillus niger*. Afr. J. Biotechnol. 5, 472–476.

Ncube, T., Howard, R.L., Abotsi, E.K., et al., 2012. *Jatropha curcas* seed cake as substrate for production of xylanase and cellulase by *Aspergillus niger* FGSCA733 in solid-state fermentation. Ind. Crop. Prod. 37, 118–123. https://doi.org/10.1016/j.indcrop.2011.11.024.

Neethu, K., Rubeena, M., Sajith, S., et al., 2012. A novel strain of *Trichoderma viride* shows complete lignocellulolytic activities. Adv. Biosci. Biotechnol. 3, 1160.

Poletto, P., Borsói, C., Zeni, M., da, S.M.M., 2015. Downstream processing of pectinase produced by *Aspergillus niger* in solid state cultivation and its application to fruit juices clarification. Food Sci. Technol. 35, 391–397.

Pothiraj, C., Balaji, P., Eyini, M., 2006. Enhanced production of cellulases by various fungal cultures in solid state fermentation of cassava waste. Afr. J. Biotechnol. 5.

Rastogi, G., Bhalla, A., Adhikari, A., et al., 2010. Characterization of thermostable cellulases produced by *Bacillus* and *Geobacillus* strains. Bioresour. Technol. 101, 8798–8806. https://doi.org/10.1016/j.biortech.2010.06.001.

Rathnan, R.K.A.M., 2011. Cellulase enzyme production by streptomyces sp using fruit waste as substrate. Aust. J. Basic Appl. Sci. 5, 1114–1118.

Reczey, K., Szengyel, Z., Eklund, R., Zacchi, G., 1996. Cellulase production by *T. reesei*. Bioresour. Technol. 57, 25–30. https://doi.org/10.1016/0960-8524(96)00038-7.

Rezaei, F., Joh, L.D., Kashima, H., et al., 2011. Selection of conditions for cellulase and xylanase extraction from switchgrass colonized by *Acidothermus cellulolyticus*. Appl. Biochem. Biotechnol. 164, 793–803.

Rezania, S., Din, M.F.M., Mohamad, S.E., et al., 2017. Review on pretreatment methods and ethanol production from cellulosic water hyacinth. Bioresources 12, 2108–2124.

Rubeena, M., Neethu, K., Sajith, S., et al., 2013. Lignocellulolytic activities of a novel strain of *Trichoderma harzianum*. Adv. Biosci. Biotechnol. 4, 214.

Sajith, S., Priji, P., Sreedevi, S., Benjamin, S., 2016. An overview on fungal cellulases with an industrial perspective. J. Nutr. Food Sci. 6, 1–13. https://doi.org/10.4172/2155-9600.1000461.

Salmon, D.N.X., Walter, A., Porto, T.S., et al., 2014. Aqueous two-phase extraction for partial purification of phytase produced under solid-state fermentation. Biocatal. Biotransform. 32, 45–52. https://doi.org/10.3109/10242422.2013.872633.

Saqib, A.A., Hassan, M., Khan, N.F., Baig, S., 2010. Thermostability of crude endoglucanase from *Aspergillus fumigatus* grown under solid state fermentation (SSF) and submerged fermentation (SmF). Process Biochem. 45, 641–646.

Saravanan, P., Muthuvelayudham, R., Viruthagiri, T., 2013. Enhanced production of cellulase from pineapple waste by response surface methodology. J. Eng. (United States) 2013, https://doi.org/10.1155/2013/979547.

Sasmal, S., Mohanty, K., 2018. Pretreatment of lignocellulosic biomass toward biofuel production. In: Biorefining of Biomass to Biofuels. Springer, Cham, pp. 203–221.

Shajahan, S., Moorthy, I.G., Sivakumar, N., Selvakumar, G., 2017. Statistical modeling and optimization of cellulase production by *Bacillus licheniformis* NCIM 5556 isolated from the hot spring, Maharashtra, India. J. King Saud. Univ. Sci. 29, 302–310. https://doi.org/10.1016/j.jksus.2016.08.001.

Sharma, A., Khare, S.K., Gupta, M.N., 2001. Enzyme-assisted aqueous extraction of rice bran oil. JAOCS 78, 949–951.

Sharma, A., Mondal, K., Gupta, M.N., 2003. Separation of enzymes by sequential macroaffinity ligand-facilitated three-phase partitioning. J. Chromatogr. A 995, 127–134. https://doi.org/10.1016/S0021-9673(03)00522-3.

Sherief, A.A., El-Tanash, A.B., Atia, N., 2010. Cellulase production by *Aspergillus fumigatus* grown on mixed substrate of rice straw and wheat bran. Res. J. Microbiol. 5, 199–211.

Shin, K., Kim, Y.H., Jeya, M., et al., 2010. Purification and characterization of a thermostable cellobiohydrolase from *Fomitopsis pinicola*. J. Microbiol. Biotechnol. 20, 1681–1688. https://doi.org/10.4014/jmb.1008.08009.

Shirkavand, E., Baroutian, S., Gapes, D.J., Young, B.R., 2016. Combination of fungal and physicochemical processes for lignocellulosic biomass pretreatment—a review. Renew. Sust. Energ. Rev. 54, 217–234.

Silvério, S.C., Rodríguez, O., Teixeira, J.A., Macedo, E.A., 2013. The effect of salts on the liquid–liquid phase equilibria of PEG600+ salt aqueous two-phase systems. J. Chem. Eng. Data 58, 3528–3535.

Sindhu, R., Kuttiraja, M., Binod, P., et al., 2011. Dilute acid pretreatment and enzymatic saccharification of sugarcane tops for bioethanol production. Bioresour. Technol. 102, 10915–10921. https://doi.org/10.1016/j.biortech.2011.09.066.

Singh, R., Kumar, R., Bishnoi, K., Bishnoi, N.R., 2009. Optimization of synergistic parameters for thermostable cellulase activity of *Aspergillus heteromorphus* using response surface methodology. Biochem. Eng. J. 48, 28–35.

Singhania, R.R., Sukumaran, R.K., Pandey, A., 2007. Improved cellulase production by *Trichoderma reesei* RUT C30 under SSF through process optimization. Appl. Biochem. Biotechnol. 142, 60–70. https://doi.org/10.1007/s12010-007-0019-2.

Sohail, M., Siddiqi, R., Ahmad, A., Khan, S.A., 2009. Cellulase production from *Aspergillus niger* MS82: effect of temperature and pH. New Biotechnol. 25, 437–441.

Sreena, C.P., Sebastian, D., 2018. Augmented cellulase production by *Bacillus subtilis* strain MU S1 using different statistical experimental designs. J. Genet. Eng. Biotechnol. 16, 9–16. https://doi.org/10.1016/j.jgeb.2017.12.005.

Sugumaran, K.R., Chakravarthi, S.P., Ponnusami, V., 2013. Cassava bagasse: a potential and low cost substrate for cellulase production in an economical fermentation. Res. J. Pharm., Biol. Chem. Sci. 4, 1168–1175.

Sukumaran, R.K., Singhania, R.R., Mathew, G.M., Pandey, A., 2009. Cellulase production using biomass feed stock and its application in lignocellulose saccharification for bio-ethanol production. Renew. Energy 34, 421–424.

Sun, F., Mukasekuru, M.R., Tan, L., et al., 2018. Optimization of on-site cellulase preparation for efficient hydrolysis of atmospheric glycerol organosolv pretreated wheat straw. J. Chem. Technol. Biotechnol.

Sun, H.-Y., Li, J., Zhao, P., Peng, M., 2011. Banana peel: a novel substrate for cellulase production under solid-state fermentation. Afr. J. Biotechnol. 10, 17887–17890. https://doi.org/10.5897/AJB10.1825.

Szabo, O.E., Csiszar, E., Koczka, B., Kiss, K., 2015. Ultrasonically assisted single stage and multiple extraction of enzymes produced by *Aspergillus oryzae* on a lignocellulosic substrate with solid-state fermentation. Biomass Bioenergy 75, 161–169.

Tabssum, F., Irfan, M., Shakir, H.A., Qazi, J.I., 2018. RSM based optimization of nutritional conditions for cellulase mediated Saccharification by *Bacillus cereus*. J. Biol. Eng. 12, https://doi.org/10.1186/s13036-018-0097-4.

Tang, Z., Li, P., Sun, R., et al., 2018. Optimum mixture and synergy analysis of three main cellulases. Int. J. Agric. Biol. 20, 255–262. https://doi.org/10.17957/IJAB/15.0485.

Tantayotai, P., Pornwongthong, P., Muenmuang, C., et al., 2017. Effect of cellulase-producing microbial consortium on biogas production from lignocellulosic biomass. Energy Procedia 141, 180–183. https://doi.org/10.1016/j.egypro.2017.11.034.

Tariq, R., Ansari, I., Qadir, F., et al., 2018. Optimization of endoglucanase production from thermophilic strain of *Bacillus licheniformis* RT-17 and its application for saccharification of sugarcane bagasse. Pak. J. Bot. 50, 807–816.

Taylor, L.E., Knott, B.C., Baker, J.O., et al., 2018. Engineering enhanced cellobiohydrolase activity. Nat. Commun. 9: 1186. https://doi.org/10.1038/s41467-018-03501-8.

Tuohy, M.G., Walsh, D.J., Murray, P.G., et al., 2002. Kinetic parameters and mode of action of the cellobiohydrolases produced by *Talaromyces emersonii*. Biochim. Biophys. Acta Protein Struct. Mol. Enzymol. 1596, 366–380. https://doi.org/10.1016/S0167-4838(01)00308-9.

Ülger, C., Salam, N., 2001. Partitioning of industrial cellulase in aqueous two-phase systems from trichoderma viride QM9414. Process Biochem. 36, 1075–1080. https://doi.org/10.1016/S0032-9592(01)00144-3.

Vaishnav, N., Singh, A., Adsul, M., et al., 2018. Penicillium: the next emerging champion for cellulase production. Bioresour. Technol. Rep. 2, 131–140. https://doi.org/10.1016/j.biteb.2018.04.003.

Valášková, V., Baldrian, P., 2006. Estimation of bound and free fractions of lignocellulose-degrading enzymes of wood-rotting fungi *Pleurotus ostreatus*, *Trametes versicolor* and *Piptoporus betulinus*. Res. Microbiol. 157, 119–124.

van, R.R., Nieves, I.U., Shanmugam, K.T., et al., 2018. Techno-economic evaluation of cellulosic ethanol production based on pilot biorefinery data: a case study of sweet sorghum bagasse processed via L+SScF. BioEnergy Res 11, 414–425. https://doi.org/10.1007/s12155-018-9906-3.

Vyas, A., Vyas, D., Vyas, K.M., 2005. Production and optimization of cellulases on pretreated groundnut shell by *Aspergillus terreus* AV49. J. Sci. Ind. Res. (India) 64, 281–286.

Wang, B., Jiang, L., Bai, H., et al., 2017. Regular enzyme recovery enhances cellulase production by *Trichoderma reesei* in fed-batch culture. Biotechnol. Lett. 39, 1493–1498. https://doi.org/10.1007/s10529-017-2379-1.

Wang, G., Zhang, X., Wang, L., et al., 2012. The activity and kinetic properties of cellulases in substrates containing metal ions and acid radicals. Adv. Biol. Chem. 2, 390–395.

Wang, J.S., Wang, J., Gulfraz, M., 2005. Efficient cellulase production from corn straw by *Trichoderma reesei* LW1 through solid state fermentation process. Ethnobot. Leafl. 2005, 7.

Wati, L., Kumari, S., Kundu, B.S., 2007. Paddy straw as substrate for ethanol production. Indian J. Microbiol. 47, 26–29. https://doi.org/10.1007/s12088-007-0005-y.

Wyman, C.E., 2013. Aqueous pretreatment of plant biomass for biological and chemical conversion to fuels and chemicals.

Xiao, C., Anderson, C.T., 2013. Roles of pectin in biomass yield and processing for biofuels. Front. Plant Sci. 4, https://doi.org/10.3389/fpls.2013.00067.

Yang, Y., Du, J., Zuo, R., et al., 2018. Synergistic effect of cell surface-displayed xylanase on cellulase during the hydrolysis of lignocellulosic substrates. Chinese J. Appl. Environ. Biol. 24, 362–366. https://doi.org/10.19675/j.cnki.1006-687x.2017.05041.

Yuan, X.Z., Peng, X., Huang, H.J., et al., 2014. Precipitation and recovery of cellulase using biosurfactant. Sep. Sci. Technol. 49, 2249–2254. https://doi.org/10.1080/01496395.2014.910525.

Znameroski, E.A., Coradetti, S.T., Roche, C.M., et al., 2012. Induction of lignocellulose-degrading enzymes in Neurospora crassa by cellodextrins. Proc. Natl. Acad. Sci. 109, 6012–6017.

Zoglowek, M., Lübeck, P.S., Ahring, B.K., Lübeck, M., 2015. Heterologous expression of cellobiohydrolases in filamentous fungi: an update on the current challenges, achievements and perspectives. Process Biochem. 50, 211–220. https://doi.org/10.1016/j.procbio.2014.12.018.

CHAPTER

13

Genetic Engineering Applications to Improve Cellulase Production and Efficiency: Part I

Enosh Phillips

Department of Biotechnology, St. Aloysius College (Autonomous), Jabalpur, India

13.1 INTRODUCTION

Cellulose, a biomass produced by the process of photosynthesis, is an organic polymer that is produced in huge quantities and is a never-ending source of raw material for a number of processes in the paper industry, food industry, textile industry, etc. (Sukumaran et al., 2005). Cellulose is synthesized by bacteria, fungi, turnicates, and higher plants. It is synthesized by cellulose synthase, which is a conserved enzyme in all organisms with different amino acid sequences, indicating the production of cellulose under diverse conditions. Cellulose is made up of polymerizing glucan chains having a β-1,4-linkage. Cellulose is a crystalline biopolymer that contains inter- and intrahydrogen bonds and has van der Waals attraction. The *CesA* gene is responsible for the synthesis of cellulose in higher plants. Bacterial and plant *CesA* gene products show no similarity (Li et al., 2014). Cellulose is present in nature in basically two forms: cellulose I and cellulose II, which are called polymorphs. However, two other polymorphs have been detected designated cellulose III and cellulose IV. All these polymorphs are interconvertible forms on chemical treatment (Lavanya et al., 2011).

The primary and natural source of cellulose is predominantly wood, which contains a large amount of lignocellulose. Other sources include agricultural residues, grasses, water plants, etc. This cellulose in its natural form comes in combination with other materials. However, the development of sophisticated purification techniques like that of chromatography led to the isolation of a highly purified form of this protein. There have been approaches to synthesize cellulose chemically by ring-opening polymerization of 3,6-di-*O*-benzyl-α-D-glucopyranose 1,2,4-orthopivalate by Nakatsubo et al., and by β-D-glucose as 1-allyl-2,6-di-*O*-acetyl-3-benzyl-4-*O*-(*p*-methoxybenzyl)-D-glucopyranoside by Klemn et al. (2005).

Cellulose is basically present in complex form in wild plants; it is found to be implanted in the matrixes of other biopolymers like hemicelluloses and lignin. Being the leading agrowaste, a strategy to hydrolyze it with the help of microbes is a promising way to produce desired metabolites or biofuel. Although lignocellulosic biomass is ideal for microbial degradation, pretreatment for disrupting the structure of lignin increases the susceptibility of cellulose to cellulases for biodegradation (Sukumaran et al., 2005).

13.1.1 Isolation and Purification of Cellulose

Biomass from which cellulose is to be extracted is collected in the required quantity depending on the final product to be formed. It is then sun dried and treated with NaOH (alkaline treatment is carried out to remove waxes, resins, and pectins). After alkaline treatment the biomass is washed in water and cooled, shredded, and dried. Cellulose can then be extracted from the biomass by a pulping technique that is completed in four steps:

- Acid hydrolysis with 0.4% H_2SO_4 for an hour and then washing.
- Chlorination with NaClO (3.5%) in a water bath and then washing it with distilled water until neutral pH is achieved.
- Alkaline extraction with 20% NaOH.
- Bleaching with 0.5% NaClO and again washing to neutral pH.

From Cellulose to Cellulase: Strategies to Improve Biofuel Production
https://doi.org/10.1016/B978-0-444-64223-3.00013-8

(if lipid is present in the sample, then the sample is added to toluene and subjected to Soxhlet extraction prior to the aforementioned four steps (Hu et al., 2017)). The treated biomass is shredded and placed in an aluminum pan for 24h at 60°C. Fourier transform infrared spectroscopy is then used for chemical characterization of the sample (Mora'n et al., 2008).

Cellulose so produced needs to be completely degraded by a group of enzymes, i.e., endoglucanases, exoglucanases, and β-glucocsidases. Together they are known as cellulosome. They work in a synchronous manner to bring about the degradation of cellulose. Exoglucanases find crystalline cellulose as their substrate and cleave its reducing and nonreducing ends, i.e., cleaving externally, whereas endoglucanases attack the amorphous cellulose bringing about its degradation. β-Glucosidases, on the other hand, cleave down the cellobiose produced in due time into glucose.

It is now evident that cellulase is produced by a number of microbes such as bacteria and fungi; however, for industrial enzyme production, filamentous fungi have been the preference. This is because the filamentous fungi produce a large quantity of fungi in industrial conditions as compared to any other microbe. Among all the known fungi, *Aspergillus* spp. and *Trichoderma* spp. have been used the most. Conventionally in industries, submerged fermentation (SmF) is used for the production of enzymes, but solid-state fermentation (SSF) has shown its fruitfulness in the process. Moreover, the filamentous fungi used in production develops better in SSF than in SmF, since they require less water. Although there are no such comparative studies to show the benefit of one technique over the other, the fermentation technique used for enzyme production depends purely on cost effectiveness and culture development (Mrudula and Murugammal, 2011; Zhang and Zhang, 2013).

Current strategies involved in this are not economically encouraging. Among the various options available, genetic engineering is becoming the best way to view the various shortcomings of the conventional strategies, which is discussed later in the chapter.

13.2 MICROBIAL CELLULASE PRODUCTION

The substrates used for cellulase enzyme production are generally corn cob, rice straw, or sugarcane bagasse because they are a rich and good source of cellulose. Raw material is dried at 70°C overnight and is then pretreated with acid (0.1N HCl) for an hour and then washed and neutralized. *Trichoderma reesei* is the commonly used fungus for enzyme production, but with the efforts of researchers worldwide new strains of *T. reesei* and new fungi and bacteria are being isolated for enhanced functionalized cellulase. Fungal culture is developed over potato agar slant and is subcultured. For inocula preparation, Tween 80 is added so that the spores are dislodged in the liquid media, which is then used as an inoculum (Sukumaran et al., 2009).

SSF is one of the cheapest methods of enzyme production. There is evidence that SSF gives a conducive environment for fungal enzyme production (Imran et al., 2016). Enzyme production by SSF has been known for quite some time and is used a lot in the case of fungal enzyme production. With the upgradation of microbiology/biotechnology, SSF has attracted a number of researchers. SSF has been used to make agrofeed such as animal feed more digestible for animals. SSF can be defined as a system having a liquid phase attached to a solid matrix and gas entrapped within these particles. Interest in SSF comes from its relationship with microbes' natural living conditions. Since less water is required, its removal at the end of the process is minimized and an approach for manipulation of system working becomes easier. SSF provides less mechanical sheering as compared to other fermenters, which is good for the cultivation of shear-sensitive fungal mycelia. SSF also provides the benefit of low-cost recovery, and is therefore economically the best option (Raibault, 1998; Singh et al., 1999).

Mineral salt medium is now used to moisten the pretreated substrate in SSF and then inoculum is prepared and added to the mixture, which is incubated for 96h at 30°C for cellulase production. Cellulase is extracted in 0.1N citrate buffer and centrifuged at 10,000rpm for 10min at 4°C as crude sample. It is generally concentrated in ice-cold acetone and resuspended in 50mM citrate buffer for hydrolysis (Sukumaran et al., 2009) (Fig. 13.1).

Presently, for increasing the production of enzymes by traditional methods, optimization of the process is taken into account. A number of optimizations take place starting with the substrate itself:

- Substrate is pretreated in alkaline conditions. NaOH is generally used to maintain the alkaline conditions. Substrate is incubated in alkaline conditions for approximately 12h and then this pretreated substrate is used in fermentation media as the carbon source for the growing fungi.
- Optimization is also performed by taking substrate in different concentrations. It has been reported that cellulase is yielded in high quantity at a particular concentration. It bypasses the idea that the quantity of

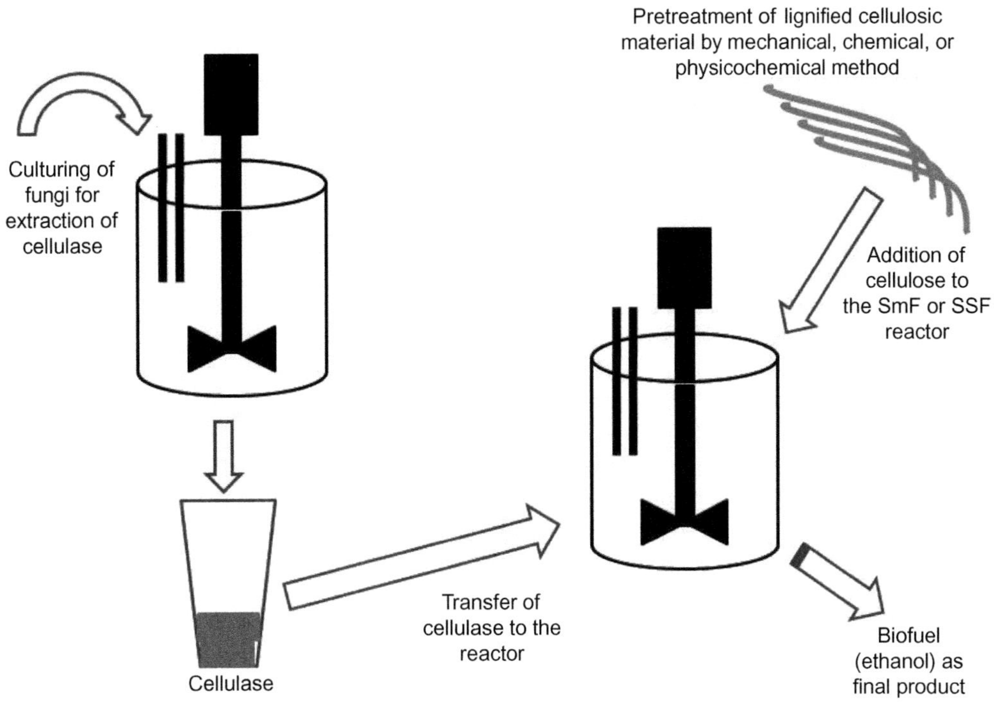

FIG. 13.1 A generalized depiction of how cellulase is produced and its use for the production of ethanol. The process starts with the culturing of fungi to produce cellulase enzyme and then this cellulase is transferred to another reactor where pretreated cellulose is present. This reaction mixture produces glucose, which is used for making ethanol.

cellulase produced is proportional to the substrate concentration. This means that increasing the concentration of substrate does not affect enzyme production on the same scale; rather an optimum concentration is required of the substrate in which the fungi produce cellulase at a high level.

- pH of the media also affects the production of enzyme. It is quite evident from the works of researchers so far that cellulase activity is seen over a pH range of 3.0–9.0, but most of the fungi, primarily *Aspergillus* spp. and *Trichoderma* spp., show maximum cellulase activity in the range of pH 4.0–4.5, i.e., in the acidic range. Therefore during the process, pH is optimized around 4–5 to get the maximum yield.
- Optimization of temperature also helps in increasing the production of cellulase. The optimal temperature for fungi is between 28°C and 30°C. The production process is kept around this temperature to achieve a good yield.
- The amount of inoculum added affects the production of cellulase. It has been seen that by increasing the inoculum load, less enzyme is produced, whereas by decreasing the amount of inoculum the production of enzyme increases. Therefore care is always taken to add less inoculum to get maximum yield.
- The time required for incubation in fermentation greatly affects the quantity of enzyme produced. It is mostly seen that the maximum activity for most species is on the fourth day of fermentation. Therefore fermentation is carried out for 96h to achieve a good yield.
- The nitrogen source used also affects the production process. Peptone, urea, and NH_4SO_4 are commonly used as nitrogen sources. Although the result depends on the species used in the process, every species has its own choice of nitrogen source depending on its metabolic activities. So, efforts are taken to understand which nitrogen source can be utilized for a process and also what quantity will give the maximum output (Sukumaran et al., 2009; Raibault, 1998; Singh et al., 1999; Acharya et al., 2008a).

13.2.1 Analysis of Cellulase Activity

A number of ways are available to analyze the activity of cellulase:

1. *Filter paper-based technique*: This is a simple technique in which cellulase enzyme activity is assayed by means of glucose released from the filter paper. The reaction mixture contains just two components, i.e., Whatman filter paper (WFP) and culture filtrate. In one of the processes about 50mg of WFP is incubated with 0.1mL of culture filtrate and the glucose released during the incubation due to the action of cellulase over cellulose

is measured by the glucose oxidase-peroxidase method (Liming and Xeuliang, 2003). In this, H_2O_2 formed by the oxidation of glucose by glucose oxidase is used to oxidize *o*-tolidine by peroxidase that produces a blue color. The color so produced is estimated using a glucose concentration colorimeter (Kingsley and Getchell, 1960). The activity is also assayed by taking the culture filtrate and reaction mixture and incubating for 1h and thereafter measuring the presence of reducing sugar by the dinitrosalysilic acid method (Liming and Xeuliang, 2003).

2. *Agar plate-based method*: In this method, agar (1%) mixed with 0.2% carboxymethyl cellulose (CMC) is poured into a Petri dish. The cellulase enzyme dissolved in 1× TBS buffer is plated over the solidified plate and incubated at 27°C for 12–16h. The zone of hydrolysis is visualized by means of Gram's iodine. In some of the methods, Congo red is used instead of Gram's iodine. Some researchers omit CMC with Gelrite. Plates containing no CMC are taken as nonsubstrate control (Johnsen and Krause, 2014).
3. *CMC plate-based method*: This is a current method of analyzing cellulase activity, which overcomes the drawbacks of the agar-based method. This method utilizes the ability of CMC to form a gel-like surface. About 7% of CMC is prepared at 70°C, poured into a Petri dish, and left overnight to be polymerized. Cellulase is then plated over this overnight plate and incubated at 27°C for 12–16h. The hydrolysis zone was examined by Gram's iodine. Another way is to pour about 200mL of CMC into a 96 well titer plate and allow it to polymerize; 20mL of cellulase is added and incubated at 27°C for 12–16h. Hydrolysis is visualized by Gram's iodine and then washed with ethanol. Absorption at 575nm is measured after staining (Johnsen and Krause, 2014).

13.3 CHALLENGES IN ENZYME PRODUCTION

Microbial cellulases have been promising in degrading cellulose into desired products like biofuel. However, the traditional way (using wild strains for cellulase production) of production is not cost effective. The present scenario for the production of cellulases is not economically feasible. Two of the major concerns of present-day enzyme production are:

- Economic feasibility of enzyme production.
- Making enzymes efficient so that fewer enzymes are required for the conversion of a large amount of substrate.

The economic feasibility of cellulase production is presently taken care of by the phenomenon known as the process of optimization, such as altering physical and chemical parameters, namely temperature, pH, etc., that enhance microbial growth and hence increase the production of cellulase enzymes. However, these attempts have not encouraged increased production. The efficiency and increased expression of enzymes can be made by means of protein engineering and knowledge of microbial genetics, because it is now evident that DNA holds the key for coding every protein, and enzymes are protein. Therefore any chance of manipulating the DNA code can help in eradicating the shortcomings of conventional cellulase production. Hence genetic engineering has become a handy tool in solving these vital issues (Sukumaran et al., 2005).

13.4 GENETIC ENGINEERING: A TOOL FOR ENZYME PRODUCTION

DNA is the hereditary material in almost all organisms, from complex eukaryotic humans to the simplest prokaryotic bacteria. Apart from being located in the cell nucleus it is also found in mitochondria inside a cell. Looking at DNA we find that it is primarily made up of just four bases: adenine (A), guanine (G), cytosine (C), and thymine (T), where G makes a triple hydrogen bond with C and A makes a double hydrogen bonded with T. The entire information is coded in just these four bases. In a similar way to where alphabets structure words that give rise to sentences, the sequence of DNA helps in structuring and maintaining an organism. It is this DNA that controls the activity of an organism by coding proteins by the intermediate processes of transcription and translation. It is these proteins that act in several ways in structuring and maintaining an organism. A way in which it does this is by acting like an enzyme, which brings about a biochemical reaction (Albert, 2002; Druzhinina and Kubicek, 2017).

Therefore increasing the efficiency of an enzyme or increasing its expression depends on how much one is able to understand and work on these sequences of DNA that could be a tool to eradicate much of the problems of protein expression.

13.4.1 Structure of Genes

The functional unit of DNA that contains the code for the synthesis of protein is the gene. A gene consists of many regions that have different functions from each other during the expression of genes. Broadly, there are two main regions:

- *Promoter*: the region that initiates the transcription of genes targeted for expression.
- *Coding region*: the region that codes for functional genes.

The promoter region dictates the point of deciding whether a gene has to be expressed or not in a particular cell and when, i.e., under inducing conditions. It contains the structural element, a specific nucleotide sequence that undertakes this task. These sequences are called *cis*-acting sequences because they reside in the same molecule of DNA as that of the gene. Sometimes these *cis*-acting elements are also known as enhancers but are far away from the gene. These *cis*-acting elements bind to the transcriptional machinery, the proteins involved in the initiation of transcription, and thus regulate the expression of the gene under appropriate conditions. The coding region contains the sequences that code for the correct amino acids required to make a functional protein. Apart from this structure the gene is found to be associated with a number of accessory proteins, which form a structure known as chromatin. These proteins also affect the expression of a gene by virtue of holding tightly to the gene to silent it or by loosening the gene from its association to make the expression possible (Polyak, 2003).

13.4.2 Genetic engineering

Genetic engineering has become a powerful tool for placing multiple desirable traits in an organism. This practice can be used to increase enzyme production and imply its cost effectiveness on the one hand and on the other hand to engineer such an enzyme that is not only efficient but can adapt to conditions involved in a particular process. Since enzymes are protein and protein is directly coded from gene(s), we can say that altering the genetics of the organism will give the desired product, here in the form of an enzyme. Such alteration is also termed protein engineering (Singh et al., 2017). The search for a microorganism that has the inbuilt above-stated capacity, from varied environmental conditions, is a time-consuming process.

Genetic engineering includes recombinant DNA technology (rDNA), mutation, protoplast fusion, etc. for manipulation of a gene. This may look quite simple but requires a large amount of knowledge and expertise because of the genomic complexity of the organism. An understanding of the whole genome sequence makes things simpler when deciding on the target to be manipulated (Ray and Rosell, 2013). The steps involved in genetic engineering are (Fig. 13.2):

- Isolation of DNA fragment from donor organism.
- Insertion of isolated fragment into an appropriate vector system.

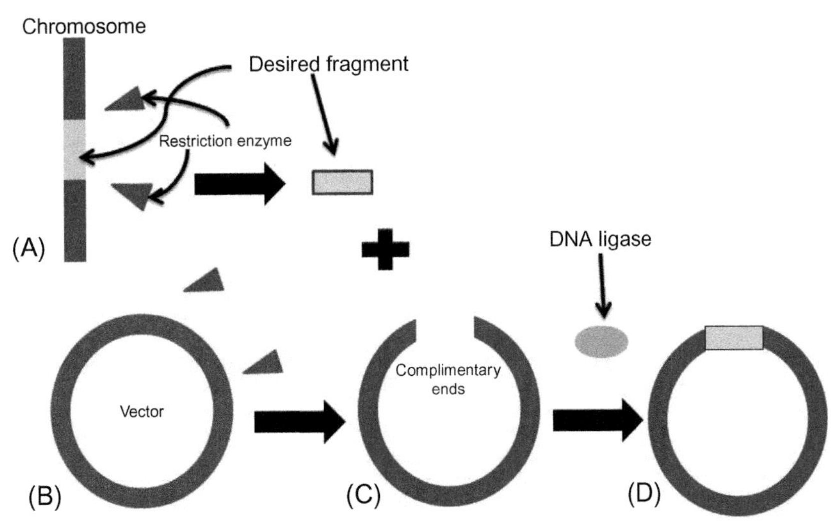

FIG. 13.2 A generalized view of the steps involved in genetic engineering. (A) Using the specific restriction enzyme (RE) to cut the desired fragment containing the gene of interest from the chromosome. (B) Taking an appropriate vector by cutting the same RE to create complementary ends. (C). Mixing the desired gene (DG) fragment and RE-digested vector with ligase enzyme. (D). Joining ligase with the DG and the vector, which is then multiplied and expressed in the appropriate host.

- Growth of the vector in the host.
- Expression of gene and purification of product.

The use of rDNA technology is already in action in many biotechnology industries involved in enzyme/protein production. Engineering a gene enables the design of such an enzyme that is quite apt for a process. In most industrial processes where enzymes are used, the process has to be optimized to such an extent to make the enzyme do its work that the cost of product production increases or yield is decreased if optimization fails. Manipulation helps in a way to make the enzyme optimized for product formation, therefore increasing the efficiency and fidelity of the enzyme (Nichol, 2008).

As in the case of cellulase production from fungi, the challenge for genetic alteration becomes an uphill task due to the presence of introns in the genes and complexities of glycosylation, which generates fewer transformants by any of the methods (Singh et al., 2017; Greenway, 1920). Before going into details of genetic manipulation, let us first look at the cellulase expression system, and then we can look at how manipulation can be performed.

Site-directed mutagenesis is another way by which the sequence within a gene can be changed by will and thus cause production of a mutant protein product with a differential function, which may or may not enhance protein activity. It is an in vitro method for inculcating mutations in the known or given sequence. In this a single-stranded oligonucleotide is taken, which is complementary to the host DNA. This oligonucleotide contains certain changed sequences that may bring about a mutation in the host DNA when it is inserted. The mutation may be a point mutation or deletion or addition. It becomes targeted in nature due to complementarity. There are several ways to achieve site-directed mutagenesis. One way is where a mutated primer is annealed to a klenow fragment and this arrangement is added to DNA POL I with deoxynucleotide triphosphates to construct the double-stranded DNA. DNA ligase is present to join the two ends at the end of polymerization. This heteroduplex is made to transfect *Escherichia coli*, giving rise to transformed cells. If the frequency of the mutants in the culture is more than 50%, then sequencing the clones can be used to identify the mutants. However, if it is less, then radioactive hybridization is a better way of identifying (Carter, 1986; Bachman, 2013). There are many other ways to embed a mutation in a given sequence. It is a priceless tool that helps in modifying a gene to understand the structural and functional properties of the protein. The properties are based on the structure, sequence of amino acids, function, and catalytic mechanism (Yang et al., 2017; Cavicchioli et al., 2006). It is used to insert a single amino acid or many amino acids, which are different to the wild-type protein or even the entire structural feature. All these make it a useful tool for the future of genetic engineering.

13.5 CELLULASE EXPRESSION SYSTEM

The most studied organism for understanding the expression of cellulosome (because cellulase is a multienzyme complex it is also referred to as cellulosome) is *T. reesei*. The cellulosome of *T. reesei* is composed of the following components:

- **Cellobiohydrolase (CBH)**, which is an exo-acting enzyme that breaks down crystalline cellulose into cellobiose.
- **Endoglucanase**, which acts on soluble cellulose and is an endo-acting cellulase.
- **β-Glucosidase**, which converts celloligosaccharide and disaccharide into glucose, which is then used by the fungi as a C source.

A number of experiments suggest that *T. reesei* has the genes listed in Table 13.1 that code for different components of cellulosome (Ilmén et al., 1997) (Fig. 13.3).

Like every gene that has an external inducer for its activation, cellulase is also expressed only in the presence of cellulose, a complex plant material, but production drops in the presence of other carbon sources such as glucose,

TABLE 13.1 Different Genes Involved in Expressing Different Components of Cellulosome (Druzhinina and Kubicek, 2017)

Enzyme part of cellulosome	Gene expressed from
CBH	*cbh1* and *cbh2*
EG	*egl1*, *egl2*, *egl3*, and *egl5*
β-Glucosidase	*bgl1* and *bgl2*

CBH, cellobiohydrolase; *EG*, endoglucanase.

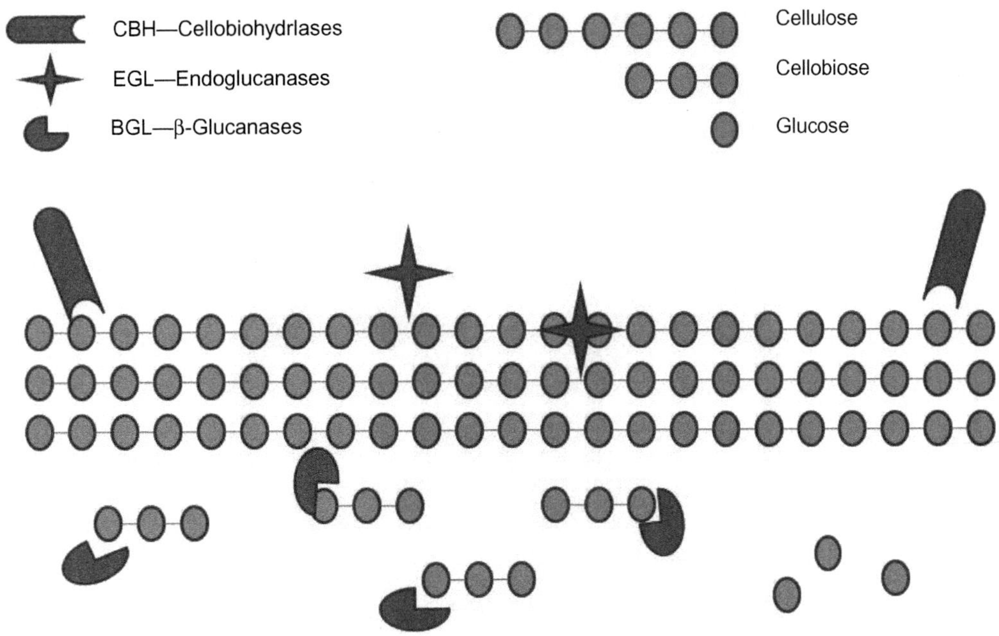

FIG. 13.3 Cellulase is a combination of enzymes that belong to the glycosyl hydrolase family. The cellobiohydrolase breaks down the cellulose from the end areas, whereas the endoglucanase breaks down cellulose from the middle region. The action of both of these produces cellobiose, which is acted on by the third enzyme of cellulase, i.e., β-glucanase, which breaks down cellobiose into glucose. This glucose is then used for various industrial purposes.

fructose, and glycerol. Inducers of cellulase formation are transglycolated. The *cbh* gene is expressed by the presence of cellobiose (a disaccharide containing carbohydrate), whereas its expression reduces in the presence of glucose (the end product of cellulolytic hydrolysis of cellulose) indicating cellulase is regulated by feedback inhibition. *Cbh2* and *egl2* gene products are the ones that give rise to these inducers, but neither of them transglycolate it. In an experiment, it is been said that the deletion of *cbh1* and *cbh2* generates no inducer for *egl1* and *egl2* genes and the fungi is unable to grow, especially when an inducer analog called sophorose is added, then *egl1* and *egl2* start accumulating, which suggests that *egl1* and *egl2* are regulated by the products formed by the expression of the *cbh* gene. It is a probability now that the inducer formed by *cbh2* and *egl2* is transglycolated by other enzymes. *Cbh1* has also shown glycosylating activity, which makes it a candidate for transglycolation of the inducer. Therefore we can say that the *cbh1/2* gene product is essential for cellulase formation, whereas *cbh2* and *egl2* are essential for the development of inducers. However, all this remains a hypothesis and much work is yet to be done (Ilmén et al., 1997; Seiboth et al., 1997).

13.5.1 Why Only in the Presence of Cellulose?

An interesting fact is that *T. reesei* has fewer cellulase and hemicellulase genes as compared to other genes, which raises the question: in spite of its fewer genes, why does cellulase have such a high level of secretion? This was answered by Nakari et al. who said that these genes are under the control of carbon catabolite repression (CCR) (CCR allows an organism to be specific for its substrate choice as a carbon source), which is regulated by a gene known as *cre1*. In a work conducted by Nakari Setala et al. *cre1* was replaced by a mutant subtype designated as *cre1-1*. Due to this truncated gene there was a difference in the colony morphology of the mutant fungus as compared with the wild type and it showed increased enzyme production. The studies thus reported that the genes coding cellulases are under the modulation of *cre1*, under both inducing and noninducing conditions, and explains how expression of cellulases increases in the presence of celluloses compared to any other carbohydrate(s). Therefore *cre1* can be the target to be engineered in a way that increases enzyme production (Raibault, 1998) (Fig. 13.4).

13.5.2 Degradation of Cellulose

The following is a short discussion of how the cellulose present in the biomass is degraded by the cellulososme. It is now known that cellulase is a mixture of enzymes, which act in a stepwise manner to covert cellulose into glucose, cellobiose, and cellooligosaccharide, which are used for various product formations. But how is this accomplished?

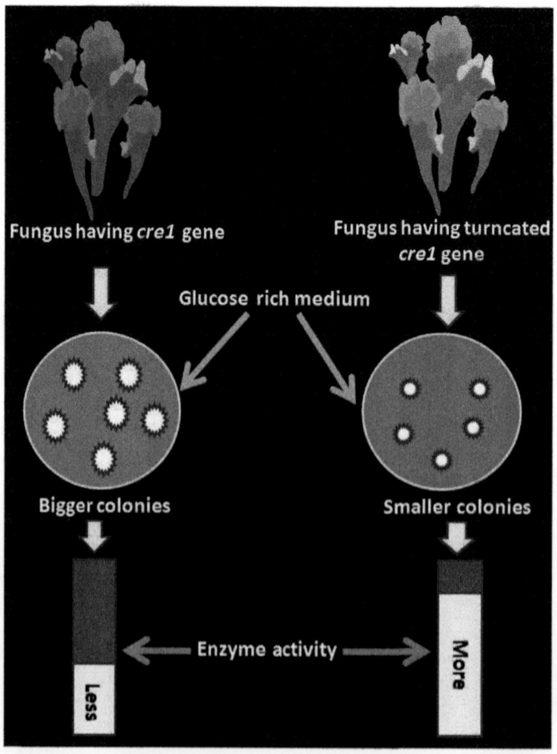

FIG. 13.4 A diagrammatic representation of the experiment carried on the role of the *cre1* gene for cellulase production. It can be clearly seen that in the absence of *cre1*, enzyme activity increases due to increase in production of the cellulases.

The cellulosome acts in perfect synergism to degrade cellulose. The enzyme degrades the β-1,4-linkages present in the cellulose structure. The endoglucanases come into action first by breaking the polymer from the center and produce reducing and nonreducing ends that become the substrate for CBHs. CBHs act on these reducing and nonreducing ends to convert them into cellooligosaccharides and cellobiose. Cellobiose is then acted on by β-glucosidases to be digested into glucose. Crystalline cellulose is also degraded by lytic polysaccharide monooxygenase, which is nonhydrolytic in nature and brings about a nick over the polymer that helps endoglucanases to produce more reducing and nonreducing ends, thus accelerating the process (Singh et al., 2017). Cellulases are secreted out of the cell.

Sequence analysis of the cellulase system has revealed that this enzyme system is composed of many domains. These domains are found to be homologous in related species and have the same function (Beguin and Aubert, 1994).

13.6 MICROBIAL GENETIC MODIFICATION FOR CELLULASE PRODUCTION

Since much of the work has been done on fungi, here we will discuss a number of ways by which it can be understood how different strains can be developed to increase the production of cellulase.

13.6.1 Disruption of *cre1*

In a work on *Acremonium cellulolyticus*, the role of the transcription factor *cre1*, discussed earlier, has been worked out. *Cre1* regulates the expression of *cbh1* by binding to two close motifs having a sequence 5′-CCCCAC-3′ in the promoter region in *Hypocrea jecorina*. *Cbh2* is also indirectly regulated by *cre1* by controlling the expression of *XYR1* (Mach-Aigner et al., 2008), since it is now known that the expression of cellulases is under the tight regulation of *cre1*. The expression of *cre1* depends on the presence and absence of cellulose. Moreover, *cre1* is not expressed when glucose is present; glucose is now the end product of cellulose degradation by cellulases, which means that there is an inhibition of the *cre1* expression by the end product of that enzyme's activity whose expression *cre1* controls. Due to this fact the production of cellulases is widely affected.

Therefore in an approach to solve this problem an attempt was made by Fuji et al. in which a plasmid was constructed containing the 5′ and 3′ sequence of the *cre1* into the upstream and downstream region of the *pyrF* gene (this is a gene that codes an enzyme known as orotidine-5′-monophosphate decarboxylase, which is involved in pyrimidine biosynthesis) in pbs-pyrF. This plasmid was then incorporated in *A. cellulolyticus*. This set-up caused homologous recombination with the *cre1* gene present in the host chromosome and converted the fungi into a transformant that contained the disrupted *cre1* gene and hence avoided its expression. On measuring the activity of FPase (total cellulase activity) in disruptant and nondisruptant, the disruptant showed more activity comparatively, which indicates that when *cre1* is disrupted the production of cellulase increases. On performing reverse transcription polymerase chain reaction (RT-PCR) for examining the gene expression the disruptant showed higher gene expression of cellulosome than the nondisruptant, which tells us the reason for increased activity of FPase in the disruptant. Thus on removing or disrupting the essential *cre1* transcription factor the production of cellulases can be increased (Fujii et al., 2013, 2014; Turnbough et al., 1987).

In another experiment, an RNA interference (RNAi) technique was used to see the effect of silencing the *cre1* gene in a thermophilic fungus. The *cre1* sequence is taken and a small interfering RNA (siRNA) was made based on the sequence of *cre1*. This siRNA sequence was then loaded onto the *pUC-19 pdc* plasmid containing the Amp^r. The transformant was prepared by making the protoplast and inserting the plasmid into it. On the successful silencing of *cre1* by RNAi the *cre1* silenced strain gave higher hydrolyzing activity than that of the strain in which *cre1* was not silenced but kept active. In the quantitative estimation by RT-PCR the *cre1*-silenced strains showed higher expression of *egl* and *cbh* genes. Moreover, matrix-assisted laser desorption/ionization/time of flight analysis of the protein secreted in the medium was also performed and it was seen that endoglucanases and CBHs were produced in a different manner in the transformants (Yang et al., 2015; Acharya et al., 2008b). Both these works clearly suggest that on silencing or disrupting the *cre1* gene, which acts as a carbon catabolite repressor, production of cellulase enzyme was increased.

13.6.2 By Altering the Promoter

Enzyme production can be enhanced by either inactivating the gene or changing the promoter. Cellulases are under the tight regulation of many elements, including promoters. The promoter of *cbh1* is a highly inducible promoter. The hypercellulolytic fungus *T. reesei* has been utilized for the production of different components of cellulases by the use of *cbh1* promoter. For instance, pAMH110 vector is used in this purpose for the production of endoglucanases. Therefore altering the promoter can be used for the production of a single type of cellulase (Singh et al., 2017).

The regulation of cellulase expression by feedback inhibition, influenced by the presence of glucose and cellobiose, and the end products of cellulolytic degradation of cellulose are a hindrance in cellulase production. Cellobiose is a major inhibitor of *cbh* and *egl*. β-Glucosidase is produced in very low quantities in *T. reesei* because cellobiose is not converted into glucose and therefore inhibits cellulase production and hence the degradation of cellulose. In an approach to solve this issue, Rahman et al. showed that when *egl3* and *xyn1* promoters are used to express the *bgl1* gene encoding β-glucosidase, the synthesis of this enzyme is increased many fold. The transformed strain overexpressed the *bgl1* gene. This also confirms that *egl3* and *xyn1* genes are expressed in higher quantities than *bgl1* in the wild-type *T.reesei* strain, explaining the reason behind accumulation of cellobiose in *Trichoderma* (Rahman et al., 2014).

Engineering of the binding sites of transcription regulators is also being done to increase the yield. Zou et al. changed the *cre1* binding site for transcription activators within the *cbh1* promoter with *ACE2* and *HAP3/HAP2/HAP5*. This increased the expression of cellulase up to sevenfold. Zou said that the *cbh1* promoter is under carbon catabolite repressor *cre1* regulation, especially in the presence of glucose. Promoter efficiency could be increased under the shadow of CCR if the binding sites of *cre1* can be replaced by the sequence that binds the activator factors. Therefore the idea was to replace the repressor binding site with the activator binding site, which could increase the expression of the cellulase gene. Using site-directed mutagenesis they changed the binding sites of the *cbh1* promoter. The three *cre1* motifs were changed, one with *ACE2* sequence and the other two with *HAP3/2/5* protein-binding sites. *ACE2* binds to a particular site in the *cbh1* promoter and positively regulates the expression of cellulases. It binds to the 5′-GGCTAA-3′ sequence over the promoter. Deletion of *ACE2* showed a decrease in the expression of cellulase. *HAP3/HAP2/HAP5*, which is a protein complex, has been found to regulate the *cbh2* promoter by binding to its CAAT box. To compare the efficiency of mutant promoter and wild-type promoter, enhanced green fluorescent protein (*egfp*) was attached to each promoter. When the wild and mutated strains were grown on inducing culture, both showed good fluorescence, but due to the presence of cellulose in the medium, which inhibited the expression of *cre1* in the wild type, they also showed cellulase expression. However, when both strains were cultured in a repressing medium, the one that contained the glucose, the expression of *egfp* in the wild type was retarded due to the action of *cre1* over the *cbh1* promoter because it contained the binding site for *cre1*. When fluorescence was checked in the mutant strain it showed good fluorescence because it did not contain the binding site for *cre1* on the *cbh1* promoter; rather it had sequences for the binding of activators *ACE2* and *HAP3/2/5* that kept the promoter activated even in the presence of the repressor. This move toward increasing cellulase expression, however, is beneficial only when a few genes are targeted for its overexpression and not the entire cellulosome (Zou et al., 2012).

13.6.3 Genetically Altering the Regulatory Pathway

In one of the more interesting works it has been proposed that knocking out certain regulatory genes could be a good way of increasing cellulase production. For instance, cellobiose, which is a substrate for *bgl2* gene product, converts cellobiose into glucose. Now, cellobiose is also being used by another important transcription factor, *ClrB*, to activate the expression of cellulases. However, the expression of cellulases is inhibited by glucose produced by *bgl2* mediated by *cre1*. The deletion of *bgl2* demonstrated the accumulation of cellobiose and therefore increased cellulase expression because cellobiose is an essential inducer of the cellulase system. So, if *bgl2* and *cre1* are knocked out, then cellobiose can be directly used by *ClrB* to stimulate the synthesis of cellulases. Another study has looked into increasing the expression of *ClrB*. This is achieved by integrating the *ClrB* gene with the promoter *gpdA* from *Aspergillus nidulans*. All this led to the increased expression of the *ClrB* gene, which in turn increased the production of cellulases (Yao et al., 2015).

Another study by Coradetii and coworkers on *Neurospora crassa* and *A. nidulans* revealed another window of cellulase expression without the requirement of an inducer. *N. crassa* consists of a transcription regulator known as *Clr2*, which is an ortholog (genes from different species that have common ancestral evolution by speciation) for *ClrB* found in *A. nidulans*. *Clr2* is a zinc binuclear cluster transcriptional factor that if deleted causes the failure of cellulase expression. It is a conserved feature of filamentous fungi. The ortholog of *Clr2* (*ClrB*) in *A. nidulans* also showed the same gene function. Therefore these two genes are orthologs of each other. Coradetti tried to discover whether the expression of *Clr2* alone is sufficient for the expression of cellulase enzyme. He placed a copy of *Clr2* under the controlled clock promoter gene 1 (*Pccg1*—a constitutive promoter). This arrangement was transferred to the fungi and was allowed to grow without an inducer, which is required for cellulase expression. The culture was grown under controlled conditions of light and temperature, and the circadian rhythm was reduced. When an expression pattern was seen, it was noted that major cellulase genes were expressed under noninducing conditions. He called this misexpression of *Clr2*; this misexpressed condition also showed good carboxymethyl cellulase activity. This clearly suggests that *Clr2* has the capacity to activate the cellulase expression system in the absence of the inducer, which is not found in its ortholog *ClrB* (Coradetti et al., 2013).

In another work by Nina Aro et al. the role of *ace1* was checked. This regulatory factor contains a Cys-His type zinc finger that binds to the *cbh1* promoter having a sequence 5′-AGGCA-3′. Initially, in their work they founded that on deletion of *ace1* the growth of the transformant was hindered more than the host. Later in their work they found that strains that deleted *ace1* showed good growth of fungi, which contradicted their initial work. They discovered that in the initial experiment they had added peptone, which inhibited the growth of the deleted *ace1* strain. So now the task was to understand whether the deletion causes increased expression of the cellulase or not. Therefore under different inducing conditions the expression of different cellulase genes was examined by checking the concentration of mRNA transcripts of the respective genes. Under all inducing conditions, all the major genes of cellulase enzymes were expressed in high quantities in the deleted *ace1* strain compared to the normal strain. The deleted strain showed the expression of *cbh1, cbh2, and egl1* by two, three, and two times, respectively, as compared to the normal strain. Although this expression was found to be high only in the initial hours it later expressed the same quantity as that of the normal strain.

This finding clearly suggests that *ace1*, unlike *ace2*, negatively regulates the expression of cellulase. In the presence of *ace1*, cellulase production decreases, whereas in the absence of *ace1*, cellulase production increases. However, so far it has been shown that accumulation of glucose decreases the expression of cellulase, but now it is evident from this work that this regulatory factor also inhibits the expression of the enzyme under inducing conditions. Therefore an attempt to create a deleted *ace1* strain could increase enzyme production (Azo et al., 2003).

13.6.4 Epigenetic Remodeling

Waddington defined epigenetics as a heritable change that does not involve changes in DNA sequence. These are the factors that are closely governing the expression of genes by methylation, acetylation, and posttranscriptional modification. Currently, much emphasis is given to understanding the basic mechanism of epigenetic tagging of a particular region of the DNA that is actively involved in gene expression. Modulating these factors helps in expressing a gene in a controlled manner without bringing about any change in the gene sequence (Gibney and Nolan, 2010). This concept is also being used to develop industrially important fungi in cellulase production. Although not much work has been done, efforts are being made, e.g., Gupta and coworkers tried to show that this phenomenon could be an approach toward increasing cellulase production. They found that under inducing conditions the chromatin around *cbh1* and *cbh2* opens up so that its expression can take place. However, this condition was absent in the strains where *xyr1* was deleted. In their approach they found that *xyr1* is an important activator of transcription of cellulase gene and its epigenetic effect was worked on. Therefore they deleted the *xyr1* gene and then developed the *xyr1* deleted strain and wild-type strains on different types of media, i.e., glucose, sophorose, and media containing no carbon. After 3h of incubation they applied chromatin accessibility real time-PCR to analyze the chromatin structure around *cbh1* and *cbh2* under both inducing and noninducing conditions for the *xyr1 and cre1* binding sites on the foregoing promoter. Later, they also quantified gene expression by quantitative PCR to see whether expression is affected by changes in chromatin structure. Under inhibiting conditions on glucose media, both the strains, wild type and deleted *xyr1*, showed no gene expression. However, in the presence of a sophorose-inducing condition, gene expression was seen at a high level in the wild-type strain but was much less in the *xyr1* deleted strain, although there was some expression but not to an acceptable level. This answers the question regarding the role of *xyr1* in changing the chromatin structure. In the absence of *xyr1*, chromatin is found to be in a closed state, therefore inhibiting the expression of cellulolytic genes, whereas in the presence of *xyr1* the expression is seen, which signifies the role of *xyr1* in remodeling the chromatin structure during the inducing condition (Mello-de-Souza et al., 2015).

Other researchers later identified many genes that were controlling chromatin remodeling for cellulase expression. One such gene was histone acetyltransferase whose expression differed greatly in the *xyr* deleted strain and normal strain. Xin et al. said that GCN5, a histone acetyltransferase, is essential for cellulase expression. It modifies the K9 and K14 residues of H3 in the *cbh* promoter region, thus expressing it. They found that deletion of GCN5 greatly reduces the acetylation of the *cbh* promoter region and reduced cellulase expression was seen (Druzhinina and Kubicek, 2017).

Therefore such approaches can be very handy because epigenetics is a very powerful tool when it comes to expressing a gene. Because it is very closely related to switching on and off the expression of a gene, one way of understanding how it works by various means can help in overexpressing a gene without the input of any sequence. Moreover, because it is inheritable it does not possess the problem of being fixed in nature and changing the wild type. A combination of other genetic tools with epigenetics can be a windfall for industrial cellulase production. Although this epigenetic approach remains to be incorporated at an industrial level, it is a promising feature for the future of cellulase production.

13.7 MICROBIAL GENETIC MODIFICATION FOR INCREASING CELLULASE EFFICIENCY

To have the highest degradation of lignocellulosic biomass, optimization of enzyme mixture and adaptability toward the substrate are necessary. There is a need to increase the catalytic efficiency of the enzyme and its thermal tolerance. Approaches have been rationally based on knowledge of substrate binding and the catalytic mechanism of cellulases. An alternative is to have different combinations of enzymes having different catalytic properties and adding to them the carbohydrate binding (CB) protein, which makes sure that a large amount of substrate is accessible to be converted, e.g., cellobiose converted to glucose. Hence through genetic modifications, the production of β-glucosidase is increased as compared to other cellulases like *cbh1* and *cbh2* that produce cellobiose, thus preventing inhibition. As stated earlier regarding the addition of CB protein, such proteins are known as accessory proteins. They primarily consist of a CB domain only. They are the potential attention-getters because they are able to loosen the rigid structure of cellulose fiber. Two such proteins are identified in *T. reesei*: *CIP1* and *CIP2*. They contain a signal peptide that enables them to be secreted. Having only a CB domain they help in accessing a large number of substrates. *Swollenin* (*SWO1*) is another protein that has a CB domain and helps in loosening the cell wall. The *CEL61A* protein also found in *T. reesei* has shown to have both a CB domain and endoglucanase activity. These accessory proteins are vital tools for increasing the efficiency of cellulase (Seidl and Seiboth, 2014).

Overexpressing a particular gene of the cellulosome has shown an increase in the efficiency of the rest of the cellulolytic genes. However, this is a novel approach in which no knockout or addition of any sequence is included in the process; rather overexpressing a gene is all that is needed to increase efficiency. Quian et al. were the first to study this and showed that overexpression of one of the genes of cellulases, i.e., *egl2*, enhanced the hydrolysis activity of the cellulase enzyme. Initially, they ligated the *cbh2* gene with a *pyrG* marker gene by double joint PCR. This was then transferred to *T. reesei* by the polyethylene glycol-mediated method. The transformant was grown on Avicel- (a reference material for microcrystalline cellulose) containing media. Further quantitative real-time reverse transcription PCR (qPCR) was done to confirm the overexpression of the *cbh2* gene. Through this they were able to overexpress the *cbh2* gene 32 times more than the wild-type strain. Similarly, they ligated another gene, *egl2*, with the *pyrG* gene and transformed *T. reesei* with the same method. When qPCR was done it was found that *egl2* was found to be 46-fold increased as compared to the wild-type strain. Let us call the *cbh2* overexpressing strain T_{cbh2} and the *egl2* overexpressing strain T_{egl2}. To understand whether overexpression of either of these genes also enhances their hydrolysis activity, enzyme activity over a natural pretreated cellulose source such as corn cob residues was checked. The corn cob was pretreated as acid pretreated corn cob residue and delignified corn cob residue. When T_{cbh2} was grown on acid pretreated corn cob residue and delignified corn cob residue it showed the same hydrolytic and saccharification activity as that of the wild-type strain, whereas when T_{egl2} was grown on these two pretreated cellulose materials, it was able to convert about 42% of the cellulose into glucose when on acid-treated corn cob and was able to convert 74% of delignified corn cob cellulose media into glucose. Comparatively, the wild type was only able to convert 20% and 34% of the pretreated material, respectively, into glucose. Therefore T_{egl2} showed enhanced activity as compared to the wild-type strain. The overexpression of *egl2* can therefore be used to enhance the activity of the cellulase enzyme, which could be a promising approach (Qian et al., 2017).

Thermostable cellulases are a great asset to cellulose degradation because they provide greater specific activity and stability during a process. Apart from avoiding contamination due to high temperature, they also increase the transfer of the mass because of the decrease in the viscosity of the fluid during high substrate concentration, since a cellulase enzyme of *T. reesei* is mesophilic and can operate up to 50°C. Therefore an attempt has been made to insert the thermotolerant cellulase from a number of fungi such as *Acremonium thermophilum*, *Thermoascus aurantiacus*, and *Chaetomium thermophilum* into *T. reesei* to produce thermotolerant cellulases. The pioneering work on this aspect was done by Sandgren et al. In their study they tried to analyze a number of homologs of GH12 in various species. Their findings gave a comparative insight between *Helianthus schweinitzii* and *T. reesei*. The *Cel12A* enzyme of *H. schweinitzii* was different from *T. reesei* by just 14 amino acid residues. This difference made the enzyme of the latter more stable than the former. To verify the effects of this difference in amino acid sequence between the two species, the author and his coworkers via site-directed mutagenesis inserted these 14 amino acids into the *Cel12A* enzyme of *T. reesei* and observed the decrease in thermal stability. By looking at the T_m of the homologs, the T_m of the *Cel12A* enzyme of *T. reesei* was found to be 54.4°C and that of *H. schweinitzii* was found to be 5°C less. A homolog from *Streptomyces* sp. showed a variant that was 12.6°C higher than that of *T. reesei*. The changes in stability among the homologs was attributed to the first 63 amino acids. On examining the relative enzyme activity, the species having higher T_m showed 30% more activity than the less T_m-valued species (Sandgren et al., 2003). Through their effort they showed that by bringing a change to the amino acid sequence of the enzyme it could bring about changes to the efficiency of the cellulase enzyme, which is a novel approach toward increasing the efficiency of the enzyme. It is evident that the fermentation process leads to an increase in temperature, which may decrease the activity of the cellulases, and can solve this issue.

Lantz et al. showed that two of the CBHs, CEL6A and CEL7A, are more thermolabile than the rest of the enzyme mixture and therefore become a reason for rate limiting. Therefore Day et al. in 2007 compared the amino acid sequence of CEL7A from many fungi and found that there were about 19 sites available for making it thermotolerant. By site-directed mutagenesis they caused mutation in 18 such sites and increased the T_m value by 14.8°C (Druzhinina and Kubicek, 2017; Lantz et al., 2003).

All these methods suggest that there are specific sites over the cellulase enzymes that can be engineered to achieve the best output from the enzyme. Rather than changing the entire sequence or inserting a new one, just by changing a few regions can bring about a change in the structure and nature of bonding between the amino acids and thus affect the overall activity of the enzyme.

13.8 PLANT GENETIC MODIFICATION FOR ENZYME PRODUCTION

The previous cost of the production of cellulosic ethanol of 5$/gal is being reduced to 0.30–0.50$/gal. Efforts are continuously being made to bring the cost down even more. One way is to produce the cellulase enzyme within the crop, which brings down the cost of taking the biomass to microbial tanks (Sticklen, 2006). At present, microbes are the sites for cellulase production, whereas plants are basically used for the production of enzymes and proteins required in pharmaceutical industries. However, today attempts are being made toward expressing the cellulose hydrolyzing genes inside the plants, thereby decreasing their production costs. Genetic engineering mediated by *Agrobacterium tumefaciens* and the gene gun method are efficient tools in this respect. Plant genetic modification has emerged to be a good option because it requires less energy input when compared with microbial production. In microbial enzyme production a lot of energy has to be inculcated during the fermentation process; moreover, there is a chance of contamination that could decrease the yield or may have an inhibitory effect. Plant-based enzyme production circumvents this more so than tissue-directed expression methodology and helps in easy extraction from specific regions and even from specific organelles. It simply means that it increases the specificity of the enzyme (Hefferon, 2017).

When expressed in plants, care is always taken that enzymes are produced in subcellular compartments rather than in the cytosol, because they show correct folding in spite of being of microbial origin, and also show good activity and glycosylation. They also have a longer life as compared to the cytosol. The enzyme is collected from the dry biomass of the transgenic plant and then pretreated before use.

13.8.1 Transgene in Plants for Cellulase Production

In view of this, an approach has been made to produce a thermostable enzyme 1,4-β-endoglucanases E1 from *Acidothermus cellulolyticus* in Arabidopsis, tobacco and potato plant and the translated product targeted to the apoplast (i.e., the extracellular pathway provided by the continuous matrix of cell walls). In *Arabidopsis* it was produced at a level of 25% of total plant soluble protein. The same work is being carried out on rice and maize. When the crude extract was pretreated with ammonia fiber explosion (AFEX), 22%–30% of the biomass was converted to glucose. However, there were certain concerns such as:

- Can the harsh condition of AFEX destroy the enzyme?
- Is it possible to express enough enzymes within the plant to convert the biomass?
- Will there be any detrimental effects on plants?

If mild treatment of AFEX is given only to the extracted crude or pure form of the enzyme, then 30% of the biomass is convertible. Gene expression can be increased by regulating the transcriptional, translational, and posttranslational machinery and accumulating the enzyme away from the cytosol into the noncytosolic compartments. This approach would therefore also remove concerns regarding the interference of cellulase with other cytosolic enzymes and substrates (Sticklen, 2006).

A good source of biomass for biofuel production is rice. Production of cellulase inside the rice could be a beneficial step because the residual tissue that is left after the grain is harvested could be converted comfortably into simple sugars. So, endo-β-glucanase gene V from *T. reesei* was transferred to the rice plant. The transgenic plant carried the gene in the plastid so that it would not be expressed in the cytoplasm of the plant cell. As we have discussed already, expressing the transgene in the cell organelle is much safer than expressing it in the cytoplasm. Here the gene is expressed in plastids. In a different approach, cellulase was overexpressed with an inducible promoter inside the plant, e.g., the senescence-inducing promoter *STAY GREEN* (*SGR*). The *SGR* gene is involved in natural and dark-induced chlorophyll degradation. Since it is a constitutive promoter, insertion of a gene that disrupts its function helps on the one hand to reduce the stunted growth of the plant and on the other hand to increase the production of the inserted gene. This was clearly seen in the case of transgenic rice that was carrying the cellulase gene under the control of senescence inducible promoter *SGR*, which increased the production of cellulase and avoided the growth retardation of the plant. Overexpression of the cellulase gene in the rice plant enhanced its saccharification properties, which suggests that steps like this are quite beneficial for enhancing enzyme production and efficiency.

The fiber from sugarcane is also an excellent source of biomass for conversion into important biofuel. A synthetic endoglucanase was expressed in the sugarcane cells with an inducible promoter from the maize plant and was expressed inside the endoplasmic reticulum, chloroplast, and vacuoles. The chloroplast-derived endoglucanases were found to be better options for biomass degradation when the pretreated bagasse was subjected to saccharification in the presence and absence of commercially available cellulases. The raw material from the transgenic plant having the cellulase gene showed better degradation in the presence of commercial cellulases. This proves that transgenic plants are better substrates for cellulolytic cleavage by enzymes.

The expression of a number of genes in the same plant that could act synchronously over the substrate has been reported. For example, foot and mouth disease virus 2A peptide, which is a self-cleaving peptide, is placed between different cellulase enzymes. Such a polyprotein containing different cellulase enzymes when expressed had the enzymes cleaved automatically, which could help in the degradation of the cellulose material. β-Glucanase and endoglucanase from *Thermotoga maritima*, xynalase from *T. reesei*, and exoglucanase from *Thermobifida fusca* were all expressed together in the tobacco plant in this manner, i.e., with a 2A peptide of foot and mouth disease virus. Such an effort would be useful in degrading cellulose from a single plant source (Hefferon, 2017).

One of the major concerns related to transgenic plants producing heterologous enzymes is the increase in yield and enzyme activity. In one of the approaches like in case of E1 enzyme (a ubiquitin-activating enzyme), that is expressed in wheat and maize at 2% and 5% of total soluble protein which is quite low. In a similar manner the cellulosome produced in plants has a low yield. Yield can be increased if only the catalytic domain is expressed or the enzyme genes are engineered inside the chloroplast rather than in the nucleus.

Bioconfinement is another issue that needs to be focused on with respect to transgenic crops producing biomass. The genetically engineered plants are manipulated in such a way that the enzymes are produced in specific tissues and not in any other organs of plants like fruits, seeds, and flowers, because this step will ensure that the transgenes are not transferred from the transgenic crop to the same crop species or other closely related species through pollination (Sticklen, 2006).

13.8.2 Decreasing the Need for Pretreatment

Lignin is an important component of the plant biomass that is required for many industrial processes. Due to the presence of lignin the plant raw material is pretreated before being fed into the fermentation process to be acted on by the cellulosome. Lignin acts as a recalcitrant for cellulose degradation and therefore needs to be pretreated. It is required for maintaining the structural integrity of the cellulose and inhibits the swelling up of the crystalline cellulose. Pretreatment for avoiding lignin inhibition is done by physical, chemical, and physicochemical processes. Pretreatment poses a number of disadvantages such as:

- High power and energy consumption
- Generation of toxic compounds
- Large amount of water required
- High cost of oxygen, acid, and alkaline catalysts used in the pretreatment
- A need for corrosion-resistant equipment.

All this increases the cost of cellulose degradation (Druzhinina and Kubicek, 2017). So, *lignin genetic modification* becomes necessary for the industrial process. Therefore many steps are being taken to downregulate the biosynthesis of lignin in plants. Inhibiting the synthesis of 4-coumarate-3-hydroxylase (C3H) in alfalfa saw a change in lignin structure and increased the digestibility of the cellulose. *C3H* is an enzyme involved in lignin biosynthesis (Maurya et al., 2015). In another approach cinnamyl alcohol dehydrogenase was downregulated in alfalfa and brought about a change in lignin structure that in situ digestibility of lignocelluloses increased. However, it did not decrease the concentration of the lignin component but changed its structure so that it was digestible in low energy (Ralph et al., 2006).

In a very different approach it was attempted to divert the carbon source from lignin production in the plants, which could increase the concentration of the polysaccharides in the plant cell wall and decrease the concentration of the lignin thereby avoiding pretreatment. As in the case of aspen, 4-coumarate CoA ligase was downregulated, which increased the concentration of sugar in plant cells by 15% and decreased the lignin concentration by 45%. Similarly, when coniferaldehyde 4-hydroxylase was downregulated, the cellulose component increased by 30% and the lignin content in the plant was decreased by 52%.

Therefore when essential enzymes involved in lignin biosynthesis are downregulated the concentration of lignin goes down considerably so that its pretreatment either becomes negligible or is no longer required. This is quite advantageous at the industrial level because it removes all the former disadvantages involved in cellulose degradation. Steps like this always become an easy threshold to meet industrial expectations for an ongoing process and bring down the cost of product formation. This can be a big boon for the production of ethanol as biofuel, since pretreatment of lignin raises the cost to heights that make the use of ethanol unfeasible.

13.8.3 Substrate Disruption

Certain factors have been noted recently that act like a CB domain of the family glycosyl hydrolase. These factors are found to act synergistically to loosen the cellulosic material or increase its solubility so that the cellulolytic enzyme has easy access to the substrate and thereby increases the efficiency of the enzyme. One of such factors is *gr2* (grass pollen group 2/3 allergens). This classic invention is the work of Daniel J. Cosgrove and Takumi Takeda. *Gr2* is a nonglycosylated protein that has so far been found in grass pollens but not in any other pollen plants. The C-terminal domain sequence of *gr2* is somewhat similar to expansins. Expansins have been used to increase the digestibility of cellulose by cellulases. However, *gr2* lack cellulolytic activity by themselves but are found to enhance cellulolytic activity when combined with a cellulase system. This protein makes the substrate accessible for cellulases, thus increasing the efficiency of the enzyme. In their method they compared the degradation of the cellulose in the presence and absence of the *gr2* protein. They added *gr2* protein at a concentration that was 0.001 times the amount of cellulose present for degradation. The concentration of *gr2* as compared to the cellulase added was between 0.01 and 0.5 times. Therefore it can be predicted that even at less concentration this factor can increase the efficiency of the cellulases. This factor was found to increase the efficiency of the enzyme by increasing the accessibility of the cellulose-based material to the enzyme under mild conditions of temperature, pH, and pressure. They do so by altering the structure and mechanical criteria of the substrate. It is quite stable in nature. They were successful in inserting the *gr2* protein gene within the plant host by using a specific vector. It was stably hybridized within the host chromosome and expressed (Cosgrove and Tanada, 2007). This work is currently under the scrutiny of a US patent. Such inventions and discoveries are the need of the hour in the field of cellulose degradation because such factors increase the efficiency of the naturally occurring cellulases without bringing any change to their structure and function. Another point to note when looking at expansins, swollenin, and *gr2* is that naturally occurring cellulase has the capacity to degrade cellulose quite efficiently and there is little necessity to bring about a change in the structure or sequence of the enzyme to enhance the efficiency. Rather, the use of accessory factors in very small quantities could increase the efficiency of the wild-type enzyme. It has shown its importance in paper, textile, and other industries (Baucher et al., 1999; Cosgrove and Tanada, 2007).

13.9 TOOLS FOR THE GENETIC MODIFICATION OF A *T. REESEI* MODEL ORGANISM FOR CELLULASE PRODUCTION

Gene targeting has always been a difficult task with respect to the expression of cellulase systems. A good tool is the prerequisite for genetically modifying *T. reesei*, the most used fungi for the industrial production of cellulase enzyme. The low efficiency of gene targeting in *T. reesei* is solved by inactivating the nonhomologous end joining of a DNA repair system such as tku70. Apart from this, the CRISPR/CAS9 system has also been used to transform the fungi. It takes into account a chimeric guide RNA that causes a targeted double strand break in the DNA and thus integrates the desired gene at the target.

Other approaches were also made by Demntl and others. In his work he took *T. reesei* having turnacated versions of many genes. Due to this the fungi become an auxotroph. To transform the fungi he took the homologous version of the same gene, which reverted the original functional gene. For instance, he utilized the *ade2* locus (codes for the phosphoribosylaminoimidazole carboxylase enzyme required for purine biosynthesis) for this purpose. When the targeted gene is inserted into this locus by homology, *ade2* is disrupted causing the accumulation of polymerized 5-aminoimidazole ribonucleotide, which can be identified by a red color. He showed the same applicability of other loci such as *hah1*, *pyr4*, and *asl1*.

If transformation has been carried out, then the problem arises of how to screen out the transformant. Subramanian solved this problem by using a 2A peptide system from foot and mouth disease virus. This system allows transcription of more than one gene as a single mRNA. When the translation is taking place a 2A peptide sequence causes the "ribosomal skip" (a phenomenon used by viruses in which a peptide stops the ribosome from attaching a new amino acid to the growing chain of protein, yet continuing the translation, thereby forming a cleaved protein from a polyprotein) between the two C-terminal amino acids and produces two separate proteins at the same time. This tool is used in *T. reesei* where CEL7A and GFP markers were coexpressed. This made it possible for the desired gene and the marker to be expressed at the same level and therefore it becomes easy to screen out the transformant (Druzhinina and Kubicek, 2017).

13.10 CONCLUSION

Changing the structure of cellulolytic genes has brought about positive functional changes in the activity of cellulases. As, has been found, wild type enzymes are under strict regulations by a number of restricting factors, the changes made in the regulatory mechanism have empowered to overcome the effects of these restricting factors. It is clearly evident from the manipulation of *ClrB* and *clr2*, which are actively involved in the regulatory mechanism of cellulase expression, that they can increase the production of the enzyme. Although the CCR gene *cre1* makes sure that the cellulase gene is expressed only in the presence of cellulose, it restricts the expression level when the concentration of glucose increases, which is the final product of cellulose metabolism. Nakari and coworkers suggested through their work that knocking out the CCR gene increases the expression even in noninducing or inhibitory conditions. The *cbh1* promoter, which is a constitutive one, can be utilized for expressing one of the cellulosome enzymes by just prefixing it on the gene to be expressed. Although *T. reesei* is the promising fungus for the production of cellulase, product synthesis requires thermotolerant strains because the temperature during the process rises due to pretreatment. Isolation and incorporation of thermotolerant genes have eradicated the need to optimize the product process.

On the other hand, apart from expressing the genes in fungi, plants have shown to be the new host for cellulolytic genes. Plants that harbor cellulolytic genes are quite promising in nature in bringing down product cost. Since they do not show the necessity for lignin pretreatment, the cost and time of cellulose degradation are reduced considerably. However, much work is still needed.

One of the approaches of genetic modification is epigenetic remodeling, and has been shown by some of the works as a novel method for adjusting the expression level. Through this approach a new window opens that shows that by manipulating noncellulolytic genes, the expression level of cellulolytic genes is enhanced under both inducing and noninducing conditions. Moreover, enzymatic activity was found to be enhanced. This suggests that cellulases are efficient enough to degrade the cellulose without any major/minor genetic code change.

The activity of cellulases is of vital importance to many industries and is an input fuel for the production of ethanol—the fuel of the future. The key to future energy requirements lies largely in the production of this enzyme and its efficiency in doing so. The development and advances in genetic modulation methods directly pave the way toward achieving novel methods of increasing the production of cellulases cheaply. The lesser the amount required for enzyme production the better the fulfillment of the promises forged for industrial purposes. Cellulase has clearly defined its market value at a global level, thereby attracting both developed and underdeveloped nations to undergo robust scientific approaches for its production to supply their energy needs. With the rise in the need for green methods for accomplishing development criteria, cellulose is the promising element and cellulase is the *gem* above which it increases its utility.

Acknowledgments

I am heartily grateful to God almighty, because without Him nothing is possible. I extend my deep sense of gratitude to Dr. Pramod Ramteke, my mentor, for giving me this opportunity and all the help required to complete this task. My special thanks go to Dr. Neha Shrivastava for her constant help and all the valuable suggestions needed to write this chapter. I thank my family for their constant support and prayers for accomplishing this.

References

Acharya, P.B., et al., 2008a. Optimization of cellulase production by *Aspergillus niger* using saw dust as substrate. Afr. J. Biotechnol. 4147–4152.

Acharya, P.B., et al., 2008b. Optimization of cellulase production by *Aspergillus niger* using saw dust as substrate. Afr. J. Biotechnol. 4147–4152.

Albert, B., et al., 2002. Structure and function of DNA. In: Molecular Biology of the Cell. fourth ed...

Azo, N., et al., 2003. *ACE-I* of *T. reesei* is a repressor of cellulase xynalase expression. Appl. Environ. Microbiol. 56–65.

Bachman, J., 2013. Site directed mutagenesis. Methods Enzymol.

Baucher, M., et al., 1999. Downregulation of cinnamyl alcohol dehydrogenase in transgenic *alfalfa* (*Medicago satival*) and effect on lignins composition and digestibility. Plant Mol. Biol. 437–447.

Beguin, P., Aubert, J.P., 1994. The biological degradation of cellulose. Microbiol. Rev. 25–58.

Carter, P., 1986. Site directed mutagenesis. Biochem. J. 1–7.

Cavicchioli, R., et al., 2006. Proteins from psychrophiles. Methods Microbiol.

Coradetti, S.T., Xiong, Y., Glass, N.L., 2013. Analysis of a conserved cellulase transcriptional regulator reveals inducer independent production of cellulolytic enzymes in *Neurospora crassa*. Microbiol. Open, 595–609.

Cosgrove, D.J., Tanada, T., 2007. Use of *gr2* proteins to modify cellulosic material and to enhance enzymatic and chemical modification of cellulose. United States Patent 20070166805.

Druzhinina, I.S., Kubicek, C.P., 2017. Genetic engineering of *Trichoderma reesei* cellulases and their production. Microb. Biotechnol. 1485–1499.

Fuji, T., et al., 2014. Characterization of the *Xylanase regulatorprotein* gene, in *Talanomyces cellulolyticus* (formerly known as *Acremonium cellulolyticus*). Biosci. Biotechnol. Biochem. 178, 1564–1567.

Fujii, T., et al., 2013. Enhancing cellulase and hemicellulase production by genetic modification of the carbon catabolite repressor gene, creA, in *Acremonium cellulolyticus*. AMB Express.

Gibney, E.R., Nolan, C.M., 2010. Epigenetics and gene expression. Heredity, 4–13.

Greenway, P.J., 1920. Basic steps in genetic engineering. Int. J. Environ. Stud.

Hefferon, K., 2017. Cellulase production in transgenic plant: molecular farming with twist. Adv. Biotechnol. Microbiol.

Hu, Y., et al., 2017. Extraction and characterization of cellulose from agricultural waste argan press cake. Cellul. Chem. Technol. 1–10.

Ilmén, M., et al., 1997. Regulation of cellulase gene expression in the filamentous fungus *Trichoderma reesei*. Appl. Environ. Microbiol. 63, 1298–1306.

Imran, M., et al., 2016. Cellulase production from species of fungi and bacteria from agricultural wastes and its utilization in industry: a review. Adv. Enzym. Res. 44–55.

Johnsen, H.R., Krause, K., 2014. Cellulase activity screening using pure carboxymethylcellulose: application to soluble cellulolytic samples to plant tissue prints. Int. J. Mol. Sci. 830–838.
Kingsley, G.R., Getchell, G., 1960. Direct ultramicro glucose oxidase method for determination of glucose in biologic liquid. Clin. Chem. 6, 466–475.
Klemn, D., et al., 2005. Cellulose: fascinating biopolymer and sustainable raw material. Angew. Chem. Int. Ed. 44 (3), 3358–3393.
Lantz, S.E., et al., 2003. *Hypocrea jecorina* CEL6A protein engineering. Biotechnol. Biofuels, 1–13.
Lavanya, D., et al., 2011. Sources of cellulose and their applications—a review. Int. J. Drug Formul. Res. 2 (6), 19–34.
Li, S., et al., 2014. Cellulose synthesis and its regulation. In: The Arabidopsis Book, pp. 1–21.
Liming, X., Xeuliang, S., 2003. High yield of cellulase production by *Trichoderma reesei* ZU-02 on corn cob residue. Bio/Technology 259–262.
Mach-Aigner, A.R., et al., 2008. Transcriptional regulation of xyr1 encoding the main regulator of the xylanolytic and cellulolytic enzyme system in *Hypocrea jeconina*. Appl. Environ. Microbiol, 6554–6562.
Maurya, P.D., et al., 2015. An overview of key pretreatment processes for biological conversion of lignocellulosic biomass to ethanol. Biotech.
Mello-de-Souza, T.M., et al., 2015. The impact of chromatin remodeling on cellulase expression in *T. reesei*. BMC Genomics, 1–11.
Mora'n, J.I., et al., 2008. Extraction of cellulose and preparation of nanocellulose from sisal fibers. Cellulose, 149–159.
Mrudula, S., Murugammal, R., 2011. Production of cellulase by *Aspergillus niger* under submerged fermentation and solid state fermentation using corn waste as a substrate. Braz. J. Microbiol. 1119–1127.
Nichol, D.S.T., 2008. An Introduction of Genetic Engineering. Cambridge University Press, pp. 216–217.
Polyak, K., 2003. Overview gene structure. In: Cancer Medicine, sixth ed...
Qian, Y., et al., 2017. Production of highly efficient cellulase mixtures by genetically exploiting the potentials of *Trichoderma reesei* endogenous cellulases for hydrolysis of corncob residues. Microb. Cell Fact.
Rahman, Z., et al., 2014. Application of *Trichoderma reesei* cellulase and xylanase promoters through homologous recombination for enhanced production of extracellular β-glucosidase I. Biosci. Biotechnol. Biochem. 1083–1089.
Raibault, M., 1998. Solid state fermentation in bioconversion of agro-industrial raw materials. In: Seminar Proceedings of ORTOM-MONTPELLER.
Ralph, J., et al., 2006. Effects of coumarate-3-hydroxylase down regulation on lignin structure. J. Biol. Chem. 8843–8853.
Ray, R.C., Rosell, G.M., 2013. Microbial Enzyme Technique in Food Application, p. 13.
Sandgren, M., et al., 2003. Comparison of family 12 glucoside hydolases and recruited substitutions important for thermal stability. Protein Sci. 848–860.
Seiboth, B., et al., 1997. Role of four major cellulases in triggering of cellulase gene expression by cellulose in *Trichoderma reesei*. J. Bacteriol. 179, 5318–5320.
Seidl, V., Seiboth, B., 2014. *Trichoderma reesei*: genetic approaches to improving strain efficiency. Biofuels, 343–351.
Singh, P., et al., 1999. Solid-state fermentation technology for bioconversion of biomass and agricultural residues. Curr. Sci. 198–216.
Singh, A., et al., 2017. Genetic modification: a tool for enhancing cellulase secretion. Biofuel Res. J. 14, 600–610.
Sticklen, M., 2006. Plant genetic engineering to improve biomass characteristics for biofuels. Curr. Opin. Biotechnol. 315–318.
Sukumaran, R.K., et al., 2005. Microbial cellulases—production, applications and challenges. J. Sci. Ind. Res. 64, 832–844.
Sukumaran, R.K., et al., 2009. Cellulase production using biomass feed stock and its application in lignocellulose saccharification for bio-ethanol production. Renew. Energy, 421–442.
Turnbough, C.L., et al., 1987. Nucleotide sequence and characterization of the pyrF operon of *Escherichia coli* K12. J. Biol. Chem. 10239–10245.
Yang, F., et al., 2015. Enhancing cellulase production in thermophilic fungus *Myceliophthora thermophila* ATCC42464 by RNA interference of cre1 gene expression. J. Microbiol. Biotechnol. 1101–1107.
Yang, H., et al., 2017. Microbial production and molecular engineering of industrial enzyme. In: Biotechnology of Microbial Enzyme.
Yao, G., et al., 2015. Redesigning the regulatory pathway to enhance cellulase production in *Penicillium oxalicum*. In: Biotechnology for Biofuels.
Zhang, Z.X., Zhang, P.Y., 2013. Cellulases: characterstics, sources production and applications. In: Bioprocessing Technologies in Biorefinery for Sustainable Production of Fuels, Chemicals and Polymers, pp. 131–142.
Zou, G., et al., 2012. Construction of a cellulase hyper-expresssion system in *Trichoderma reesei* by promoter and enzyme engineering. Microb. Cell Fact. 1–12.

CHAPTER

14

Genetic Engineering Applications to Improve Cellulase Production and Efficiency: Part II

Pragati Misra, Pradeep Kumar Shukla†, K. Prasada Rao†, P.W. Ramteke†*

*Department of Molecular and Cellular Engineering, Jacob Institute of Biotechnology and Bioengineering, Sam Higginbottom University of Agriculture, Technology and Sciences, Allahabad, India †Department of Biological Sciences, Sam Higginbottom University of Agriculture Technology & Sciences (Formerly Allahabad Agricultural Institute), Allahabad, India

14.1 INTRODUCTION

Recent industrial research is converging toward biopredicated production of biochemicals and biomaterials from renewable waste biomass, as a promising route to achieve constant sustainability (Chung et al., 2015; Chen and Nielsen, 2016). Diverse genera of microorganisms have been genetically modified/engineered for production/improvement of biopredicated products comprising enzymes, metabolites, etc. (Chae et al., 2017). Microorganisms, ranging over millions of acres, causing relinquishment of fields in Asian countries like India, can be an admirable miniature source for cellulase production (Chandel et al., 2009, 2012; Scordia et al., 2010).

Several countries have initiated bioconversion-based technology innovation programs to provide biofuels from cellulosic/lignocellulosic biomaterials (Goldemberg, 2007). The production of recombinant enzymes involves remarkably progressive processes, focused on efficient development and the scaling-up of production for industries (Adrio and Demain, 2014), following coalescence between directed evolution and rational design (Zhao et al., 2002; Cherry and Fidantsef, 2003). However, the still constrained erudition of cellulase substrates, relationships, and regulation of cellulosic activity compromises rational design, leading to the utilization of directed evolution as an implement for a higher rate of prosperity (Zhang et al., 2006). Several authors have reviewed and compared recombinant cellulase expression systems (Sadhu and Maiti, 2013; Hasunuma et al., 2013; Garvey et al., 2013; Juturu and Wu, 2014; Lambertz et al., 2014).

Unfortunately, the enzymatic breakdown of biomass is narrowly studied and not standardized optimally for large-scale applications. Substantial research involvements therefore are indispensable to amend productivities and yields of cellulase. Exploration of microbial diversity to discover novel enzymes and cognizance-predicated peptide engineering may be a better strategy to amend cellulase yield for industrial applications. Endeavors to ameliorate cellulase performance are in progress, but are slow due to constraints in effective strategies and lack of appropriate expression systems (Adsul et al., 2011). Lignocellulosic biomaterial, a prime potential for fuels and range of chemicals, is biochemically composed of cellulose (35%–50%), hemicellulose (25%–30%), and lignin (25%–30%). Conversely, cellulose, the most important polysaccharide found in higher and lower plant cell walls, is synthesized in nature by photosynthesis (Fig. 14.1) (Juturu and Wu, 2014; Yoon et al., 2014; Lynd and Laser, 2005). Cellulases are employed in many industrial applications, including formulations of medicines, preparation of perfumes, development of resins, production of starch, use in baking processes, treatment of biowaste, etc. (Couto and Sanroman, 2006; FitzPatrick et al., 2010), and most prominently, production of bioethanol from lignocellulosic biomaterials (Chapple et al., 2007). Furthermore, the demand for bioethanol as the major component of renewable fuels is predicted to increase by 3.4-fold by 2035 globally (Limayem and Ricke, 2012; Baeyens et al., 2015). Many substantial experiments have been

From Cellulose to Cellulase: Strategies to Improve Biofuel Production
https://doi.org/10.1016/B978-0-444-64223-3.00014-X

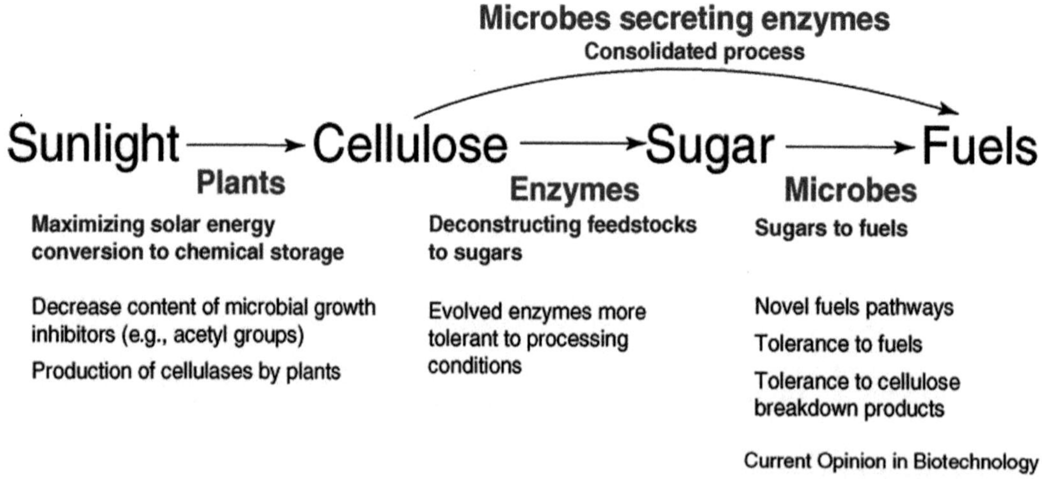

FIG. 14.1 Route from sunlight to fuels. Conversion of biomass to fuels will involve the development of dedicated energy plants that maximize solar energy conversion to chemical storage and minimize the use of water and fertilizer, enzymes that depolymerize cellulose and hemicellulose into useful sugars, and microorganisms that produce advanced biofuels that are compatible with our existing transportation infrastructure. To achieve economically viable biofuel production, all aspects of these processes must be optimized. In particular, production hosts must natively have or be endowed with several important characteristics: extension of the substrate range, elimination of cellulose hydrolysates and biofuel product toxicity, and improvement of global regulatory functions. One method that has been proposed to reduce fuel production cost is to perform cellulose hydrolysis and fermentation in one step, called consolidated bioprocessing (CBP); this alternative approach avoids costs associated with cellulase production (Lynd and Laser, 2005). *Adapted from Lee, S.K., Chou, H., Ham, T.S., Lee, T.S., Keasling, J.D., 2008b. Metabolic engineering of microorganisms for biofuels production: from bugs to synthetic biology to fuels. Curr. Opin. Biotechnol. 19, 556–563.*

focused on upgrading cellulase actions and yield through strategic introduction of desired features to enzymes by engineering metabolic circuits, strain improvement, optimization of culture conditions, and recombinant or heterologous cellulase expression (Vu et al., 2011). The enzyme-mediated hydrolysis of lignocellulosic biomaterial is a key factor to establish economically viable cellulose bioconversion processes. The gap between low enzymatic hydrolysis proficiency and the high cost of cellulase production underlines the overriding burdens in this bioconversion process. A multifarious cellulase system exhibiting enhanced hydrolytic competency together with the capability to synthesize effective inducers was developed by the genetic engineering of *Trichoderma reesei* (Gao et al., 2017).

For enzymatic degradation of biomaterials, the high production cost of cellulase is a prodigious constraint for wide usage of cellulosic materials. Utilization of cellulolytic microorganisms and alteration of their genetic makeup for cost-effective synthesis of cellulase is the prospect of modern research. However, despite various scientific efforts over the past several years, cost-effective production of cellulolytic enzymes still faces numerous technological constrictions. Furthermore, cellulase, activities, desired features, and yields can be manipulated/ameliorated by DNA recombinant technology in alliance with an enzyme engineering approach (Behera and Ray, 2016).

Genetic engineering for cellulolytic enzymes synthesis is a cognate, yet unsettled, issue: strains synthesizing prodigious cellulase quantities (approximately $120\,g\,L^{-1}$ total protein) have been noted at the industrial level (Gupta et al., 2016). Unfortunately, many of the genetic manipulations accomplished through alternations in metabolic circuits, targeted engineering of promoter or transcription factor, or redesigning of epigenetic frameworks have been examined only under laboratory normal/auxotrophic conditions, as analyzed only through the proportionate abundance of cellulase mRNA (Druzhinina and Kubicek, 2017). There is no doubt that the cost of second-generation bioethanol can only be reduced when the cost component of cellulases can be reduced. There have been combined efforts to amend the process parameters, bioengineering, as well as strains. Strain amelioration via mutation has been prosperously employed on industrial cellulase producing strains; however, the process is quite skeptical and lengthy. Whole genome sequences of important cellulase-producing fungal strains and cognizance of their genetic functions have paved the way for prosperous targeted modifications. Genetic engineering of robust industrial strains has enabled a single strain to produce a repertoire of cellulases depicting high saccharification efficiency of the biomass. Targeted modifications at gene sequences of promoter, catabolite repressor, transcriptional factor, and enzymes (e.g., β-glucosidase [BGL], endoglucanase [EG], cellobiohydrolase [CBH], xylanase, auxiliary enzymes, etc.) have led to the development of the current cellulase-producing status, and the next steps should be focused on crossing the economic barriers for economically feasible second-generation bioethanol production. Though genetic modification is a potent technique to obtain a higher and efficient cellulase-producing strain, revelation of novel potential cellulase producers should remain a priority because the hidden treasures of nature are hard to predict (Singh et al., 2017).

14.2 BIOMASS: CELLULOSE AND LIGNOCELLULOSE

Biomass typically consists of three kinds of biopolymers, i.e., cellulose, hemicellulose, and lignin, while cellulose represents about half of the total wood mass and exhibits dynamic physicochemical and biochemical connections with lignin and hemicelluloses. Biochemically, cellulose is a linearized polymer of glucose units, represented by a well-orchestrated cellobiose polymer (D-glucopyranosyl-β-1,4-D-glucopyranose). In general, in its native form, cellulose is comprised of about 10,000 glycosyl units, constituting microfibrils, stabilized through intermolecular hydrogen bonds with the neighboring molecules. Crystalline domains of cellulosic biomaterials are disseminated by poorly organized, amorphous configurations, making it a potential target for enzymatic and chemical attacks. However, high-ordered arrangement of cellodextrins chains within crystalline cellulose makes it highly resilient to enzymatic/chemical hydrolysis to its monomer, i.e., glucose (Adsul et al., 2011). Biochemical composition of biomass is plant/species dependent and dominated by cellulose (Lerouge et al., 1998) that can be depolymerized via enzymatic actions of cellulase, which can further be subcategorized into EGs and CBHs (Teeri et al., 1998).

Acid-based and/or enzyme-based conversion approaches can potentially transform lignocellulosic biomass to commodity chemicals (Hahn-Hagerdal et al., 2007). A multienzyme-centered conversion process is generally preferable to a chemical-based approach due to many reasons, including (1) moderate operational environments, (2) superior transformation competences, (3) minimum substrate loss, and (4) zero production of hazardous chemicals (Alvira et al., 2010). Enzyme-predicated biomass pretreatment can facilitate degradation of cellulosic biomaterial due to the synergistic action of hydrolytic enzymes produced by certain microorganisms (Adsul et al., 2011). Plant-based biomass, being the most considerate sustainable biomaterial on Earth, can be the future of biofuel because it can be effortlessly converted into sugar and other carbohydrates and plausibly transformed into ethanol and several other subsidiary biochemicals (Lynd et al., 2002).

To reach sustainable economic viability, genetic and biotechnological approaches can be pivotal in the depolymerization of the cell wall through altered interactions among lignocellulosic components. Due to overcompensation in a plant's metabolic perturbances, negotiated growth and developmental physiology may generate novel structures in the cell wall without negative effects (Vermerris and Abril, 2015).

Utilization of sustainable organic resources, dominated by lignocellulosic biomaterials, with annual production of approximately 1×10^{10} MT globally (Sánchez and Cardona, 2008), for energy and commodity chemicals is projected to increase in the future. Cost-effective and technologically viable metabolic conversion bioprocesses are being comprehensively reviewed to satisfy the ever-demanding production of biofuel using lignocellulosic material as primary substrate. A great deal of attention has consequently been focused on developing genetically modified strains for competent utilization of sugars and their subsequent conversion to secondary compounds with major objectives pertaining to better yields, production, and to broaden the substrate and commodity spectra (Aristidou and Penttilä, 2000). Ethanol production through metabolic transformation of varied lignocellulosic biomaterials from forest and agricultural remainders offers several advantages; however, it is steadily obstructed by technological and economic issues (Sánchez and Cardona, 2008). Cellulose and hemicellulose are the major constituents of plant cell wall polysaccharides, and the key substrate for designing novel enzymes, in biofuels research. Agriculture- and forest-predicated lignocellulosic residues are among the most available but underutilized biomaterial on the biosphere (Beckham et al., 2012; Dashtban et al., 2009; Sánchez, 2009). Ecofriendly production of biofuel using sustainable bioresources, e.g., cellulosic material, is now receiving comprehensive attention (Sun and Cheng, 2002; Lin and Tanaka, 2006). Microbes, being a biofactory, harbor cellulose-cleaving enzymes, and play a crucial role in bioconversion mechanisms. However, cost-effective production of cellulolytic enzymes is among the major hurdles for the evolution of economically viable bioethanol-producing industries (Balan, 2014). In a genetic engineering study, *cre1* was proved to be an important target gene in strain manipulation for amending enzyme synthesis in *T. reesei* (Nakari-Setälä et al., 2009).

14.3 CELLULASE: THE MOLECULAR ENTITY FOR CELLULOSE-BASED INDUSTRIES AND OTHER APPLICATIONS

Cellulase is the key industrial enzyme, accountable for the bioconversion of lignocellulosic biomaterials into simpler products pertinent for a diversity of industrial operations (Esterbauer et al., 1991). Cellulose can be efficiently hydrolyzed through the synergistic action of a multienzyme complex of cellulases, comprising three different enzymes: (1) EG to defragment cellulose fibers to liberate minuscule fragments with free-reducing and/or nonreducing ends, (2) exoglucanase-/CBH to release diminutive oligosaccharides, i.e., cellobiose through their targeted enzymatic action on free-reducing and/or nonreducing ends of minuscule fragments, and (3) BGL to cleave cellobiose units into glucose monomers, the final product of the cellulase multienzyme complex (Gadgil et al., 1995). Bioconversion of cellulose-enriched biowastes is of substantial economic interest to design competent strategies for the treatment and consumption of cellulosic biowastes as low-cost carbon material for cellulase production (Ballesteros et al., 2004; Annamalai et al., 2013; Lee et al., 2008a; Chen et al., 2010). Biosystems can efficiently hydrolyze complex cellulosic biomaterials into simpler monomers through enzymatic bioconversion processes accomplished by multienzyme cellulase complexes (Johnvesly et al., 2002). Several biosystems, including different fungi, yeasts, and bacteria, and even some plants, are efficient producers of cellulases, but are usually challenging to culture/grow on an industrial level (Table 14.1).

Recently, a number of fungal mutants have shown their potential for synthesizing cellulase and its application in industrial biotechnology for the bioconversion of lignocellulosic materials into second-generation biofuels (Gusakov, 2013). Most of the genetic modification approaches are now targeted at redesigning/improving cellulases exhibiting consistent activity, superior stability/thermostability, broadened substrate spectrum, minimized/zero inhibition rates, etc. (Voutilainen et al., 2008; Gusakov, 2011, 2014; Bommarius et al., 2014; Gusakov et al., 2017). In general, cellulolytic enzymes comprise a catalytic domain and a cellulose-binding domain (CBD) coupled through a linker complex (Linder et al., 1996). Because of their widespread utilizations in cotton-based industries, pulp and paper recycling industries, juice extraction processes, surfactant/detergent enzymes, and cattle feed supplements, etc., cellulolytic enzymes are the third most industrially applicable enzymes globally (Singhania et al., 2010). However, for industrial applications, three crucial constraints are (1) high production cost of cellulase, (2) microbial susceptibility toward toxic components produced from the hydrolysis of cellulosic material, and (3) inefficiency of some microbes to ferment certain saccharides (Zhao et al., 2016).

14.4 MICROBES: BIOFACTORIES FOR CELLULASE PRODUCTION

Several microbes, including fungi, bacteria, yeasts, etc., harbor potential cellulose-degrading properties due to the occurrence of gene-encoding cellulases and therefore could be utilized as robust and smart biofactories for the decomposition of cellulosic biomass and their enzymatic bioconversion into high-energy organic molecules (Table 14.2) (Singh et al., 2017; Wen et al., 2005). Hence globally, several research groups are trapping novel microorganisms for the isolation of competent cellulolytic enzymes with higher specific activities (Annamalai et al., 2014). Some genetically modified bacteria and some thermophilic anaerobes exhibit potential properties for definitive microbial bioconversion of cellulosic material to ethanol (Bull, 1990). On average, microbe-based bioconversion of sugar remainders in wastepaper and yard waste from US landfills alone could generate more than 400 billion liters of ethanol (Lynd et al., 1991), equivalent to 10 times the corn-derived ethanol exhausted yearly as a 10% blend with gasoline (Keim and Venkatasubramanian, 1989). Metabolic bioconversion of lignocellulosic biomass to ethanol as a primary product comprises three steps: (1) chemical/enzymatic removal of the lignin component from the lignocellulosic complex, (2) defragmentation of cellulosic and lignocellulosic polymers to liberate free sugars, and (3) fermentation to release sugars to yield ethanol. However, no microbe in its native form can satiate all obligatory functions (e.g., high yield and productivity, broad spectrum of substrate utilization, tolerance to ethanol and other moderators present in hydrolysates, and cost related to biomaterial disposal). Therefore there is a need for the deployment of contemporary genetic manipulation technologies for the development of recombinants endowed with specific characters for executing such bioprocesses (Aristidou and Penttilä, 2000). Cellulose-degrading fungi through orchestrated regulation of inducible and repressible mechanisms direct synthesis of cellulolytic enzymes for precise responses to carbon-enriched environments. Many cellulose-degrading fungi, including *T. reesei* and *Penicillium oxalicum*, were genetically modified for preferable expression of transcription regulatory factors of many cellulolytic proteins. Nevertheless, in spite of such considerable attempts, the consequence of altering a single target to ameliorate the synthesis of cellulase is highly restricted (Yao et al., 2015).

14.5 MUTATION: MOLECULAR ARCHITECTURE TRANSFORMER BUT WITH UNPREDICTABLE CONSEQUENCES

Several mutagens, whether chemicals or radiations, can be strategically utilized for genetic improvement of different microbial strains for the production of cellulolytic enzymes (Table 14.3) (Singh et al., 2017; Li et al., 2010a; El-Ghonemy et al., 2014a,b). The genetic architecture of many fungi has been modified using mutagenesis as the preferred choice for converting them into a robust biological system for the synthesis of cellulose-degrading fungi (Chandel et al., 2012; Dashtban et al., 2009; Häkkinen et al., 2014). Mutants developed through mutagenesis are able to combat

TABLE 14.1 Cellulase Expression and Activity Levels When Produced in Bacterial, Yeast, Plant, and Fungal Organisms

Recombinant expression organism	Cellulase/s expressed			Protein expression (mg/L)	Substrate degradation, specific activity under ideal conditions (U/min)				Biomass degradation		References
	Cellulase/s	Native organism	Format		4-MUC	CMC	Avicel	Xylan	Pretreated	Untreated	
NATIVE EXPRESSION SYSTEMS											
1. *Trichoderma reesei* QM9414 2. *T. reesei* RUT-C30	Cellulase mixture			1. 14,000[a] 2. 19,000[a]		1. 109[b] 2. 150[b]					Peterson and Nevalainen (2012)
BACTERIAL RECOMBINANT EXPRESSION SYSTEMS											
Escherichia coli DH5α	Unidentified endo-glucanase	*Bacillus* sp.	Free enzyme	186[a] 11.2[c]		393[d]					Afzal et al. (2010)
E. coli BL21; *E. coli* Rosetta	Unidentified cellulases	Archael derived	Free enzyme	0.02[e]		138[f]	1241[g]		M: 1002[h] CS: 1318[h]		Graham et al. (2011b)
E. coli BL21 (DE3)	1. BglA 2. Bgl-cohesin chimera	*Clostridium thermocellum*	1. Free enzyme 2. Chimeric enzyme (with native cellulosome)	1. 90[c] 2. 25.2[c]					SG: ~160[i]		Gefen et al. (2012)
Bacillus subtilis BAL2238	Cel5A; Cel48F; Cel9E; Scaf	Catalytic domains from *Clostridium cellulolyticum*	Minicellulosome	~150,000[j]					CS: ~360[k]	CS: ~23[k] S: ~16[k] SG: ~15[k]	Anderson et al. (2013)
Zymomonas mobilis ATCC39676	1. E1 2. GH12	*Acidothermus cellulolyticus*	Free enzymes	1. 1.7[l] 2. 4.6[l]	Activity shown						Linger et al. (2010)
YEAST RECOMBINANT EXPRESSION SYSTEMS											
Saccharomyces cerevisiae	CBH I; CBH II; CBH IIb	Fungal sources	Free enzymes in mixtures	~1000[a]			~22[m]				Ilmen et al. (2011)
Kluyveromyces lactis	Swo	*T. reesei*	Free protein	~30[a]			Activity shown[n]				Jäger et al. (2011)
PLANT RECOMBINANT EXPRESSION SYSTEMS											
1. *Nicotiana tabacum* (T) 2. *Zea mays* (CS)	E1	*A. cellulolyticus*	Free enzymatic catalytic domain	1. 3.1[o] 2. 0.003[o]					T: ~10[p] CS: 15[p]	CS: 8[p]	Brunecki et al. (2011)
Oryza sativa L. (R)	E1	*A. cellulolyticus*	Free enzymatic catalytic domain	6.1[q]	0.079[r]					0.135[n]	Chou et al. (2011)
N. tabacum	SSO1354	*Sulfolobus solfataricus*	Free enzyme in endoplasmic reticulum	1.1[q]	Activity shown	Activity shown			T PCWP: 35[k]		Klose et al. (2012)

Continued

TABLE 14.1 Cellulase Expression and Activity Levels When Produced in Bacterial, Yeast, Plant, and Fungal Organisms—cont'd

Recombinant expression organism	Cellulase/s expressed			Protein expression (mg/L)	Substrate degradation, specific activity under ideal conditions (U/min)				Biomass degradation		References
	Cellulase/s	Native organism	Format		4-MUC	CMC	Avicel	Xylan	Pretreated	Untreated	
N. tabacum	1. Cel9A 2. Cel6B; and others	*Thermobifida fusca*		1. 40[q] 2. 5[q]		~0.028[s]			WS: ~0.048[t]		Petersen and Bock (2011)
N. tabacum	a. Cel5A b. Cel5A chimera	*T. maritima*	Free enzyme in 1. Apoplast 2. Chloroplast 3. Cytoplasm	1. a: 2[q] 2. b: 5.2[q] 3. a: 0[q]		b: 2[u]				1.891[v]	Mahadevan et al. (2011)
FUNGAL AND OTHER RECOMBINANT EXPRESSION SYSTEMS											
T. reesei QM9414	XylII	*T. reesei*	Free enzyme	1.94[a]				8866[w]			Harris et al. (2013)
T. reesei RUT-C30	Bgl1	*Penicillium decumbens*	Free and native enzymes	10 710[a]					CS: 5.01[x]		Ma et al. (2011b)
Chlamydomas reinhardtii	Xyl1	*T. reesei*	Free enzyme	0.25[q]				~2300[y]			Kim et al. (2010)

[a] *Crude enzyme solution.*
[b] *U = mmol glucose or glucose equivalent/mL enzyme solution.*
[c] *Purified enzyme solution.*
[d] *U = g glucose or glucose equivalent/g enzyme.*
[e] *mg/g cell pellet.*
[f] *U = mmol glucose or glucose equivalent/mmol enzyme.*
[g] *Hydrolysis of 1 mmol glucose equivalent per mmol enzyme/15 min.*
[h] *Glucose/glucose equivalent released/15 min.*
[i] *Glucose equivalent released (mM).*
[j] *Number of units/cell.*
[k] *Glucose or glucose equivalent released (mg/L).*
[l] *Percent of total cellular protein.*
[m] *Percentage of substrate hydrolyzed.*
[n] *Activity as cellulose swelling (pretreatment) molecule.*
[o] *mg/g biomass.*
[p] *Increase in biomass digestibility (%).*
[q] *Percent of total soluble protein.*
[r] *U = hydrolysis of 1 mmol substrate/mmol enzyme.*
[s] *U = mmol of reducing sugars/mL enzyme solution.*
[t] *mmol glucose equivalents.*
[u] *U = 1 mM glucose reducing sugar/mg total soluble protein.*
[v] *mg/mL of solubilized glucose by autohydrolysis in 24 h.*
[w] *U = mmol xylan released from birchwood xylan/mL enzyme solution.*
[x] *U = mg/mL glucose after 12 h.*
[y] *U = mmol 6,8-difluoro-4-methylumbelliferyl b-D-xylobioside degraded/mL enzyme solution.*

Abbreviations: *4-MUC*, 4-methylumbelliferyl-D-cellobiose; *CMC*, carboxymethyl cellulose; *CS*, corn stover; *M*, *Miscanthus*; *PCWP*, plant cell wall polysaccharides; *R*, rice; *S*, straw; *SC*, sugarcane; *SG*, switch grass; *T*, tobacco; *W*, wheat.

Adapted from Garvey, M., Klose, H., Fischer, R., Lambertz, C., Commandeur, U., 2013. Cellulases for biomass degradation: comparing recombinant cellulose expression platforms. Trends Biotechnol. 31(10), 581–593.

TABLE 14.2 Cellulase- or Cellulase-Encoding Genes Present in Microorganisms

Microorganisms	Cellulase/genes	References
Trichoderma reesei	2 CBHs, 8EGs, 7 BGLs, 16 hemicellulases, 9 LPMOs genes	Aro et al. (2005) and Törrönen et al. (1994)
Humicola insolens	2 CBHs, 5 EGs lacking CBM	Schulein (1997)
Neurospora crassa	3 CBHs, 4 EGs, and 1 BGL	Yazdi et al. (1990)
Penicillium decumbens	3 CBHs, 11 EGs, and 11 BGLs	Liu et al. (2013b)
Myceliophthora thermophila	8 EGs, 7 CBHs, 9 BGLs, and 25 LPMOs genes	Karnaouri et al. (2014b)
Aspergillus niger	2 CBHs, 2 EGs, 5 BGLs, and 7 LPMOs genes	de Vries and Visser (2001) and Singhania et al. (2011)
Clostridium thermocellum	12 endo- and exocellulases	Bayer et al. (1998b)

Adapted from Singh, A., Patel, A.K., Adsul, M., Singhania, R.R., 2017. Genetic modification: a tool for enhancing cellulase secretion. Biofuel Res. J. 14, 600–610.

TABLE 14.3 Cellulase Production Improvement via Mutagenesis Among Microorganisms

Microorganism	Type of mutation	Recombinant properties	References
Pencillium oxalicum	Functional mutation	Enhanced cellulase production (160 FPU/L/h)	Liu et al. (2013b)
Aspergillus species	Sequential random mutation	Enhanced cellulase production	Vu et al. (2009)
Acremonium cellulolyticus	Random mutation	Enhanced β-glucosidase activity and FPase activity	Fujii et al. (2013)
Celluloytic fungi	Random mutation	Stable and enhanced cellulase production	Chand et al. (2005)
Penicillium janthinellum NCIM1171	Random mutation	Enhanced cellulase production and hydrolysis	Adsul et al. (2007)
Trichoderma atroviride	Random mutation	Enhanced cellulase production	Li et al. (2010b)
Humicola insolens	Random mutation	Enhanced cellulase production	Javed et al. (2011)
Aspergillus sp. Su 14	Random mutation	Enhanced cellulase production	Vu et al. (2011)
Penicillium decumbens 114	Random mutation	Enhanced cellulase and hemicellulase production	Liu et al. (2013a)
Trichoderma viride	Random mutation	Enhanced cellulase production	Li et al. (2010b)
Aspergillus terreus AUMC 10138	Random mutation	Enhanced alkaline cellulase production	Isaac and Abu-Tahon (2015)
Fusarium oxysporum	Random mutation	Enhanced cellulase production	Kuhad et al. (1994)
Cellulomonas	Random mutation	Enhanced xylanolytic activity	Mayorga-Reyes and Ponce-Noyola (1998)
P. funiculosum	Site-directed mutagenesis	Increased enzyme activity against crystalline cellulose	Adney et al. (2009)
Thermotoga maritima	Site-directed mutagenesis	Increased thermostability and activity toward Avicel	Mahadevan et al. (2008)
Macrophomina phaseolina	Site-directed mutagenesis	Modified endoglucanase for hydrolysis of β-glucans	Wang and Jones (1997)

Adapted from Singh, A., Patel, A.K., Adsul, M., Singhania, R.R., 2017. Genetic modification: a tool for enhancing cellulase secretion. Biofuel Res. J. 14, 600–610.

more astringent environments or acquire superior resistance to environmental fluctuations through the repeated cycles of their cultivation (Talbot, 2001). Some alkylating chemicals, e.g., *N*-methyl-*N*′-nitro-*N*-nitrosoguanidine and ethyl methanesulfonate (EMS), can be potential mutagens to effectively alter the genetic architecture of fungi (Talbot, 2001; Chand et al., 2005). The mutagenesis approach, to augment cellulase synthesis in brown rot fungi and white rot fungi, is not usually preferred, in spite of being a first-rate and conducive genetic modification technique. Several fungal strains have been exposed to different chemical mutagens, e.g., EMS, *N*-methyl-*N*-nitro-*N*-nitrosoguanidine, and radiations, namely ultraviolet (UV) and gamma irradiation (through Co^{60}), to amend cellulose production (Fawzi and Hamdy, 2011; Mostafa, 2014; El-Ghonemy et al., 2014b). Dhillon et al. (2012) experimentally established higher extracellular cellulase production from NRRL 3484, a mutant strain of *Aspergillus oryzae*, through a strategic multistep mutation approach. The fungus *Chaetomium cellulolyticum* NRRL 18756, when exposed to 0.5 kGy of gamma irradiation, exhibited 1.6-fold higher extracellular carboxymethyl cellulase production compared to the wild type (Fawzi and Hamdy, 2011). An indigenous strain of *Trichoderma viride* FCBP-142, in an experiment of induced mutation, was exposed to UV and EMS to produce a sufficient quantity of cellulases (Shafique et al., 2011). Mutant strains have produced approximately $35\,g\,L^{-1}$ of extracellular protein (Singhania et al., 2012), and all consisting of cellulases and hemicellulases (Gritzali and Brown, 1979).

14.6 GENE MINING FOR EFFICIENT CELLULASE PRODUCTION

Biomass processing can commence only under an environment of slow kinetics of biomaterial decomposition to relinquish products with high yields. Contemporary advancements in synthetic biology (Pleiss, 2006) alone or in permutation with metabolic engineering (Lee et al., 2006; Keasling and Chou, 2008) and a system biology approach (Mukhopadhyay et al., 2008) have discovered microbial cell factories to synthesize commodity chemicals.

Genome-wide analysis of cellulose-degrading microbes has been accomplished over the last few years and the DNA sequences have contributed consequential emergent tips about the genetic basis of microorganism-mediated cellulose degradation. Genome sequences of some of the aerobic microorganisms, *Hypocrea jecorina* (*T. reesei*), *Phanerochaete chrysosporium*, and *Thermofida fusca*, reveal that all harbor cellulase-coding multigene complexes, mostly encoding a CBM, and also confirm the presence of many presumptive cellulase genes in each organism (Martinez et al., 2004; Lykidis et al., 2007). However, genome-wide sequence analysis of some anaerobic bacteria, namely *Clostridium thermocellum*, *Ruminococcus albus*, and *Ruminococcus flavifaciens*, proclaim the presence of scaffoldin genes in all genomes and cellulase-coding multigenes complexes encoding docerin domains. Furthermore, these domains are also thought to be consistent with their habitation of cellulosomes and many operative cellulase genes are reported among the docerin-transcribing genes (Bayer et al., 2008). Some microorganisms, in spite of being cellulolytic in nature, do not confirm the presence of processive cellulase, scaffoldin, or docerin domains. *Cytophiga hutchinsonii* (aerobic cellulolytic bacterium) (Xie et al., 2007) and *Fibrobacter succinogenes* (anaerobic cellulolytic bacterium) (Qi et al., 2007) show strong affinity to cellulose fibers during growth and multiplication; however, their genome sequence analyses do not support utilization of free cellulase/cellulose-degrading cellulosomal mechanisms, therefore, hypothetically, they have adapted a novel route for cellulose degradation (Wilson, 2008, 2009).

Transcriptional substratum, medium composition, and dynamics are the key stimulants and regulators of cellulase. A cellulose-based medium can be the positive regulator for cellulase production, whereas utilization of glucose and cellobiose-based media show negative regulation. Cellulase production is reported to be positively regulated by xyr, Ace2, and HAP2/3/5 factor complexes on transcriptional substratum; however, regulatory factors Cre1 and Ace1 show negative repressible regulation (Singhania et al., 2012). Enzymatic analysis of CBHs1) EGs, and BGLs has confirmed that strong synergism among these is obligatory for efficient cellulose hydrolysis. The *Trichoderma* genetic system entails a complex set of hemicellulolytic enzymes comprising two side chain cleaving enzymes, namely endo-β-xylanases and β-mannanase. In *Trichoderma*, cellulase production is controlled at the transcriptional level and is reliant on the type of carbon source available (Kubicek and Penttilä, 1998). The genes are strongly suppressed by glucose and persuaded up to several thousand-fold by cellulose/sophorose (Mandels et al., 1962). Glucose-mediated suppression of cellulase expression is regulated by the gene *Cre1* of *Trichoderma* (Ilmén et al., 1996). Under variable conditions, expression of principal cellulase genes, *cbh1*, *cbh2*, *egl1*, and *egl2*, has been reported to be harmonized, showing utmost expression of the *cbh1* gene encoding CBH (Ilmén et al., 1997). Many carbon substrates and several inducers were utilized to analyze operative regulatory mechanisms of comparative expression levels of different genes encoding cellulase (Margolles-Clark et al., 1997). However, the precise molecular mechanism confirming the facts of cellulase genes activation and regulation needs to be explored (Ali et al., 2014).

14.7 GENETIC ENGINEERING: CONTEMPORARY AND COMPETENT TECHNOLOGY FOR AUGMENTING CELLULASE PRODUCTION AND EFFICIENCY

Genetic alterations in different microorganisms have been accomplished either to broaden the spectrum of substrate utilization, dominantly cellulose and hemicellulose, or to channelize intermediary metabolites typically toward ethanol (Ladisch et al., 1983). Modern tools of genetic engineering have empowered many scientific groups to

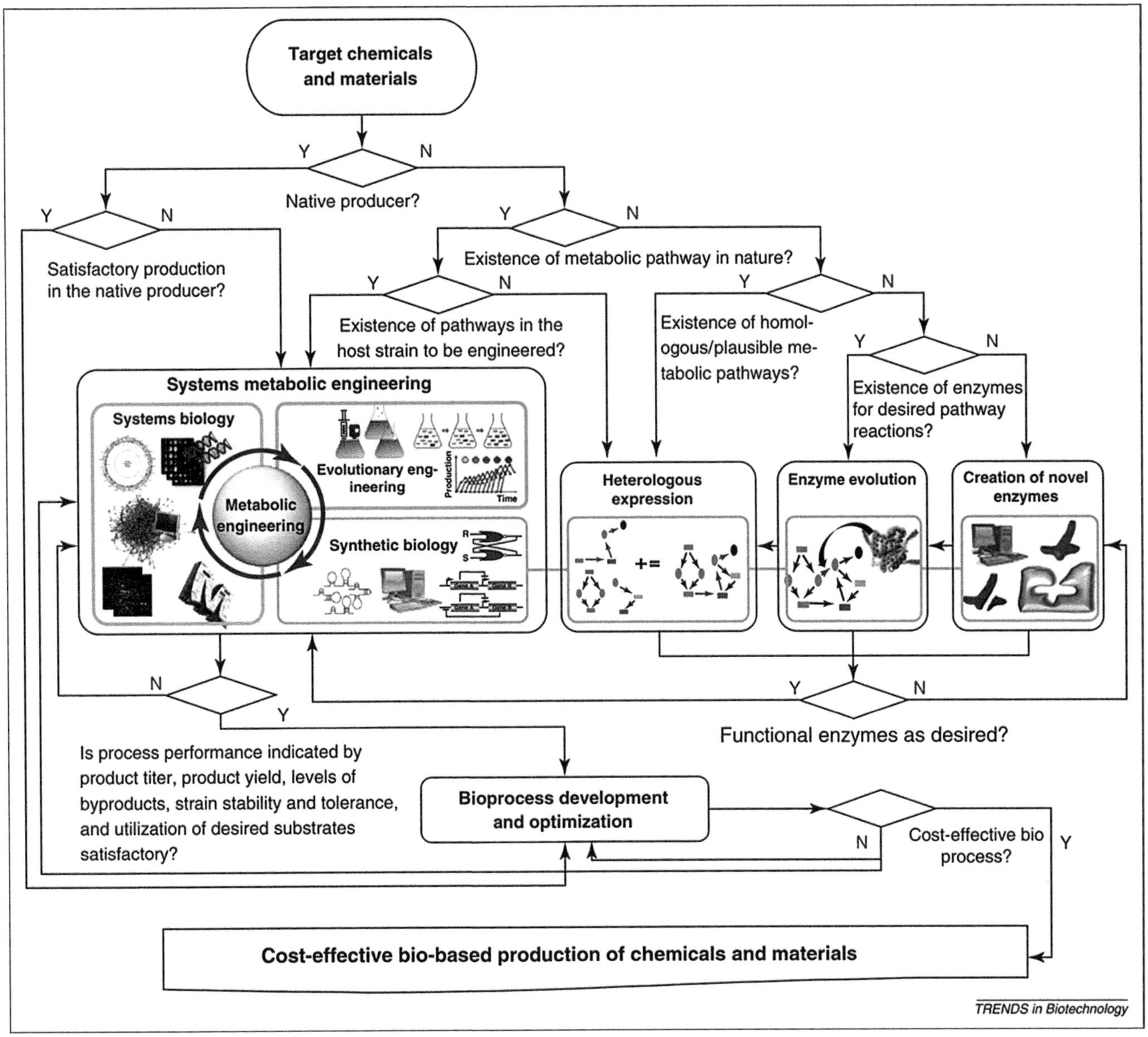

FIG. 14.2 Systems metabolic engineering for the production of chemicals and materials. Systems biology, synthetic biology, and evolutionary engineering have been integrated into metabolic engineering. Combined use of these disciplines that complementarily and synergistically overcome the weaknesses of each technology will increase the efficiency of strain development by metabolic engineering, by allowing the creation of novel enzymes and metabolic and gene regulatory circuits; fine-tuning and optimization of the metabolic fluxes that lead to increased yield and concentration of a desired product; increased tolerance of cells to the product and harmful medium components; formation of fewer or no by-products; enhanced utilization of desired carbon substrates; and even development of cost-effective fermentation and downstream processes. Based on several criteria such as the availability of native producer, biosynthetic pathways, enzymes, and others, the starting point of systems metabolic engineering can be decided. Several cycles of systems metabolic engineering can be performed until the efficiencies of production of the desired chemicals and materials become satisfactory. *Adapted from Lee, J.W., Kim, T.Y., Jang, Y., Choi, S., Lee, S.Y., 2011a. Systems metabolic engineering for chemicals and materials. Trends Biotechnol. 29(8), 370–378.*

redesign/reconstruct microorganisms (yeasts or bacteria) for efficient conversion of biomaterial to effective products. This stratagem can be executed via two main approaches: (1) to genetically design organisms to utilize a broadened spectrum of substrate and (2) to genetically manipulate an organism concealing all possible routes to convert intermediary metabolites of the central carbon metabolism. The first approach is focused on efficient ethanologenic microorganisms (generally, yeast and the bacterium *Zymomonas mobilis*) to introduce the metabolic pathways for xylose or arabinose catabolism (Kötter et al., 1990; Kötter and Ciriacy, 1993). The second approach is to commence with microorganisms consuming a broad range of substrates, mostly hexoses and pentoses (e.g., *Escherichia coli*), and to introduce metabolic routes for their conversion to ethanol (Carey et al., 1983; Alterthum and Ingram, 1989; Ohta et al., 1991a,b; Wood and Ingram, 1992; Beall and Ingram, 1993; Yomano et al., 1998).

Metabolic engineering (or a homogeneous field, synthetic biology) provides implements for genetic alteration of nonnative organisms to produce a wide range of fuel-grade molecules (Fig. 14.2). Regular advancements in redesigning the metabolic pathway along with strain optimization, using experimental and computational tools, are prerequisites for economically feasible and highly efficient production (Shota Atsumi and Liao, 2008; Lee et al., 2011a). Expertise regarding manipulation of plant genetic makeup to produce a transgenic crop, their farming and harvesting, and postharvest processing the plant matter are accessible (Howard and Hood, 2005). Analysis of heterologous expression of specific enzymes responsible for plant cell wall degradation with enhanced production is under consideration for prudent hydrolysis of cellulosic materials. Most of the cell wall depolymerizing enzymes recognized so far are of bacterial lineage, and codon modification in the open reading frame is obligatory to ascertain competent expression in plants; this is an outright strategy to be broadly employed for heterologous expression of prokaryotic proteins in eukaryotic systems (Sticklen, 2008). A more preeminent diversity of protein engineering implements can be helpful to tightly couple complex enzymes with higher prospects of success in amending their activities (Bommarius et al., 2014). Microorganisms have an inherent distinct genetic architecture comprising information to synthesize enzymes for production of alternative fuels through a metabolic engineering approach. However, targeted gene manipulation is not possible through traditional procedures of genetic strain improvement, but industrial biotechnology has played a pivotal role to design novel pathways to produce commodity chemicals, which are typically not naturally synthesized by microorganisms (Chandel and Singh, 2011). Genetic engineering offers utmost potential for modifying the broad spectrum of lignocellulolytic enzymes to construct designer enzymes for a diversity of applications. However, miraculously, it may additionally be plausible to merge different lignocellulolytic genes or segments of genes from different organisms to generate naïve chimeric proteins or enzymes with desired properties (Kumar et al., 2008). With the dawn of advanced biotechnological and computational biological tools, novel enzymes can be explored through metagenomics approaches. Metagenomic analysis of the *Nasutitermes* hindgut shows an affluent miscellany of bacterial genes comprising up to 100 families of proteins with the ability to hydrolyze glycosidic linkages between carbohydrates or between carbohydrate and noncarbohydrate molecules (Warnecke et al., 2007).

A metabolic engineering approach genetically manipulates cellular metabolic networks by implementing biological system-based information to amend its properties pertinent for industrial application (Bailey, 1991; Stephanopoulos and Vallino, 1991; Nielsen, 2001). The complexities of cellular metabolisms and their cognate regulation network are major challenges for the development of a robust cellulase-producing system. However, recent advancements in omics technology and computational modeling are progressively being utilized to decrypt the genetic information of microorganisms. Systems biology can expedite analysis of metabolic networks in an organized and comprehensive manner to accelerate the genetic engineering process more efficiently. Moreover, systems biology is principally pertinent to decipher genetic circuits and mechanisms to amend phenotypes of the strains derived through mutation-based approaches and screening, or through adaptive laboratory-based evolution (Hong et al., 2011; Caspeta et al., 2014). Systematic integration of these state-of-the-art technologies have been proved to be smart tools for altering metabolic circuitry to optimize de novo metabolic pathways and strain optimization, typically represented by the cycle of metabolic engineering (Dai and Nielsen, 2015).

Lucrative cloning and efficient expression of cellulases and polygalacturonase genes can provide a solution for the development of a cost-effective biorefinery industry to realize the decline in the utilization of unsustainable resources. Consequent laboratory research related to optimization of enzyme-mediated bioconservation of lignocellulosic waste can investigate to support maximum biomaterial conversions into fermentable sugars (Ibrahim et al., 2013).

Furthermore, integration among evolutionary engineering, synthetic biology, and systems biology has already transmuted and elevated metabolic engineering to an inimitable level (Park et al., 2008; Jung et al., 2010; Lee et al., 2007). Due to universal acceptance and accuracy in the alteration of gene regulatory metabolic circuits, systems metabolic engineering has proven to be the foremost strategy for the improvement of industrially competitive strains. Furthermore, systems metabolic engineering also sanctions the probing of many target genes; the synthesis of biomolecules with amended functions; and the adaptive evolution of cells (Lee et al., 2011a).

Rewiring of cellular metabolisms can redirect metabolic pathways for augmented production of desired metabolites, with the proficiency to generate novel products. However, transformation of cells into competent biofactories is challenging due to hardwired, robust, cellular metabolic networks, connected via tightly coupled communication lines to regulate molecular networks (Nielsen and Keasling, 2016). Metabolic engineering of noncomplexed cellulase systems, as produced by *T. reesei* and other filamentous fungi, involves three key research dimensions: (1) coherent restructuring of each cellulase, predicated on cellulase molecular structure and mechanism of catalytic action

(Schulein, 2000; Wilson, 2004); (2) designing of expression cassettes and directed evolution approaches for each cellulase, for screening competent enzymes with novel characters through arbitrary mutations approaches and/or molecular recombination (Cherry and Fidantsef, 2003; Arnold, 2001; Hibbert et al., 2005; Shoemaker et al., 2003); and (3) optimization of different cellulase cocktails to evaluate their activity on insoluble cellulosic materials with upgraded hydrolysis rate for superior cellulose digestibility (Sukumaran et al., 2009; Boisset et al., 2001; Himmel et al., 1999; Kim et al., 1998; Zhang and Lynd, 2004). Designing of an artificial cellulase system seems to open new vistas to metabolic engineering and appears to be more favorable for producing competent cellulases with desired characteristics. Structure-directed reconstitution of three fungal class II CBHs (CBH II) assisted development of profoundly thermotolerant CBH II recombinants (Heinelman et al., 2009). Inventories of robust cellulase systems, as optimized through proficient mixtures of cellulases, can set a blueprint for successful biomass conversion (Singhania et al., 2010). Genetically recombinant fungal strains have already been sanctioned for cellulase production at an industrial level (Holker et al., 2004). Furthermore, genetic alteration of fungal strains is intended to attain a dual purpose: (1) amendment to cellulase yield and (2) enhancement in thermotolerance, tolerance to chemical inhibitors, and resistance to environmental fluctuations of cellulase systems (Chandel et al., 2012; Dashtban et al., 2009). Improvements in these and other quality parameters in cellulase can elevate the hydrolysis rate for conversion of cellulose to glucose (Salinas et al., 2011; Lee et al., 2001). In spite of its potential advantages, genomic alteration is a comparatively intricate process, hence the biosafety aspect of genetically altered fungal strains should be precisely prioritized (Yoon et al., 2014).

14.7.1 Fungi: The Most Preferred Genetic Platform for Cellulase Production

Fungi have inherent multienzyme complexes to synthesize hydrolytic and oxidative enzymes for sufficient degradation of lignocellulosic biomaterials (Bon and Picataggio, 2002; Ferreira-Leitao et al., 2003). In plants, hydrolysis of fundamental carbohydrates, especially the cell wall, to sugars is vital to generate lignocellulosic biofuels (Chundawat et al., 2011; Himmel et al., 2007). Several enzyme engineering approaches are reliant on enzymatic cocktails developed from effective cellulase-producing filamentous fungi for decomposition of biomaterial (Himmel et al., 2007; Bu et al., 2011), particularly accelerated by the discovery and cellulase characterization of *T. reesei* in the 1940s (Martinez et al., 2008; Reese, 1976). Expression of chimeric cellulase in fungal, yeast, bacterial, and plant systems was successfully demonstrated by the targeted insertion of the *T. reesei Cel5A* gene into different hosts for protein synthesis (Fig. 14.3). Enzymatic components, synthesized by *T. reesei*, e.g., BGLs, EGs, and CBHs (Kumar et al., 2008; Martinez et al., 2008), were experimentally expressed in yeasts to establish augmented levels of enzymes and ethanol production simultaneously. Some carbon sources or disaccharides such as lactose or sophorose were reported as strong inducers for the expression of cellulase enzyme complexes and of the majority of hemicellulases in an orchestrated manner in *T. reesei*. Coordinated expression of cellulase enzyme complexes can be manipulated via genetic engineering of regulators responsible for their expression (Gao et al., 2017; Stricker et al., 2006). Key enzymatic machinery of cellulose hydrolysis contains three enzymatic components: (1) EGs to cleave cell wall cellulose in the middle, (2) CBHs (exoglucanases) to liberate cellobiose units, and (3) BGLs to catalyze cellobiose and cello-oligosaccharides to glucose. This key enzymatic complex is assisted by swollenins, a protein family, exhibiting sequence resemblance to plant expansins and having inherent competency to disintegrate the plant cell wall without showing any damage to its components (Saloheimo et al., 2002; Verma et al., 2013; Sampedro and Cosgrove, 2005). The swollenin-expressing gene *swo1* was induced in a coordinated manner under cellulolytic enzyme-inducing culture environments in *T. reesei* (Foreman et al., 2003). Further cellulose-mediated induction of the swollenin gene sustained overproduction of swollenin in *Trichoderma asperellum* and reported improved root colonization (Brotman et al., 2008). Heterologous expression of swollenin Afswo1 from *Aspergillus fumigatus* exhibited enzyme-mediated degradation of Avicel cellulose to simpler molecules without apparently reducing sugar production. Fast-rated decomposition of Avicel by a cellulose admixture was also validated with the inclusion of Afswo1 (Chen et al., 2010). Similar consequences with the swollenin addition of *P. oxalicum* were also noticed (Kang et al., 2013). Jäger et al. (Jäger et al., 2011) expressed *T. reesei* swollenin in a *Kluyveromyces lactis* system and reported swollenin-regulated disruption of filter paper surface, evaluated through microscopic analysis. Furthermore, cellulose crystallinity and filter paper index were also inhibited by swollenin treatment. They also revealed that pretreatment of cellulosic substrate or filter paper with swollenin could be supportive for abrupt degradation of these substrates through accelerated activity of the cellulase complex (Gourlay et al., 2013). Fungi-based swollenins have been reported to degrade the structure of many lignocellulosic complex materials such as Avicel cellulosic materials (Kang et al., 2013; Jäger et al., 2011), filter paper (Saloheimo et al., 2002; Verma et al., 2013; Jäger et al., 2011), and corn stover (Andberg et al., 2015). Swollenin can efficiently cleave β-1,4 linkages as a major key catalytic activity and hence designate itself analogous to EG. Nevertheless, the

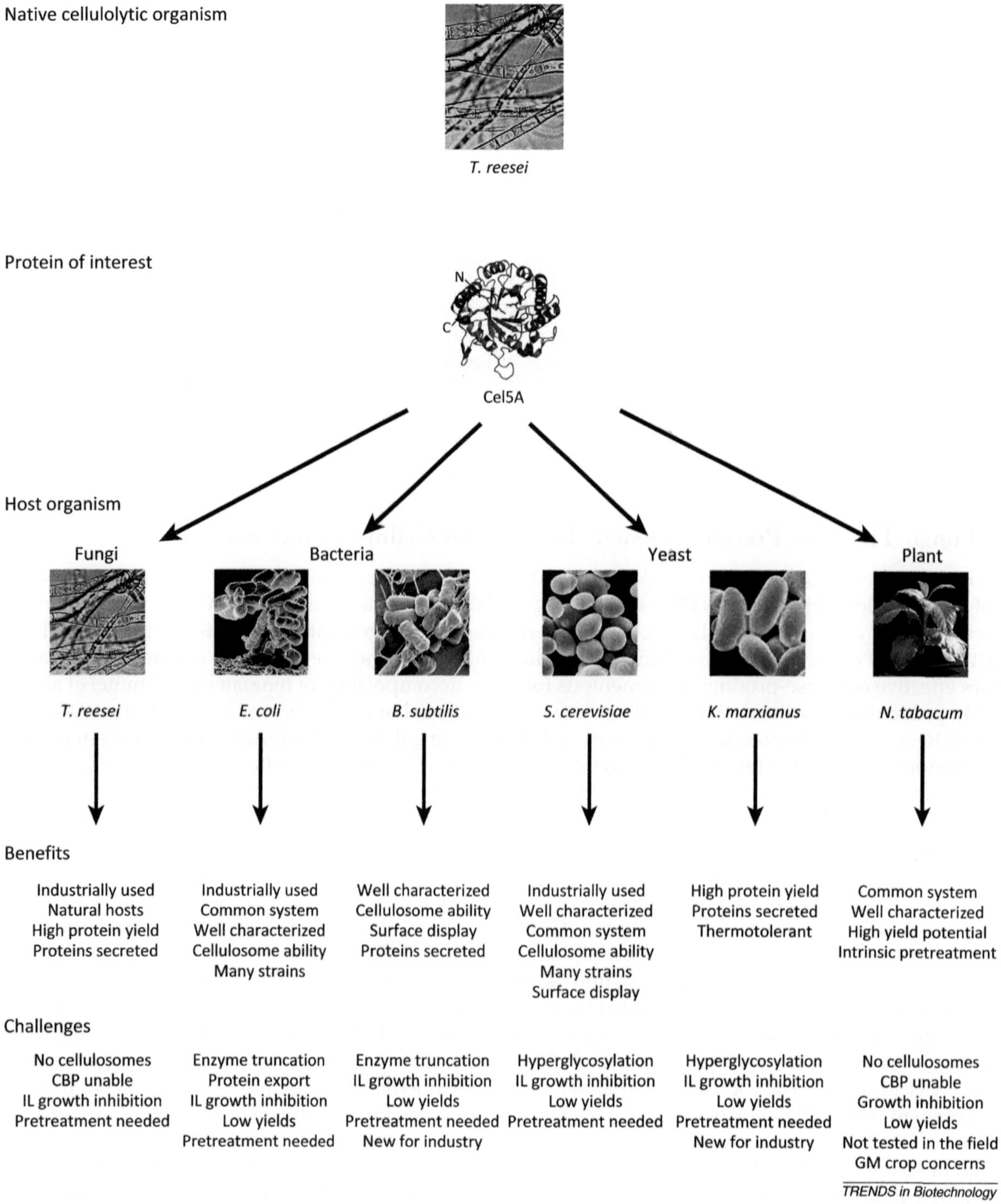

FIG. 14.3 Recombinant cellulase expression in bacterial, yeast, plant, and fungal systems, exemplified by the transfer of the *Trichoderma reesei Cel5A* gene into different host organisms for protein production, and outlining some of the foremost benefits and drawbacks from each expression system. The examples selected are commonly used for cellulase expression in laboratories, with the aim of increasing protein yields or activities from the protein when expressed through its native (unmodified) host. Although each recombinant host displays benefits for cellulase expression, only synthetic biological modifications to the native host (*T. reesei*) have been able to improve upon the yields and activity found natively. Organism images are adapted from: *T. reesei* (Górka-Nieć et al., 2010; Lee et al., 2011b) (micrograph and crystal structure, respectively), *Escherichia coli* (E. Erbe, Agricultural Research Service, https://www.ars.usda.gov/is/graphics/photos/mar05/K11077-1.htm), *Bacillus subtilis* (Zweers et al., 2008), *Saccharomyces cerevisiae* (Suh et al., 2003), and *Kluyveromyces marxianus* (Odronitz et al., 2007). The image of *Nicotiana tabacum* was taken in-house. *CBP*, Consolidated bioprocessing; *GM*, genetically modified; *IL*, ionic liquid. *Adapted from Garvey, M., Klose, H., Fischer, R., Lambertz, C., Commandeur, U., 2013. Cellulases for biomass degradation: comparing recombinant cellulose expression platforms. Trends Biotechnol. 31(10), 581–593.*

main catalytic product of swollenin action is cellobiose, a characteristic feature of CBHs. Therefore these disparities in mechanism of action do not support swollenin or either of the typical cellulase categories (Andberg et al., 2015).

Alriksson et al. (Alriksson et al., 2009) successfully developed a genetically recombinant *Aspergillus niger* D15 (egl) strain heterologously expressing EG Cel7B of *H. jecorina*, which demonstrated higher EG activity (2100 nkat/mL) when allowed to grow on the spent hydrolysates. DNA sequence analysis of the potential cellulose-degrading fungus *H. jecorina* (*T. reesei*) was also performed (Martinez et al., 2008). In *T. reesei*, unlike in *Neurospora crassa*, *Fusarium graminearum*, or *Aspergillus* spp., the XYR1 represents a fungi-specific class of main activators of transcriptional machinery of both cellulase and xylanase genes, comprising Zn^{2+} binding domains by six cysteine residues (Zn2Cys6) (Gao et al., 2017; Furukawa et al., 2008). Cellulase induction is positively regulated by *xyr1* because deletion of *xyr1* eliminates cellulase induction by all possible inducers; however, *xyr1* overexpression boosts it (Gupta et al., 2016; Bischof et al., 2016; Druzhinina and Kubicek, 2016). Genetic engineering of XYR1 revealed that 26-fold constitutive overexpression of *xyr1* in *Trichoderma* cf. *harzianum* consequently resulted in the induction of an enzymatic complex catalyzing 25% improved digestion of sugarcane bagasse during the initial 24 h of saccharification (da Silva Delabona et al., 2017). Additionally, reduced glucose repression was noticed through constitutively expressing the XYR1 (V821F) mutant (Derntl et al., 2013) in *T. reesei* and achieved improved cellulase and xylanase synthesis on sugarcane bagasse (Ellilä et al., 2017). In *T. reesei*, cellulase encoding gene expression is potentially regulated through the Zn2Cys6 transcriptional activators, namely ACE2 and ACE3, and CCAAT-binding protein complex (HAP2/HAP3/HAP5) (Häkkinen et al., 2014). Furthermore, the cellulase gene expression machinery in *N. crassa* shows participation of many transcription factors, from which CLR2 and VIB1 (Coradetti et al., 2012; Xiong et al., 2014) have orthologs in *T. reesei* as revealed through genome annotation (Druzhinina et al., 2016). Nonetheless, their role in the regulation of cellulose gene expression in *T. reesei* is not evident, and consequently their structural and functional manipulation could be adapted to amend cellulose production. Two fascinating alternatives for manipulation of gene encoding cellulase have recently been described: (1) a hybrid cellulose regulator, consisting of the DNA-binding domain of the glucose-repressor CRE1 fused with XYR1 effector binding domains, was constructed (Zhang et al., 2017). Overexpression of this hybrid cellulose regulator elevated the synthesis of constitutive cellulase and hemicellulose on glucose as primary substrate by approximately 30-fold in *T. reesei*. (2) One of the transformants obtained from an artificial zinc finger protein library (expressed in *T. reesei*) exhibited 55% upsurge in cellulase activity against filter paper and about 8.1-fold augmented BGL activity (Zhang et al., 2016). Carbon catabolite repressor protein CRE1 has also been shown to suppress cellulase gene expression by rapidly assimilated carbohydrates (Nakari-Setälä et al., 2009). CRE1 restricts both induced and constitutive cellulase and hemicellulose gene expression; however, the degree of this repression differs for different varieties of cellulase and hemicellulose genes (Druzhinina and Kubicek, 2016). For example, constitutive expression and noninductive expression of the CBH gene *cel6A* and of the xylanase gene *xyn1* is suppressed by CRE1 (Mach et al., 1996; Zeilinger et al., 2003). However, in contrast, CRE1 suppresses both constitutive and induced expression of the CBH I gene *cel7a* (Nakari-Setälä et al., 2009).

Mutation-based silencing of CRE1 resulted in a mutant confirming improved cellulase synthesis induced via higher quantities of carbon sources (cellulose or lactose) (Eveleigh and Montenecourt, 1979). This mutant, recognized as *T. reesei* RUT-C30, inherits a truncated variant of *cre1*, and has the potential to express only the zinc finger but not trans-activating domains (CRE1-96) (Ilmen et al., 1996). It is worth noting that some mutants/strains utilized for synthesis of cellulases and hemicellulases for industrial applications are scions of RUT-C30. Genetically engineered strains of *T. reesei* with eliminated *cre1* function have also been analyzed, but in comparison to RUT-C30 they exhibit a pleomorphic phenotype (Nakari-Setälä et al., 2009) and therefore their use for cellulose production is restricted. Alternatively, Zn2Cys6 transcription factor ACE1 acts as a co/partial repressor of genes encoding cellulase and xylanase enzymes, and strains carrying a truncated allele exhibit improved cellulase formation (Aro et al., 2003). In another investigation, deactivation of the *RCE1* gene improved the induced expression of cellulose genes and did not contribute in CRE1-regulated catabolite repression; however, it provoked the binding of XYR1 to the cellulase promoters (Cao et al., 2017).

Expression of cellulase genes can also be strategically modulated by engineering the binding sites for transcriptional regulators. Nonetheless, substituting the CRE1 binding sites from ACE2 and the HAP2/HAP3/HAP5 complex (transcriptional activator binding sites) within the cbh1 promoter increased transcription of a green fluorescent protein in cellulase-inducing environments by sevenfold (Zou et al., 2012). However, manipulation in the xyr1 promoter indicates an upfront approach for manipulation of cellulase production (Lichius et al., 2014). Zhang et al. (Zhang et al., 2017), for the first time, reported cellulase production from glucose as a solitary carbon source by genetically engineered *T. reesei* RUT-C30 with the artificially incorporated transcription factor. In this experiment, constitutive overexpression of artificial transcription activator *xyr1-cre1b* was analyzed in *T. reesei* RUT-C30 using glucose as the sole carbon substrate. Results confirmed that FPase activity (2.63 IU/mL) and xylanase activities (108.72 IU/mL) were attained in batch fermentation through a short fermentation time (48 h) (Zhang et al., 2017).

Gao et al. (2017) employed a combinatorial genetic manipulation technique in *T. reesei* by three specific genes to introduce a versatile cellulase enzymatic complex for competent biomaterial saccharification and a potential cellulase inducer to attain tailored cellulase production. A systematic genetic engineering approach was employed to design strain *T. reesei* SCB18 through strategic truncation of *cre1* (catabolite repressor gene) for producing a cellulase multienzyme complex exhibiting a dynamic cellulolytic activity for biomass saccharification and an unusually high BGL efficiency for cellulase-inducing disaccharides production. Elimination of *cre1* with the *pyrG* (selectable marker) showed improvement in total cellulase activity by 72.6%; nonetheless, a slight reduction in saccharification proficiency was also reported. Furthermore, strain *T. reesei* SCB18 displayed a 29.8% upsurge in total cellulase activity and 51.3-fold improvement in BGL activity (equivalent to 103.9 IU/mL) after overexpression of the *A. niger* BGLA-encoding gene *bglA*. Moreover, the crude enzymatic complex, with high BGL activity, prepared from SCB18 also exhibited robust transglycosylation capability to produce β-disaccharides from glucose. This transglycosylation derivative was conclusively used as the inducer for cellulase synthesis, displaying a 63.0% upsurge in total cellulase activity in comparison to the regularly used soluble inducer lactose (Gao et al., 2017).

Recently, homologous and heterogenous overexpression of *bgl* genes was adapted as the preferred strategy for the designing of a novel strain for enhanced cellulase production (Chauve et al., 2010; Murray et al., 2004; Rahman et al., 2009). The endogenous *bgl1* gene, regulated through various promoters, was genetically transformed into *T. reesei*, exhibiting BGL activities ranging between 0.48 and 8.30 IU/mL (Rahman et al., 2009; Zhang et al., 2010; Qian et al., 2017). *Aspergillus* and *Penicillium* species-based BGL enzymes exhibited relatively superior activities, and furthermore, heterogenous expression of these *bgl* genes in *T. reesei* significantly increased BGL activity (34.31 IU/mL), subsequently showing the improved competence for cellulose hydrolysis (Ma et al., 2011a; Wang and Xia, 2011). A core cellulolytic enzyme complex strengthened with accessory hemicellulases and pectinase displayed strong synergism for hydrolysis of rice straw. The synergistic relationship of three different recombinant enzymes, namely GH62 α-L-arabinofuranosidase (ARA), CE8 pectin esterase (PET), and GH10 endo-1,4-beta-xylanase (XYL) from *Aspergillus aculeatus*, was demonstrated in *Pichia pastoris* to a commercial *T. reesei* cellulase (i.e., Accellerase 1500; ACR). The optimal proportion of the quaternary enzymatic complex (i.e., ACR:ARA:PET:XYL of 0.171:0.079:0.100:0.150) displayed glucose-releasing proficiency of 0.173 g_{glc}/FPU, superior to the binary ACR:XYL amalgamation (0.122 g_{glc}/FPU) and ACR alone (0.081 g_{glc}/FPU), ultimately leading to a 47.3% growth in glucose yield paralleled with ACR alone at the same cellulase dosage. These results established a synergistic relation between core cellulases and auxiliary hydrolytic/nonhydrolytic enzymes. However, a variable degree of synergism of auxiliary enzymes to cellulases would be suitable for designing customized enzyme systems for bio-industrial applications (Laothanachareon et al., 2015). For substantial improvement of cellulolytic enzyme production, genetic combinational manipulation of three key genes was adapted as a novel approach in *P. oxalicum* for sustainable industrial applications. An RE-10, a trigenic recombinant strain of *P. oxalicum*, was genetically engineered through elimination of *bgl2* and *creA*, and with overexpression of the gene *clrB* to improve cellulase yields. RE-10 showed consequential improvement in cellulolytic activity; the extracellular protein concentration and filter paper activity (FPA) improved by up to 10- and 20-fold, respectively, higher than the wild types using pure cellulose and composite wheat bran as substrate. Additionally, the cellulolytic activity of RE-10 was equivalent to industrial *P. oxalicum* strain JU-A10-T, developed from arbitrary mutagenesis. Proportional proteomics and secretomes analysis between RE-10 and wild-type strains confirmed the roles of the enzymes and auxiliary proteins for dramatically elevated decomposition of lignocellulosic matter through recombinant strains, thereby attesting their significance for biomaterial decomposition and suggesting a promising coregulatory strategy (Yao et al., 2015).

Cellulase genes of fungal origin can be cloned and expressed in other microorganisms through heterologous expression (Table 14.4) (Singh et al., 2017; Dashtban et al., 2009). Protein sequencing and gene isolation are prerequisites for heterologous expression for any cellular system (Lee et al., 2001; Hamada et al., 1999a,b). Low molecular weight proteins, contributing cellulolytic activity, are of prime concern for heterologous expression (Quiroz-Castaneda et al., 2011). One of the EG genes from *Tinea versicolor* was successfully expressed in a heterologous manner in *P. pastoris*, a yeast strain, and produced a functional recombinant enzyme; however, the yield was low (Salinas et al., 2011). A cellobiose dehydrogenase (CDH) gene from *Pycnoporus cinnabarinus* (a white rot fungi) was heterologously expressed into *P. pastoris* (Bey et al., 2011) and an enzymatic cocktail was prepared with the purified recombinant enzymes, complemented with an enzymatic complex of commercial *T. reesei*. Outcomes from this study revealed that higher CDH activities were produced through heterologous expression of cellulase in *P. pastoris* as validated through effective performance of enzymatic cocktails in the saccharification of the wheat straw process. Furthermore, the recombinant CDH exhibited thermotolerance and was found to be thermostable at a higher temperature (up to 70°C) during saccharification (Bey et al., 2011).

TABLE 14.4 Heterologous Expression for Enhanced Cellulase Production

Source	Host	Strategy used	Recombinant properties	References
Promoter *gpdA* from *Aspergillus*	*Pencillium oxalicum*	Amplifying induction along with depression (knockout strategy)	Enhancement of cellulase production	Yao et al. (2015)
Penicillium decumbens	*Trichoderma reesei* (RUT-C30)	Agrobacterium-mediated transformation	Enhancement of β-glucosidase activity	Ma et al. (2011c)
E. coli β-glucosidase and acetamidase gene (*amdS*) or the *argB* gene of *A. nidulans*	*T. reesei*	Transformation via plasmid-carrying selectable marker	Enhancement of cellulase production with higher β-glucosidase content	Penttilä et al. (1987)
Chymosin (remin) isolated from the stomach of young calves	*T. reesei*	Transformation via plasmid constructed by prochymosin cDNA fused with promoter, signal sequence, and terminator of the *cbhl* gene	Production of eukaryotic heterologous proteins for enhanced cellulase production	Uusitalo et al. (1991)
Expression vector PAMH10	*T. reesei*	Cellobiohydrolase gene inactivation	Enhanced cellulase production	Harkki et al. (1991)
Myceliophthora thermophila	*Pichia pastoris*	Cloned and expressed *eg7*	Thermostable cellulase	Karnaouri et al. (2014a))
Penicillium funiculosum NCLI	*P. pastoris*	Overexpressed β-glucosidase under constitutive and inducible promoter	Glucose-tolerant β-glucosidase	Ramani et al. (2015)
Penicillium oxalicum	*P. oxalicum*	Heterologous expression of *bgl* and *lpmo* under the control of gla 1 promoter	65-Fold higher yield compared with wild type	Yao et al. (2016)
bgl from *Aspergillus niger.* Tr*LPMO* from *T. reesei*	*P. verrucosum*	Gene *cre I* knocked out and replaced by a *crel-1*	Production of heterologous enzyme simultaneously	Bulakhov et al. (2017)
RUT-C30	*T. reesei*	Gene *cre I* knocked out and replaced by a *crel-I*	Enhanced cellulase production	Nakari-Setälä et al. (2009)
AN3046 LPMO from *A. nidulans*	*A. nidulans*	Cloned and expressed in protein-secreting vector pEXPYR using ligation-free cloning	Enhanced LPMO level in cellulase produced	Jagadeeswaran et al. (2016)

Adapted from Singh, A., Patel, A.K., Adsul, M., Singhania, R.R., 2017. Genetic modification: a tool for enhancing cellulase secretion. Biofuel Res. J. 14, 600–610.

In another study, *T. reesei* ZU-02 strain was used for overexpression analysis of the *cbh2* gene, and one of the mutants, C10, showed the highest production level of FPA up to 4.3-fold in comparison to the parental strain (Fang and Xia, 2013). In another genetic engineering approach, the FPA was enhanced by 130% through the exclusive mutant B12, as developed via genetic manipulation of BGL-encoding gene *bgl1* under artificial constructs of the *cbh1* promoter (Zhang et al., 2010). Several factors have been identified to regulate cellulase induction in *T. reesei*; however, more novel regulators remain to be explored. In a previous study, overexpression of 28 putative regulatory genes was identified in *T. reesei* after induction by various cellulose-dependent substrates, out of which only seven genes showed enhanced cellulase production (Häkkinen et al., 2014). Interestingly, arbitrary T-DNA insertion in the fungal chromosome, in a process of constructing a T-DNA-tagged mutant library, conferred enhanced cellulolytic enzyme activity (Zhong et al., 2012).

A novel hypercellulolytic *T. reesei* variant, SN1, was developed and characterized, and was found to be different from other cellulase producers, e.g., QM9414 and RUT-C30. A uracil auxotrophic strain, SP4, a uracil auxotroph, was engineered through a gene deletion approach to achieve competent genetic transformation. Furthermore, the *bgl1* gene was overexpressed in SP4 and showed significant enhancement in BGL activity in the cellulolytic multienzyme complex (Qian et al., 2016).

With the help of an orchestrated genetic strategy based on the *pyrG* marker, Qian et al. (2017) demonstrated that utilizing the genetic potentials of *T. reesei* endogenous cellulases to synthesize competent cellulase combinations was a significant approach to endorse saccharification efficacy, eventually resulting in cost reduction for lignocellulose-centered biofuels. Results revealed 1.5-fold enrichment in overall cellulase activity (approximately 5.8 U/mL FPA) and a substantial rise in saccharification competence toward differently pretreated corncob biomass in the engineered strain QPE36 when EGs (EG2) were overexpressed with a 46-fold increase in the transcription level and an equivalent protein level to CBH2. Furthermore, an 11.6-fold elevation in BGL activity was observed when BGL1 was overexpressed in the EG2 overexpression strain QE51 (*pyrG*-excised). The EG2-BGL1 twin overexpression strain QEB4 showed a notable improvement in cellulolytic capacity on pretreated corncob residues (Qian et al., 2017).

An important role of cellulase, from glycoside hydrolase family 12 (GH12), was observed in cellulose defragmentation and cell wall deformity, being extensively used in a variety of bio-industrial practices. A gene for efficient synthesis of GH12 was cloned from *Aspergillus terreus* (AtGH12), and expressed in *Aspergillus nidulans* strain A773. The physicochemical properties of purified protein confirmed that the optimum temperature and pH for the enzyme were 55°C and 5.0, respectively, at which it displayed the maximum activity against β-glucan and xyloglucan and also toward glucomannan and carbomethoxy cellulose (CMC) (Segato et al., 2017).

In one study, *Bacillus amyloliquefaciens* cellulase exhibiting both exoglucanase and EG actions was expressed in *A. niger*. Diminished exoglucanase and EG enzymatic activities caused a reduction in hydrolyzing capacity of cellulases derived from marine *A. niger*. A cellulase with exhibiting superior EG and exoglucanase activities was proficiently expressed with promoter glaA to boost the efficiency of halostable cellulase derived from marine *A. niger*. As a result, exoglucanase and EG efficiencies augmented from 0.21 and 4.51 to 0.89 and 15.12 U/mL of the transformant, correspondingly. A 7.1-fold (0.63–4.47 U/mL) increase in FPA was reported. The cellulase obtained from the transformant not only displayed higher activity, but moreover also retained halostability (Xue et al., 2017).

14.7.2 Bacteria: A Classical and Reliable Genetic Source for Cellulase Production

An extensive series of bacterial expression systems are generally used for the remarkable advancements in recombinant cellulase protein production. These bacteriological systems are comprised of *B. subtilis*, *E. coli*, *Z. mobilis*, *Corynebacterium glutamicum*, etc., while *E. coli* is still the utmost and frequently employed microbial system for the production of recombinant cellulase enzymes, which is generally found in harsh environments or in animal digestive tracts (Kim and Kim, 2012). In addition, *B. subtilis* or *Clostridium* species, being Gram-positive, are favorable organisms because of their significant secretion capacities along with surface-display properties (Olson et al., 2012).

Characterization and overexpression of novel bacterial cellulases have already been documented with special features such as thermostability, halotolerance, and synergy with other enzymes, confirming high potential for industrial applications (Table 14.5) (Garvey et al., 2013; Menendez et al., 2015).

Cellulases, hemicellulases, and peroxidases are generally found in bacterial secretions having enzyme complexes, for example, *Streptomyces* secretes lignocellulase, which digests lignocelluloses. Cellulolytic enzymes of bacterial origin are found to be more thermostable and perform well under alkaline conditions than those produced by fungi (Bon and Picataggio, 2002; Ferreira-Leitao et al., 2003). For multiple protein expression, more consistent gene expression systems are already established for cellulases (Mutalik et al., 2013). Codon mismatching is another possible hurdle to cellulase expression in a heterologous manner, for which further possibilities for codon optimization need to be investigated (Linger et al., 2010). Moreover, surface replications are explored to study and optimize production of recombinant cellulase (Lee et al., 2012). In contradiction to all these barriers, production of recombinant cellulases in bacterial systems generally result in an increment in enzymatic yield, in comparison to the pristine host, which is supposed to be an outcome of the differences in the expression systems comprising promoters, gene copy number, and inducer types (Afzal et al., 2010). Directed evolution techniques in combination with a cogent strategy to overexpress cellulases in their own bacterial source are already reported. Genera such as *Bacillus* (*B. subtilis*) and *Clostridium* (*C. thermocellum*) were exploited as homologous cellulase production platforms, due to their easy genetic variation and other appropriate characters (Lambertz et al., 2014). An ethanologenic *E. coli* strain was used to express β-1,4-endoglucanase (EGase) and a β-1,4-glucosidase from other *E. coli* strains under a constitutive promoter, with the potential for fermenting biomass hydrolysates (Munjal et al., 2015). Chung et al. (2015) cloned and engineered *Caldicellulosiruptor bescii*, a bacterium able to degrade lignocellulosic biomass by itself, and executed a homologous expression of a multimodular cellulase called CelA, which harbors GH9 and GH48 domains. This work displayed an efficacious approach to improve and augment the cellulolytic activity of this important thermophilic bacterium with a potential for industrial biotechnological applications.

Moreover, a classical method to overproduce an appealing protein or enzyme by cloning its codifying genes in a high copy number plasmid, compatible with its host, is still commonly utilized. A homologous overexpression of cellulase CelC2 was obtained from *Rhizobium leguminosarum* bv *trifolii* ANU843, which incremented threefold its cellulolytic activity (Robledo et al., 2011). Similarly, *Clostridium* species harboring cellulosomes (cell-bound complexes of cell wall-degrading enzymes), act synergistically in heterologous manner using yeasts as preferred platform (Koeck et al., 2014; Bayer et al., 1998a, 2004; Fontes and Gilbert, 2010). The strategies predicated in heterologous expression are focused on the utilization of noncellulolytic microorganisms that have a high production ratio for expressing

TABLE 14.5 Significant Developments in the Production of Recombinant Cellulolytic Proteins Through Bacterial Expression Systems

Host organism	Protein/s expressed	Source organism	Features	References
ESCHERICHIA COLI				
BL21	Cel6B and Cel5A; stable protein 1	*Thermobifida fusca; Populus tremula*	Synergism of scaffold-linked (up to 12 subunits) free enzymes	Morais et al. (2010)
BL21	1. EglA, EglC-CBD 2. EglD, CBM3	1. *Aspergillus niger; Cellulomonas fimi* 2. *Clostridium thermocellum*	Nanoparticle scaffolds, with up to 30 biotin linkage-associated enzymes or protein domains	1. Kim et al. (2011) 2. Kim et al. (2012)
BL21	Egl-I-IA-F, Egl-I-IA-R; Egl-III-3A-F, EglIII-3A-R; amylase-F, amylase-R	*Bacillus* sp. I-IA; *Bacillus akibai; Bacillus subtilis* 32356	GHs with enhanced secretion, via chemical influences on the cell wall	Liu et al. (2012b)
BL21	BGIA chimera	*C. thermocellum*	Attachment of a chimeric (originally free) enzyme to a native cellulosome	Gefen et al. (2012)
BACILLUS SUBTILIS				
TAD-derived strains	1. Cel8A and Coh 2. Scaffoldin, Cel48F, Cel9E, Cel5A	1. *C. thermocellum* 2. *Clostridium cellulolyticum*	Spontaneously assembling, surface-displayed endoglucanase, and minicellulosomes	1. Anderson et al. (2011) 2. Anderson et al. (2013)
MGB874	Egl237	Alkaliphilic *Bacillus* sp. strain KSM-S237	Genetically depleted host organism strain with enhanced enzyme production	Morimoto et al. (2008)
WB600 and WB800	CbhA, CelK, CelO, Cel48Y, Cel48S, CelA	*C. thermocellum*	Secretively expressed GHs with significant synergistic effects seen once combined	Liu et al. (2012a)
CORYNEBACTERIUM GLUTAMICUM				
	Minicellulosome and chimeric components	*Clostridium cellulovorans; C. thermocellum*	Enhanced industrial viability of GHs and minicellulosome	Hyeon et al. (2011)
ZYMOMONAS MOBILIS				
	E1 and GH12	*Acidothermus cellulolyticus*	Soluble, active enzymes at high levels without significant effect on bacterial growth rate	Linger et al. (2010)

Adapted from Garvey, M., Klose, H., Fischer, R., Lambertz, C., Commandeur, U., 2013. Cellulases for biomass degradation: comparing recombinant cellulose expression platforms. Trends Biotechnol. 31(10), 581–593.

microbial cellulases (Batthacharya et al., 2015). Bacteria such as *E. coli*, different species from the genus *Bacillus*, *Pseudomonas fluorescens*, *Ralstonia eutropha*, and *Z. mobilis*, yeasts such as *Saccharomyces cerevisiae* and *P. pastoris*, and filamentous fungi from *Aspergillus* and *Trichoderma* genera are the most utilized in research and industry, and are considered as host systems for producing recombinant enzymes. Furthermore, cell cultures of mammals, plants, or insects, and transgenic plants and/or animals, are utilized for protein expression (Demain and Vaishnav, 2009).

E. coli and *Bacillus* are often utilized as bacterial systems for expressing recombinant proteins. Moreover, other bacteria are also utilized as platforms, including *Z. mobilis* and *Streptomyces lividans*. *E. coli* is a prevalent expression system for cellulases for industrial applications. Additionally, *Z. mobilis* presented itself as an alternative to the utilization of yeast due to its multifariousness, fermenting a wide range of sugars, and also as an alternative to *E. coli* due to its higher yields and the ability to express recombinant proteins intra- and extracellularly. *S. lividans* has proved its potential for voluminous-scale biopharmaceutical production of recombinant proteins (Juturu and Wu, 2014). Nonetheless, the vast majority of researchers preferred *E. coli* because of its impressive expression system. Expression of an EG, Cel5G, identified by metagenomics was demonstrated (Liu et al., 2011) and presented the closest similarity with the myxobacterium *Plesiocystis pacifica*. A strain of *Paenibacillus cookii* synthesizing a novel EG with chitosanase activity was isolated and cloned into an *E. coli* strain (Shinoda et al., 2012). Heterologous expression of the novel thermohalotolerant EG Cel5H from *Dyctioglomus thermophilum* in *E. coli* was also described (Shi et al., 2013). There are a number of reasons that still make *E. coli* a striking and preferential expression system for recombinant cellulase protein production predominantly to express and characterize new cellulolytic proteins, comprising proteins from extreme habitats or animal digestive systems and for the bioconversion of sustainable biomass to expedient products (Adlakha et al., 2011; Girfoglio et al., 2012; Graham et al., 2011a). CBD is supposed to be comprised of a unique sequence found in cellulose-binding protein A, an integral of the cellulase involute of *Clostridium cellulovorans*. An *E. coli* expression vector pET-8c was used for the subcloning of DNA coding for this presumed CBD. High affinity for crystalline cellulose was recorded in proteins synthesized through the guidance of the genetically engineered plasmid pET-CBD (Goldstein et al., 1993). *E. coli* has been considered an expedient host for biocatalysis in the production of biofuel and glucose fermentation into a diverse array of simpler alcohols and generation of deoxygenated hydrocarbons. The bacterium *Pectobacterium carotovorum* subsp. *carotovorum* is known for its property to macerate plant cell walls, triggering soft rot. Its potential to degrade plants is due to synthesis of a vast array of biomass-degrading enzymes, including cellulase and polygalacturonase. The DNA of *P. carotovorum* ATCC no. 15359 was used for the magnification of genes encoding two cellulases and a polygalacturonase, i.e., *cel*B, *cel*C, and *peh*, respectively. Cloning of amplified polymerase chain reaction products was done in the pTAC-MAT-2 expression vector and *E. coli* was used for transformation. Agar diffusion assays were generally used for the qualitative examination of the enzymatic activities of cloned genes, expressing cellulase and polygalacturonase. Comparative variations in the activity of cellulase may be ascribed to the heterogeneous nature of their catalytic actions. The construed amino acid sequences of proteins encoded by *cel*B and *cel*C displayed a close resemblance to β-(1,4)-glucan glucanohydrolase precursor and endo-1,4-D-glucanase of *P. carotovorum*, correspondingly (Ibrahim et al., 2013).

Typically, glucose production from cellulose hydrolysis is a consequence of synergism among EG, exoglucanase, and BGL. A complete cellulase system was coproduced by using a multiple promoter and polycistronic gene construct concealing an EG (CenA), an exoglucanase (CbhA), and a BGL in *E. coli*, and examined for coproduction and harmonious response of the three enzymes. The outcomes established that synergistic action of CenA and BGL was precarious in the decomposition of filter paper to glucose. A multiple promoter and polycistronic gene construct concealing three enzymes exhibited considerably superior glucose production compared with monocistronic constructs. The categorical activity of the multiple promoter construct was 22.5 and 6.9 times and the polycistronic construct was 33.6 and 10.3 times greater than CbhA and CenA, respectively. Conclusively, a complete cellulolytic enzymatic complex may possibly be coexpressed in *E. coli* for the conversion of cellulose to fermentable sugar. The approach to restructure a whole cellulase enzymatic system proved to be very operative and pragmatic to many other cellulolytic enzymes from several organisms, along with other multienzymatic systems (Liu and Yu, 2012).

B. subtilis additionally showed importance in these regions due to its secretion and surface exhibit characteristics (Anderson et al., 2011; Liu et al., 2012a). *B. subtilis* has a minutely minuscular amount of native cellulolytic activity (Anderson et al., 2011). *B. subtilis* was utilized pristinely as a cellulose producer through recombinant DNA technology. Various modifications have been done in bacterial systems to efficiently monitor heterogeneous genes for cellulase enzymes (Pottkamper et al., 2009) and microorganisms, namely *Z. mobilis* (Linger et al., 2010) and *B. subtilis* (Liu et al., 2012a), and are being explored for their potentials in consolidated bioprocessing (CBP). Genome curtailment in *B. subtilis*, accomplished by phase-wise gene expunction, has additionally been displayed to positively affect the yield of cellulose enzymes and ameliorate organism magnification proficiency (Morimoto et al., 2008). Pathogenic bacteria of plants *Erwinia carotovora* and *Erwinia chrysanthemi* have developed refined methods of lyases and hydrolases that support the dissolution of lignocellulosic biomass and let them soften and enter the plant cell (Kado, 1992).

Genome modification of these bacteria for ethanol formation signifies a captivating substitute to the dissolution of lignocellulose via enzymatic or chemical methods. Genetically engineered *E. carotovora* SR38 and *E. chrysanthemi* EC16 harboring PET operon were reported to proficiently metabolize cellobiose and glucose into ethanol and CO_2 as the main fermentation product (Beall and Ingram, 1993). *S. cerevisiae* is used as a model organism for cellulase and hemicellulase expression in active form, indicating the feasibility of construction of hydrolytic yeast strains (Penttilä et al., 1988; Béguin and Aubert, 1994). Two ethanologenic strains of *Erwinia* produced around $50\,g\,L^{-1}$ ethanol from fermentation of $100\,g\,L^{-1}$ cellobiose within 48 h showing a volume output of $1.5\,g\,L^{-1}/h$, and a twofold higher speed than for the cellobiose-digesting yeast *Brettanomyces custersii* in close culture (Spindler et al., 1992). Hydrolytic activity of a cellulase (*C. thermocellum*) expressed in *E. coli* through genetic modification was reported. The performance of recombinant cellulolytic cellulase was found as good as that of cellulases isolated in natural conditions. The outcomes showed that the activity of recombinant cellulolytic cellulases in shake flask simultaneous saccharifcation and fermentation was seven times the upsurge of ethanol titer ($8.8\,g\,L^{-1}$) in comparison to cellulases (*B. subtilis* and *T. reesei*) isolated in natural conditions (Das et al., 2013).

A suitable host is prerequisite to overexpress thermostable microbial cellulase applications in various industries. Thermostable *B. subtilis* and *E. coli* BL21 (DE3) were successfully used for the cloning and expression of thermostable endocellulase (CelDR), which showed almost three times the activity (0.78 U/mL) (Chandel and Singh, 2011). A recombinant multidomain cellulase (90 kDa) was expressed and reported as cellulase, classified under the glycosyl hydrolase super family (Graham et al., 2011a), with an optimum activity at 109°C, displaying a half-life of 5 h at 100°C, and also denatured in concentration extremes of salts, robust detergents, and ion-enriched liquids. Moreover, these thermostable cellulase enzymes, active at more than 100°C, may enable biofuel formation from lignocellulose biomass. A GH10 endoxylanase-encoded gene was cloned from *Geobacillus* sp. WSUCF1 and expressed heterologously in *E. coli* with proficient expression (Bhalla et al., 2013). The outcomes confirmed higher specific activity (461.0 U/mg), thermotolerance, and hydrolytic digestion (92%) of recombinant enzymes and suggested its possible applications in biofuels, pulp, and bleaching industries for improved production. One perpetual problem related to the expression of recombinant protein in a bacterial system is the incidence of enzyme attenuation, which occurs even in single bacterial families, for example, in *Clostridium* sp. (Mingardon et al., 2011). This may possibly be connected to the absence of defensive glycosylation of cellulases expressed in bacterial systems and being natively fungal in inception (Xia et al., 2013). Actually, the production of consummate, active enzymes, concretely those of fungal inception, in bacteria species is still a consequential challenge, supported by different instances of recombinant cellulase exhibiting expression failures or profoundly low yields in bacterial systems (Girfoglio et al., 2012; Xia et al., 2013; Abdeljabbar et al., 2012). The synergistic effect witnessed in organisms with various cellulases and/or cellulosomes has initiated attention in repeating these outcomes through genetic engineering (Fontes and Gilbert, 2010). Initially, more than one bacterial species was required to effectively produce the desired recombinant proteins (Morais et al., 2010; Tsai et al., 2009). Moreover, the latest developments in expression practices, including protein transfer and localization, sanction these diverse minicellulosome constituents to be formed within one particular organism, mainly in *E. coli*, *Corynebacterium* species, and *B. subtilis* (Anderson et al., 2011; Hyeon et al., 2011; Ding et al., 2012). Expression of standard-sized, auto-assembled cellulase and/or cellulosomes recombinantly has been shown to be a challenging task (Kim et al., 2011). As a consequence, efforts have been initiated on synthetic cellulosomes, such as comprising chimeric cellulases with a nanoscaffold backbone (Kim et al., 2011, 2012). The activity of GH-12 cellulase from *Thermotoga neapolitana* on cellotriose and cellotetrose produced minuscule amounts of glucose under protracted incubation with cellobiose as the key product. Furthermore, the activity of the *cel*B enzyme on CMC substrate proposes both endo- and exocellulase activities of this enzyme (Bok et al., 1998).

In spite of outstanding biocompatibility and mechanical properties, cellulose has restricted applications in biomedical and biomaterial conversion due to its poor degradability under in vitro and in vivo conditions. Therefore a metabolic engineering-based strategy was adapted to reform cellular metabolites for transforming *N*-acetylglucosamine (GlcNAc) remainders into cellulosic polymers via de novo synthetic pathways from *Gluconacetobacter xylinus*. Cellulose from such metabolically engineered systems (modified bacterial cellulose [MBC]) was assessed and matched with cellulose from typical standard cells (bacterial cellulose [BC]). Due to higher GlcNAc content and reduced crystallinity, MBC in comparison to BC, showed susceptibility towards lysozyme and, also to rapid hydrolysis by cellulase. Degradation capacity was evaluated under in vivo conditions using subcutaneous grafts in mice, showing degradation of altered cellulose within a time period of 20 days. This finding offers a new direction in the production of a family of modifiable cellulosic polymers that can sort out the long-term restriction connected with the reduced degradation capacity of cellulose for its potential applications in different fields (Yadav et al., 2010). A novel cellulose-cleaving strain of *Streptomyces* sp. G12 was extracted from industrial biowaste. The cloning and chimeric expression of the gene encoding the *Streptomyces* sp. G12 cellulase *CelStrep* displayed competent thermostability subsequently improving its efficiency for cellulose bioconversion (Amore et al., 2012).

14.7.3 Yeast: A New and Less Explored Genetic System for Cellulase Production

Even though yeast is not the natural host of cellulases, recombinant expression in yeast systems is also gaining popularity on an industrial scale (Mattanovich et al., 2012). A number of yeast expression systems have already been developed for their usage in the biofuel industry, including *S. cerevisiae*, *Kluyveromyces marxianus*, and *P. pastoris* (Akcapinar et al., 2011; Tanaka et al., 2012; Yanase et al., 2010). Among these, *S. cerevisiae* is immensely characterized and typically considered as a model organism for heterologous gene expression and is particularly chosen as a CBP organism (Hasunuma and Kondo, 2012; Amerik et al., 2010). To optimize chimeric yeast systems to achieve product yields and titers adequate for economically feasible biomass degradation is still a great challenge, possibly through approaches like producing minicellulosomes with enzymes exhibiting a diverse array of metabolic reactions (Fontes and Gilbert, 2010), or immensely massive cellulosomes by integrating numerous scaffolds in proteins (Fan et al., 2012). Moreover, a five-gene cassette exhibiting single-step cellulase coexpression was produced in thermotolerant *K. marxianus* (Chang et al., 2012), presenting the system for competent biodegradation of cellulose and ethanol fermentation at higher temperatures (Chang et al., 2013). Yeast strains are among the most used fungal expression systems, namely *S. cerevisiae*, but *P. pastoris* and *Kluyveromyces* sp., especially the last one, are important due to their high yields in heterologous cellulase production. However, nowadays comparisons between those systems are arduous, because there is not enough knowledge to surmount the low cellulolytic activity of the expressed cellulases (Garvey et al., 2013). It is quite mundane to express fungal cellulases in *Saccharomyces*, such as *T. reseei* and other fungal sources (Garvey et al., 2013), but there are some studies reporting the expression of clostridial minicellulosomes in this yeast (Lilly et al., 2009).

Recombinant strains of *S. cerevisiae* Y294[SFI] were constructed and compared for the expression of BGLs in *Thermoascus aurantiacus* (Tabgl1) and *P. chrysosporium* (PcbglB and Pccbgl1), and results confirmed a 2.6-fold higher BGL activity (under aerobic and anaerobic environments) in *S. cerevisiae* Y294 [Pccbgl1] strain than the other two strains (Njokweni et al., 2012). Two strains of *S. cerevisiae* were genetically altered to express the *BGL1* (BGL) gene from *Saccharomycopsis fibuligera*, an industrialized strain of *S. cerevisiae*, for improvement in the production of heterologous cellulase at growth incompetence on nonnative substrates (Yeoman et al., 2010). A plasmid construct with an EG gene originating from *C. thermocellum* (Ctcel8A) was utilized for enhancing the possibility of cellulase production. Genetic engineering of *S. cerevisiae* to use cellobiose by coexpression of a cellobiose transporter and BGL is an innovative approach. A cellulase gene identified from *C. thermocellum* (Ctcel8A) was transformed into a homozygous diploid yeast strain, and when heterologously expressed an improvement in BGL activity was reported (Kitagawa et al., 2011).

A recombinant cellulolytic yeast strain established for cell-surface presentation of fungal EG, BGL, and CBH showed 1.4 times more ethanol (89% of proposed yield) production from rice straw (200 g-dry weight/L) in less than 72 h than the wild-type strain supplemented with 10 FPU/g-biomass cellulase. Engineering of the cell surface leads to successful reduction in the bulk of commercial enzyme needed for the fermentation of cellulose. This finding reveals that cellulases exhibited on the yeast cell surface have the tendency to hydrolyze cellulose that remained unhydrolyzed by commercial cellulases, subsequently leading to an increase in sugar consumption for amended ethanol production (Matano et al., 2012).

Two newly isolated genes, *cbh6A* and *egGH45*, coding cellulase were cloned from *Phialophora* sp. G5, and showed efficient expression in *P. pastoris*. The presumed polypeptide of CBH6A comprises a catalytic domain of glycosyl hydrolase family 6 CBHs and a family 1 CBM, whereas construed EgGH45 comprises merely a catalytic domain of family 45 EGs. Optimum activities of CBH6A and EgGH45 were observed at 65°C and pH 7.0 and 60°C and pH 6.0, correspondingly. Both the enzymes were highly active and stable over a wide-ranging pH range and showed high thermostability at 70°C. These enzymes were significantly resistant to detergents like sodium dodecyl sulfate (10 mM), and displayed residual activities of 35% and 54%, respectively. Both of these enzymes displayed synergism in the degradation of filter paper, and had maximum proficiency in the ratio of CBH6A to EgGH45 at 80:20. The enzymatic mixtures of CBH6A and EgGH45 also proved to be highly significant for commercialization to be utilized in the textile and detergents industries (Zhao et al., 2012). Cellulase enzymes CBH6A and EgGH45, showing activity at neutral pH, were isolated from *Phialophora* sp. G5 and proficiently expressed in *P. pastoris* (Zhao et al., 2012).

14.7.4 Plant: An Alternate Genetic Source for Cellulase Production

Plants are considered as nonsumptuous alternative systems for the production of recombinant enzymes. The approaches to achieve an efficient recombinant enzyme expression involve their expression from the plant nuclear genetic material, chloroplast genome, or plant tissues harboring viral particles. Additionally, there are several strategies to enhance recombinant enzyme production yields (Streatfield, 2007). Plants are perhaps the perfect system for profitable cellulase production, possibly due to their potential to be synthesized cost effectively on an industrial level (Shen et al., 2012; Jung et al., 2013). The most common model plants used for recombinant enzyme expression studies are *Arabidopsis thaliana*, *Zea mays*, *Medicago sativa/truncatula*, *Nicotiana tabacum*, and *Populus alba*.

Initially, production of cellulase and several other glycoside hydrolase enzymes in plants was believed to be impractical, due to the probable harmful effects caused by cellulases on the polysaccharides of the plant cell wall (Taylor et al., 2008). In fact, cellulases, which are present endogenously in plants, perform a significant role in plant growth (Xie et al., 2013) and their effect on cell wall constituents and structure is mainly species, tissue, and enzyme specific (Abramson et al., 2013). The promising and damaging nature of cellulases, which are overexpressed in plants, have been studied (Jung et al., 2012). Tobacco plant is generally used as a model system for research into heterologous cellulase production in plants (Taylor et al., 2008; Sainz, 2011). Moreover, a few investigations on the expression of cellulose have additionally involved more economically significant crops like sugarcane, rice (Chou et al., 2011; Harrison et al., 2011), and feedstocks, for instance, woody plants and grasses, which can be developed in large masses (Jung et al., 2013). Various cellulases, concretely EGs, along with other glycoside hydrolases, for example,

TABLE 14.6 Heterologous Expression of Cell Wall-Deconstructing Enzymes in Plants

Plant	Transgenic enzyme	Subcellular storage compartment	References
Arabidopsis thaliana	*Acidothermus cellulolyticus* El$_{CAT}$	Apoplast	Ziegler et al. (2000)
Tobacco	*A. cellulolyticus* El and El$_{CAT}$	Apoplast	Ziegelhoffer et al. (2001) and Dai et al. (2005)
Tobacco	*A. cellulolyticus* El	Endoplasmic reticulum	Dai et al. (2005)
Tobacco	*Clostridium thermocellum* XynZ	Apoplast	Herbers et al. (1995)
Tobacco	*A. cellulolyticus* E1$_{CAT}$	Chloroplast	Dai et al. (2005, 2000b)
Tobacco	Maize *p*-glucosidase	Chloroplast	Ralph et al. (2006)
Tobacco	*A. cellulolyticus* E1$_{CAT}$	Chloroplast	Ziegelhoffer et al. (2001)
Tobacco	*Thermomonospora fusca* E1 and El$_{CAT}$ E3	Cytosol	Ziegelhoffer et al. (1998)
Tobacco	*Trichoderma reesei* CBH1	Cytosol	Dai et al. (1999)
Tobacco	Human *p*-glucosidase	Cytosol	Reggi et al. (2005)
Tobacco	*C. thermocellum* XynA	Cytosol	Kimura et al. (2003a)
Tobacco	*A. cellulolyticus* El	Cytosol	Teymouri et al. (2004) and Dai et al. (2000a)
Potato	*A. cellulolyticus* El	Apoplast	Dai et al. (2000a)
Potato	*Streptomyces olivaceoviridis* XynB	Apoplast	Yang et al. (2007)
Potato	*A. cellulolyticus* El	Chloroplast	Dai et al. (2000b)
Potato	*A. cellulolyticus* El	Vacuole	Dai et al. (2000b)
Potato	*T. fusca* E2	Cytosol	Smidansky et al. (2003)
Potato	*T. fusca* E3	Cytosol	Ziegelhoffer et al. (1998)
Potato	*S. olivaceoviridis* XynB	Cytosol	Yang et al. (2007)
Alfalfa	*T. fusca* E2 and E3	Cytosol	Ziegelhoffer et al. (1998)
Rice	*A. cellulolyticus* El$_{CAT}$	Apoplast	Oraby et al. (2007)
Rice	*C. thermocellum* XynA$_{CAT}$	Cytosol	Kimura et al. (2003b)
Barley	Rumen *Neocallimastix patriciarum* XynA	Cytosol	Patel et al. (2000)
Maize	*A. cellulolyticus* E1$_{CAT}$	Apoplast	Biswas et al. (2006)
Maize	*A. cellulolyticus* El$_{CAT}$	Apoplast	Balan et al. (2007)

CBH1, cellulobiohydrolase 1; *CAT*, catalytic domain; *E1*, *E2*, and *E3*, endoglucanases (endocellulases); *XynaA*, *XynB*, and *XynZ*, xylanases (hemicellulases).

Adapted from Sticklen, M.B., 2008. Plant genetic engineering for biofuel production: towards affordable cellulosic ethanol. Nat. Rev. Genet. 9, 433–443.

xylanases, have now been well expressed in plants utilizing different regulatory components and indicating diverse tissue and subcellular locations (Table 14.6) (Sticklen, 2008; Sainz, 2011).

Three important issues are related to the synthesis and utilization of cellulolytic enzymes in agricultural produce: (1) whether the biological activity of these enzymes would be eliminated by rigorous pretreatment conditions (acidic, basic, and/or temperature); (2) whether the biomass is expressing sufficient enzymes to convert carbohydrates into simple sugars without the requirement to integrate more enzymes; and (3) whether incrementing the capacity of production of these cellulolytic enzymes inside plant cells is a source of impairment in plant magnification and development. The mildest method of ammonia fiber explosion pretreatment was utilized on thermostable *Acidothermus cellulolyticus* EG E1-producing tobacco biomaterial to address the first concern. Almost 65% activity of this heterologous enzyme was disoriented in the experiment (Teymouri et al., 2004). It was reported consequently that it is a better approach to extract heterologous enzyme either in crude or pristine form, and then integrate it into the material, which is already treated for the formation of fermentable sugars (Oraby et al., 2007). To address the second issue, regulation of transcriptional, posttranscriptional, and posttranslational elements can plausibly raise the potential of gene expression. In fact, the finest means of improvement in heterologous proteins production is to mark them far from the cytosolic environment for their collection in cellular compartments not found in cytosol (Hood, 2004).

Additionally, this approach would state the third concern, that enzyme accumulation within cytosolic compartments resulted in constant plant cytosolic metabolic activities. Distinct molecular chaperone assemblies exist in targeted compartments for translocation or folding of certain proteins, which is another benefit of this approach.

A number of possibilities have been studied comprising the genetic alterations of plants for the production of cellulase and/or ligninase enzymes on their own, the improvement of plant architecture with lower lignin content, and the manipulation of crops with incremented cellulose or total bioproduct. Many pathways have been developed in all areas, as mentioned. For instance, there is vast scope for the high-level production of cellulolytic enzymes within the plant produce. Additionally, the probability of superseding microbial mass in reactors with plants can act as a bio-industrial unit for the production of industrially important enzymes along with cellulases resulting in an economically viable approach. However, genetic manipulation of plants for cost-effective biofuel production is at an early stage (Sticklen, 2006).

However, one further possibility to curtail the price is to produce cellulolytic enzymes using plant metabolic networks (Sticklen et al., 2004), reinstating cellulase production in microbial tanks. Laterally, the expression of a gene that codes for the catalytic site of the thermotolerant enzyme *A. cellulolyticus* 1,4-β-endoglucanase E1 was effectively reported in certain plants, namely *Arabidopsis*, potato, and tobacco, and the expressed produce was directed to the apoplast (extracellular path generated by the cell wall matrix). In the case of *Arabidopsis*, there is a possibility of producing this enzyme at one-fourth the concentration of a plant's total soluble protein (TSP) within the biomass. In recent times the catalytic domain of the enzyme *A. cellulolyticus* 1,4-β-endoglucanase E1 was efficiently constitutively expressed in rice (Oraby et al., 2007) and maize.

Expression of cellulases in plants to elucidate the mode of action and effects of bacterial/fungal/plant cellulases in cellulose biosynthesis has been reported by many researchers. Thermostable cellulases E2 and E3 from *Thermobifida* (formerly *Thermomonospora*) *fusca* were expressed in *M. sativa*, *Solanum tuberosum*, and *N. tabacum*, obtaining low but favorable expression patterns (Ziegelhoffer et al., 1998). Members of a GH5 family endocellulase (E1 endoglucanase) from *A. cellulolyticus* were expressed in *Z. mays* and *N. tabacum* plants. Results confirmed a reduction of cell wall recalcitrance to chemical pretreatments in both plants, proposing the plausibility of an application for improving plant biomass bioconversion (Brunecki et al., 2011). High expression levels of a thermostable cellulase cocktail in plastids of *Nicotiana* plants were reported (Petersen and Bock, 2011).

Cell wall hydrolysis enzymes are preferred to accumulate in subcellular compartments than in cytosol when expressed in plants. These enzymes affirm precise folding and activity, glycosylation, reduced degradation, and incremented stability, in comparison to enzymes that are produced and accumulated in cytosol (Sticklen, 2006; Horn et al., 2004). Heterologous enzymes can be isolated from fresh and/or dry biomass of transgenic crops as an amount of their TSP, and further used for the conversion of pretreated crop produce into fermentable sugars (Sticklen, 2008; Teymouri et al., 2004; Oraby et al., 2007; Ransom et al., 2007).

Another vital aspect is the probable reduction in the expression of cellulolytic enzymes in plant systems. An alternate strategy to deal with the probable decline in enzyme production is to express single domain enzymes, e.g., hyperthermophilic archeon *Sulfolobus solfataricus* showed successful expression of its endocellulases in the endoplasmic reticulum of tobacco plants in active form. Transgenic rice seeds are ideal as a promising bioreactor system for enzyme production to convert cellulosic biomass. Rice (*Oryza sativa* L. subsp. *japonica*), transformed with the codon-optimized synthetic gene "endo-1,4-β-glucanase (E1)" (from *A. cellulolyticus*) regulated through a Gt1 promoter (seed storage protein in rice) was successfully established and encoded enzymes perfectly (Zhang et al., 2013).

Thermophilic cellulases, which show abridged or insignificant activity below 50°C, are reported to be successfully expressed in plants deprived of antagonistic effects. However, postharvest activation in enzymes was reported by a temperature shift (Chou et al., 2011; Klose et al., 2013, 2012; Lebel et al., 2008) and also notable compensations in lignocellulose hydrolysis were observed in some (Taylor et al., 2008; Sainz, 2011). Some reports have established the prospective expression of cellulase enzymes in plants and have contributed to enhancing appropriate growth environments, for instance, during the impact of constitutive expression contrasted with alcohol-inducible expression of EG TrCel5A from *T. reesei*, a mesophilic fungus was evaluated on tobacco plant development. Results showed substantial production of cellulose in significant amounts in the plant leaves devoid of any negative impact on plant development. These outcomes reveal the possible viability of an inducible expression system for in planta production of biomass-degrading enzymes (Klose et al., 2012).

It is worth bearing in mind the theoretical, technical, or fiscal hurdles that potentially hinder the efficient expression of heterologous thermophilic cellulase in plants. For the efficient transformation of cellulose-based crop biomass into fermentable sugars, which is a very complicated procedure, a synergistic action of several major and related enzymes is required. In plants an efficient expression system for the total destruction of lignocellulosic substrate is thus liable to demand proficient expression of various heterologous genes in a particular feedstock crop, with specific expression levels personalized to reach the best standards of efficient cooperation (Mir et al., 2014).

14.7.5 Insects: A Symbiotic Genetic Association for Cellulase Production

Cellulolytic ability is also observed in some insects. Insects themselves produce cellulase that specifies its inherent character of cellulose digestion (Girard and Jouanin, 1999; Lee et al., 2005; Nakashima et al., 2002; Tokuda et al., 1999; Watanabe et al., 1998), from symbiotic actions of organisms harbored in their guts (Ohtoko et al., 2000), or both (Watanabe et al., 1998; Ohtoko et al., 2000; Breznak and Brune, 1994; Scharf et al., 2003, 2005). Genes coded for endogenous insect cellulases have been successfully isolated from three beetles and a few termites (Girard and Jouanin, 1999; Lee et al., 2005; Nakashima et al., 2002; Tokuda et al., 1999; Watanabe et al., 1998; Sugimura et al., 2003). These cellulolytic enzymes belong to three structurally distinct glycosyl hydrolase families (GHFs), namely GHF 5 (*Psacothea hilaris* beetle), GHF 45 (*Phaedon cochleariae* beetle), and GHF 9 (termites and cockroaches), and are expected to have autonomous evolutionary roots (Sugimura et al., 2003). The beetle *Batocera horsfieldi* was used for the characterization of a new cellulase (EGase), and cDNA (Bh-EGase II), which belongs to GHF 45. The protein sequence of the *Bh-EGase II* gene showed 93.72% similarity to that of the *Ag-EGase II* gene obtained from the beetle *Apriona germari*, consisting of a single exon coding for 239 amino acid residues and which spans 720bp. Bh-EGase II has conserved the GHF 45 catalytic site and has three putative N-glycosylation sites at 56–58 (N-K-S), 99–101 (N-S-T), and 237–239 (N-Y-S), respectively. The baculovirus-infected insect BmN cells and *Bombyx mori* larvae expressed cDNA encoding Bh-EGase II. The highest activity of these recombinant Bh-EGase II enzymes was nearly 928U/mg at 50°C and pH 6.0 (Xia et al., 2013). More recently, Wei et al. (Wei et al., 2015) cloned an EG from a *B. subtilis* strain, isolated from termite gut, in *E. coli.*

14.7.6 Other Genetic Systems for Cellulase Production

Epigenetic engineering has recently been discussed regarding the transcriptional activators for chromatin remodeling, which was highlighted as an evolving tactic for improved cellulase production (Gupta et al., 2016). The chromatin packing was released from *cbh1* and *cbh2* (cellulase-encoding genes) in cellulose-persuading environments, while it remained attached in an *xyr1*-deletion strain (Mello-de-Sousa et al., 2015, 2016). Consequently, numerous genes were identified that encoded chromatin remodeling enzymes (Mello-de-Sousa et al., 2015), as well as histone acetyltransferases, but had considerably diverse expression in Δxyr1 and its parent strain. In integration, chromatin was also modified by fungal genes of different groups resulting in the modulation of cellulase and hemicellulose production: the Velvet-Lae A/LAE1 involute (Seiboth et al., 2012; Aghcheh et al., 2014; Liu et al., 2016). Lae A of *A. nidulans* has features of the *S*-adenosyl-L-methionine arginine protein methyltransferase, which regulates the expression of genes for secondary metabolite production (Sarikaya-Bayram et al., 2015). However, Lae A neither methylates histones nor does it seem to interact with the transcription factors of Velvet protein involute. Cellulase expression and secretion were enhanced by constitutive overexpression of both *lae1* and *vel1* in *T. reesei* (Seiboth et al., 2012; Aghcheh et al., 2014). An impressive study described the metabolic steps included in the signaling of cellulase gene expression. The present state of information regarding cellulose production has been reviewed (Gupta et al., 2016; Druzhinina and Kubicek, 2016). In spite of reasonable advancement in this field, a necessary amendment to the production of cellulase enzyme by the management of different genes convoluted in this procedure has not yet been done. Whole-genome committed strategies, such as, for example, the utilization of whole-cell metabolic models, possibly propose the best existing strategy for the documentation of the circumscribing steps involved in the production of cellulase enzymes. Castillo et al. (2016) used the CoReCo (comparative metabolic reconstruction framework) pipeline (Pitkänen et al., 2014) for the development of an excellent metabolic model of *T. reesei* (BIOMODELS database). This model is suitable for simulating protein production processes because it has a biomass equation, uptake/export reactions, and reaction limitations.

Further, the important and correlated outlook for advanced research is increasing the levels of production and the total activity of heterologous enzymes. Genetic engineering of chloroplast genome provides a better solution for escalating enzyme production in comparison to nuclear genomes. Additionally, the chloroplast genome displays maternal inheritance, and chloroplast genome engineering also offers the advantage of transgene suppression. In the case of hyperthermophilic cellulase Bg1B, expressed in chloroplast, the maximum activity of enzyme and its accumulation, i.e., 5.8% of total soluble protein, was attained when Bg1B was targeted at the chloroplast utilizing the Rubisco activase (RA) signal peptide along with RbcS promoter (Jung et al., 2010). Hyperthermophilic EG (SSO1354) expression has been positively directed to the apoplast utilizing the plant codon-enhanced murine antibody mAb24 (LPH) signal peptide conditioned by the commonly used CaMV dual 35S promoter, having a polyadenylation signal (pA35S) at the C-terminus and a 50-untranslated region translational enhancer from the *Petroselinum hortense* chalcone synthase gene amalgamated with an endoplasmic reticulum retaining sequence (Klose et al., 2012).

Regulation of amylase and cellulase expression in *P. decumbens*, a fast-growing cellulolytic fungus, was examined as a role for the Gα subunit 3 (called PGA3 for *Penicillium* G protein alpha subunit 3). Results revealed that reduced production of amylase and considerably decremented transcription of the main amylase gene *amy15A* are due to the deletion of *pga3*. Amylase production was improved in the Dpga3 strain by the addition of exogenous cAMP or its analog dibutyryl-cAMP, signifying an important function of PGA3 in amylase production by controlling the levels of cAMP. On the other hand, cellulose enzyme activity in the medium was unaffected by the increment in the transcription of the major cellulase gene *cel7A-2* in Dpga3. Regulation of the cAMP-mediated expression level of transcription factor AmyR was helpful in achieving the regulatory effects of PGA3, which was carbon source independent. The role of PGA3 shown by gene effacement was partly fortified by the investigation of the mutant having dominantly activated PGA3. The outcomes throw light on the explanation of the physiological roles of the G protein-cAMP pathway in filamentous fungi (Hu et al., 2013).

14.8 CONCLUSION

Shifting of the focus of energy generation toward biofuel use was due to elevations in the consumption of energy, incremented environmental anxieties, and diminution of fossil fuels. Biomass is the most abundant and also the most renewable biomaterial on our planet; as a result, bioconversion of biomass has consequential advantages over other alternative energy strategies. The most abundant organic compound on the planet as well as the most abundant waste product produced by human activity, and which is found in almost every ecosystem, is cellulose. The hydrolysis of cellulose in the presence of cellulase enzymes is referred to as cellulolysis. Cellulase is a member of certain glycoside hydrolase families. However, lower activity and higher cost of cellulase are the bottlenecks for cellulase industrial production. Henceforth, low-cost cellulase hyperproduction with improved efficiency is the need of the hour. Therefore strategies to develop competent cellulase-producing microbial strains and enzyme systems must be able to (1) comprehend the biology underlying cellulase hyperproduction, (2) optimize conditions for enzyme induction, (3) improve genetic engineering approaches in combination with metabolic engineering and system biology approaches, and (4) establish appropriate and versatile gene assembly for cellulase production. The competency to contrive involute polygenic phenotypes is fundamental for genetic engineering endeavors because it entails targeted manipulation in gene circuits to maximize anticipated products, including cellulases. The amalgamation of recent synthetic biology implements with directed evolution and metabolic engineering can sanction the engineering of biological systems holistically via manipulation of increasingly intricate traits. Recent research in systems and synthetic biology have expedited competency to construct and optimize cell factories for applications of metabolic engineering. Advancements in modeling and estimating genome-scale metabolic networks, multifarious platforms for gene assembly, and optimization strategies for delicate synthetic pathways have provided us with exhilarating prospects to utilize the full potential of cell metabolisms. The present state of molecular biology, genetic engineering, metabolic engineering, and synthetic biology techniques, combined with powerful simulation techniques, can establish a naïve blueprint for benign, efficient, versatile, and low-cost cellulase production.

14.9 FUTURE PROSPECTS

The diverse array of complexities of biological systems may encounter serious hurdles to designing novel strategies for robust cellular feasibilities for cellulase production. Stratagems to develop recombinatorial strain and enzyme systems for exclusively producing significant cellulase, and the production of cellulase synthesizers based on efficient scrutiny under conditions that are immune to using biological systems, are still unsolved. Metagenomic approach technologies should be optimized to trap and culture uncultivable microbes, and should be exploited for their potentially desirable enzymes. It is considered to be the most viable method for searching new metabolic pathways. Nevertheless, a synergistic association of genetic and metabolic engineering, as well as systems and synthetic biology, would help to address the issue of developing new strains along with novel, efficient, and sturdy pathways for the production of cellulase without by-products.

References

Abdeljabbar, D.M., Song, H.J., Link, A.J., 2012. *Trichoderma reesei* cellobiohydrolase II is associated with the outer membrane when overexpressed in *Escherichia coli*. Biotechnol. Lett. 34, 91–96.

Abramson, M., Shoseyov, O., Hirsch, S., Shani, Z., 2013. Genetic modifications of plant cell walls to increase biomass and bioethanol production. In: Lee, J.W. (Ed.), Advanced Biofuels and Bioproducts. Springer, pp. 315–338.

Adlakha, N., Rajagopal, R., Kumar, S., Reddy, V.S., Yazdani, S.S., 2011. Synthesis and characterization of chimeric proteins based on cellulase and xylanase from an insect gut bacterium. Appl. Environ. Microbiol. 77, 4859–4866.

Adney, W.S., Jeoh, T., Beckham, G.T., Chou, Y.C., Baker, J.O., Michener, W., Brunecky, R., Himmel, M.E., 2009. Probing the role of *N*-linked glycans in the stability and activity of fungal cellobiohydrolases by mutational analysis. Cellulose 16 (4), 699–709.

Adrio, J.L., Demain, A.L., 2014. Microbial enzymes: tools for biotechnological processes. Biomolecules 4, 117–139.

Adsul, M.G., Bastawde, K.B., Varma, A.J., Gokhale, D.V., 2007. Strain improvement of *Penicillium janthinellum* NCIM 1171 for increased cellulase production. Bioresour. Technol. 98 (7), 1467–1473.

Adsul, M.G., Singhvi, M.S., Gaikaiwari, S.A., Gokhale, D.V., 2011. Development of biocatalysts for production of commodity chemicals from lignocellulosic biomass. Bioresour. Technol. 102 (6), 4304–4312.

Afzal, S., Saleem, M., Yasmin, R., Naz, M., Imran, M., 2010. Pre and post cloning characterization of a beta-1,4-endoglucanase from Bacillus sp. Mol. Biol. Rep. 37, 1717–1723.

Aghcheh, R.K., Néemeth, Z., Atanasova, L., Fekete, E., Paholcsek, M., Sandor, E., Aquino, B., Druzhinina, I.S., Karaffa, L., Kubicek, C.P., 2014. The VELVET A orthologue VEL1 of *Trichoderma reesei* regulates fungal development and is essential for cellulase gene expression. PLoS One 9, e112799.

Akcapinar, G.B., Gul, O., Sezerman, U., 2011. Effect of codon optimization on the expression of *Trichoderma reesei* endoglucanase 1 in *Pichia pastoris*. Biotechnol. Prog. 27, 1257–1263.

Ali, N., Athar, M.A., Khan, Y.H., Idrees, M., Ahmad, D., 2014. Regulation and improvement of cellulase production: recent advances. Nat. Resour. 5, 857–863.

Alriksson, B., Rose, S.H., van Zyl, W.H., Sjöde, A., Nilvebrant, N.O., Jönsson, L.J., 2009. Cellulase production from spent lignocellulose hydrolysates by recombinant *Aspergillus niger*. Appl. Environ. Microbiol. 75, 2366–2374.

Alterthum, F., Ingram, L.O., 1989. Efficient ethanol production from glucose, lactose, and xylose by recombinant *Escherichia coli*. Appl. Environ. Microbiol. 55, 1943–1948.

Alvira, P., Thomas-Pejo, E., Ballesteros, M., Negro, M.J., 2010. Pretreatment technologies for an efficient bioethanol production process based on enzymatic hydrolysis. Bioresour. Technol. 101, 4851–4861.

Amerik, A., et al., 2010. New Yeast Strain Comprises a Yeast Cell Comprising Heterologous Genes That Encode the Cellulase Enzymes, Useful for Producing Ethanol From Hemicellulose. Arbor Fuel Inc. (ARBO-Non-standard). (WO2010062332-A1).

Amore, A., Pepe, O., Ventorino, V., Birolo, L., Giangrande, C., Faraco, V., 2012. Cloning and recombinant expression of a cellulase from the cellulolytic strain *Streptomyces* sp. G12 isolated from compost. Microb. Cell Fact. 11 (164), 1–12.

Andberg, M., Penttilä, M., Saloheimo, M., 2015. Swollenin from *Trichoderma reesei* exhibits hydrolytic activity against cellulosic substrates with features of both endoglucanases and cellobiohydrolases. Bioresour. Technol. 181, 105–113.

Anderson, T.D., Robson, S.A., Jiang, X.W., Malmirchegini, G.R., Fierobe, H.P., Lazazzera, B.A., Clubb, R.T., 2011. Assembly of mini cellulosomes on the surface of *Bacillus subtilis*. Appl. Environ. Microbiol. 77, 4849–4858.

Anderson, T.D., Miller, J.I., Fierobe, H.P., Clubb, R.T., 2013. Recombinant *Bacillus subtilis* that grows on untreated plant biomass. Appl. Environ. Microbiol. 79, 867–876.

Annamalai, N., Rajeswari, M.V., Elayaraja, S., Balasubramanian, T., 2013. Thermostable, haloalkaline cellulose from *Bacillus halodurans* CAS 1 by conversion of lignocellulosic wastes. Carbohydr. Polym. 94, 409–415.

Annamalai, N., Rajeswari, M.V., Balasubramanian, T., 2014. Enzymatic saccharification of pretreated rice straw by cellulase produced from *Bacillus carboniphilus* CAS 3 utilizing lignocellulosic wastes through statistical optimization. Biomass Bioenergy 68, 151–160.

Aristidou, A., Penttilä, M., 2000. Metabolic engineering applications to renewable resource utilization. Curr. Opin. Biotechnol. 11, 187–198.

Arnold, F.H., 2001. Combinatorial and computational challenges for biocatalyst design. Nature 409, 253–257.

Aro, N., Ilmén, M., Saloheimo, A., Penttilä, M., 2003. ACEI of *Trichoderma reesei* is a repressor of cellulose and xylanase expression. Appl. Environ. Microbiol. 69, 56–65.

Aro, N., Pakula, T., Penttila, M., 2005. Transcriptional regulation of plant cell wall degradation by filamentous fungi. FEMS Microbiol. Rev. 29 (4), 719–739.

Baeyens, J., Kang, Q., Appels, L., Dewil, R., Lv, Y., Tan, T., 2015. Challenges and opportunities in improving the production of bio-ethanol. Prog. Energy Combust. Sci. 47, 60–88.

Bailey, J.E., 1991. Toward a science of metabolic engineering. Science 252, 1668–1675.

Balan, V., 2014. Current challenges in commercially producing biofuels from lignocellulosic biomass. ISRN Biotechnol. 463074.

Balan, V., Biswas, G., Dale, B., Crockett, E., Sticklen, M., 2007. Heterologous *Acidothermus cellulolyticus* 1,4-β-endoglucanase E1 produced within the corn biomass converts corn stover into glucose. Appl. Biochem. Biotechnol. 36, 207–220.

Ballesteros, M., Oliva, J.M., Negro, M.J., Manzannares, P., Ballesteros, L., 2004. Ethanol from lignocellulosic materials by a simultaneous saccharification and fermentation process (SSF) with *Kluyveromeces marxianus* CECT 10875. Process Biochem. 39, 1843–1848.

Batthacharya, A.S., Batthacharya, A., Pletschke, B.I., 2015. Synergism of fungal and bacterial cellulases and hemicellulases: a novel perspective for enhanced bio-ethanol production. Biotechnol. Lett. 37 (6), 1117–1129.

Bayer, E.A., Shimon, L.J.W., Lamed, R., 1998a. Cellulosomes: structure and ultrastructure. J. Struct. Biol. 124, 221–234.

Bayer, E.A., Chanzy, H., Lamed, R., Shoham, Y., 1998b. Cellulose, cellulases and cellulosomes. Curr. Opin. Struct. Biol. 8 (5), 548–557.

Bayer, E.A., Belaich, J.P., Sholam, Y., Lamed, R., 2004. The cellulosomes: multienzyme machines for degradation of plant cell wall polysaccharides. Annu. Rev. Microbiol. 58, 521–554.

Bayer, E.A., Lamed, R., White, B.A., Flint, H.J., 2008. From cellulosomes to cellulosomics. Chem. Rec. 8, 364–377.

Beall, D.S., Ingram, L.O., 1993. Genetic engineering of soft-rot bacteria for ethanol production from lignocellulose. J. Ind. Microbiol. 11, 151–155.

Beckham, G.T., Dai, Z., Matthews, J.F., Momany, M., Payne, C.M., Adney, W.S., Baker, S.E., Himmel, M.E., 2012. Harnessing glycosylation to improve cellulase activity. Curr. Opin. Biotechnol. 23 (3), 338–345.

Béguin, P., Aubert, J.P., 1994. The biological degradation of cellulose. FEMS Microbiol. Rev. 13, 25–58.

Behera, S.S., Ray, R.C., 2016. Solid state fermentation for production of microbial cellulases: recent advances and improvement strategies. Int. J. Biol. Macromol. 86, 656–669.

Bey, M., Berrin, J.-G., Poidevin, L., Sigoillot, J.-C., 2011. Heterologous expression of *Pycnoporus cinnabarinus* cellobiose dehydrogenase in *Pichia pastoris* and involvement in saccharification processes. Microb. Cell Fact. 10 (113), 1–15.

Bhalla, A., Bansal, N., Kumar, S., Bischoff, K.M., Sani, R.K., 2013. Improved lignocellulose conversion to biofuels with thermophilic bacteria and thermostable enzymes. Bioresour. Technol. 128, 751–759.

Bischof, R.H., Ramoni, J., Seiboth, B., 2016. Cellulases and beyond: the first 70 years of the enzyme producer *Trichoderma reesei*. Microb. Cell Fact. 15, 106.

Biswas, G.C.G., Ransom, C., Sticklen, M., 2006. Expression of biologically active *Acidothermus cellulolyticus* endoglucanase in transgenic maize plants. Plant Sci. 171, 617–623.

Boisset, C., Petrequin, C., Chanzy, H., Henrissat, B., Schulein, M., 2001. Optimized mixtures of recombinant *Humicola insolens* cellulases for biodegradation of crystalline cellulose. Biotechnol. Bioeng. 72, 339–345.

Bok, J.D., Yernool, D.A., Eveleigh, D.E., 1998. Purification, characterization, and molecular analysis of thermostable cellulases CelA and CelB from *Thermotoga neapolitana*. Appl. Environ. Microbiol. 64, 4774–4781.

Bommarius, A.S., Sohn, M., Kang, Y., Lee, J.H., Realff, M.J., 2014. Protein engineering of cellulases. Curr. Opin. Biotechnol. 29, 139–145.

Bon, E.P.S., Picataggio, S., 2002. Enzyme and microbial biocatalysts. Appl. Biochem. Biotechnol. 98, 163.

Breznak, J.A., Brune, A., 1994. Role of microorganisms in the digestion of lignocellulose by termites. Annu. Rev. Entomol. 39, 453–487.

Brotman, Y., Briff, E., Viterbo, A., Chet, I., 2008. Role of swollen in, an expansion-like protein from Trichoderma, in plant root colonization. Plant Physiol. 147, 779–789.

Brunecki, R., Selig, M.J., Vinzant, T.B., Himmel, M.E., Lee, D., Blaylock, M.J., Decker, S.R., 2011. In planta expression of *A. cellulolyticus* Cel5A endocellulase reduces cell wall recalcitrance in tobacco and maize. Biotechnol. Biofuels 4 (1).

Bu, L.T., Beckham, G.T., Shirts, M.R., Nimlos, M.R., Adney, W.S., Himmel, M.E., Crowley, M.F., 2011. Probing carbohydrate product expulsion from a processive cellulase with multiple absolute binding free energy methods. J. Biol. Chem. 286, 18161–18169.

Bulakhov, A.G., Volkov, P.V., Rozhkova, A.M., Gusakov, A.V., Nemashkalov, V.A., Satrutdinov, A.D., Sinitsyn, A.P., 2017. Using an inducible promoter of a gene encoding *Penicillium verruculosum* glucoamylase for production of enzyme preparations with enhanced cellulase performance. PLoS One 12 (1), e0170404.

Bull, S.R., 1990. Energy From Biomass & Wastes XIV. Institute of Gas Technology, Chicago, IL.

Cao, Y., Zheng, F., Wang, L., Zhao, G., Chen, G., Zhang, W., Liu, W., 2017. Rce1, a novel transcriptional repressor, regulates cellulase gene expression by antagonizing the transactivator Xyr1 in *Trichoderma reesei*. Mol. Microbiol. https://doi.org/10.1111/mmi.13685. (Epub ahead of print).

Carey, V.C., Walia, S.K., Ingram, L.O., 1983. Expression of a lactose transposon (Tn951) in *Zymomonas mobilis*. Appl. Environ. Microbiol. 465, 1163.

Caspeta, L., Chen, Y., Ghiaci, P., Feizi, A., Buskov, S., Hallstrom, B.M., Petranovic, D., Nielsen, J., 2014. Altered sterol composition renders yeast thermotolerant. Science 346, 75–78.

Castillo, S., Barth, D., Arvas, M., Pakula, T.M., Pitkänen, E., Blomberg, P., Seppanen-Laakso, T., Nygren, H., Sivasiddarthan, D., Penttila, M., Oja, M., 2016. Whole-genome metabolic model of *Trichoderma reesei* built by comparative reconstruction. Biotechnol. Biofuels 9, 252.

Chae, T.U., Ko, Y., Hwang, K., Lee, S.Y., 2017. Metabolic engineering of *Escherichia coli* for the production of four-five and six-carbon lactams. Metab. Eng. 41, 82–91.

Chand, P., Aruna, A., Maqsood, A., Rao, L., 2005. Novel mutation method for increased cellulase production. J. Appl. Microbiol. 98 (2), 318–323.

Chandel, A.K., Singh, O.V., 2011. Weedy lignocellulosic feedstock and microbial metabolic engineering: advancing the generation of 'Biofuel'. Appl. Microbiol. Biotechnol. 89, 1289–1303.

Chandel, A.K., Narasu, M.L., Chandrasekhar, G., Manikeyam, A., Rao, L.V., 2009. Use of *Saccharum spontaneum* (wild sugarcane) as biomaterial for cell immobilization and modulated ethanol production by thermotolerant *Saccharomyces cerevisiae* VS3. Bioresour. Technol. 100, 2404–2410.

Chandel, A.K., Chandrasekhar, G., Silva, M.B., da Silva, S.S., 2012. The realm of cellulases in biorefinery development. Crit. Rev. Biotechnol. 32 (3), 187–202.

Chang, J.J., Ho, C.Y., Ho, F.J., Tsai, T.Y., Ke, H.M., Wang, C.H.T., Chen, H.L., Shin, M.C., Huang, C.C., Li, W.H., 2012. PGASO: a synthetic biology tool for engineering a cellulolytic yeast. Biotechnol. Biofuels 5, 53.

Chang, J.J., Ho, F.J., Ho, C.Y., Wu, Y.C., Hou, Y.H., Huang, C.C., Shin, M.C., Li, W.H., 2013. Assembling a cellulase cocktail and a cellodextrin transporter into a yeast host for CBP ethanol production. Biotechnol. Biofuels 6, 19.

Chapple, C., Ladisch, M., Meilan, R., 2007. Loosening lignin's grip on bio fuel production. Nat. Biotechnol. 25, 746–748.

Chauve, M., Mathis, H., Huc, D., Casanave, D., Monot, F., Lopes Ferreira, N., 2010. Comparative kinetic analysis of two fungal beta-glucosidases. Biotechnol. Biofuels 3 (1), 3.

Chen, Y., Nielsen, J., 2016. Biobased organic acid production by metabolically engineered microorganisms. Curr. Opin. Biotechnol. 37, 165–172.

Chen, X., Ishida, N., Todaka, N., Nakamura, R., Maruyama, J., Takahashi, H., Kitamoto, K., 2010. Promotion of efficient saccharification of crystalline cellulose by *Aspergillus fumigatus* Swo1. Appl. Environ. Microbiol. 76, 2556–2561.

Cherry, J.R., Fidantsef, A.L., 2003. Directed evolution of industrial enzymes: an update. Curr. Opin. Biotechnol. 14, 438–443.

Chou, H.L., Dai, Z., Hsieh, C.W., Ku, M.S.B., 2011. High level expression of *Acidothermus cellulolyticus* beta-1,4-endoglucanase in transgenic rice enhances the hydrolysis of its straw by cultured cow gastric fluid. Biotechnol. Biofuels 4, 58.

Chundawat, S.P.S., Beckham, G.T., Himmel, M.E., Dale, B.E., 2011. Deconstruction of lignocellulosic biomass to fuels and chemicals. In: Prausnitz, J.M. (Ed.), Annual Review of Chemical and Biomolecular Engineering, pp. 121–145.

Chung, D., Cha, M., Guss, A.M., Westpheling, J., 2015. Direct conversion of plant biomass to ethanol by engineered *Caldicellulosiruptor bescii*. Proc. Natl. Acad. Sci. U. S. A. 111, 8931–8936.

Coradetti, S.T., Craig, J.P., Xiong, Y., Shock, T., Tian, C., Glass, N.L., 2012. Conserved and essential transcription factors for cellulase gene expression in ascomycete fungi. Proc. Natl. Acad. Sci. U. S. A. 109, 7397–7402.

Couto, S.R., Sanroman, M.A., 2006. Application of solid-state fermentation to food industry—a review. J. Food Eng. 76, 291–302.

da Silva Delabona, P., Rodrigues, G.N., Zubieta, M.P., Ramoni, J., Codima, C.A., Lima, D.J., Farinas, C.S., da cruz Pradella, D.G., 2017. The relation between xyr1 overexpression in *Trichoderma harzianum* and sugarcane bagasse saccharification performance. J. Biotechnol. 246, 24–32.

Dai, Z., Nielsen, J., 2015. Advancing metabolic engineering through systems biology of industrial microorganisms. Curr. Opin. Biotechnol. 36, 8–15.

Dai, Z., Hooker, B.S., Quesenberry, R.D., Gao, J., 1999. Expression of *Trichoderma reesei* exo-cellobiohydrolase I in transgenic tobacco leaves and calli. Appl. Biochem. Biotechnol. 77, 689–699.

Dai, Z., Hooker, B.S., Anderson, D.B., Thomas, S.R., 2000a. Improved plant-based production of E1 endoglucanase using potato: expression optimization and tissue targeting. Mol. Breed. 6, 277–285.

Dai, Z., Hooker, B.S., Anderson, D.B., Thomas, S.R., 2000b. Expression of *Acidothermus cellulolyticus* endoglucanase E1 in transgenic tobacco: biochemical characteristics and physiological effects. Transgenic Res. 9, 43–54.

Dai, Z., Hooker, B.S., Quesenberry, R.D., Thomas, S.R., 2005. Optimization of *Acidothermus cellulolyticus* endoglucanase (e1) production in transgenic tobacco plants by transcriptional, post-transcription and post-translational modification. Transgenic Res. 14, 627–643.

Das, S.P., Deka, D., Ghosh, A., Das, D., Jawed, M., Goyal, A., 2013. Scale up and efficient bioethanol production involving recombinant cellulase (Glycosidehydrolase family 5) from *Clostridium thermocellum*. Sustain. Chem. Processes 1 (1), 19.

Dashtban, M., Schraft, H., Qin, W., 2009. Fungal bioconversion of lignocellulosic residues; opportunities & perspectives. Int. J. Biol. Sci. 5 (6), 578–595.

de Vries, R.P., Visser, J., 2001. *Aspergillus* enzymes involved in degradation of plant cell wall polysaccharides. Microbiol. Mol. Boil. Rev. 65 (4), 497–522.

Demain, A.L., Vaishnav, P., 2009. Production of recombinant proteins by microbes and higher organisms. Biotechnol. Adv. 27, 297–306.

Derntl, C., Gudynaite-Savitch, L., Calixte, S., White, T., Mach, R.L., Mach-Aigner, A.R., 2013. Mutation of the xylanase regulator 1 causes a glucose blind hydrolase expressing phenotype in industrially used Trichoderma strains. Biotechnol. Biofuels 6, 62.

Dhillon, G.S., Kaur, S., Brar, S.K., Verma, M., 2012. Potential of apple pomace as a solid substrate for fungal cellulase and hemi cellulase bioproduction through solid-state fermentation. Ind. Crops Prod. 38, 6–13.

Ding, S.Y., Liu, Y.S., Zeng, Y., Himmel, M.E., Baker, J.O., Bayer, E.A., 2012. How does plant cell wall nanoscale architecture correlate with enzymatic digestibility? Science 338, 1055–1060.

Druzhinina, S., Kubicek, C.P., 2016. Familiar stranger: ecologicagenomics of the model saprotroph and industrial enzyme producer *Trichoderma reesei* breaks the stereotypes. Adv. Appl. Microbiol. 95, 69–147.

Druzhinina, I.S., Kubicek, C.P., 2017. Genetic engineering of *Trichoderma reesei* cellulases and their production. Microb. Biotechnol. 10 (6), 1485–1499.

Druzhinina, I.S., Kopchinskiy, A.G., Kubicek, E.M., Kubicek, C.P., 2016. A complete annotation of the chromosomes of the cellulase producer *Trichoderma reesei* provides insights in gene clusters, their expression and reveals genes required for fitness. Biotechnol. Biofuels 9, 75.

El-Ghonemy, D.H., Ali, T.H., El-Bondkly, A.M., Moharam, M.E.S., Talkhan, F.N., 2014a. Improvement of *Aspergillus oryzae* NRRL 3484 by mutagenesis and optimization of culture conditions in solid-state fermentation for the hyper-production of extracellular cellulase. Antonie Van Leeuwenhoek 106, 853–864.

El-Ghonemy, D.H., Ali, T.H., Moharam, M.E., 2014b. Optimization of culture conditions for the production of extracellular cellulases via solid state fermentation. Br. Microbiol. Res. J. 4, 698–714.

Ellilä, S., Fonseca, L., Uchima, C., Cota, J., Goldman, G.H., Saloheimo, M., Sacon, V., Siika-aho, M., 2017. Development of a low-cost cellulase production process using *Trichoderma reesei* for Brazilian biorefineries. Biotechnol. Biofuels 10, 30.

Esterbauer, H., Steiner, W., Labudova, I., Hermann, A., Hayn, M., 1991. Production of Trichoderma cellulose in laboratory and pilot plant. Bioresour. Technol. 36, 67–76.

Eveleigh, D.E., Montenecourt, B.S., 1979. Increasing yields of extracellular enzymes. Adv. Appl. Microbiol. 25, 57–74.

Fan, L.H., Zhang, Z.J., Yu, X.Y., Xue, Y.X., Tan, T.W., 2012. Self-surface assembly of cellulosomes with two miniscaffoldins on *Saccharomyces cerevisiae* for cellulosic ethanol production. Proc. Natl. Acad. Sci. U. S. A. 109, 13260–13265.

Fang, H., Xia, L., 2013. High activity cellulase production by recombinant *Trichoderma reesei* ZU-02 with the enhanced cellobiohydrolase production. Bioresour. Technol. 144, 693–697.

Fawzi, E.M., Hamdy, H.S., 2011. Improvement of carboxymethyl cellulase production from *Chaetomium cellulolyticum* NRRL 18756 by mutation and optimization of solid state fermentation. Bangladesh J. Bot. 40, 139–147.

Ferreira-Leitao, V.S., da Silva, J.G., Bon, E.P.S., 2003. Methylene blue and azure B oxidation by horseradish peroxidase: a comparative evaluation of class II and class III peroxidases. Appl. Catal. B: Environ. 42 (2), 213–221.

FitzPatrick, M., Champagne, P., Cunningham, M.F., Whitney, R.A., 2010. A bio refinery processing perspective: treatment of lignocellulosic materials for the production of value-added products. Bioresour. Technol. 101, 8915–8922.

Fontes, C.M.G.A., Gilbert, H.J., 2010. Cellulosomes: highly efficient nanomachines designed to deconstruct plant cell wall complex carbohydrates. Annu. Rev. Biochem. 79, 655–681.

Foreman, P.K., Brown, D., Dankmeyer, L., Dean, R., Diener, S., Dunn-Coleman, N.S., Goedegebuur, F., Houfek, T.D., England, G.J., Kelley, A.S., Meerman, H.J., Mitchell, T., Mitchinson, C., YaoWard, H.M., 2003. Transcriptional regulation of biomass-degrading enzymes in the filamentous fungus *Trichoderma reesei*. J. Biol. Chem. 278, 31988–31997.

Fujii, T., Inoue, H., Ishikawa, K., 2013. Enhancing cellulase and hemicellulase production by genetic modification of the carbon catabolite repressor gene, creA, in *Acremonium cellulolyticus*. AMB Express 3 (1), 73.

Furukawa, T., Shida, Y., Kitagami, N., Ota, Y., Adachi, M., Nakagawa, S., Shimada, R., Kato, M., Kobayashi, T., Okada, H., Ogasawara, W., Morikawa, Y., 2008. Identification of the cis-acting elements involved in regulation of xylanase III gene expression in *Trichoderma reesei* PC-3-7. Fungal Genet. Biol. 45, 1094–1102.

Gadgil, N.J., Daginawala, H.F., Chakarabarti, T., Khanna, P., 1995. Enhanced cellulase production by a mutant of *Trichoderma reesei*. Enzyme Microb. Technol. 17, 942–946.

Gao, J., Qian, Y., Wang, Y., Qu, Y., Zhong, Y., 2017. Production of the versatile cellulose for cellulose bioconversion and cellulose inducer synthesis by genetic improvement of *Trichoderma reesei*. Biotechnol. Biofuels 10, 272.

Garvey, M., Klose, H., Fischer, R., Lambertz, C., Commandeur, U., 2013. Cellulases for biomass degradation: comparing recombinant cellulose expression platforms. Trends Biotechnol. 31 (10), 581–593.

Gefen, G., Anbar, M., Morag, E., Lamed, R., Bayer, E.A., 2012. Enhanced cellulose degradation by targeted integration of a cohesin-fused β-glucosidase into the *Clostridium thermocellum* cellulosome. Proc. Natl. Acad. Sci. U.S.A. 109, 10298–10303.

Girard, C., Jouanin, L., 1999. Molecular cloning of cDNAs encoding a range of digestive enzymes from a phytophagous beetle, *Phaedon cochleariae*. Insect Biochem. Mol. Biol. 29, 1129–1142.

Girfoglio, M., Rossi, M., Cannio, R., 2012. Cellulose degradation by Sulfolobus solfataricus requires a cell-anchored endo-beta-1-4-glucanase. J. Bacteriol. 194, 5091–5100.

Goldemberg, J., 2007. Ethanol for a sustainable energy future. Science 315, 808–810.

Goldstein, M.A., Takagi, M., Hashida, S., Shoseyov, O., Doi, R.H., Segel, I.H., 1993. Characterization of the cellulose-binding domain of the clostridium cellulovorans cellulose-binding protein A. J. Bacteriol. 175 (18), 5762–5768.

Górka-Nieć, W., Perlinska-Lenart, U., Zemberk, P., Palamarczyk, G., Kruszewska, J.S., 2010. Influence of sorbitol on protein production and glycosylation and cell wall formation in Trichoderma reesei. Fungal Biol. 114, 855–862.

Gourlay, K., Hu, J., Arantes, V., Andberg, M., Saloheimo, M., Penttilä, M., Saddler, J., 2013. Swollen in aids in the a morphogenesis step during the enzymatic hydrolysis of pretreated biomass. Bioresour. Technol. 142, 498–503.

Graham, J.E., Clark, M.E., Nadler, D.C., Huffer, S., Chokhawala, H.A., Rowland, S.E., Robb, F.T., 2011a. Identification and characterization of a multi domain hyper thermophilic cellulase from an archaeal enrichment. Nat. Commun. https://doi.org/10.1038/ncomms1373.

Graham, J.E., Clark, M.E., Nadler, D.C., Huffer, S., Chokhawala, H.A., Rowland, S.E., Blanch, H.W., Clark, D.S., Robb, F.T., 2011b. Identification and characterization of a multidomain hyperthermophilic cellulase from an archaeal enrichment. Nat. Commun. 2, 375.

Gritzali, M., Brown, R.D.J., 1979. The cellulase system of *Trichoderma* relationships between purified extracellular enzymes from induced or cellulose-grown cells. Adv. Chem. 181, 237–260.

Gupta, V.K., Steindorff, A.S., de Paula, R.G., Silva-Rocha, R., Mach-Aigner, A.R., Mach, R.L., Silva, R.N., 2016. The post-genomic era of *Trichoderma reesei*: what's next? Trends Biotechnol. 34, 970–982.

Gusakov, A.V., 2011. Alternatives to *Trichoderma reesei* in biofuel production. Trends Biotechnol. 29, 419–425.

Gusakov, A.V., 2013. Cellulases and hemicellulases in the 21st century race for cellulosic ethanol. Biofuels 4, 567–569.

Gusakov, A.V., 2014. Comment on "revealing nature's cellulase diversity: the digestion mechanism of *Caldicellulosiruptor bescii* CelA". Science 344, 578-a.

Gusakov, A.V., Dotsenko, A.S., Rozhkova, A.M., Sinitsyn, A.P., 2017. N-linked glycans are an important component of the processive machinery of cellobiohydrolases. Biochimie 132, 102–108.

Hahn-Hagerdal, B., Karhumaa, K., Fonseca, C., Spencer-Martins, I., Gorwa-Grauslund, M.F., 2007. Towards industrial pentose-fermenting yeast strains. Appl. Microbiol. Biotechnol. 74 (5), 937–953.

Häkkinen, M., Valkonen, M.J., Westerholm-Parvinen, A., Aro, N., Arvas, M., Vitikainen, M., Penttila, M., Saloheimo, M., Pakula, T.M., 2014. Screening of candidate regulators for cellulase and hemicellulase production in *Trichoderma reesei* and identification of a factor essential for cellulase production. Biotechnol. Biofuels 7, 14.

Hamada, N., Fuse, N., Shimosaka, M., Kodaira, R., Amano, Y., Kanda, T., Okazaki, M., 1999a. Cloning and characterization of a new exocellulase gene, cel3, in Irpex lacteus. FEMS Microbiol. Lett. 172 (2), 231–237.

Hamada, N., Okumura, R., Fuse, N., Kodaira, R., Shimosaka, M., Kanda, T., 1999b. Isolation and transcriptional analysis of a cellulase gene (cel1) from the basidiomycete Irpex lacteus. J. Biosci. Bioeng. 87 (1), 97–102.

Harkki, A., Mäntylä, A., Penttilä, M., Muttilainen, S., Bühler, R., Suominen, P., Knowles, J., Nevalainen, H., 1991. Genetic engineering of *Trichoderma* to produce strains with novel cellulase profiles. Enzyme Microb. Technol. 13 (3), 227–233.

Harris, D., Petti, C., DeBolt, S., 2013. The synthesis, regulation and modification of lignocellulosic biomass as a resource for biofuels and bioproducts. In: Lee, J.W. (Ed.), Advanced Biofuels and Bioproducts. Springer, pp. 281–314.

Harrison, M.D., Geijskes, J., Coleman, H.D., Shand, K., Kinkema, M., Palupe, A., Hassall, R., Sainz, M., Lloyd, R., Miles, S., Dale, J.L., 2011. Accumulation of recombinant cellobiohydrolase and endoglucanase in the leaves of mature transgenic sugar cane. Plant Biotechnol. J. 9, 884–896.

Hasunuma, T., Kondo, A., 2012. Development of yeast cell factories for consolidated bioprocessing of lignocellulose to bioethanol through cell surface engineering. Biotechnol. Adv. 30, 1207–1218.

Hasunuma, T., Okazaki, F., Okai, N., Hara, K.Y., Ishii, J., Kondo, A., 2013. A review of enzymes and microbes for lignocellulosic biorefinery and the possibility of their application to consolidated bioprocessing technology. Bioresour. Technol. 135, 513–522.

Heinelman, P.C., Snow, D., Wu, I., Nguyen, C., Villalobos, A., Govindarajan, S., Minshull, J., Arnold, F.H., 2009. A family of thermostable fungal cellulases created by structure guided recombination. PNAS 106 (14), 5610–5615.

Herbers, K., Wilke, I., Sonnewald, U., 1995. A thermostable xylanase from *Clostridium thermocellum* expressed at high levels in the apoplast of transgenic tobacco has no detrimental effects and is easily purified. Nat. Biotechnol. 13, 63–66.

Hibbert, E.G., Baganz, F., Hailes, H.C., Ward, J.M., Lye, G.J., Woodley, J.M., Dalby, P.A., 2005. Directed evolution of biocatalytic processes. Biomol. Eng. 22, 11–19.

Himmel, M.E., Ruth, M.F., Wyman, C.E., 1999. Cellulase for commodity products from cellosic biomass. Curr. Opin. Biotechnol. 10, 358–364.

Himmel, M.E., Ding, S.Y., Johnson, D.K., Adney, W.S., Nimlos, M.R., Brady, J.W., Foust, T.D., 2007. Biomass recalcitrance: engineering plants and enzymes for biofuels production. Science 315, 804–807.

Holker, U., Hofer, M., Lenz, J., 2004. Biotechnological advantages of laboratory-scale solid-state fermentation with fungi. Appl. Microbiol. Biotechnol. 64 (2), 175–186.

Hong, K.K., Vongsangnak, W., Vemuri, G.N., Nielsen, J., 2011. Unravelling evolutionary strategies of yeast for improving galactose utilization through integrated systems level analysis. Proc. Natl. Acad. Sci. U. S. A. 108, 12179–12184.

Hood, E.E., 2004. Bioindustrial and biopharmaceutical products from plants. In: New Directions for a Diverse Planet. Proceedings of the 4th International Crop Science Congress, Brisbane, Australia. ISBN 1-920842-20-9.

Horn, M.E., Woodard, S.L., Howard, J.A., 2004. Plant molecular farming: systems and products. Plant Cell Rep. 22, 711–720.

Howard, J., Hood, E., 2005. Bioindustrial and biopharmaceutical products produced in plants. Adv. Agron. 85, 91–124.

Hu, Y., Liu, G., Li, Z., Qin, Y., Qu, Y., Song, X., 2013. G protein-cAMP signaling pathway mediated by PGA3 plays different roles in regulating the expressions of amylases and cellulases in *Penicillium decumbens*. Fungal Genet. Biol. 58–59, 62–70.

Hyeon, J.E., Jeon, W.J., Whang, S.Y., Han, S.O., 2011. Production of minicellulosomes for the enhanced hydrolysis of cellulosic substrates by recombinant *Corynebacterium glutamicum*. Enzyme Microb. Technol. 48, 371–377.

Ibrahim, E., Jones, K.D., Hosseny, E.N., Escudero, J., 2013. Molecular cloning and expression of cellulase and polygalacturonase genes in *E. coli* as a promising application for biofuel production. J. Pet. Environ. Biotechnol. 4, 3. https://doi.org/10.4172/2157-7463.1000147.

Ilmén, M., Thrane, C., Penttilä, M., 1996. The glucose repressor gene cre1 of *Trichoderma*: isolation and expression of a full-length and a truncated mutant form. Mol. Genet. Genomics 251, 451–460.

Ilmen, M., Thrane, C., Penttilä, M., 1996. The glucose repressor gene cre1 of Trichoderma: isolation and expression of a full-length and a truncated mutant form. Mol. Gen. Genet. 251, 451–460.

Ilmén, M., Saloheimo, A., Onnela, M.L., Penttilä, M.E., 1997. Regulation of cellulase gene expression in the filamentous fungus *Trichoderma ressei*. Appl. Environ. Microbiol. 63, 1298–1306.

Ilmen, M., den Haan, R., Brevnova, E., McBride, J., Wiswall, E., Froehlich, A., Koivula, A., Voutilainen, S.P., Siika-aho, M., l Grange, D.C., Thorngren, N., Ahlgren, S., Mellon, M., Deleault, K., Rajgarhia, V., van Zyl, W.H., Penttila, M., 2011. High level secretion of cellobiohydrolases by *Saccharomyces cerevisiae*. Biotechnol. Biofuels 4, 30.

Isaac, G.S., Abu-Tahon, M.A., 2015. Enhanced alkaline cellulases production by the thermohalophilic *Aspergillus terreus* AUMC 10138 mutated by physical and chemical mutagens using corn stover as substrate. Braz. J. Microbiol. 46 (4), 1269–1277.

Jagadeeswaran, G., Gainey, L., Prade, R., Mort, A.J., 2016. A family of AA9 lytic polysaccharide monooxygenases in *Aspergillus nidulans* is differentially regulated by multiple substrates and at least one is active on cellulose and xyloglucan. Appl. Microbiol. Biotechnol. 100 (10), 4535–4547.

Jäger, G., Girfoglio, M., Dollo, F., Rinaldi, R., Bongard, H., Commandeur, U., Fischer, R., Spiess, A.C., Büchs, J., 2011. How recombinant swollenin from *Kluyveromyces lactis* affects cellulosic substrates and accelerates their hydrolysis. Biotechnol. Biofuels 4, 33.

Javed, M.M., Haq, I., Mariyam, I., 2011. Multistep mutagenesis for the over-expression of cellulase in *Humicola insolens*. Pak. J. Bot. 43 (1), 669–677.

Johnvesly, B., Virupakshi, S., Patil, G.N., Ramalingam, A., Naik, G.R., 2002. Cellulase-free thermostable alkaline xylanase from thermophilic and alkalophilic Bacillus sp. JB-99. J. Microbiol. Biotechnol. 12, 153–156.

Jung, S., Kim, S., Bae, H., Lim, H.S., Bae, H.J., 2010. Expression of thermostable bacterial b-glucosidase (BglB) in transgenic tobacco plants. Bioresour. Technol. 101, 7144–7150.

Jung, S.K., Parisutham, V., Jeong, S.H., Lee, S.K., 2012. Heterologous expression of plant cell wall degrading enzymes for effective production of cellulosic biofuels. J. Biomed. Biotechnol. https://doi.org/10.1155/2012/405842.

Jung, Y.H., Cho, H.J., Lee, J.S., Noh, E.W., Park, O.K., Kim, K.H., 2013. Evaluation of a transgenic poplar as a potential biomass crop for biofuel production. Bioresour. Technol. 129, 639–641.

Juturu, V., Wu, J.C., 2014. Microbial cellulases: engineering, production and applications. Renew. Sustain. Energy Rev. 33, 188–203.

Kado, C.I., 1992. Plant pathogenic bacteria. In: Balows, A., Truper, H.G., Dworkin, M., Harder, W., Scheifer, K.-H. (Eds.), The Prokaryotes. vol. 1. Springer-Verlag, New York, pp. 659–674.

Kang, K., Wang, S., Lai, G., Liu, G., Xing, M., 2013. Characterization of a novel swollen in from *Penicillium oxalicum* in facilitating enzymatic saccharification of cellulose. BMC Biotechnol. 13, 42.

Karnaouri, A.C., Topakas, E., Christakopoulos, P., 2014a. Cloning, expression, and characterization of a thermostable GH7 endoglucanase from *Myceliophthora thermophila* capable of high-consistency enzymatic liquefaction. Appl. Microbiol. Biotechnol. 98 (1), 231–242.

Karnaouri, A., Topakas, E., Antonopoulou, I., Christakopoulos, P., 2014b. Genomic insights into the fungal lignocellulolytic system of *Myceliophthora thermophila*. Front. Microbiol. 5, 281.

Keasling, J.D., Chou, H., 2008. Metabolic engineering delivers next-generation biofuels. Nat. Biotechnol. 26 (3), 298–299.

Keim, C.R., Venkatasubramanian, K., 1989. Economics of current biotechnological methods of producing ethanol. Trends Biotechnol. 7, 22–29.

Kim, S., Kim, C.H., 2012. Production of cellulase enzymes during the solid-state fermentation of empty palm fruit bunch fiber. Bioprocess Biosyst. Eng. 35, 61–67.

Kim, E., Irwin, D.C., Walker, L.P., Wilson, D.B., 1998. Factorial optimization of a six-cellulase mixture. Biotechnol. Bioeng. 58, 494–501.

Kim, T.W., Chokhawala, H.A., Nadler, D., Blanch, H.W., Clark, D.S., 2010. Binding modules alter the activity of chimeric cellulases: effects of biomass pretreatment and enzyme source. Biotechnol. Bioeng. 107, 601–611.

Kim, D.M., Umetsu, M., Takai, K., Matsuyama, T., Ishida, N., Takahashi, H., Asano, R., Kumagai, I., 2011. Enhancement of cellulolytic enzyme activity by clustering cellulose binding domains on nanoscaffolds. Small 7, 656–664.

Kim, D.M., Nakazawa, H., Umetsu, M., Matsuyama, T., Ishida, N., Ikeuchi, A., Takahashi, H., Asano, R., Kumagai, I., 2012. A nanocluster design for the construction of artificial cellulosomes. Catal. Sci. Technol. 2, 499–503.

Kimura, T., Mizutani, T., Sakka, K., Ohmiya, K., 2003a. Stable expression of a thermostable xylanase of *Clostridium thermocellum* in cultured tobacco cells. J. Biosci. Bioeng. 95, 397–400.

Kimura, T., Mizutani, T., Tanaka, T., Koyama, T., Sakka, K., Ohima, K., 2003b. Molecular breeding of transgenic rice expressing a xylanase domain of the *nnA* gene from *Clostridium thermocellum*. Appl. Microbiol. Biotechnol. 62, 374–379.

Kitagawa, T., Kohda, K., Tokuhiro, K., Hoshida, H., Akada, R., Takahashi, H., Imaeda, T., 2011. Identification of genes that enhance cellulase protein production in yeast. J. Biotechnol. 151, 194–203.

Klose, H., Roder, J., Girfoglio, M., Fischer, R., Commendeur, U., 2012. Hyperthermophilic endoglucanase for in planta lignocellulose conversion. Biotechnol. Biofuels 5, 63.

Klose, H., Gunl, M., Usadel, B., Fischer, R., Commandeur, U., 2013. Ethanol inducible expression of a mesophilic cellulase avoids adverse effects on plant development. Biotechnol. Biofuels 6, 53.

Koeck, D.E., Pechtl, A., Zerlov, V.V., Schwarz, W.H., 2014. Genomics of cellulolytic bacteria. Curr. Opin. Biotechnol. 29, 171–183.

Kötter, P., Ciriacy, M., 1993. Xylose fermentation by *Saccharomyces cerevisiae*. Appl. Microbiol. Biotechnol. 38, 776–783.

Kötter, P., Amore, R., Hollenberg, C.P., Ciriacy, M., 1990. Isolation and characterization of the *Pichia stipitis* xylitol dehydrogenase gene, *XYL2*, and construction of a xylose-utilizing *Saccharomyces cerevisiae* transformant. Curr. Genet. 18, 493.

Kubicek, C.P., Penttilä, M., 1998. Regulation of production of plant polysaccharide degrading enzymes by Trichoderma. In: Harman, G.E., Kubicek, C.P. (Eds.), Trichoderma and Gliocladium. Taylor & Francis Ltd., London.

Kuhad, R.C., Kumar, M., Singh, A., 1994. A hypercellololytic mutant of *Fusarium oxysporum*. Lett. Appl. Microbiol. 19 (5), 397–400.

Kumar, R., Singh, S., Singh, O.V., 2008. Bioconversion of lignocellulosic biomass: biochemical and molecular perspectives. J. Ind. Microbiol. Biotechnol. 35, 377–391.

Ladisch, M.R., Lin, K.W., Voloch, M., Tsao, G.T., 1983. Process considerations in the enzymatic hydrolysis of biomass. Enzyme Microb. Technol. 5, 82–102.

Lambertz, C., Garvey, M., Klinger, J., Heesel, D., Klose, H., Fischer, R., Commandeur, U., 2014. Challenges and advances in the heterologous expression of cellulolytic enzymes: a review. Biotechnol. Biofuels 7, 135.

Laothanachareon, T., Bunterngsook, B., Suwannarangsee, S., Eurwilaichitr, L., Champreda, V., 2015. Synergistic action of recombinant accessory hemicellulolytic and pectinolytic enzymes to Trichoderma reesei cellulase on rice straw degradation. Bioresour. Technol. 198, 682–690.

Lebel, E.G., Heifetz, P.B., Ward, E.R., Uknes, S.J., 2008. Syngenta Participations AG, Transgenic Plants Expressing a Cellulase. . (US 07361806).

Lee, C.C., Wong, D.W.S., Robertson, G.H., 2001. Cloning and characterization of two cellulase genes from *Lentinula edodes*. FEMS Microbiol. Lett. 205 (2), 355–360.

Lee, S.J., Lee, K.S., Kim, S.R., Gui, Z.Z., Kim, Y.S., Yoon, H.J., Kim, I., Kang, P.D., Sohn, H.D., Jin, B.R., 2005. A novel cellulase gene from the mulberry longicorn beetle, *Apriona germari*: gene structure, expression, and enzymatic activity. Comp. Biochem. Physiol. B: Biochem. Mol. Biol. 140, 551–560.

Lee, S.J., Song, H., Lee, S.Y., 2006. Genome-based metabolic engineering of *Mannheimia succiniciproducens* for succinic acid production. Appl. Environ. Microbiol. 72 (3), 1939–1948.

Lee, K.H., Park, J.H., Kim, T.Y., kim, H.U., Lee, S.Y., 2007. Systems metabolic engineering of Escherichia coli for L-threonine production. Mol. Syst. Biol. 3, 149.

Lee, S.Y., Park, J.H., Jang, S.H., Nielsen, L.K., Kim, J., Jung, K.S., 2008a. Fermentative butanol production by Clostridia. Biotechnol. Bioeng. 101 (2), 209–228.

Lee, J.W., Kim, T.Y., Jang, Y., Choi, S., Lee, S.Y., 2011a. Systems metabolic engineering for chemicals and materials. Trends Biotechnol. 29 (8), 370–378.

Lee, T.M., Farrow, M.F., Arnold, F.H., Mayo, S.L., 2011b. A structural study of *Hypocrea jecorina* Cel5A. Protein Sci. 20, 1935–1940.

Lee, Y.J., Kim, H.J., Gao, W., Chung, C.H., Lee, J.W., 2012. Statistical optimization for production of carboxy methyl cellulase of *Bacillus amyloliquefaciens* DL-3 by a recombinant *Escherichia coli* JM109/DL-3 from rice bran using response surface method. Biotechnol. Bioprocess Eng. 17, 227–235.

Lerouge, P., Cabanes-Macheteau, M., Rayon, C., Fischette-Lainé, A.C., Gomord, V., Faye, L., 1998. N-glycoprotein biosynthesis in plants: recent developments and future trends. Plant Mol. Biol. 38, 31–48.

Li, X.H., Yang, H.J., Roy, B., Park, E.Y., Jiang, L.J., Wang, D., Miao, Y.G., 2010a. Enhanced cellulase production of the *Trichoderma viride* mutated by microwave andultraviolet. Microbiol. Res. 165, 190–198.

Li, X.H., Yang, H.J., Roy, B., Park, E.Y., Jiang, L.J., Wang, D., Miao, Y.G., 2010b. Enhanced cellulase production of the *Trichoderma viride* mutated by microwave and ultraviolet. Microbiol. Res. 165 (3), 190–198.

Lichius, A., Seidl-Seiboth, V., Seiboth, B., Kubicek, C.P., 2014. Nucleo-cytoplasmic shuttling dynamics of the transcriptional regulators XYR1 and CRE1 under conditions of cellulase and xylanase gene expression in *Trichoderma reesei*. Mol. Microbiol. 94, 1162–1178.

Lilly, M., Fierobe, H.P., van Zyl, W.H., Volschenk, H., 2009. Heterologous expression of a *Clostridium* minicellulosome in *Saccharomyces cerevisiae*. FEMS Yeast Res. 9, 1236–1249.

Limayem, A., Ricke, S.C., 2012. Lignocellulosic biomass for bioethanol production: current perspectives, potential issues and future prospects. Prog. Energy Combust. Sci. 38, 449–467.

Lin, Y., Tanaka, S., 2006. Ethanol fermentation from biomass resources: current state and prospects. Appl. Microbiol. Biotechnol. 69 (6), 627–642.

Linder, M., Salovuori, I., Ruohonen, L., Teeri, T.T., 1996. Characterization of a double cellulose-binding domain: synergistic high affinity binding to crystalline cellulose. J. Biol. Chem. 271 (35), 21268–21272.

Linger, J.G., Adney, W.S., Darzins, A.I., 2010. Heterologous expression and extracellular secretion of cellulolytic enzymes by *Zymomonas mobilis*. Appl. Environ. Microbiol. 76, 6360–6369.

Liu, M., Yu, H., 2012. Co-production of a whole cellulase system in *Escherichia coli*. Biochem. Eng. J. 69, 204–210.

Liu, J., Liu, W.D., Zhao, X.L., Shen, W.J., Cao, H., Cui, Z.L., 2011. Cloning and functional characterization of a novel endo-β-1,4-glucanase gene from a soil-derived metagenomic library. Appl. Microbiol. Biotechnol. 89, 1083–1092.

Liu, J.M., Xin, X.J., Li, C.X., Xu, J.H., Bao, J., 2012a. Cloning of thermostable cellulase genes of *Clostridium thermocellum* and their secretive expression in *Bacillus subtilis*. Appl. Biochem. Biotechnol. 166, 652–662.

Liu, S.L., Wei-Zhao Chen, K.D., Liu, G., Xing, M., 2012b. Effective approach to greatly enhancing selective secretion and expression of three cytoplasmic enzymes in *Escherichia coli* through synergistic effect of EDTA and lysozyme. J. Ind. Microbiol. Biotechnol. 39, 1301–1307.

Liu, G., Zhang, L., Qin, Y., Zou, G., Li, Z., Yan, X., Wei, X., Chen, M., Chen, L., Zheng, K., Zhang, J., Ma, L., Li, J., Liu, R., Xu, H., Bao, X., Fang, X., Wang, L., Zhong, Y., Lio, W., Zheng, H., Wang, S., Wang, C., Xun, C., Zhao, G.P., 2013a. Long-term strain improvements accumulate mutations in regulatory elements responsible for hyper-production of cellulolytic enzymes. Sci. Rep. 3, 1569.

Liu, G., Zhang, L., Wei, X., Zou, G., Qin, Y., Ma, L., Li, J., Zheng, H., Wang, S., Wang, C., Xun, L., Zhao, G.P., Zhou, Z., Qu, Y., 2013b. Genomic and secretomic analyses reveal unique features of the lignocellulolytic enzyme system of *Penicillium decumbens*. PLoS One 8 (2), e55185.

Liu, K., Dong, Y., Wang, F., Jiang, B., Wang, M., Fang, X., 2016. Regulation of cellulase expression, sporulation, and morphogenesis by velvet family proteins in *Trichoderma reesei*. Appl. Microbiol. Biotechnol. 100, 769–779.

Lykidis, A., Mavromatis, K., Ivanova, N., Anderson, I., Land, M., DiBartolo, G., Martinez, M., Lapidus, A., Lucas, S., Copeland, A., Richardson, P., Wilson, D.B., Kyrpides, N., 2007. Genome sequence and analysis of the soil cellulolytic actinomycete *Thermobifida fusca* YX. J. Bacteriol. 189, 2477–2486.

Lynd, L.R., Laser, W.H.v.Z.J.E.M.B.M., 2005. Consolidated bioprocessing of cellulosic biomass: an update. Curr. Opin. Biotechnol. 16, 577–583.

Lynd, L.R., Cushman, J.H., Nichols, R.J., Wyman, C.E., 1991. Fuel ethanol from cellulosic biomass. Science 251, 1318–1323.

Lynd, L.R., Weimer, P.J., vanZyl, W.H., Pretorius, I.S., 2002. Microbial cellulose utilization: fundamentals and biotechnology. Microbiol. Mol. Biol. Rev. 66, 506–577.

Ma, L., Zhang, J., Zou, G., Wang, C., Zhou, Z., 2011a. Improvement of cellulose activity in *Trichoderma reesei* by heterologous expression of a beta-glucosidase gene from *Penicillium decumbens*. Enzyme Microb. Technol. 49, 366–371.

Ma, L., Zhang, J., Zou, G., Wang, C., Zhou, Z., 2011b. Improvement of cellulase activity in *Trichoderma reesei* by heterologous expression of a beta-glucosidase gene from *Penicillium decumbens*. Enzyme Microb. Technol. 49, 366–371.

Ma, L., Zhang, J., Zou, G., Wang, C., Zhou, Z., 2011c. Improvement of cellulase activity in *Trichoderma reesei* by heterologous expression of a beta-glucosidase gene from *Penicillium decumbens*. Enzyme Microb. Technol. 49 (4), 366–371.

Mach, R.L., Strauss, J., Zeilinger, S., Schindler, M., Kubicek, C.P., 1996. Carbon catabolite repression of xylanase I (xyn1) gene expression in *Trichoderma reesei*. Mol. Microbiol. 21, 1273–1281.

Mahadevan, S.A., Wi, S.G., Lee, D.S., Bae, H.J., 2008. Site-directed mutagenesis and CBM engineering of Cel5A (*Thermotoga maritima*). FEMS Microbiol. Lett. 287 (2), 205–211.

Mahadevan, S.A., Wi, S.G., Kim, Y.O., Lee, K.H., Bae, H.J., 2011. In planta differential targeting analysis of *Thermotoga maritima* Cel5A and CBM6-engineered Cel5A for autohydrolysis. Transgenic Res. 20, 877–886.

Mandels, M., Parrish, F.W., Reese, E.T., 1962. Sophorose as an inducer of cellulase in *Trichoderma viride*. J. Bacteriol. 83, 400–408.

Margolles-Clark, M., Ilmén, M., Penttilä, M., 1997. Expression patterns of ten hemicellulase genes of the filamentous fungus *Trichoderma reesei* on various carbon sources. J. Biotechnol. 57, 167–179. https://doi.org/10.1016/S0168-1656(97)00097-7.

Martinez, D., Larrondo, L.F., Putnam, N., Gelpke, M.D., Huang, K., Chapman, J., Helfenbein, K.G., Ramaiya, P., Detter, J.C., Larimer, F., Coutinho, P.M., Henrissat, B., Berka, R., Cullen, D., Rokhsar, D., 2004. Genome sequence of the lignocellulose degrading fungus *Phanerochaete chrysosporium* strain RP78. Nat. Biotechnol. 22, 679–680.

Martinez, D., Berka, R.M., Henrissat, B., Saloheimo, M., Arvas, M., Baker, S.E., Chapman, J., Chertkov, O., Coutinho, P.M., Cullen, D., Brettin, T.S., 2008. Genome sequencing and analysis of the biomass-degrading fungus *Trichoderma reesei* (syn *Hypocrea jecorina*). Nat. Biotechnol. 26, 553–560.

Matano, Y., Hasunuma, T., Kondo, A., 2012. Display of cellulases on the cell surface of Saccharomyces cerevisiae for high yield ethanol production from high-solid lignocellulosic biomass. Bioresour. Technol. 108, 128–133.

Mattanovich, D., Branduardi, P., Dato, L., Gasser, B., Sauer, M., Porro, D., 2012. Recombinant protein production in yeasts. Methods Mol. Biol. 824, 329–358.

Mayorga-Reyes, L., Ponce-Noyola, T., 1998. Isolation of a hyperxylanolytic cellulomonas xavigena mutant growing on continuous culture on sugar cane bagasse. Biotechnol. Lett. 20 (5), 443–446.

Mello-de-Sousa, T.M., Rassinger, A., Pucher, M.E., dos Santos Castro, L., Persinoti, G.F., Silva-Rocha, R., Pocas-Fonseca, M.J., Mach, R.L., Silva, R.N., Mach-Aigner, A.R., 2015. The impact of chromatin remodelling on cellulose expression in *Trichoderma reesei*. BMC Genom. 16, 588.

Mello-de-Sousa, T.M., Rassinger, A., Derntl, C., Poc_as-Fonseca, M.J., Mach, R.L., Mach-Aigner, R.A., 2016. The relation between promoter chromatin status, XYR1 and cellulase expression in *Trichoderma reesei*. Curr. Genomics 17, 145–152.

Menendez, E., Garcia-Fraile, P., Rivas, R., 2015. Biotechnological applications of bacterial cellulases. Bioengineering 2 (3), 163–182.

Mingardon, F., Chanal, A., Tardif, C., Fierobe, H.P., 2011. The issue of secretion in heterologous expression of *Clostridium cellulolyticum* cellulase-encoding genes in *Clostridium acetobutylicum* ATCC 824. Appl. Environ. Microbiol. 77, 2831–2838.

Mir, B.A., Mewalal, R., Mizrachi, E., Myburg, A.A., Cowan, D.A., 2014. Recombinant hyperthermophilic enzyme expression in plants: a novel approach for lignocellulose digestion. Trends Biotechnol. 32 (5), 281–289.

Morais, S., Heyman, A., Barak, Y., Caspi, J., Wilson, D.B., Lamed, R., Shoseyov, O., Bayer, E.A., 2010. Enhanced cellulose degradation by nanocomplexed enzymes: synergism between a scaffold-linked exoglucanase and a free endoglucanase. J. Biotechnol. 147, 205–211.

Morimoto, T., Kadoya, R., Endo, K., Tohata, M., Sawada, K., Liu, S., Ozawa, T., Kodama, T., Kakeshita, H., Kageyama, Y., 2008. Enhanced recombinant protein productivity by genome reduction in *Bacillus subtilis*. DNA Res. 15, 73–81.

Mostafa, A.A., 2014. Effect of gamma irradiation on *Aspergillus niger* DNA and production of cellulases enzymes. J. Am. Sci. 10, 152–160.

Mukhopadhyay, A., Redding, A.M., Rutherford, B.J., Keasling, J.D., 2008. Importance of systems biology in engineering microbes for biofuel production. Curr. Opin. Biotechnol. 19 (3), 228–234.

Munjal, N., Jaewd, K., Wajid, S., Yazdani, S.S., 2015. A Constitutive expression system for cellulase secretion in *Escherichia coli* and its use in bioethanol production. PLoS One 10, e0119917.

Murray, P., Aro, N., Collins, C., Grassick, A., Penttila, M., Saloheimo, M., Tuohy, M., 2004. Expression in *Trichoderma reesei* and characterisation of a thermostable family 3 beta-glucosidase from the moderately thermophilic fungus *Talaromyces emersonii*. Protein Expr Purif. 38, 248–257.

Mutalik, V.K., Guimaraes, J.C., Cambray, G., Lam, C., Christoffersen, M.J., Mai, Q.A., Tran, A.B., Paul, M., Keasling, J.D., Arkin, A.P., Endy, D., 2013. Precise and reliable gene expression via standard transcription and translation initiation elements. Nat. Methods 10, 354–360.

Nakari-Setälä, T., Paloheimo, M., Kallio, J., Vehmaanperä, J., Penttiläa, M., Saloheimo, M., 2009. Genetic modification of carbon catabolite repression in *Trichoderma reesei* for improved protein production. Appl. Environ. Microbiol. 75, 4853–4860.

Nakashima, K., Watanabe, H., Saitoh, H., Tokuda, G., Azuma, J.I., 2002. Dual cellulosedigesting system of the wood-feeding termite, *Coptotermes formosanus* Shiraki. Insect Biochem. Mol. Biol. 3, 777–784.

Nielsen, J., 2001. Metabolic engineering. Appl. Microbiol. Biotechnol. 55, 263–283.

Nielsen, J., Keasling, J.D., 2016. Engineering cellular metabolism. Cell 164, 1185–1197.

Njokweni, A.P., Rose, S.H., Van Zyl, W.H., 2012. Fungal-glucosidase expression in *Saccharomyces cerevisiae*. J. Ind. Microb. Biotechnol. 39, 1445–1452.

Odronitz, F., Hellkamp, M., Kollmar, M., 2007. diArk—a resource for eukaryotic genome research. BMC Genomics 8, 103.

Ohta, K., Beall, D.S., Mejia, J.P., Shanmugam, K.T., Ingram, L.O., 1991a. Genetic improvement of *Escherichia coli* for ethanol production: chromosomal integration of *Zymomonas mobilis* genes encoding pyruvate decarboxylase and alcohol dehydrogenase II. Appl. Environ. Microbiol. 57, 893–900.

Ohta, K., Beall, D.S., Mejia, J.P., Shanmugam, K.T., Ingram, L.O., 1991b. Metabolic engineering of *Klebsiella oxytoca* M5A1 for ethanol production from xylose and glucose. Appl. Environ. Microbiol. 57, 2810–2815.

Ohtoko, K., Ohkuma, M., Moriya, S., Inoue, T., Usami, R., Kudo, T., 2000. Diverse genes of cellulase homologues of glycosyl hydrolase family 45 from the symbiotic protists in the hindgut of the termite *Reticulitermes speratus*. Extremophiles 4, 343–349.

Olson, D.G., McBride, J.E., Shaw, A.J., Lynd, L.R., 2012. Recent progress in consolidated bioprocessing. Curr. Opin. Biotechnol. 23, 396–405.

Oraby, H., Venkatesh, B., Dale, B., Ahmad, R., Ransom, C., Oehmke, J., Sticklen, M., 2007. Enhanced conversion of plant biomass into glucose using transgenic rice-produced endoglucanase for cellulosic ethanol. Transgenic Res. (6)739–749.

Park, J.H., Lee, S.Y., Kim, T.Y., Kim, H.U., 2008. Application of systems biology for bioprocess development. Trends Biotechnol. 26, 404–412.

Patel, M., Johnson, J.S., Brettell, R.I.S., Jacobsen, J., Xue, G.P., 2000. Transgenic barley expressing a fungal xylanase gene in the endosperm of the developing grains. Mol. Breed. 6, 113–124.

Penttilä, M., Nevalainen, H., Rättö, M., Salminen, E., Knowles, J., 1987. A versatile transformation system for the cellulolytic filamentous fungus *Trichoderma reesei*. Gene 61 (2), 155–164.

Penttilä, M.E., André, L., Lehtovaara, P., Bailey, M., Teeri, T.T., Knowles, J.K., 1988. Efficient secretion of two fungal cellobiohydrolases by *Saccharomyces cerevisiae*. Gene 63, 103–112.

Petersen, K., Bock, R., 2011. High-level expression of a suite of thermostable cell wall-degrading enzymes from the chloroplast genome. Plant Mol. Biol. 76, 311–321.

Peterson, R., Nevalainen, H., 2012. Trichoderma reesei RUT-C30—thirty years of strain improvement. Microbiology 158, 58–68.

Pitkänen, E., Jouhten, P., Hou, J., Syed, M.F., Blomberg, P., Kludas, J., Oja, M., Hoim, L., Penttila, M., Rousu, J., Arvas, M., 2014. Comparative genome-scale reconstruction of gapless metabolic networks for present and ancestral species. PLoS Comput. Biol. 10, 1003465.

Pleiss, J., 2006. The promise of synthetic biology. Appl. Microbiol. Biotechnol. 73 (4), 735–739.

Pottkamper, J., Barthen, P., Ilmberger, N., Schwaneberg, U., Schenk, A., Schulte, M., Ignatiev, N., Streit, W.R., 2009. Applying metagenomics for the identification of bacterial cellulases that are stable in ionic liquids. Green Chem. 11, 957–965.

Qi, M., Jun, H.S., Forsberg, C.W., 2007. Characterization and synergistic interactions of *Fibrobacter succinogenes* glycoside hydrolases. Appl. Environ. Microbiol. 73, 6098–6105.

Qian, Y., Zhong, L., Hou, Y., Qu, Y., Zhong, Y., 2016. Characterization and strain improvement of a hypercellulolytic variant, *Trichoderma reesei* SN1, by genetic engineering for optimized cellulase production in biomass conversion improvement. Front. Microbiol. 7, 1349.

Qian, Y., Zhong, L., Gao, J., Sun, N., Wang, Y., Sun, G., Qu, Y., Zhong, Y., 2017. Production of highly efficient cellulose mixtures by genetically exploiting the potentials of *Trichoderma reesei* endogenous cellulases for hydrolysis of corncob residues. Microb. Cell Fact. 16, 207.

Quiroz-Castaneda, R.E., Perez-Mejia, N., Martinez-Anaya, C., Acosta-Urdapilleta, L., Folch-Mallol, J., 2011. Evaluation of different lignocellulosic substrates for the production of cellulases and xylanases by the basidiomycete fungi *Bjerkandera adusta* and *Pycnoporus sanguineus*. Biodegradation 22 (3), 565–572.

Rahman, Z., Shida, Y., Furukawa, T., Suzuki, Y., Okada, H., Ogasawara, W., 2009. Application of *Trichoderma reesei* cellulase and xylanase promoters through homologous recombination for enhanced production of extracellular beta-glucosidase I. Biosci. Biotechnol. Biochem. 73, 1083–1089.

Ralph, J., Akiyama, T., Kim, H., Lu, F., Schatz, P.F., Marita, J.M., Ralph, S.A., Srinivasa Reddy, M.S., Chen, F., Dixo, R.A., 2006. Effects of coumarate 3-hydroxylase down-regulation on lignin structure. J. Biol. Chem. 281, 8843–8853.

Ramani, G., Meera, B., Vanitha, C., Rajendhran, J., Gunasekaran, P., 2015. Molecular cloning and expression of thermostable glucose-tolerant β-glucosidase of *Penicillium funiculosum* NCL1 in *Pichia pastoris* and its characterization. J. Ind. Microbiol. Biotechnol. 42 (4), 553–565.

Ransom, C., Venkatesh, B., Biswas, G., Dale, B., Crockett, E., 2007. Heterologous *Acidothermus cellulolyticus* 1,4-β-endoglucanase E1 produced within the corn biomass converts corn stover into glucose. Appl. Biochem. Biotechnol. 36, 207–220.

Reese, E.T., 1976. History of cellulose program at US Army Natick Development Center. Biotechnol. Bioeng, 9–20.

Reggi, S., Marchetti, S., Patti, T., De Amicis, F., Cariati, R., Bembi, B., Fogher, C., 2005. Recombinant human acid β-glucosidase stored in tobacco seed is stable, active and taken up by human fibroblasts. Plant Mol. Biol. 57, 101–113.

Robledo, M., Jimenez-Zurdo, J.I., Velazquez, M.J.S.E., Dazzo, F., Molina, E.M., Mateos, P.F., 2011. Development of functional symbiotic white clover root hairs and nodules requires tightly regulated production of rhizobial cellulose CelC2. Mol. Plant Microbe Interact. 24, 798–807.

Sadhu, S., Maiti, T.K., 2013. Cellulase production by bacteria: a review. Br. Microbiol. Res. J. 3, 235–258.

Sainz, M.B., 2011. Commercial cellulosic ethanol: the role of plantexpressed enzymes. In: Tomes, D., et al. (Eds.), Biofuels. Springer, pp. 237–264.

Salinas, A., Vega, M., Lienqueo, M.E., Garcia, A., Carmona, R., Salazar, O., 2011. Cloning of novel cellulases from cellulolytic fungi: heterologous expression of a family 5 glycoside hydrolase from Trametes versicolor in *Pichia pastoris*. Enzyme Microb. Technol. 49 (6e7), 485–491.

Saloheimo, M., Paloheimo, M., Hakola, S., Pere, J., Swanson, B., Nyyssönen, E., Bhatia, A., Ward, M., Penttilä, M., 2002. Swollenin, a *Trichoderma reesei* protein with sequence similarity to the plant expansins, exhibits disruption activity on cellulosic materials. Eur. J. Biochem. 269, 4202–4211.

Sampedro, J., Cosgrove, D.J., 2005. The expansin superfamily. Genome Biol. 6, 242.

Sánchez, C., 2009. Lignocellulosic residues: biodegradation and bioconversion by fungi. Biotechnol. Adv. 27 (2), 185–194.

Sánchez, O.J., Cardona, C.A., 2008. Trends in biotechnological production of fuel ethanol from different feedstocks. Bioresour. Technol. 99, 5270–5295.

Sarikaya-Bayram, Ö., Palmer, J.M., Keller, N., Braus, G.H., Bayram, Ö., 2015. One Juliet and four Romeos: VeA and its methyltransferases. Front. Microbiol. 6, 1.

Scharf, M.E., Wu-Scharf, D., Pittendrigh, B.R., Bennett, G.W., 2003. Caste-and development associated gene expression in a lower termite. Genome Biol. 4, R62.1–R62.11.

Scharf, M.E., Wu-Scharf, D., Zhou, X., Pittendrigh, B.R., Bennett, G.W., 2005. Gene expression profiles among immature and adult reproductive castes of the termite Reticulitermes flavipes. Insect Mol. Biol. 14, 31–44.

Schulein, M., 1997. Enzymatic properties of cellulases from *Humicola insolens*. J. Biotechnol. 57 (1-3), 71–81.

Schulein, M., 2000. Protein engineering of cellulases. Biochem. Biophys. Acta Protein Struct. Mol. Enzymol. 1543 (2), 239–252.

Scordia, D., Cosentino, S.L., Jeffries, T.W., 2010. Second generation bioethanol production from *Saccharum spontaneum* L. ssp. aegyptiacum (Willd.) Hack. Bioresour. Technol. 101, 5358–5365.

Segato, F., Dias, B., Berto, G.L., de Oliveira, D.M., De Souza, F.H.M., Citadini, A.P., Murakami, M.T., Damásio, A.R.L., Squina, F.M., Polikarpov, I., 2017. Cloning, heterologous expression and biochemical characterization of a non-specific endoglucanase family 12 from *Aspergillus terreus* NIH2624. Biochim. Biophys. Acta 1865, 395–403.

Seiboth, B., Karimi, R.A., Phatale, P.A., Linke, R., Hartl, L., Sauer, D.G., Smith, K.M., Baker, S.E., Freitag, M., Kubicek, C.P., 2012. The putative protein methyltransferase LAE1 controls cellulase gene expression in *Trichoderma reesei*. Mol. Microbiol. 84, 1150–1164.

Shafique, S., Bajwa, R., Shafique, S., 2011. Strain improvement in *Trichoderma viride* through mutation for overexpression of cellulase and characterization of mutants using random amplified polymorphic DNA (RAPD). Afr. J. Biotechnol. 10, 19590–19597.

Shen, B., Sun, X., Zuo, X., Shilling, T., Apgar, J., Ross, M., Bougri, O., Samoylov, V., Parker, M., Hancock, E., Lucero, H., Gray, B., Ekborg, N.A., Zhang, D., Johnson, J.C.S., Lazar, G., Raab, R.M., 2012. Engineering a thermoregulated intein-modified xylanase into maize for consolidated lignocellulosic biomass processing. Nat. Biotechnol. 30, 1131–1136.

Shi, R., Li, Z., Ye, Q., Xu, J., Liu, Y., 2013. Heterologous expression and characterization of a novel thermohalotolerant endoglucanase Cel5H from *Dictyoglomus thermophilum*. Bioresour. Technol. 142, 338–344.

Shinoda, S., Kanamasa, S., Arai, M., 2012. Cloning of an endoglycanase gene from *Paenibacillus cookii* and characterization of the recombinant enzyme. Biotechnol. Lett. 34, 281–286.

Shoemaker, H.E., Mink, D., Wubbolts, M.G., 2003. Dispelling the mythsbiocatalysis in industrial synthesis. Science 299, 1694–1697.

Shota Atsumi, S., Liao, J.C., 2008. Metabolic engineering for advanced biofuels production from *Escherichia coli*. Curr. Opin. Biotechnol. 19, 414–419.

Singh, A., Patel, A.K., Adsul, M., Singhania, R.R., 2017. Genetic modification: a tool for enhancing cellulase secretion. Biofuel Res. J. 14, 600–610.

Singhania, R.R., Sukumaran, R.K., Patel, A.K., Larroche, C., Pandey, A., 2010. Advancement and comparative profiles in the production technologies using solid-state and submerged fermentation for microbial cellulases. Enzyme Microb. Technol. 46, 541–549.

Singhania, R.R., Sukumaran, R.K., Rajasree, K.P., Mathew, A., Gottumukkala, L., Pandey, A., 2011. Properties of a major β-glucosidase-BGL1 from *Aspergillus niger* NII-08121 expressed differentially in response to carbon sources. Process Biochem. 46 (7), 1521–1524.

Singhania, R.R., Patel, A.K., Sukumaran, R.K., Larrocheand, C., Pandey, A., 2012. Role and significance of β-glucosidases in the hydrolysis of cellulose for bioethanol production. Bioresour. Technol. 127, 500–507.

Smidansky, E.D., Martin, J.M., Hannah, C.L., Fischer, A.M., Giroux, M.J., 2003. Seed yield and plant biomass increases in rice are conferred by deregulation of endosperm ADP-glucose pyrophosphorylase. Planta 216, 656–664.

Spindler, D.D., Wyman, C.E., Grohmann, K., Philippidis, G.P., 1992. Evaluation of the cellobiose-fermenting yeast *Brettanomyces custersii* in the simultaneous saccharification and fermentation of cellulose. Biotechnol. Lett. 14, 403–407.

Stephanopoulos, G., Vallino, J.J., 1991. Network rigidity and metabolic engineering in metabolite overproduction. Science 252, 1675–1681.

Sticklen, M.B., 2006. Plant genetic engineering to improve biomass characteristics for biofuels. Curr. Opin. Biotechnol. 17, 315–319.

Sticklen, M.B., 2008. Plant genetic engineering for biofuel production: towards affordable cellulosic ethanol. Nat. Rev. Genet. 9, 433–443.

Sticklen, M., Teymouri, F., Maqbool, S., Salehi, H., Ransom, C., Biswas, G., Ahmad, R., Dale, B., 2004. Production of microbial hydrolysis enzymes in biomass crops via genetic engineering. In: 2nd International Ukrainian Conference on Biomass for Energy, September 20–22, pp. 133.

Streatfield, A.J., 2007. Approaches to achieve high-level heterologous protein production in plants. Plant Biotechnol. J 5, 2–15.

Stricker, R., Grosstessner-Hain, K., W€urleitner, E., Mach, R.L., 2006. Xyr1 (xylanase regulator 1) regulates both the hydrolytic enzyme system and D-xylose metabolism in *Hypocrea jecorina*. Eukaryot. Cell 5, 2128–2137.

Sugimura, M., Watanabe, H., Lo, N., Saito, H., 2003. Purification, characterization, cDNA cloning and nucleotide sequencing of a cellulase from the yellow-spotted longicorn beetle, *Psacothea hilaris*. Eur. J. Biochem. 270, 3455–3460.

Suh, K.J., Hong, Y.S., Lee, C.H., 2003. Water self-diffusion behavior in yeast cells studied by pulsed field gradient NMR. Biophys. Chem. 104, 121–130.

Sukumaran, R.K., Singhania, R.R., Mathew, G.M., Pandey, A., 2009. Cellulase production using biomass feed stock and its application in lignocellulose saccharification for bioethanol production. Renew. Energy 34 (2), 421–424.

Sun, Y., Cheng, J., 2002. Hydrolysis of lignocellulosic materials for ethanol production: a review. Bioresour. Technol. 83 (1), 1–11.

Talbot, N.J., 2001. Molecular and Cellular Biology of Filamentous Fungi. Oxford University Press Inc., New York.

Tanaka, T., Yamada, R., Ogino, C., Kondo, A., 2012. Recent developments in yeast cell surface display toward extended applications in biotechnology. Appl. Microbiol. Biotechnol. 95, 577–591.

Taylor, L.E., Dai, Z., Decker, S.R., Bruneckt, R., Adney, W.S., Ding, S.Y., Himmel, M.E., 2008. Heterologous expression of glycosyl hydrolases in planta: a new departure for biofuels. Trends Biotechnol. 26, 413–424.

Teeri, T.T., Koivula, A., Linder, M., Wohlfahrt, G., Divne, C., Jones, T.A., 1998. *Trichoderma reesei* cello bio hydrolases: why so efficient on crystalline cellulose? Biochem. Soc. Trans. 26, 173–178.

Teymouri, F., Alizadeh, H., Laureano-Perez, L., Dale, B.E., Sticklen, M.B., 2004. Effects of ammonia fiber explosion treatment on activity of endoglucanase from *Acidothermus cellulolyticus* in transgenic plant. Appl. Biochem. Biotechnol. 116, 1183–1192.

Tokuda, G., Lo, N., Watanabe, H., Slaytor, M., Matsumoto, T., Noda, H., 1999. Metazoan cellulase genes from termites: intron/exon structures and sites of expression. Biochim. Biophys. Acta 1447, 146–159.

Törrönen, A., Harkki, A., Rouvinen, J., 1994. Three-dimensional structure of endo-1,4-beta-xylanase II from *Trichoderma reesei*: two conformational states in the active site. EMBO J. 13 (11), 2493–2501.

Tsai, S.L., Oh, J., Singh, S., Chen, R., Chen, W., 2009. Functional assembly of minicellulosomes on the *Saccharomyces cerevisiae* cell surface for cellulose hydrolysis and ethanol production. Appl. Environ. Microbiol. 75, 6087–6093.

Uusitalo, J.M., Nevalainen, K.M.H., Harkki, A.M., Knowles, J.K.C., Penttilä, M.E., 1991. Enzyme production by recombinant *Trichoderma reesei* strains. J. Biotechnol. 17 (1), 35–49.

Verma, D., Jin, S., Kanagaraj, A., Singh, N.D., Daniel, J., Kolattukudy, P.E., Miller, M., Daniell, H., 2013. Expression of fungal cutinase and swollenin in tobacco chloroplasts reveals novel enzyme functions and/or substrates. PLoS One 8, e57187.

Vermerris, W., Abril, A., 2015. Enhancing cellulose utilization for fuels and chemicals by genetic modification of plant cell wall architecture. Curr. Opin. Biotechnol. 32, 104–112.

Voutilainen, S.P., Puranen, T., Siika-aho, M., Lappalainen, A., Alapuranen, M., Kallio, J., Hooman, S., Viikri, L., Vehmaanperä, J., Koivula, A., 2008. Cloning, expression, and characterization of novel thermostable family 7 cellobiohydrolases. Biotechnol. Bioeng. 101, 515–528.

Vu, V.H., Pham, T.A., Kim, K., 2009. Fungal strain improvement for cellulase production using repeated and sequential mutagenesis. Mycobiology 37 (4), 267–271.

Vu, V.H., Pham, T.A., Kim, K., 2011. Improvement of fungal cellulase production by mutation and optimization of solid state fermentation. Mycobiology 39, 20–25.

Wang, H., Jones, R.W., 1997. Site-directed mutagenesis of a fungal β-1,4-endoglucanase increases the minimum size required for the substrate. Appl. Microbiol. Biotechnol. 48 (2), 225–231.

Wang, B., Xia, L., 2011. High efficient expression of cellobiase gene from *Aspergillus niger* in the cells of *Trichoderma reesei*. Bioresour. Technol. 102 (6), 4568–4572.

Warnecke, F., Luginbühl, P., Ivanova, N., Ghassemian, M., Richardson, T.H., Stege, J.T., Cayouette, M., McHardy, A.C., Djordjevic, G., Aboushadi, N., Sorek, R., Tringe, S.G., Podar, M., Martin, H.G., Kunin, V., Dalevi, D., Madejska, J., Kirton, E., Platt, D., Szeto, E., Salamov, A., Barry, K., Mikhailova, N., Kyrpides, N.C., Matson, E.G., Ottesen, E.A., Zhang, X., Hernández, M., Murillo, C., Acosta, L.G., Rigoutsos, I., Tamayo, G., Green, B.D., Chang, C., Rubin, E.M., Mathur, E.J., Robertson, D.E., Hugenholtz, P., Leadbetter, J.R., 2007. Metagenomic and functional analysis of hindgut microbiota of a wood-feeding higher termite. Nature 450, 560–565.

Watanabe, H., Noda, H., Tokuda, G., Lo, N., 1998. A cellulase gene of termite origin. Nature 394, 330–331.

Wei, K.S.C., Teoh, T.C., Koshy, P., Salmah, I., Zainudin, A., 2015. Cloning, expression and characterization of the endoglucanase gene from *Bacillus subtilis* UMC7 isolated from the gut of the indigenous termite *Macrotermes malaccensis* in *Escherichia coli*. Electron. J. Biotechnol. 18, 103–109.

Wen, Z., Liao, W., Chen, S., 2005. Production of cellulase by *Trichoderma reesei* from dairy manure. Bioresour. Technol. 96, 491–499.

Wilson, D.B., 2004. Studies of *Thermobifida fusca* plant cell wall degrading enzymes. Chem. Rec. 4, 72–82.

Wilson, D.B., 2008. Three microbial strategies for plant cell wall degradation. Ann. N. Y. Acad. Sci. 1125, 289–297.

Wilson, D.B., 2009. Cellulases and biofuels. Curr. Opin. Biotechnol. 20, 295–299.

Wood, B.E., Ingram, L.O., 1992. Ethanol production from cellobiose, amorphous cellulose, and crystalline cellulose by recombinant *Klebsiella oxytoca* containing chromosomally integrated *Zymomonas mobilis* genes for ethanol production and plasmids expressing thermostable cellulase genes from *Clostridium thermocellum*. Appl. Environ. Microbiol. 58, 2103–2110.

Xia, D., Wei, Y., Zhang, G., Zhao, Q., Zhang, Y., Xiang, Z., Lu, C., 2013. cDNA cloning, expression, and enzymatic activity of a novel endogenous cellulase from the beetle *Batocera horsfieldi*. Gene 514, 62–68.

Xie, G., Bruce, D.C., Challacombe, J.F., Chertkov, O., Detter, J.C., Gilna, P., Han, C.S., Lucas, S., Misra, M., Myers, G.L., Richardson, P., Tapia, R., Thayer, N., Thompson, L.S., Brettin, T.S., Henrissat, B., Wilson, D.B., McBride, M.J., 2007. Genome sequence of the cellulolytic gliding bacterium *Cytophaga hutchinsonii*. Appl. Environ. Microbiol. 73, 3536–3546.

Xie, G., Yang, B., Xu, Z., Li, F., Guo, K., Zhang, M., Wang, L., Zou, W., Wang, Y., Peng, L., 2013. Global identification of multiple OsGH9 family members and their involvement in cellulose crystallinity modification in rice. PLoS One 8, e50171.

Xiong, Y., Sun, J., Glass, N.L., 2014. VIB1, a link between glucose signaling and carbon catabolite repression, is essential for plant cell wall degradation by *Neurospora crassa*. PLoS Genet. 10, e1004500.

Xue, D., Lin, D., Gong, C., Peng, C., Yao, S., 2017. Expression of a bifunctional cellulase with exoglucanase and endoglucanase activities to enhance the hydrolysis ability of cellulose from a marine *Aspergillus niger*. Process Biochem. 52, 115–122.

Yadav, V., Paniliatis, B.J., Shi, H., Lee, K., Cebe, P., Kaplan, D.L., 2010. Novel *in vivo*-degradable cellulose-chitin copolymer from metabolically engineered *Gluconacetobacter xylinus*. Appl. Environ. Microbiol. 76 (18), 6257–6265.

Yanase, S., Hasunuma, T., Yamada, R., Tanaka, T., Ogino, C., Fukuda, H., Kondo, A., 2010. Direct ethanol production from cellulosic materials at high temperature using the thermotolerant yeast *Kluyveromyces marxianus* displaying cellulolytic enzymes. Appl. Microbiol. Biotechnol. 88, 381–388.

Yang, P., Wang, Y., Bai, Y., Meng, K., Luo, H., Yuan, T., Fan, Y., Yao, B., 2007. Expression of xylanase with high specific activity from *Streptomyces olivaceoviridis* A1 in transgenic potato plants (*Solanum tuberosum* L). Biotechnol. Lett. 29, 659–667.

Yao, G., Li, Z., Gao, L., Wu, R., Kan, Q., Liu, G., Qu, Y., 2015. Redesigning the regulatory pathway to enhance cellulase production in *Penicillium oxalicum*. Biotechnol. Biofuels 8, 71.

Yao, G., Wu, R., Kan, Q., Gao, L., Liu, M., Yang, P., Du, J., Li, Z., Qu, Y., 2016. Production of a high-efficiency cellulase complex via β-glucosidase engineering in *Penicillium oxalicum*. Biotechnol. Biofuels 9, 78.

Yazdi, M.T., Woodward, J.R., Radford, A., 1990. Cellulase production by *Neurospora crassa*: the enzyme of the complex and their regulation. Enzyme Microb. Technol. 12 (2), 116–119.

Yeoman, C.J., Han, Y., Dodd, D., Schroeder, C.M., Mackie, R.I., Cann, I.K., 2010. Thermostable enzymes as biocatalysts in the biofuel industry. Adv. Appl. Microbiol. 70, 1–55.

Yomano, L.P., York, S.W., Ingram, L.O., 1998. Isolation and characterization of ethanol-tolerant mutants of *Escherichia coli* KO11 for fuel ethanol production. J. Ind. Microbiol. Biotechnol. 20, 132–138.

Yoon, L.W., Ang, T.N., Ngoh, G.C., Chua, A.S.M., 2014. Fungal solid-state fermentation and various methods of enhancement in cellulase production. Biomass Bioenergy 67, 319–338.

Zeilinger, S., Schmoll, M., Pail, M., Mach, R.L., Kubicek, C.P., 2003. Nucleosome transactions on the *Hypocrea jecorina* (*Trichoderma reesei*) cellulase promoter cbh2 associated with cellulase induction. Mol. Genet. Genomics 270, 46–55.

Zhang, Y.-H.P., Lynd, L.R., 2004. Toward an aggregated understanding of enzymatic hydrolysis of cellulose: noncomplexed cellulase systems. Biotechnol. Bioeng. 88, 797–824.

Zhang, Y.H.P., Himmel, M.E., Mielenz, J.R., 2006. Outlook of cellulase improvement: screening and selection strategies. Biotechnol. Adv. 24, 452–481.

Zhang, J., Zhong, Y., Zhao, X., Wang, T., 2010. Development of the cellulolytic fungus *Trichoderma reesei* strain with enhanced beta-glucosidase and filter paper activity using strong artificial cellobiohydrolase 1 promoter. Bioresour. Technol. 101 (24), 9815–9818.

Zhang, W., Kou, Y., Xu, J., Cao, Y., Zhao, G., Shao, J., Wang, H., Wang, Z., Bao, X., Chen, G., Liu, W., 2013. Two major facilitator superfamily sugar transporters from *Trichoderma reesei* and their roles in induction of cellulase biosynthesis. J. Biol. Chem. 288, 32861–32872.

Zhang, F., Bai, F., Zhao, X., 2016. Enhanced cellulose production from *Trichoderma reesei* Rut-C30 by engineering with an artificial zinc finger protein library. Biotechnol. J. 11, 1282–1290.

Zhang, X., Li, Y., Zhao, X., Bai, F., 2017. Constitutive cellulase production from glucose using the recombinant *Trichoderma reesei* strain overexpressing an artificial transcription activator. Bioresour. Technol. 223, 317–322.

Zhao, H., Chockalingam, K., Chen, Z., 2002. Directed evolution of enzymes and pathways for industrial biocatalysis. Curr. Opin. Biotechnol. 13, 104–110.

Zhao, J., Shi, P., Li, Z., Yang, P., Luo, H., Bai, Y., Wang, Y., Yao, B., 2012. Two neutral thermostable cellulases from *Phialophora* sp. G5 act synergistically in the hydrolysis of filter paper. Bioresour. Technol. 121, 404–410.

Zhao, X., Xiong, L., Zhang, M., Bai, F., 2016. Towards efficient bioethanol production from agricultural and forestry residues: exploration of unique natural microorganisms in combination with advanced strain engineering. Bioresour. Technol. 215, 84–91.

Zhong, Y., Wang, X., Yu, H., Liang, S., Wang, T., 2012. Application of T-DNA insertional mutagenesis for improving cellulase production in the filamentous fungus *Trichoderma reesei*. Bioresour. Technol. 110, 572–577.

Ziegelhoffer, T., Will, J., Austin-Phillips, S., 1998. Expression of bacterial cellulase genes in transgenic alfalfa (*Medicago sativa* L.), potato (*Solanum tuberosum* L.) and tobacco (*Nicotiana tabacum* L.). Mol. Breed. 5, 309–318.

Ziegelhoffer, T., Raasch, J.A., Austin-Phillips, S., 2001. Dramatic effects of truncation and sub-cellular targeting on the accumulation of recombinant microbial cellulase in tobacco. Mol. Breed. 8, 147–158.

Ziegler, M.T., Thomas, S.R., Danna, K.J., 2000. Accumulation of a thermostable endo-1,4-D-glucanase in the apoplast of *Arabidopsis thaliana* leaves. Mol. Breed. 6, 37–46.

Zou, G., Shi, S., Jiang, Y., van den Brink, J., de Vries, R.P., Chen, L., Ma, L., Wang, C., Zhou, Z., 2012. Construction of a cellulase hyperexpression system in *Trichoderma reesei* by promoter and enzyme engineering. Microb. Cell Fact. 11, 21.

Zweers, J.C., Barak, I., Becher, D., Driessen, A.J.M., Hecker, M., Kontinen, V.P., Saller, M.J., Vavrova, L., Dijl, J.M.V., 2008. Towards the development of Bacillus subtilis as a cell factory for membrane proteins and protein complexes. Microb. Cell Fact. 7, 10.

Further Reading

Lee, S.K., Chou, H., Ham, T.S., Lee, T.S., Keasling, J.D., 2008b. Metabolic engineering of microorganisms for biofuels production: from bugs to synthetic biology to fuels. Curr. Opin. Biotechnol. 19, 556–563.

CHAPTER

15

Biofuel Cells With Enzymes as a Catalyst

Kelvii Wei Guo

Department of Mechanical and Biomedical Engineering, City University of Hong Kong, Kowloon Tong, Hong Kong

15.1 INTRODUCTION

Energy is the currency that makes the worldwide economy of the future. Under the influences of global economic development and population growth, world primary energy use continues to increase. It is reported that energy usage limits economic activity directly and varies greatly by nation. The increase in the consumption of energy results in increasing wealth: there exists a definitely clear connection between energy use and income, and how energy consumption is highly correlated with higher GDP per capita.

Moreover, because of the rapid growth of the worldwide economy, the requirements of energy increase are extreme, particular for critical countries. Currently, the resources of fossil fuel needed for energy generation are becoming increasingly scarce and climate variation connected with carbon emissions to the atmosphere make the saving of energy and protection of the environment attractive.

In recent years, with climate variation and energy requirements, huge price fluctuations of international oil and other issues have attracted worldwide attention. Now, the demand for renewables is the focus of the energy mix, and provides superior methods to solve the relevant problems of the security and sustainability of energy. Subsequent to countries around the world attaining the Kyoto Protocol, reducing greenhouse gases (GHGs), and the security and sustainability of energy became key policy points, and are the current mainstream of energy development. Today, the use of energy is explored unrestrictedly, including the extinction of global stocks, the large amount of controllable and uncontrollable waste generated during energy conversion processes, and even the serious deterioration of the global environment. To maintain a stable supply of energy, improve its security, and decrease both the demand for traditional fossil fuels and GHG emissions, exploring the new kinds of renewable energy is the biggest issue of the contemporary era so that economic development and environmental resources can be sustained.

In the 21st century, many countries are actively involved in green technology and renewable energy R&D improvements. Especially after the second energy crisis in 1970, the United States and Europe recognized the importance of energy diversification and are dedicated to a variety of alternative energy development and scenarios, coupled with environmental awareness. Also, clean and diverse renewable energy is being revalued, e.g., Europe, the United States, and other countries are keen on the development of modern wind turbines to promote wind turbine performance and their efficiency has increased dramatically. After 1990, Europe, America, Japan, and other countries, depending on the growth of solar energy power generation systems as replacements for petrochemicals and nuclear energy, continued to promote the granting of incentives to encourage people to install solar power generation systems.

Energy can be provided from many different sources and in many different forms. Energy sources can be divided into renewable sources (sun, wind, hydropower, geothermal, bioenergy, etc.) and nonrenewable sources (oil, coal, gas, petroleum, etc.).

The distinction between renewable (nonfossil) and nonrenewable (fossil) sources is made on the basis of the time scale of the reliable energy being stored. Generally, energy mostly comes from the sun and the uptake of CO_2 is driven by the effect of photosynthesis. By this method, energy will be stored in the manner of carbohydrates. For fossil sources, the sun's energy was utilized millions of years ago, while for renewable sources it was utilized on a shorter time scale (for bioenergy only in the order of decades). In addition to bioenergy, the sun's energy can also be used indirectly as wind and hydropower, or directly as solar power. Most of the energy we use comes from fossil

From Cellulose to Cellulase: Strategies to Improve Biofuel Production
https://doi.org/10.1016/B978-0-444-64223-3.00015-1

fuels, such as coal, natural gas, petroleum, and uranium. Uranium is another nonrenewable source, but it is not a fossil fuel. Uranium is converted to a fuel and used in nuclear power plants. Once these natural resources are used up, they are gone forever. However, renewable sources of energy can be used repeatedly. Renewable resources include solar energy, wind, geothermal energy, bioenergy, and hydropower. They generate much less pollution, both in gathering and in production, than nonrenewable sources. Before the industrial revolution, most goods and services were produced with energy from animals, the sun, wood, other bioenergies, and early forms of hydropower. At present, most of the world's energy is still supplied by fossil fuels, and the rest is supplied by nonfossil energy sources such as nuclear, hydroelectric, solar, bioenergy, wind power, and others. When considering the energy sources used for electricity, it is evident that 82% of electricity is produced from fossil sources such as coal, oil, and gas.

Energy sources and the form in which energy is used are mainly dependent on income. Energy consumed in cities by a high-income population will be in the form of electricity, oil, or gas, and heat is produced from a variety of sources. Bioenergy, coal, and gas can be directly used for heating and cooking, while oil is converted into liquid fuel, and part of coal, gas, and bioenergy is converted into electricity before use. Whereas low-income populations depend on the use of bioenergy for cooking and heating, higher-income populations switch their energy use from bioenergy to gas, coal, oil, and finally electricity because they are the most modern and advanced forms of energy.

The efficiency of energy production is another determinant of the degree to which energy consumption contributes to GHG emissions. Some energy is always lost between production and end use, but the amount lost (often dependent on the infrastructure quality) varies greatly depending on the efficiency of production and the quality of the transmission infrastructure. Urban areas relying on inefficient or wasteful energy sources contribute more GHG emissions than those that consume the same amount from more efficient sources. For the production of electricity, efficiency is an issue of major interest, because it indicates which part of the energy present in the source ends up as electricity. While most electricity is derived from fossil fuels (such as coal), conversion into electricity via combustion has the drawback of low efficiency since a large part of the energy is lost as unused heat.

Fossil fuels are essentially a nonrenewable energy source. Fossil fuels are the remains of plants and animals from the prehistoric era, which have now been reduced to mere hydrocarbon chains with a very high combustion rate and which release tremendous amounts of energy. Fossil fuels have provided a great impetus to the industrial revolution that took place in the 20th century. However, the irrational consumption of fossil fuels has led to several problems all over the world. Burning fossil fuels (such as oil, coal, and gas) results in the production of carbon dioxide and other GHGs, which cause global warming. Leakage of some fossil fuels, such as natural gas and crude oil, can lead to severe hazards. Hence the transportation of these fuels is very risky. Fossil fuels have contributed in several ways to global warming, an issue that has been discussed all over the world for many years. This is a stark truth, and the scientific facts point to significant warming, indicated by sea levels rising, hotter temperatures, and freakish weather patterns, like tsunamis and hurricanes, becoming more and more regular. Though fossil fuels have efficiently met our growing energy demands over the years, it is crucial that we start looking for alternative sources of energy.

15.1.1 Renewable Energy

No single solution can meet society's future energy needs. The answer lies instead in a family of diverse energy technologies that share a common thread: they do not deplete our natural resources or destroy our environment. Renewable energy sources such as solar, wind, geothermal, wave, hydropower, and biomass are not at risk of depletion from their use for energy production. Bioenergy, however, is a renewable resource only if its rate of consumption does not exceed its rate of regeneration. Renewable energy is abundant, especially for solar and wind energy.

Renewable energy technologies in natural cycles and systems are used to direct the energy we use in daily life into usable forms. The movement of wind and water, the heat and light of the sun, the heat in the ground, the carbohydrates in plants are all natural energy sources that can supply our needs in a sustainable way. By comparing renewable sources with fossil sources, the major advantage is that, ideally, they do not emit GHGs (CO_2), which makes them carbon neutral. For wind and water this may be evident, but also biomass, where CO_2 is released when energy is used, is considered carbon neutral. The reason for this is that the CO_2 they emit has been taken up by photosynthesis, and thus has a short recycle time compared to fossil fuels, where the CO_2 was taken up millions of years ago. This carbon neutrality, however, does not make renewable sources directly advantageous or more environmentally sustainable than fossil sources. To assess if the use of renewable sources is more environmentally sustainable than the use of fossil sources, each source should be judged separately based on its adverse effects. To understand just how alternative energy sources can help in preserving the earth's delicate ecological balance, one must have the relevant knowledge to realize what kinds of alternative energy resources there are and which can be taken in our ordinary life, especially for advantages and disadvantages of the relevant energy resources.

Up to now, the resources of renewable alternative energy combined with cutting-edge techniques have played an important part in the future prospects of the consumption of energy, and the attractive properties of renewable energy with environmental and technical characteristics make them a promising alternative to traditional energy resources.

15.1.1.1 Solar Energy

It is well known that solar energy is the most ancient energy source, and it is a substantial resource for almost all fossil and renewable types. Solar energy is formed by the radiation of the sun. For parts of the world with plentiful access to solar radiation, one can take solar energy as one of the most promising energy sources, along with the techniques of active and passive solar design to convert solar radiation into practical applications. In order to make the designed solar energy systems run efficiently, the selected area should be investigated with the relevant radiation data fixed, where the radiometric station nets with a low spatial resolution are generally taken to measure the solar radiation correspondingly, for instance interpolation and extrapolation techniques. However, it should be noted that the adopted techniques may be valid for places where the significance of spatial variability of radiation is not crucial and is less accurate if some complex areas of the terrain exist between the radiometric stations. The freely available characteristics and easily harnessed solar energy will make us less dependent on traditional energy resources. Moreover, solar power plants are environmentally friendly.

As is known, a solar cell is sensitive to incident light resulting in variable efficiency of the solar cell. Different areas have different amounts of solar radiation. Therefore the design of a solar cell should consider various parameters, such as the distance between earth and sun, the activity of the sun, the radiation time of the sun, atmospheric conditions, etc. For different solar devices, it is essential to use a standard spectrum and power densities to compare the efficiency precisely. Infrared, visible, and ultraviolet regions of the electromagnetic spectrum are included in the sun's emission wavelength with a maximum peak at about 500 nm, and the relevant spectrum is similar to that of a blackbody at 5670 K and is affected by the absorption of molecules in the atmosphere, such as O_3, O_2, CO_2, and H_2O. The surface of the sun's power density is 62 MW/m^2 and decreases to 1353 W/m^2 (known as the solar constant) at the point just above the earth's atmosphere. Through the solar constant, the amount of incoming perpendicular solar radiation can be measured on the outer surface of earth's atmosphere. Before solar radiation arrives on the ground, when the radiation passes through the atmosphere, the solar spectrum should be modified due to the angle of incidence of the sunrays, where the higher the tilt of sunrays, the lower the energy that lands on the soil, and clouds that absorb a large amount of radiation and atmospheric mass.

15.1.1.2 Wind Power

Wind is also one of the most popular energy sources to replace conventional energy. In recent years, wind power has become more and more important. For many parts of the world, increasingly popular wind power applications have been distributed in various geographical locations, along with grid-connected turbines on a small scale. Meanwhile, the design and operation of wind energy systems with state-of-the-art strategies is continuously being explored. Similar to solar systems, wind energy systems are unsuitable for all locations because of the lack of wind speeds; these systems are also more unpredictable and complex than solar systems. Obviously, high-altitude and offshore locations where winds are constant and strong are ideal sites for wind energy. To get the best from wind energy, selection of location, estimation of wind speed, and conversion of wind energy are extremely significant.

Nowadays, wind turbines are constructed to allow the air to flow through them to push the power generators mechanically to generate the electricity. As a promising replacement for traditional energy, widely distributed wind energy is environmentally friendly with no GHG emissions; it is also plentiful, unobtrusive, consumes no water, is renewable, and especially is far less problematic than nonrenewable power sources.

15.1.1.3 Hydropower

Hydropower, also known as water power/hydraulic power, generates energy from moving water. An attractive characteristic of hydropower is that water drives the energy systems. Because the density of water is higher than air, sea tides or a slow flowing stream of water can produce a large amount of energy. Up to now, ocean waves, tides, and falling and flowing water are general used to generate hydroelectricity. With the development of hydropower plants and their control techniques, the size of a small hydropower plant of the run-of-river type is key to the cost effectiveness of the investment. As for wave power and tides, it is necessary to study systematically various locations in the world because such energy is not easily harnessed effectively. It is known that ancient civilizations used hydropower to irrigate the soil and as a renewable energy to operate many kinds of mechanical devices, for example, watermills, sawmills, textile mills, gristmills, dock cranes, and domestic lifts.

15.1.2 Bioenergy

Bioenergy, another promising renewable energy, is generated from materials derived from biological sources. As a kind of renewable energy source, biomass is obtained from the biological materials of both animals and plants. To date, although biomass is taken to be one of the most beneficial renewable energy sources, the bottleneck of practical applications of biomass still exists from technical and economical viewpoints. Energy such as heat or electricity can be generated by burning biomass to produce steam. Furthermore, biomass can be converted to other useful things such as ethanol, methane, and so on. At present, there are many biomass power plants all over the world that are producing increasing amounts of electricity.

Countries around the world are focusing on augmenting their pool of renewable energy resources, which play extremely significant roles in economic development and political perspectives. In the past decade, huge advances have been made in the methods of energy harvesting and renewable energy. Prediction in the field of renewable energy resources shows that there is still a huge growth potential for the future.

15.1.2.1 Biomass

Broadly speaking, "biomass" refers to organic material generated under the effect of light photosynthesis. The chemical materials (organic compounds of carbons) are stored and then used to produce energy. This can be achieved, e.g., by using animal waste and converting it for everyday human use (Arregi et al., 2018; Correa and Kruse, 2018; Md Isa et al., 2018; Guedes et al., 2018). Biomass has to be considered in the search for an alternative source of energy that is abundant yet nondisruptive, since it is capable of being implemented at all levels of society. Biomass or biofuels from renewable sources have become some of the most plentiful and well-utilized sources of renewable energy in the world over the last two decades. Biomass is used in solid, liquid, and gas forms to produce electricity and heat, and to manufacture biofuels. Many of the biomass fuels used today come in the form of wood products, dried vegetation, crop residues, compost, rapeseed oil, and forestry and agricultural products. In addition, residual wastes of biogenic origin are also used, in particular biowaste, sewage sludge, solid manure, and straw (Widjaya et al., 2018; Rios et al., 2018).

The most common biomass used for energy is wood (Danish and Ahmad, 2018). Wood has been used by humans for producing energy for heating and cooking for a very long time. Wood is still used extensively for energy in both household situations primarily for cooking and space heating, and in industry, particularly in the timber, paper and pulp, and other forestry-related industries. Woody biomass accounts for much of the primary energy consumed in most of the developing world. It is used to raise steam, which, in turn, is used as a by-product to generate electricity. Through anaerobic digestion, methane can be produced from solid landfill waste or other biomass materials, such as sewage, manure, and agricultural waste. Sugars can be extracted from agricultural crops and alcohols can be produced by distillation for use as transportation fuels. Also, numerous other technologies exist or are being developed to take advantage of other biomass feedstocks. Considerable research and development work is currently under way to develop smaller gasifiers that can produce electricity on a small scale.

When comparing biomass with fossil sources, the sustainability of its use strongly depends on the type of biomass feedstock. It has been argued that some of these biomass feedstocks might even be less sustainable than fossil fuels because of competition with food production and CO_2 emissions as a result of land clearing for crop growth. There are two issues that affect the evaluation of biomass as a viable solution to the current energy problem: the effects of farming and production of biomass, and the effects of the factory conversion of biomass into usable energy or electricity. There are many environmental and economic benefits along with disadvantages for each issue, which present a difficult challenge in evaluating the potential success of biomass as an alternative fuel. For instance, the replacement of coal by biomass could result in a considerable reduction in net carbon dioxide emissions that contribute to the greenhouse effect. On the other hand, the use of wood and other plant material for fuel may mean deforestation. The problems associated with denuding forests are evident, and widespread tree cutting can lead to groundwater contamination and irreversible erosion patterns that could literally change the structure of the world ecology. Although tree plantations have considerable advantages in supplying an energy source, the commercial use of plantation-grown fuels for power generation is limited to a few isolated experiences. It is not possible for plantations to be implemented at this scale, not to mention that soil exhaustion would eventually occur. Biomass cannot replace the current dependence on coal, oil, and natural gas, but it can complement other renewables. At present, some technologies are cost inhibitive, but the rate of technological development is making them more affordable with time. Energy recovery from biogenic residues and waste helps to solve or mitigate the problem of the conflicting uses of biomass as an energy source or material.

It is generally accepted that biomass that is considered to be waste, such as municipal and industrial organic wastes, are an attractive and sustainable biomass feedstock, because these cannot be used for any food or feed production. Municipal and industrial organic wastes are wastewater streams consisting of a mixture of proteins, lipids, carbohydrates, and organic acids. These organic materials can be broken down by microorganisms under anaerobic conditions into smaller organic components. Nevertheless, other biological processes have also been proposed for the conversion of chemical energy in wastewater into other useful forms of chemical energy, e.g., biogas (methane + CO_2), hydrogen gas, by: (1) methanogenic anaerobic digestion and (2) biological fermentative hydrogen production (also known as dark hydrogen fermentation). The by-products of these two bioprocesses (i.e., biogas/hydrogen) are eventually converted into electrical energy using heat combustion engines or hydrogen fuel cells, respectively (Johnson et al., 2013; Krzyżaniak et al., 2015; Ghani et al., 2018; Çepelioğullar et al., 2018). The challenge is to convert these wet organic waste streams in an efficient manner into electricity. Wet waste streams need low-temperature conversion if they are to be efficiently converted, because heating the water would lead to major energy losses. Currently, the only available technology for converting these wet streams into electricity is anaerobic digestion. Anaerobic digestion is a process that uses microbes to convert biomass under anaerobic conditions to produce methane gas, which can be used to produce the electricity in a gas motor. Overall, the conversion from biomass into electricity via anaerobic digestion occurs in several steps: (1) biochemical conversion of biomass into methane and (2) cleaning of methane gas from H_2S before being combusted. This makes the overall conversion from biomass to electricity via anaerobic digestion complex and expensive.

Fermentative hydrogen production, on the other hand, is an emerging waste-to-energy process mainly targeted toward the concept of hydrogen economy, a hypothetical future economy in which the primary form of stored energy is hydrogen. The underlying principle of fermentative hydrogen production is very much like methanogenic anaerobic processes except that the former process requires inhibition of hydrogen-consuming microorganisms. Although there is a significant industrial demand for hydrogen, producing hydrogen from wastewater treatment via fermentation with the aim of recovering renewable energy is not appropriate. Compared to methane, hydrogen has a much lower density and boiling point, only 0.09 kg/Nm^3 and −252.9°C, respectively (vs. 0.717 kg/Nm^3 and −182.5°C, respectively, for methane). These physical properties make hydrogen gas less suitable for transport and storage compared to methane.

From a point where hydrogen gas is produced to a point where it is converted into electricity, many energy-consuming intermediate steps are involved. These extra energies can only be obtained from other energy sources (either fossil fuels or other renewables), but certainly not from the primary hydrogen production step. In other words, a biologically produced hydrogen-rich biogas requires energy-intensive posttreatment such as purification and pressurization before it is finally delivered to a fuel cell for electricity generation.

It should be noted that biomass is present in many forms, such as wood, crop residues, animal manure, and municipal waste. Not all of these sources have the same suitability for electricity production in a fuel cell. While wet organic waste streams (municipal organic wastes, animal slurries) are particularly interesting, they likely need pretreatment before being used for electricity generation. At the same time, pretreatment may also enable the use of other forms of biomass for conversion into electricity.

15.1.2.2 *Biofuel*

Biofuels can be fuels produced from biomass for the purposes of transport, heating, electricity generation, and cooking. Among the liquid biofuels, ethanol and biodiesel are the most used, accounting for more than 90% of global biofuel usage, especially for biodiesel usage. In total, biofuels now provide 1% of the world's liquid transport fuels, mainly in the transportation sector. Despite the different types of biofuels, biodiesel is a more attractive resource in practical applications.

15.1.2.2.1 Biodiesel

Biodiesel, from the Greek "bio" = "life" and "diesel" from Rudolf Diesel, refers to a diesel-equivalent, processed fuel obtained from biological sources. In chemical terms, it designates a group of esterified vegetable oils produced from different oil-containing crops, such as sunflower or palm, rapeseed, soybean, and animal fats or waste oil. Because oil crops are renewable, biodiesel is considered as a potentially inexhaustible energy resource, with an energy content close to that of fossil fuel. Historically, the first attempt to use biodiesel as a fuel was made by Rudolf Diesel, the German inventor of the diesel engine, who in 1893 fueled his first engine using 100% peanut oil, and suggested that biomass could be used as a fuel. His objective was to give an opportunity to small-scale industries, farmers, and communities in isolated areas to generate their own energy. At the same time, they could compete with the large monopolies, which dominated all energy production at that time, or reduce their control to some extent. Although worldwide usage of petroleum consumed from diesel fuel, the captivating results of vegetable oil as a fuel was shown in some areas in the 1920s and 1930s and later during World War II. However, due to energy supply and environmental concerns the production of biodiesel has been intensively boosted since the last decade.

In comparison with petroleum-derived diesel fuel, biodiesel stands out from diesel because of its superior environmental characteristics because it reduces the GHG emissions by 41% and because it recycles the carbon dioxide extracted from the atmosphere in the biomass. In addition, biodiesel presents less danger to the environment; all of its chemical components only need about 1 month to be biodegraded. Compared with ethanol, biodiesel stands out first because of its minimal impact on human and environmental health through nitrogen (N), phosphorus (P), and pesticide release, and second because it has a positive net energy balance, which means that it does not require more energy to make it than it does to yield it.

From a technical viewpoint, biodiesel can be used directly in engines without requiring substantial modifications. In addition, biodiesel's energy account is nearly 10% lower than diesel; however, it has a higher fuel density and better ignition, which extends the life of the engine. Finally, because of this, biodiesel has one more advantage over petrol/diesel: it is safe to use, handle, and store. On the negative side, biodiesel cannot be stored for prolonged periods of time due to its biodegradability. In addition, pure biodiesel does not have a good flow property at low temperatures because of its "cloud point" and "pour point," which consequently could bring problems for those who use outdoor storage tanks in colder climates. Lastly, because of its solvent property, biodiesel cannot be used in conjunction with seals applied in the older machinery and vehicles. As a result, seals used in these fuel systems should be replaced by items suitable for the blends.

15.1.2.2.2 Ethanol

Nowadays, the sugar industry of modern society is still plagued by governments' many levels of interference. It is estimated that sugar consumption per annum is up by around 1 million tons and is increasing at a rate of around 2 million tons every year. The European Union, Brazil, and India are the top producers, with China catching up quickly, and together they account for about 45% of the yield per year. The majority of sugar is used internally for production and only about 31% is exported.

Ethanol and sugar are traditionally linked to the by-product "molasses," derived from both beet and cane conversion. This is a syrup with a high sucrose content, but process technology does not allow it to be extracted into the final product. Ethanol can also stand alone when all sucrose of the sugar plant is directly converted into alcohol. Ethanol can also be produced from maize by converting the starch into alcohol, and this method is used for all ethanol production in the United States.

Sugar is produced entirely from two crops: sugar beet and sugarcane. The final product, "refined sugar," is identical regardless of the origin of the raw material. Sugar is formed in beet/cane as a result of photosynthesis, combining carbon dioxide and water under the influence of sunlight to make sucrose (oxygen, another result of the reaction, is emitted into the atmosphere). The production of sugar is principally a physical extraction of sucrose from the roots (in the case of beet) or the stems (in the case of cane). Agricultural conditions and extraction processes differ significantly, so they are dealt with separately. In both cases the raw material (beet, cane) will deteriorate in sugar content when stored, thus processing is preferably done close to the growing areas. Sugar content by weight of beets (15%–20%) is higher than that for cane (10%–15%). Plant yield per area cultivated is higher for cane (75 tons of cane per hectare vs. 50 tons of beet). Actual values depend on soil, climate conditions, cultivar, and also on the extent and type of chemicals used as fertilizers and pesticides. For both plants a large variety of grades and hybrids have been developed for matching conditions and subsequent process technology.

Sugarcane, botanically a grass, is the ancient source of sugar. It grows to a height of 3–6 m and has stems 2–5 cm thick. The small spikelets are borne in panicles and are surrounded by long, silky fibers. Several horticultural varieties are known, differing in stem color and height. It is a perennial plant, cultivated as a monoculture (no crop rotation), distributed over the world's tropical and subtropical regions. It grows for 6 months up to 2 years and harvesting time is 3 months up to 11 months. As a general estimate the yield of sugar is 7 tons per hectare, and about 75% of all sugar is made from cane. The increase in cane percentage is also enhanced by a significant reduction in sugar production in Europe, which has been solely based on beets.

Harvesting is mostly done by a simple and laborious process, usually after flash burning the fields to remove snakes and scorpions, which populate cane fields in huge numbers. The stems are chopped off from the roots, which stay in the ground to regrow for the next harvest. The cane is free from leaves and is then taken to the factory for extraction. The cane is mechanically crushed to squeeze out the sugar juice. The remaining fibers (termed "bagasse") are separated and dried to serve as the burning fuel to produce process steam. The juice is cleaned and filtered, concentrated in various evaporation steps, and finally centrifuged to obtain raw sugar. The brown raw sugar is ready for use immediately, but the characteristic taste is undesirable. So, it is refined to white sugar on site or at the location of ultimate use. Side products are: (1) bagasse, which is a material separated from the juice. In dried form it is used as fuel for producing steam for both the requirement of the process and the generation of electricity, which is close to about 3 tons per ton of raw sugar. (2) Molasses, which is an enriched sugar syrup without the possibility of further extraction. It is divided from the thick juice in the final cleaning step (termed "centrifugation"). It is usable for various purposes. Because it contains sugar, it can be fermented to ethanol, which is close to more than 200 kg per ton of raw sugar.

Sugar beet, botanically a plant belonging to the goosefoot family, looks like giant-sized turnips. It grows in temperate climate conditions. Intensive agriculture and increasingly mild winters allow cultivation in areas previously unsuitable (England, Sweden, and Canada). The plan of the rotation of the beet crop along with other crops should be careful, and the systematic investigation of accurate field fertilization should be explored; this has recently been a concern. In the production of commercial beet seed, beets with around 15% sugar are selected, laid aside in pits in the winter period, and taken out again during spring to develop the seed crop. The small plants are individually enclosed in a ball of fertilizer and planted in early spring into plowed and smoothed ground. Soil conditions determine a sophisticated pattern of crop rotation (usually 4 years) with other agricultural products. In some cases, fields are also left idle within the crop rotation cycle to recover. Depending on the local situations, fields need to be weeded and irrigated during the growing period.

As a general estimate, the yield of sugar is 8 tons per hectare. However, the area required is significantly bigger when taking crop rotation into consideration. Marginal planting conditions, like poor soil or unreliable climate, always cause fluctuations. About 25% of global sugar production is mainly from beet. To date, decreasing yields appear mainly due to radical reduction of acreage in the European Union.

Harvesting sugar based on beet is usually done in the fall during the "campaign," which lasts from October until early January. Both planting and harvesting are entirely mechanized. On harvesting, the root is taken out of the ground, the leaves and the top of the beet are removed on site, and an initial cleaning is performed. The cut off materials are preserved by light fermentation and used as animal feed during the winter (termed "silage") or left on the field as fertilizer to be plowed in.

Because the sugar content deteriorates over time, the beets are processed right after harvesting, usually in sites close to the growing area. Sugar juice is extracted, filtered, and concentrated in several steps, and finally refined to white sugar. Side products are: (1) pulp, which consists of remnants of the sliced beet, separated from the raw juice after the first extraction step (termed "diffusion"). It is dried and used for animal feed, straight or blended as additive, which is nearly 500 kg per ton of white sugar. (2) Molasses, which enriches sugar syrup without the possibility of further extraction due to limitations in the process technology. It is separated from the thick juice in the final cleaning step (termed "centrifugation). It is usable for various purposes, mostly for mixing into animal feed (together with the pulp). Because it still contains sugar, it can also be fermented to ethanol, which is more than 200 kg per ton of sugar; the sucrose content is 49%.

Sugar beet acreage will stay stable, although there will be some shifting in regional situations. Sugarcane harvesting areas (for producing sugar) will increase by 10%–12% with regional differences. Yields for beet are affected by reductions of harvests in Europe; the high efficiency (up to 80 tons/ha) cannot be matched by new players in marginal areas. High values do not necessarily imply effective agriculture, because they might be achieved via the excessive use of chemicals or irrigation.

Additionally, there are no plans in the United States to produce alcohol from cane or beet, except for by-products when converting the plants into raw or white sugar. Acreage for both crops will remain unchanged. Cane harvests are concentrated in South Texas, Louisiana, Florida, and Hawaii. Beet areas are well distributed over the territory; with the exception of California, all of them are in the northern half. Beet production resists dynamic changes because it is part of traditionally established crop rotation with other plants. Moreover, sugar consumption is mainly determined by local diet habits. Elasticity values with local prices, income levels, and alternative sweeteners are low. Per capita intake in some developed countries is going down slightly due to concerns of health, body weight, and availability of improved alternatives (artificial sweeteners).

The shipment of bulk sugar is often comprised of both raw and white (refined) forms. Lately, nearly half of the imports of sugar have been transported in the raw form based on cane. There is no intermediate "raw sugar" if the production is based on beets. A key factor for the trading of raw sugar is better utilization of the available refining capacities. Raw sugar imports consist of a significant feedstock for sugar mills (refineries) in several countries. Nearly all consumption is in the white sugar form. The trade of raw sugar is accomplished with long-term supply agreements (because refiners need to ensure capacity utilization), whereas white sugar is traded on an auction basis to the highest bidder.

Sugar and ethanol production have been traditionally linked, because the production process does not allow the entire quantity of sucrose to be extracted from the plant. Until ethanol became a mainstream fuel for combustion engines, it was regarded as a by-product. As an excise item it enjoyed attractive prices, thus enhancing the balance sheets of sugar manufacturers. The increasing commercial nonfood use of ethanol moved it to a key commodity alongside sugar. It is encouraging that manufacturers adjust quantities between sugar and ethanol in an opportunistic manner. Ethanol can be produced from starch, carbohydrate, byproducts, and cellulose being the most common alternatives to sugar. The transition from starch (as contained, e.g., in grains) to ethanol is well supported by state-of-the-art process technology. The conversion of cellulose and other raw materials (e.g., waste) is a highly attractive concept, but the process technology is still at an early stage to make it feasible for large-scale industrial use. In the United States, ethanol production from cellulose would cost about five times the present corn conversion. The long-term future will definitely rest on the so-called "second-generation ethanol," including improved corn (or other grains) conversion concepts.

The plants for handling sugar beets can make them into the refined sugar directly. Sugar beet molasses and beet pulp are the traditional by-products of sugar beet plants, where the beet pulp can be used as feed for animals. When sugar beet plant machinery is located near to dairy and livestock factories, some of the beet pulp can be sold as wet pulp. Meanwhile, as an effective livestock feed additive, sugar beet molasses can also be traded. The sugarcane generally transforms into raw sugar in sugarcane mills, and must then be delivered to the refinery to make it into white refined sugar. During the process of converting the sugarcane to raw sugar, bagasse and cane molasses are typical by-products of sugarcane mills. Sugarcane molasses can be used for fuel alcohol, alcoholic beverage production, and human consumption directly. More attractively, sugarcane bagasse is used for steam production and electricity generation in the sugarcane mills. In the meantime, the generated excess electricity is sold to the regional utility grid. Sugar, unlike corn, usually exists as raw sugar and refined sugar, and in the form of molasses it does not need prehydrolysis before fermentation. Therefore making ethanol from sugar is definitely simpler than converting corn into ethanol.

In the United States, two typical processes (wet milling and dry milling/dry ground) are used to make ethanol from corn. During the production of ethanol, wet milling makes up 21% of the total production of ethanol. The remainder of about 79% is used for dry milling. Corn kernels are often mixed with starch, corn germ, protein, and fiber during the wet-milling procedure. It should be noted that only pure starch is adopted in ethanol production. Moreover, the process of wet milling is very similar to the process used in biorefineries. CO_2, corn oil, corn gluten feed, and corn gluten meal are the typical by-products of the ethanol production process. Moreover, food and feed additives, vitamins, aquaculture, and hydroponic vegetable production are produced in some large wet-milling ethanol plants. In the fermentation procedure, CO_2 is generated. Corn oil as the premium vegetable oil is often ready for human consumption. In the United States, corn oil is not used for biodiesel production because of its high cost. Most corn gluten meal is used in poultry feed rations because it is more than 60% protein with no fiber. Interestingly, some ethanol plants refine raw CO_2 and use it for the growth of vegetables in greenhouses, for carbonated beverages, and/or for dry ice production. CO_2, condensed syrup, and distillers dried grain with solubles are the classic by-products of dry milling. In the traditional process of dry milling, corn kernels are usually fractioned with flowing water. Next, the corn mash is cooked along with enzymes added to convert starch into glucose. By fermentation, the glucose changes to alcohol. Then the alcohol is removed, and the liquid separates by centrifuge and forms into thin and thick stillage or wet distillers grains. Wet distillers grains have 33% solids. Dairy and beef cattle have 40% of wet distillers grains in their rations. However, one disadvantage is the shelf life of wet distillers grains. It is always very short, about 4 days. For the sake of improving the shelf life of wet distillers grains, the water must be removed. Distillers dried grains contain 27% protein and 87%solids. The thin stillage is evaporated and sprayed over the distillers dried grains to make distillers dried grains with various solubles or is sold as condensed syrup for use as a kind of feed additive. In the newly developed dry fractionation process, before the starch is converted into glucose, fiber and corn germ are separated from the corn in the dry-milling plants. Besides CO_2 and distillers dried grains, corn germ and fiber are the typically valuable by-products among other existing plants for ethanol. Also, some plants for the ethanol can partly get the corn oil from the stillage.

It is well known that there is 70%–72% starch in corn. The content of starch in corn relies on the weather, type of soil, varieties of seed, and various cultural practices. Even for the same corn varieties, starch output changes for different years. The variation of starch in the corn and other sources is not obvious, and is similar to sorghum, which is used in relatively small quantities for the same purpose. Sugar crops contain sucrose, a kind of the sugar, which can be converted into ethanol directly. However, in contrast, grains that contain starch must be fermented to make sugars. The variation in sugar content in sugar beet and sugarcane is affected by the weather, climate, seasonal growth, type, and cultural practices. In fact, sugar beet has more sugar (sucrose) than sugarcane. The range of sucrose content in sugar beet is from 15% to 20%. However, the sucrose content of sugarcane ranges from 10% to 15%. The recovery rates of sugar for both sugarcane and sugar beet rely on the facilities used in processing as well as weather conditions, the varieties of the plants, and cultural practices. Recently, the adjustment of the genetic code of all agricultural crops will produce tangible results for sugar plants and maize. New grades with increased starch and sucrose contents will help to attain higher yields, combined with pest resistance and improved adjustment to specific climatic conditions. Therefore the dynamics of all aspects of the market, from production to processing, distribution and marketing, will prevail. Additional dynamics are added because raw materials compete for dual use (cane for sugar and ethanol and corn for ethanol and food/animal feedstock).

Biochemistry allows every carbohydrate to be converted into alcohol. The most popular processes for industrial purposes utilize either sugar (from cane or beet) or starch (from maize). The production of alcoholic beverages from starch goes back a long time in history. The brewing of beer dates back even before the Egyptians, and whisky in Scotland (based on barley) and vodka in Russia (based on potatoes) have been around for centuries. Conversions from cellulose (another carbohydrate) into ethanol are technically feasible, but so far technically and commercially they are not yet suitable for large-scale applications. Judgment on the future of such technology is greatly biased by local or lobby interests. There might even be a smarter fuel than alcohol that can be derived one day from carbohydrates. The search is intense, because ethanol is not a very smart fuel, because it is hygroscopic and has lower energy content as compared to traditional fuels. Molasses, a side product of the process, with a high content of sucrose is converted into ethanol. Different to the beet process, molasses is formed in two different phases. Per 100 kg of white sugar, the derived quantity is approximately 7 L of ethanol. This has been regarded as a profitable effect on the bottom line of manufacturers. In the conversion from raw sugar to white sugar, there are an additional 2–3 L. All data on acreage vary by harvesting year and region. Acreages required for the production of 1 L of ethanol should not be directly used to draw conclusions on the environmental impact of the different crops or processes. A variety of other variables (e.g., pesticides, fertilizers, transportation, etc.) are needed for accurate calculations.

State-of-the-art technologies related to energy generation must be pushed forward continuously due to the factors of economic and environmental effects. Fuel cells used as a clean, renewable, and efficient technique can directly produce chemical energy and turn fuel into electricity. Consequently, the fuel cell is beneficial to technology. Electrons flow through the external circuit driven by different potentials attained by fuel oxidation on the anode and by oxygen reduction on the cathode. As a result, electricity is realized. Generally, electrochemical reactions are classified as faradic and nonfaradaic reactions. A faradic current results when oxidation or reduction occurs at the interface of the electrode solution due to the charge transfer reaction. These electrons are the true representation of bulk electroactive species present in the solution, which are, in general, used for quantitative purposes in the electroanalytical field. An electrode is a solid material, except a dropping mercury electrode, where electrochemical reactions take place. Initially, the mercury electrode, which is highly toxic, was mainly used, but was then replaced by other electrode materials like carbon, gold, platinum, boron-doped diamond, bismuth, paper, and screen-printed electrodes (SPEs). SPEs are disposable, have surfaces that are easily modified, and are cheap, which is quite impressive for use in real field applications. Instruments employed for electrochemical measurements were initially quite large but with the surge in technology, size and complexity of handling have been tremendously reduced. Such improvements have made analysis facile and enabled even unskilled people to operate simple electrochemical devices in their day-to-day lives (i.e., glucose sensors). It is well known that for most traditional fuel cells, platinum-based catalysts are often used to decrease oxygen. However, although research has shown beneficial results, fuel cells with platinum-based catalysts still have several drawbacks for large-scale commercial applications. The main drawback that impedes practical application of this technique is the high cost of platinum-based catalysts. Fuel flow to the cathode and large overpotential lead to passivation of the electrode.

To address these drawbacks, cutting-edge techniques are being explored to overcome these challenges by using bioelectrocatalytic enzymes to replace platinum-based catalysts. Using enzymes as catalysts on the cathode has various benefits. (1) The selective of enzyme is excellent. Enzymes can endure fuel flow to the cathode because of its selectivity toward oxygen, and the fuel cell configuration therefore can be simplified. (2) The catalytic activity of enzymes is superior to that of platinum-based catalysts. Detailed research on the energy calculation of each step in oxygen reduction shows that oxygen electroreduction of the enzyme-based fuel cell possesses lower overvoltage. (3) For the living cell body, enzymes are clean, environmentally friendly, biodegradable, nontoxic, and compatible. In addition, for improving electronic devices on the microscale, bioelectrodes can also be implanted into an animal's body. (4) The cost of enzymes is cheaper than most platinum-based catalysts. Through the fermentation process with a low-cost growth medium, enzyme mass production can be successfully realized.

A fuel cell is a kind of electrochemical cell. For a typical fuel cell, chemical energy is converted from fuel into electricity through hydrogen fuel by an electrochemical reaction with oxygen or another oxidizing agent (Secanell et al., 2011).

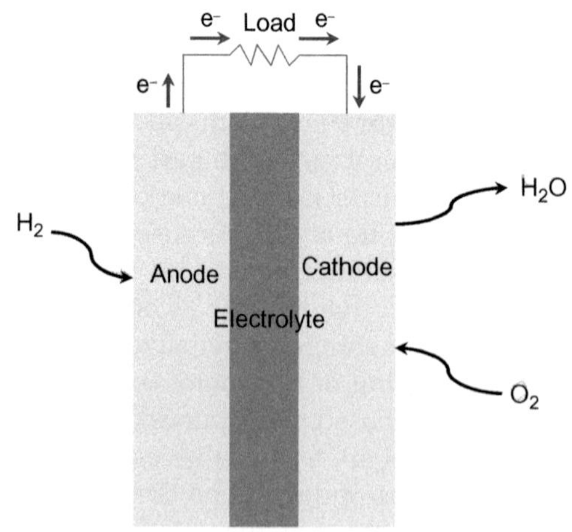

FIG. 15.1 Diagram of a fuel cell.

The characteristics of fuel cells are different to batteries when maintaining the chemical reaction by obtaining a continuous source of fuel and oxygen (usually from air). However, for a battery, chemical energy is obtained directly from chemicals already inside the battery. Fuel cells can retain electricity continuously if fuel and oxygen are provided. Therefore fuel cells come in many varieties but have the same general working manner. They have three adjacent segments: the cathode, the electrolyte, and the anode, as shown in Fig. 15.1. Two chemical reactions take place at the interfaces of the three different segments. The two reactions' net result is consistent with the consumed fuel: CO_2 or H_2O is generated. Consequently, an electric current is generated, normally referred to as the load, and can be used for powering electrical devices. A catalyst (usually hydrogen) oxidizes the fuel at the anode, and converts the fuel into a negatively charged electron and a positively charged ion. The specifically designed electrolyte is a substance that allows ions to pass through it but prevents electrons from passing through it. Under such conditions, the electric current is created by the free electrons traveling through a wire, where the ions travel to the cathode. By reaching the cathode through the electrolyte, the ions are reunited with the electrons and the two react with a third chemical, usually oxygen, to generate CO_2 or H_2O.

A biofuel cell comprises an anode and a cathode separated by a polymeric membrane like that shown in Fig. 15.1, and is similar to a proton exchange membrane fuel cell (PEMFC). PEMFCs are also used as polymer electrolyte membrane (PEM) fuel cells. These developed fuel cells are mainly used for applications such as transportation, along with applications that use stationary fuel cells and portable fuel cells. The relevant attractive characteristics are a special proton-conducting PEM and lower temperature ranges (50–100°C). The principle of PEMFC-produced electricity is opposite to PEM electrolysis (Taner, 2018).

By means of biochemical pathways, the biofuel cell makes chemical energy obtained directly from the fuel and turns it into electrical energy (https://en.wikipedia.org/wiki/Biofuel). Fuels, such as glucose, fructose, methanol, ethanol, or hydrogen, are oxidized on the anode side, while oxidizing agents, mostly oxygen, are reduced on the cathode side. The potential difference between cathode and anode makes electrons flow from anode to cathode through the external circuit and electricity is finally generated. Charges on the anode and cathode are then balanced by moving cations (usually H^+) from anode to cathode or moving anions (usually OH^-) in the opposite direction.

It is well known that the major difference between a biofuel cell and a conventional fuel cell is the catalyst, and a biofuel cell utilizes the microorganism or enzyme as the catalyst instead of conventional metal catalysts because the microorganism or enzyme is renewable and environmentally clean. Moreover, numerous chemicals can be used as fuels through the appropriate enzyme or microorganism. In addition, the biofuel cell can be operated in mild conditions, such as low temperatures around 20–40°C and under the physiological pH range, which is a more interesting alternative to powering up electronic devices that are suitable for working in mild conditions (Neto et al., 2010).

The enzymatic biofuel cell was discovered in 1964 by Yahiro et al. (1964). They constructed a biofuel cell using platinum as electrodes, while glucose and enzyme glucose oxidase were dissolved in the anolyte. As a result, a very low current density, about 30 nA/cm^2 at 330 mV, was generated. However, the lack of power output and stability in the biofuel cell encouraged research in the late 1960s and 1970s to focus more on metal catalysts. Revitalization of biofuel research occurred in the early 1980s. Plotkin et al. (1981) introduced a mediator, phenazin methasulfate, to assist electron transfer in the anolyte containing methanol and methanol dehydrogenase. Maximum current density increased three orders higher than the first biofuel cell to 30 μA/cm^2, achieved at 300 mV. In 1998 Palmore (Palmore et al., 1998) made improvements by using cascade enzymes instead of a single enzyme. Methanol oxidation in the anode could be achieved completely in the cascade reaction. Power output increased to 0.67 mW/cm^2 at 0.45 V. Significant progress was made by Heler and coworkers with an immobilization technique in the fuel cell (Chen et al., 2001). They immobilized the glucose oxidase on the redox polymer, and successfully extended the lifetime of the biofuel cell from 8 h to 7–10 days. This technique also allowed simplification and miniaturization of the biofuel cell. Their biofuel cell dimensions were 7 μm in diameter and 2 cm in length made of carbon fiber. The power density was 64 μW/cm^2 at 23°C and 137 μW/cm^2 at 37°C (Chen et al., 2001; Moehlenbrock and Minteer, 2008).

15.2 DRAWBACKS IN DEVELOPING ENZYMATIC BIOFUEL CELLS

Two defects in making enzymatic biofuel cells are short lifetime and poor power density. These problems originate from major issues in biofuel research:

1. Electron conduction between enzyme and electrode is very poor. Only a few enzymes have been recorded that are capable of establishing direct electron transfer to electrodes, such as cytochrome C, laccase, hydrogenase, and several peroxidases (Lu et al., 2006; Kihara et al., 2011; Alwarappan et al., 2010; Shleev et al., 2005). In direct electron transfer, the distance between the enzyme active center and the electrode surface is crucial. Unfortunately, the active center in the enzyme is usually buried under the nonconductive carbon shell (Lu et al., 2006; Kim et al., 2006). Therefore the critical distance between enzyme and electrode becomes a fundamental criterion, and the critical distance varies for each enzyme: laccase critical distance is about 20 Å, while horseradish peroxidase critical distance is about 18 Å. To solve this issue, enzymes can be improved by chemical modification or by adding a redox mediator to shuttle the electron from the active site of the enzyme to the electrode or in the opposite direction (Gutiérrez-Sánchez et al., 2012). The crucial issue related to mediators capable of small and easy diffusions is to keep them in the biofuel cells (Kim et al., 2006).
2. High enzyme loading is important for the biofuel cell, but enzyme loading can limit power output. For instance, when glucose oxidase was distributed stochastically as a monolayer on a flat surface, the loading of an enzyme was only 1.7×10^{-12} mol/cm^2 (0.27 μg/cm^2). It is assumed that enzyme activity in the immobilized form and free in the solution remained the same as well as a typical turnover number was 600 s^{-1}, the calculation of the current density at the upper limit was only about 0.2 mA/cm^2. The theoretical current density should be even lower for other enzymes possessing lower specific activity. Because the typical working voltage is lower than 1 V for enzymatic biofuel cells, the current density based on theoretical calculation can only be obtained with a power density of less than 0.2 mW/cm^2 (Kim et al., 2006).
3. Mass transfer can limit reaction in the biofuel cell. Three mass processes related to transfer limitations are included in the reactions of the biofuel cell: (1) fuel diffusion or oxidation to the biocatalysts' active sites; (2) membrane passage by the proton; and (3) redox mediators' diffusion between biocatalysts and electrodes, or electron passage via the catalysts' active sites to the electrodes. A concentration difference between the bulk phase and the sites of reaction is set up under the effect of high resistance to the mass transfer process of fuels, which slows the reactions down and leads to electrode polarization. It will be helpful to solve this kind of issue by classical engineering methods, as in the design of the patterned electrode and the convective transport introduced by forced stirring or relevant flow (Kim et al., 2006; Thomas et al., 2011).
4. Enzyme stability affects biofuel cell durability and power output. Enzymes can be easily denatured by both pH extremes and heat, by acetone, alcohol, and other miscible organic solvents, by guanidine hydrochloride and urea solutes, or by detergents, which allow their activity to diminish partially or totally (Nelson et al., 2008). At room temperature, enzymes can only last for a few hours or a few days. This also reflects on the lifetime of the biofuel cell, which mostly lasts for only a few days (Babu et al., 2012). Attractively, enzyme immobilization can help extend the lifetime of the biofuel cell (Moehlenbrock and Minteer, 2008).

15.3 ENZYME-BASED CATHODES FOR OXYGEN ELECTROREDUCTION

Enzymes can be constructed as catalysts on cathodes to reduce oxygen to water. Enzymes form copper oxidoreductase, such as laccase and bilirubin oxidase, which are superior, and they become favorable subjects for most studies on biocathodes. The advantages associated with employing enzymes as catalysts for biocathodes are: (1) low overpotential often near to zero; (2) excellent specificity, they are able to reduce oxygen from the ambient environment; (3) infinitely renewable; (4) conveniently applied in devices that are disposable and have completely biodegradable properties. The relevant disadvantages are: (1) per catalyst volume with low density current; (2) purification and isolation are time and cost dependent; (3) narrow temperature range and instability; and (4) hard to control the electronic transfer in the electrode with high quality in some conditions.

Biocathodes with enzymes are preferable because they show lower overpotential than that of platinum cathodes (Santoro et al., 2017). Overpotential in oxygen reduction using the enzyme-based biocathode is 0.19 V, which is lower than that of 0.45 V using the platinum-based cathode. Low overpotential of an enzyme-based biocathode is derived from its high turnover frequency (k_{cat}). This is the ultimate rate for the relevant catalyst to cycle its substrate (either an oxidant or a fuel), and the potential essential to gain catalysis. As an example, laccase k_{cat} is about 100 s^{-1}, while platinum's is only 15 s^{-1}. The formal potential of laccase is also very close to the formal potential of oxygen reduction. The reasons for overpotential in enzymes are: (1) Oxygen in enzymes is reduced by four electrons, which transfer locally in potential independent reactions. This is in sharp contrast to platinum-based oxygen reduction catalysts, where the number of electrons transferred is usually lower than 4, and ranges from 2 to 3.5. The overpotential of the catalytic cycle is kept at a minimum because any large release of free energy will have to be overcome in the following charge transfer steps. (2) Enzyme reaction always involves multiatomic reaction sites. (3) The dynamic structure of enzymes facilitates a given reaction. (4) Metal ions in the enzyme active center are assisted by nearby peptide residues.

15.4 ELECTRON TRANSFER IN BIOCATHODES

Based on the way in which electrons are shuttled to the electrode, the biocathode can be divided into two major categories. These are the biocathode employing mediated electron transfer and direct electron transfer. In direct electron transfer biocathodes, an electrical connection from electrode to enzyme active center is necessary. However, in mediated electron transfer, electrical charge transfer between electrode and enzyme is assisted by a sequential redox reaction of the mediator (Smolander et al., 2008a; Pereira et al., 2016; Macazo and Minteer, 2017). Mediators are usually used to assist the rapid transfer of electrons between cathode and enzyme. In a simplified model, mediators are oxidized by the enzymes that can diffuse to the electrode to be reduced at the cathode surface. The mediator cycles the electrons continuously between enzyme and electrode, because it is repeatedly oxidized and reduced, generating a catalytic current, while the enzyme reacts with the substrate to give its product. An ideal mediator for the biocathode requires several characteristics. It must be a good enzyme substrate, it should not inhibit the enzymatic reaction, its oxidized and reduced forms must be stable, and its redox conversion must be cyclic. At first, mediators were considered to be low-molecular-weight enzyme substrates whose enzymatic oxidation gave rise to stable high-potential intermediates. The intermediates were then reduced to their initial form by the electrode, thereby creating a reversible cycle (Borole et al., 2009; Liew et al., 2014; Rimboud et al., 2016; Bajracharya et al., 2017; Gadkari et al., 2018). The mediator in the biocathode can be inorganic, small organic, or polymer molecules (Milner et al., 2016; Ichi-Ribault et al., 2016; Baek et al., 2017; Bandapati et al., 2017; Lim et al., 2017), and there are advantages and disadvantages associated with either a direct or mediated electron transfer system. Mediated electron transfer is better in terms of electrochemical performance. Research indicates that by increasing current density, all cathodes show a distinctive drop in electrode potential commonly attributed to the occurrence of mass-transfer limitations. However, in cases when mediator 2,2′-azino-bis(3-ethylbenzothiazoline-6-sulfonic acid) (ABTS) occurs in the solution for testing, all adsorbed enzyme electrode materials show considerable ameliorated performance. Mediator molecules are easy to interact with the enzyme active center and donate their electrons, while without the mediator, electrons must travel along nonconductive carbon chains around the enzyme active center. Therefore electron transfer is more efficient in mediated electron transfer systems, which results in higher current density. However, the presence of a mediator results in decreasing the open circuit potential to a value around the mediator formal potential. In addition, the mediator can be a limiting factor on biocathode durability. If the mediator leaks from the biocathode, it will result in the declination of biocathode performance. Therefore strategies have been suggested to circumvent mediator leakage from the biocathode, for example, attaching the mediator covalently to support the material or by applying the polymer as the mediator.

15.5 IMMOBILIZATION METHODS FOR BIOCATHODES

The main concern with biocathode construction is immobilizing the enzyme permanently on the cathode. At the same time, it still retains enzyme activity. To supply electrochemical communication between the enzyme active center and the electrode, immobilization of the enzyme has also been utilized. Enzyme immobilization can be achieved by either chemical or physical attachment. Physical methods include adsorption and gel entrapment. Although simple and nonadditional reagents are required, the physical association between enzymes and electrode surface is weak. On the other hand, chemical methods involve covalent binding between enzyme and support material. Interaction between enzyme and carbon support is strong, but concern grows because covalent binding can disrupt enzyme activity (Cracknell et al., 2008).

It should be noted that as a kind of self-contained integrated device an electrochemical biosensor is able to analyze information special with quantity or semiquantity by the biochemical receptor with quantity or semiquantity specially, namely a biological recognition element, which can be retained with an electrochemical transduction element in direct spatial contact. Many of the developed biosensors rely on the detection of either coreactant or enzyme-catalyzed by-product such as NADH or H_2O_2 depending on the kind of enzyme employed. Alternatively, a redox mediator can be employed to transfer the electron between the enzyme and the electrode, or a direct electrical contact with the enzyme can be made to measure the associated current. Achieving direct electron transfer is highly desirable because it helps to avoid the use of external mediators and paves the way for developing highly efficient biofuel cells. Enzymes are so selective that they make it possible to detect the analyte of interest even in complex matrices. Though enzymes are highly useful in constructing biosensors, their stable immobilization on the electrode surface in its native state is an emergent issue that hinders the miniaturized biosensor's commercial application. Various immobilization techniques that are being employed to fix the enzyme on the electrode surface are adsorption, copolymerization/entrapment, affinity, and covalent binding.

15.5.1 Adsorption

This is the simplest way to adsorb an enzyme onto a solid on the basis of the nonspecific physical interaction between the surface of the matrix and the enzyme protein. A main benefit is that it does not require any reagents and only needs the steps of activation. Binding relies on weaker interactions and thus protein desorption is always detected resulting from the variations of ionic strength, pH, temperature, and even the mere presence of substrate.

15.5.2 Entrapment or Copolymerization

Entrapment or copolymerization is another technique for enzyme immobilization and enables the confinement of both the enzyme and the redox mediator during the polymerization process, making construction simpler. However, the formation of a thick polymer layer may obstruct the diffusion of substrate molecules, the formation of free radicals during the polymerization process, the leakage of enzyme/mediator during operation, difficulty dissolving monomers in aqueous buffer, and acidic conditions required for polymerization, all of which undermine the advantages of this method.

15.5.3 Affinity

The affinity of enzymes toward electrode surfaces can be created through the establishment of a kind of avidin-biotin bridge along with avidin-conjugated biomolecules or biotinylated biomolecules. The electrogeneration of biotinylated polymer films through electropolymerization of a biotin derivative by the functionalization of a phenol group and biotin linked by a long hydrophilic arm to an electropolymerizable pyrrole group have been investigated. The resulting copolymer electrodes are used for biotinylated glucose oxide immobilization via the avidin-biotin bridge formation. The method is more suitable for immobilization because it can help retain the origin of the enzyme by providing a facile environment; however, the requirement of synthesizing the biotin derivative with appropriate monomers is tedious and cumbersome.

15.5.4 Covalent Binding

Covalent binding is one of the immobilization methods that has been most intensely researched for many years, resulting in steady covalent bond formation between the support matrix and the enzyme. During the formation of a covalent bond, reaction of the binding can be done under various states, which cannot induce enzymic activity loss, and the usage of the reagents must not affect the enzyme active site. With the appropriate functional group such as the alpha amino and the chain carboxylic groups, usage with various coupling reagents such as 1-ethyl-3-(3-dimethylaminopropyl)carbodiimide hydrochloride/*N*-hydroxysuccinimide, glutaraldehyde, and dicyclohexylcarbodiimide affects the covalent binding. Under the effect of cross-linking, to prevent the enzyme leaching into the solution of the surroundings, covalent bonding should supply the enzyme derivatives with steady immobilization. It is a generally applicable immobilization technique, even though the structure of the protein or the enzyme active site to be coupled is not very well known because of the vast variety of binding reactions and the insoluble carriers with functional groups with abilities of covalent coupling, or such groups being activated.

The enzyme immobilization methods for constructing the biocathode are listed in Table 15.1, which illustrates that not only the method of immobilization but also the support material determines the performance of the biocathode. Mostly, support materials are carbon materials, for example, carbon nanotube (CNT) or graphene. The special characteristics of conductivity and biocompatibility of carbon materials with enzymes are preferable (Wu et al., 2010; Wang et al., 2011). The transfer of the electron to the enzyme can be assisted by the modified carbon materials.

TABLE 15.1 Enzyme Immobilization Method for Constructing a Biocathode

No.	Biocathode composite	Enzyme	Electron transfer	Immobilization method	Current density	References
1	Printable carbon-based or silver-based inks containing enzymes in a mediator	Laccase from *Trametes hirsuta*			0.04 mA/cm^2	(Smolander et al., 2008b)
2	Gold-carbon three-dimensional film electrode containing adsorbed BOD	BOD			0.05 mA/cm^2	(Lesniewski et al., 2010)
3	Film electrode consisting of phenyl sulfonate-funtionalized nanoparticles encapsulated into a hydrophilic sol-gel-processed silicate and further modified with adsorbed laccase	Laccase from *Cerrena unicolor*	DET	Physical	0.1 mA/cm^2	(Szot et al., 2009)
4	Carbon black XC72R in Nafion film	BOD from *Myrothecium verrucaria*	DET MET	Physical Physical	0.25 mA/cm^2 1.5 mA/cm^2	(Habrioux et al., 2010)
5	Adsorbed enzyme on carbon material BM-4 and coated with Nafion	Laccase from *Trametes hirsuta*	DET	Physical	0.3 mA/cm^2	(Shleev et al., 2010)
6	Pressed hydrophobized carbon black with Teflon	BOD from *Myrothecium verrucaria*			0.3 mA/cm^2	(Gupta et al., 2011)
7	Micro/macrocellular foams from silica monolith for BOD adsorption	BOD			2 mA/cm^2	(Flexer et al., 2011)
8	Polytetrafluoroethylene Ketjen black composite deposited on carbon paper	CueO from *Escherichia coli*			20 mA/cm^2	(Kontani et al., 2009)
9	Enzyme adsorbed on CNT-carbon porous suspension	Laccase	MET	Physical	0.01 mA/cm^2	(Liu et al., 2005)
10	Inks composed of laccase mixed with a dispersion of ABTS-adsorbed CNTs and stabilized with Nafion	Laccase from *Cerrena unicolor*	MET	Physical	0.1 mA/cm^2	(Karnicka et al., 2008)
	Inks composed of laccase mixed with a dispersion of ABTS covalently modified CNTs and stabilized with Nafion				0.4 mA/cm^2	(Sadowska et al., 2010)
	ABTS covalently modified CNTs covered with cubic phase film containing laccase				0.4–0.6 mA/cm^2	(Nazaruk et al., 2009)
11	Laccase covalently attached to PBSE-modified CNT	Laccase or BOD	DET	Covalent	0.4 mA/cm^2	(Ramasamy et al., 2010)
	Laccase covalently attached to amine-modified CNT through glutaraldehyde	Laccase			0.5 mA/cm^2	(Nazaruk et al., 2012)

ABTS, 2,2′-azino-bis(3-ethylbenzothiazoline-6-sulfonic acid); BOD, biochemical oxygen demand; CNT, carbon nanotube; DET, direct electron transfer; MET, mediated electron transfer; PBSE, 1-pyrene butanoic acid succinimidyl ester.

Sadowska et al. (2010) meditated the transfer of the electron by attaching ABTS to the CNT on the end tube or the side wall. Another approach is also achieved by covalently attaching an enzyme to the carbon support. Using covalent attachment, a carbon support can be located closer to the active site of the enzyme with the effective occurrence of the communication of electricity. The utilization of conjugates (glutaraldehyde, 1-pyrene butanoic acid succinimidyl ester [PBSE], or an aminophenyl layer) is needed for attachment of the enzyme (Ramasamy et al., 2010; Nazaruk et al., 2012; Vaz-Dominguez et al., 2008). By irreversible *p-p* stacking in the interface of the CNT and PBSE, the interaction between the CNT walls' aromatic-like structure and the aromatic pyrenyl moiety occurs. After PBSE with CNT incubation, laccase enables the amines on its surface to form covalent bonds of the amide-attached laccase to PBSE and subsequently to the modified CNT (Ramasamy et al., 2010). In addition, through the link to the nucleophile on the surface of the laccase, glutaraldehyde can also be utilized as conjugates, especially from lysine residue. To make the glutaraldehyde link to CNT successful, CNTs need to be modified initially by the group of amine (Wang et al., 2011).

15.6 BIOFUEL-BIOETHANOL

Up to now, as an outstanding alternative source to meet the requirements of the present and future fuels, biomass is an ideal candidate. The kinds of fuel that are produced by biomass are biofuels, where bioethanol and biodiesel are the two most general and achievable biofuels, which can be used instead of traditional liquid fuels: petroleum and diesel. Ethanol, a kind of biofuel, is expected to be utilized extensively all over the world and can be manufactured from the abundant cellulose biomass starch supplies.

Liquid biofuel-bioethanol—with its ecofriendly and renewable properties is able to be manufactured from various kinds of biomass feedstocks: (1) starch crops or sugar (sugar beet, sugarcane, wheat, and corn) and (2) lignocellulosic biomass. Based on the type of feedstock used, biofuels are generally defined as "First-generation fuels," "second-generation fuels," and "third-generation fuels" (Nigam and Singh, 2010; Chakraborty et al., 2016; Nguyen et al., 2017; Mohapatra et al., 2017; Lee et al., 2017).

First-generation biofuels comprise bioethanol from starch or sugar crops by conventional manufacturing methods and second-generation biofuels are the bioethanol generated from the lignocellulosic biomass. The output of biofuels related to oil crops and grains is restricted because the land with sufficient cultivable properties is unavailable on earth. Meanwhile, if energy crops are grown instead of food crops, this will lead to an increase in food prices resulting in a heavier burden on the poor.

Because of the increasing population, urgent resolution of the utilization of land to produce various crops set aside for bioethanol conversion or to satisfy nutritional requirements should be explored immediately. If the scale of the production of ethanol increases food will increasingly be in short supply. Furthermore, serious land contamination/pollution will be caused when more energy crops are cultivated with fertilizers and pesticides. The ecosystem of biocontrol, the erosion of soil, the biodiversity of crops, and the emissions of the GHG will be the main issues (Aditiya et al., 2016; Jambo et al., 2016; Tgarguifa et al., 2017; Gavião et al., 2017; Akram et al., 2018; Firouzi et al., 2018).

As alternatives, low-cost feedstocks, lignocellulosic biomass, and starchy wastes (crop residues, sawdust, grasses, livestock manure, sludge, and woodchips) can be hydrolyzed by enzymes to generate fermentable sugars for the production of biofuels. However, due to the drawbacks of the methods currently applied, the cost of hydrolysis is very high and the yield is very low. As a result, the feasibility of taking these materials as feedstock is restricted (Limayem and Ricke, 2012; Anwar et al., 2014; Behera et al., 2014; Urbaniec and Bakker, 2015; Zabed et al., 2016; Hafid et al., 2017; Stephen and Periyasamy, 2018).

Feedstock must be hydrolyzed into sugars for other carbohydrates before metabolization. The main factor is the overall cost of cellulase enzymes during enzymatic saccharification of the relevant agricultural biomass, including lignin. Consequently, the conversion of cellulosic materials or starch to fermentable sugars is still a main drawback despite such materials being cheaper than raw materials containing sugars. In addition, as a kind of lignocellulosic feedstock, lignin is extremely difficult to biologically decompose and ferment.

15.6.1 Algae

15.6.1.1 Algae Potential for Bioethanol

Algae are attractive due to their little impact on agriculture to act as an ideal renewable energy with proposed platforms. As a kind of alternative renewable in "third-generation biofuels," increasing attention is being paid to algae for bioethanol production. Photosynthetic algae utilize Sun energy to rapidly accumulate biomass with hydrocarbons, proteins, carbohydrates, lipids, and small molecules and pigments. Large amounts of phosphate, nitrogen, and

CO_2 are also used by microalgae. Therefore for conversion, biomass is attractive because it reduces toxic chemicals and waterways pollution, and mitigates CO_2 (Sirajunnisa and Surendhiran, 2016; Debiagi et al., 2017; Adeniyi et al., 2018; Kumar et al., 2018; Sivaramakrishnan and Incharoensakdi, 2018).

Like other biomass resources it is theoretically a carbon neutral source of energy. This is promising because there is an urgent need to decrease CO_2 emissions to the atmosphere, considering the serious issues relevant to climatic variation. Algal (autotrophic or heterotrophic) feedstock, as an alternative renewable source, is suitable for economically viable products and services, and because it alleviates CO_2 emissions from fossil fuel-driven power plants. Studies of autotrophic algae show that just like carbohydrate, it can set the inorganic carbon from CO_2 in the atmosphere assimilated as the reserve food materials afterward and utilize photosynthesis to harness sunlight (Prajapati et al., 2013; Voloshin et al., 2016; Kōnst et al., 2017; Fortier et al., 2017).

For heterotrophic algal species, small organic molecules in the environment can be absorbed and converted into primary oil or fat and proteins. A mixotroph, an algal species, can take advantage of CO_2 in the atmosphere or environmental organic carbon. No matter what methods are utilized, lipids, carbohydrates, and proteins can be manufactured by algae in a very short time and then used to make biofuels. Moreover, for the algae under anaerobic conditions, it can serve as a self-biorefinery for the output of ethanol through relevant photosynthates. Compared to other feedstocks, algal cells are easily acquired in a very short period of time. Therefore they are suitable candidates for satisfying the output of ethanol related to rocket-boosting feedstock requirements. Also, the demand for algae formation is very simple. They can efficiently form with high densities by CO_2, and use light and other inorganic nutrients (Burkholder et al., 2008; Vassilev and Vassileva, 2016; Elegbede et al., 2017; Barbera et al., 2018).

15.6.1.2 Algae Classification

Algae are generally classified as macroalgae and microalgae on the basis of their size and morphology, where microalgae are organisms with microscopic photosynthetic characteristics and many are unicellular. Contrarily, multiple cells organized into structures resembling stems, leaves, and roots of higher plants are the main components of macroalgae.

15.6.2 Bioethanol From Macroalgae

Because of their large size, macroalgae can be used as storage material to ferment sugars to produce ethanol. The enzymatic hydrolysis of algal cellulose with or without lignin is easy. For *Alaria*, *Laminaria*, and *Saccorhiza*, part of the brown algal group that can form in meters, types of macroalgal genera, mannitol, and laminarin are used for their primary food material. Similar to microalgae, macroalgae can also form in a very short time and produce large quantities of biomass. Because more energy is needed by macroalgae to generate supporting tissue than that of plants grown on land, and all the needed nutrients can be absorbed over the entire surface, output is higher. It is well known that red algae can generate more bioenergy than any other biomass source. Furthermore, macroalgae can not only incubate in 3D than that of 2D on land, but also cultivate on strings or nets and seed onto thin weighed strings suspended over a larger horizontal rope (Borines et al., 2013; Seghetta et al., 2014; Sudhakar et al., 2018; Abomohra et al., 2018).

15.6.3 Algae Pretreatment

Pretreatment of hemicellulose, cellulose, and lignin for generating sugars from algae biomass is a necessary process. Therefore hydrolysis with an enzyme catalyst (along with the addition of water) will decompose the complex carbohydrate molecules in the algae cells into simple sugars. Subsequently, the formed sugars are fermented into ethanol by relevant microorganisms. Finally, the formed ethanol is purified to agree with the specifications of the fuel. Also, a number of recovered separated solids can be used as a fuel to provide heat and electricity for the process (Raj and Krishnan, 2018; Díaz et al., 2017; Bussamra et al., 2015; Aguiar et al., 2013).

Characteristics, high moisture content, and organization with the weak connecting force are peculiar for algae biomass. Consequently, algae pretreatment is easier than that for lignocellulosic biomasses. It commonly utilizes chemical, physical-chemical, physical, and biological processes for algae material pretreatment. For saccharification, enzymes are needed to hydrolyze carbohydrates before fermentation. In case a terrestrially sourced enzyme is taken, desalinization would be needed when marine biomass is used. Consequently, it is advantageous to use a system with enzymes from a marine source (Lillebø et al., 2017; Shemer and Semiat, 2017; Sóti et al., 2018; Rempel et al., 2018; Goldbeck et al., 2016; Costanzo et al., 2015; Fasahati et al., 2015).

Acid hydrolysis is usually the most used method for various chemical pretreatment techniques. It can be utilized either as a hydrolyzing biomass technique for sugars or as a pretreatment prior to enzymatic hydrolysis.

Research results show that compared with other pretreatment techniques from the viewpoint of low energy consumption and cost effectiveness, acid pretreatment is the best pretreatment before fermentation. Acid hydrolysis is generally classified into two types: (1) concentrated acids and (2) diluted acids. To date, for algal material pretreatment, relevant methods related to dilute acid hydrolysis have been convincingly achieved. Currently, dilute sulfuric acid (H_2SO_4) is the most commonly used acid. The hydrolysis of *Enteromorpha* uses four different acids: H_2SO_4, HCl, H_3PO_4, and $C_4H_4O_4$ (maleic acid), and of these H_2SO_4 is the most effective under the same conditions (Jain and Agrawal, 2018; Paramjeet et al., 2018; Dutta and Chakraborty, 2018; Majidian et al., 2018; Chakraborty et al., 2017; Rizza et al., 2017; Martín and Grossmann, 2014; Menon and Rao, 2012).

15.6.4 *Ulva lactuca*

It is considered that the green macroalga *U. lactuca* (Chlorophyceae) is a potential aquatic energy crop because of its high carbohydrate content and rate of high potential growth. The investigated results reveal that it was not economically sustainable to use *U. lactuca* as an aquatic energy crop. Nevertheless, with increasing attention given to CO_2 neutral and renewable energy, the pressure on conventional resources of biomass increases simultaneously, especially considering the change in climate and the scarcity of land. Therefore resources with sustainable, clean, and alternative characteristics are becoming increasingly crucial, especially for aquatic biomass because the resources of land are restricted for energy harvest and food production. Consequently, macroalgae such as *U. lactuca* are becoming the highpoint of bioenergy generation. Even though restrictions are different in various areas, *U. lactuca* is common in both tropical and polar climates. Previously, the species grown in land-based systems or from natural populations in shallow coastal areas, alone or as part of integrated multitrophic aquaculture systems, have been applied. Macroalgae biochemical constituents rely highly on growth parameters and different seasons. The total solids (TS) content of *U. lactuca* is from 9.6% and 20.4%. Additionally, 0.3% lipids, 27% protein, and 62% carbohydrates are found in TS, and under the effect of high external loads of nitrogen, the content of protein in TS will exceed 40%. For *U. lactuca* in the form of carbohydrates, 8%–29% of TS—the complex hydrocolloid ulvan, a sulfated glucuronoxylormannan—is preferable for use as intracellular energy storage along with structural starch and cell wall components coupled with cellulose (Zollmann et al., 2018; Veeranan et al., 2018; Gaurav et al., 2017; Yahmed et al., 2017; Salinas and French, 2017).

Ulva contains a water-soluble sulfated polysaccharide (ulvan) and nonwater-soluble cellulose, followed by xyloglucan and gluconic acid polysaccharides (glucuronan).

15.6.4.1 *Ulva*-Sulfated Polysaccharide (ulvan)

Under the bulk of water-soluble ulvan, *Ulva* has polysaccharides of the cell wall with a dry weight basis of 38%–54%. Interesting biological and physical-chemical characteristics of the polysaccharide make it suitable for various applications. Obviously, the uncommon soft gels can be synthesized by borate and divalent cations. The repeating sequences of disaccharides comprise iduronic acid or xylose, glucuronic acid, and sulfated rhamnose can make ulvan built on it. Moreover, the repeating sequences are in a dispersed format for most of the composed sugars. Two kinds of main repeating sequences are aldobiuronic acids expressed as: (1) type A: ulvanobiuronic acid 3-sulfate and (2) type B: ulvanobiuronic acid 3-sulfate. Sulfated xylose partly residues at O^{-2} is able to take place in the uronic acids, and at O^{-2} of rhamnose 3-sulfate the glucuronic acid can also be branched. Moreover, glucose, galactose, protein, and mannose in ulvan are also usually observed at low ratios (Jung et al., 2016; Suganya et al., 2016; Karray et al., 2015; Coste et al., 2015).

15.6.4.2 Cellulose

Cellulose $(C_6H_{10}O_5)_n$ is an organic compound with a polysaccharide comprising a linear chain of over 10,000 β (1→4)-linked D-glucose units. Hemicellulose, although it is a polysaccharide relevant to cellulose, can be obtained from many sugars, such as arabinose, galactose, glucose, mannose, rhamnose, and xylose. Brief chains of about 200 sugar units are the main component of hemicellulose. In nature, cellulose is a commonly usable saccharide and is around 50% of all matter in plants. On the contrary, hemicellulose is only about 20%–30%; the rest is mainly lignin. It is noted that cellulose is broken down by hydrolysis into smaller polysaccharides (also called cellodextrin) in cellulolysis. Due to the high bonding strength of cellulose molecules attached to each other, cellulolysis is difficult compared to other broken-down polysaccharides. Because it is crystalline in nature, cellulose has two crystalline forms, α and β, where the characteristic of α-cellulose is a one-chain triclinic structure, while the characteristic of β-cellulose is a two-chain monoclinic structure.

Cellulose achieved from species of algal has a poriferous or sponge-like network, which is distinctive of higher plant cellulose (Chen and Smith, 2017; Skolrud et al., 2016; Álvarez-Murillo et al., 2016; Brown, 2015).

15.6.5 *Saccharomyces cerevisiae* Sugar Metabolism

Most as monosaccharides' main carbon and the energy sources, but there are some special yeasts that can employ sugars to use untraditional sources of carbon, the main monosaccharides such as fructose, D-stereoisomers of glucose, mannose, and galactose can be fermented to *S. cerevisiae*. Sugars with different reduction levels to glucose may be difficult to ferment to ethanol because of disturbances in the redox balance (Agu et al., 2018; Shuba and Kifle, 2018; Rodionova et al., 2017; Rodrigues et al., 2016).

15.6.6 Batch Fermentation

Fermentations can be manipulated in fed-batch, batch, or continuous reactors without any input or output flow in the experimental procedure. At the start of fermentation, all constituents are placed in the batch reactor, except for substrates, e.g., pH-controlling substances, oxygen, and antifoaming agents. In the fed-batch procedure, only one substrate should be used to control the rate of reaction by governing its concentration. In addition, the flow of input and output should be maintained continuously with constant growth of the reaction volume. Due to the combination of batch and continuous processes, fed-batch reactor applications are significantly greater, where the corresponding process first starts as a kind of batch process. After the substrate feeds into the batch at the initial glucose-consuming stage, the process will become stable, and relevant fermentation will be continuous with a steady growth rate as long as there are no limitations to hinder the growth of the cell. At present, one of the most used techniques for ethanol output in industrial applications is the fed-batch process with intermittent glucose addition and without fermentation broth removal. One of the interesting profitable properties of the fed-batch process is that substrate inhibition can be decreased, where ethanol formation and associated processes can be hindered by high sugar concentration in the fermentation environment. Other attractive properties of this technique are higher dissolved oxygen in the medium, higher productivity, shorter fermentation time, and lower toxic effects of the medium components (Assawamongkholsiri et al., 2018; bin Mohd Zain et al., 2018; Bhatia et al., 2017; Loman and Ju, 2016; Kalyani et al., 2017).

15.7 CONCLUSION

Enzymes could be an alternative catalyst for cathodes, which provides lower overpotential, lower prices, and higher selectivity. Although, some enzymes, such as laccase, are unstable and suffer from poor electron transfer, a biocathode composed of covalently attached laccase displays more effective direct electron transfer than that of physical adsorption, which results in higher current density values. Meanwhile, because there is a large amount of high energy degraded by enzymes, research should continue to find more suitable, effective, biofuel enzymes with relevant cutting-edge techniques, in-depth theories, and relevant models, which should be explored in detail.

References

Abomohra, A.E.F., El-Naggar, A.H., Baeshen, A.A., 2018. Potential of macroalgae for biodiesel production: screening and evaluation studies. J. Biosci. Bioeng. 125 (2), 231–237.

Adeniyi, O.M., Azimov, U., Burluka, A., 2018. Algae biofuel: current status and future applications. Renew. Sust. Energ. Rev. 90, 316–335.

Aditiya, H.B., Mahlia, T.M.I., Chong, W.T., Nur, H., Sebayang, A.H., 2016. Second generation bioethanol production: a critical review. Renew. Sust. Energ. Rev. 66, 631–653.

Agu, C.V., Ujor, V., Ezeji, T.C., 2018. Allopurinol supplementation of the growth medium enhances the fermentation of lignocellulosic hydrolysates to ethanol by *Saccharomyces cerevisiae*. Biocatal. Agric. Biotechnol. 14, 151–159.

Aguiar, R.S., Silveira, M.H.L., Pitarelo, A.P., Corazza, M.L., Ramos, L.P., 2013. Kinetics of enzyme-catalyzed hydrolysis of steam-exploded sugarcane bagasse. Bioresour. Technol. 147, 416–423.

Akram, F., ul Haq, I., Imran, W., Mukhtar, H., 2018. Insight perspectives of thermostable endoglucanases for bioethanol production: a review. Renew. Energy 122, 225–238.

Álvarez-Murillo, A., Sabio, E., Ledesma, B., Román, S., González-García, C.M., 2016. Generation of biofuel from hydrothermal carbonization of cellulose. Kinetics modelling. Energy 94, 600–608.

Alwarappan, S., Joshi, R.K., Ram, M.K., Kumar, A., 2010. Electron transfer mechanism of *cytochrome c* at graphene electrode. Appl. Phys. Lett. 96, 263702.

Anwar, Z., Gulfraz, M., Irshad, M., 2014. Agro-industrial lignocellulosic biomass a key to unlock the future bio-energy: a brief review. J. Radiat. Res. Appl. Sci. 7 (2), 163–173.

Arregi, A., Amutio, M., Lopez, G., Bilbao, J., Olazar, M., 2018. Evaluation of thermochemical routes for hydrogen production from biomass: a review. Energy Convers. Manag. 165, 696–719.

Assawamongkholsiri, T., Reungsang, A., Plangkang, P., Sittijunda, S., 2018. Repeated batch fermentation for photo-hydrogen and lipid production from wastewater of a sugar manufacturing plant. Int. J. Hydrog. Energy 43 (7), 3605–3617.

Babu, P.R., Pinnamaneni, R., Koona, S., 2012. Occurrences, physical and biochemical properties of laccase. Univ. J. Environ. Res. Technol. 2, 1–13.

Baek, G., Kim, J.S., Lee, S.Y., Lee, C.S., 2017. Development of biocathode during repeated cycles of bioelectrochemical conversion of carbon dioxide to methane. Bioresour. Technol. 241, 1201–1207.

Bajracharya, S., Srikanth, S., Mohanakrishna, G., Zacharia, R., PBTB Strik, D., Pant, D., 2017. Biotransformation of carbon dioxide in bioelectrochemical systems: state of the art and future prospects. J. Power Sources 356, 256–273.

Bandapati, M., Dwivedi, P.K., Krishnamurthy, B., Kim, Y.H., Kim, G.M., Goel, S., 2017. Screening various pencil leads coated with MWCNT and PANI as enzymatic biofuel cell biocathode. Int. J. Hydrog. Energy 42 (44), 27220–27229.

Barbera, E., Bertucco, A., Kumar, S., 2018. Nutrients recovery and recycling in algae processing for biofuels production. Renew. Sust. Energ. Rev. 90, 28–42.

Behera, S., Arora, R., Nandhagopal, N., Kumar, S., 2014. Importance of chemical pretreatment for bioconversion of lignocellulosic biomass. Renew. Sust. Energ. Rev. 36, 91–106.

Bhatia, S.K., Kim, S.H., Yoon, J.J., Yang, Y.Y., 2017. Current status and strategies for second generation biofuel production using microbial systems. Energy Convers. Manag. 148, 1142–1156.

bin Mohd Zain, M.Z., Kanesan, J., Kendall, G., Chuah, J.H., 2018. Optimization of fed-batch fermentation processes using the backtracking search algorithm. Expert Syst. Appl. 91, 286–297.

Borines, M.G., de Leon, R.L., Cuello, J.L., 2013. Bioethanol production from the macroalgae *Sargassum* spp. Bioresour. Technol. 138, 22–29.

Borole, A.P., LaBarge, S., Spott, B.A., 2009. Three-dimensional, gas phase fuel cell with a laccase biocathode. J. Power Sources 188 (2), 421–426.

Brown, T.R., 2015. A techno-economic review of thermochemical cellulosic biofuel pathways. Bioresour. Technol. 178, 166–176.

Burkholder, J.M., Glibert, P.M., Skelton, H.M., 2008. Mixotrophy, a major mode of nutrition for harmful algal species in eutrophic waters. Harmful Algae 8 (1), 77–93.

Bussamra, B.C., Freitas, S., da Costa, A.C., 2015. Improvement on sugar cane bagasse hydrolysis using enzymatic mixture designed cocktail. Bioresour. Technol. 187, 173–181.

Çepelioğullar, Ö., Mutlu, İ., Yaman, S., Haykiri-Acma, H., 2018. Activation energy prediction of biomass wastes based on different neural network topologies. Fuel 220, 535–545.

Chakraborty, R., Chatterjee, S., Mukhopadhyay, P., Barman, S., 2016. Progresses in waste biomass derived catalyst for production of biodiesel and bioethanol: a review. Procedia Environ. Sci. 35, 546–554.

Chakraborty, S., Singh, P.K., Paramashetti, P., 2017. Microreactor-based mixing strategy suppresses product inhibition to enhance sugar yields in enzymatic hydrolysis for cellulosic biofuel production. Bioresour. Technol. 237, 99–107.

Chen, M., Smith, P.M., 2017. The U.S. cellulosic biofuels industry: expert views on commercialization drivers and barriers. Biomass Bioenergy 102, 52–61.

Chen, T., Barton, S.C., Binyamin, G., Gao, Z., Zhang, Y., Kim, H.H., Heller, A., 2001. A miniature biofuel cell. J. Am. Chem. Soc. 123, 8630–8631.

Correa, C.R., Kruse, A., 2018. Supercritical water gasification of biomass for hydrogen production—review. J. Supercrit. Fluids 133 (Part 2), 573–590.

Costanzo, W., Jena, U., Hilten, R., Das, K.C., Kastner, J.R., 2015. Low temperature hydrothermal pretreatment of algae to reduce nitrogen heteroatoms and generate nutrient recycle streams. Algal Res. 12, 377–387.

Coste, O., Malta, E.J., López, J.C., Fernández-Díaz, C., 2015. Production of sulfated oligosaccharides from the seaweed *Ulva* sp. using a new ulvan-degrading enzymatic bacterial crude extract. Algal Res. 10, 224–231.

Cracknell, J.A., Vincent, K.A., Armstrong, F.A., 2008. Enzymes as working or inspirational electrocatalysts for fuel cells and electrolysis. Chem. Rev. 108, 2439–2461.

Danish, M., Ahmad, T., 2018. A review on utilization of wood biomass as a sustainable precursor for activated carbon production and application. Renew. Sust. Energ. Rev. 87, 1–21.

Debiagi, P.E.A., Trinchera, M., Frassoldati, A., Faravelli, T., Vinu, R., Ranzi, E., 2017. Algae characterization and multistep pyrolysis mechanism. J. Anal. Appl. Pyrolysis 128, 423–436.

Díaz, A.B., Marzo, C., Caro, I., de Ory, I., Blandino, A., 2017. Valorization of exhausted sugar beet cossettes by successive hydrolysis and two fermentations for the production of bio-products. Bioresour. Technol. 225, 225–233.

Dutta, S.K., Chakraborty, S., 2018. Mixing effects on the kinetics and the dynamics of two-phase enzymatic hydrolysis of hemicellulose for biofuel production. Bioresour. Technol. 259, 276–285.

Elegbede, I., Matemilola, S., Kies, F., Fadeyi, O., Saba, A., De Los Rios, P., Adekunbi, F., Lawal-Are, A., Fashina-Bombata, H., 2017. Risk analysis and development of algae biofuel from aquatic and terrestrial systems. Energy Procedia 128, 324–331.

Fasahati, P., Woo, H.C., Liu, J.J., 2015. Industrial-scale bioethanol production from brown algae: effects of pretreatment processes on plant economics. Appl. Energy 139, 175–187.

Firouzi, S., Nikkhah, A., Aminpanah, H., 2018. Resource use efficiency of rice production upon single cropping and ratooning agro-systems in terms of bioethanol feedstock production. Energy 150, 694–701.

Flexer, V., Brun, N., Courjean, O., Backov, R., Mano, N., 2011. Porous mediator-free enzyme carbonaceous electrodes obtained through Integrative Chemistry for biofuel cells. Energy Environ. Sci. 4, 2097–2106.

Fortier, M.O.P., Roberts, G.W., Stagg-Williams, S.M., Sturm, B.S.M., 2017. Determination of the life cycle climate change impacts of land use and albedo change in algal biofuel production. Algal Res. 28, 270–281.

Gadkari, S., Gu, S., Sadhukhan, J., 2018. Towards automated design of bioelectrochemical systems: a comprehensive review of mathematical models. Chem. Eng. J. 343, 303–316.

Gaurav, N., Sivasankari, S., Kiran, G.S., Ninawe, A., Selvin, J., 2017. Utilization of bioresources for sustainable biofuels: a review. Renew. Sust. Energ. Rev. 73, 205–214.

Gavião, L.O., Meza, L.A., Lima, G.B.A., Sant'Anna, A.P., de Mello, J.C.C.B.S., 2017. Improving discrimination in efficiency analysis of bioethanol processes. J. Clean. Prod. 168, 1525–1532.

Ghani, N.M.A.M.A., Vogiatzis, C., Szmerekovsky, J., 2018. Biomass feedstock supply chain network design with biomass conversion incentives. Energy Policy 116, 39–49.

Goldbeck, R., Gonçalves, T.A., Damásio, A.R.L., Brenelli, L.B., Wolf, L.D., Paixão, D.A.A., Rocha, G.J.M., Squina, F.M., 2016. Effect of hemicellulolytic enzymes to improve sugarcane bagasse saccharification and xylooligosaccharides production. J. Mol. Catal. B Enzym. 131, 36–46.

Guedes, R.E., Luna, A.S., Torres, A.R., 2018. Operating parameters for bio-oil production in biomass pyrolysis: a review. J. Anal. Appl. Pyrolysis 129, 134–149.

Gupta, G., Lau, C., Rajendran, V., Colon, F., Branch, B., Ivnitski, D., Atanassov, P., 2011. Direct electron transfer catalyzed by bilirubin oxidase for air breathing gas-diffusion electrodes. Electrochem. Commun. 13, 247–249.

Gutiérrez-Sánchez, C., Jia, W., Beyl, Y., Pita, M., Schuhmann, W., De Lacey, A.L., Stoica, L., 2012. Enhanced direct electron transfer between laccase and hierarchical carbon microfibers/carbon nanotubes composite electrodes. Comparison of three enzyme immobilization methods. Electrochim. Acta 82, 218–223.

Habrioux, A., Napporn, T., Servat, K., Tingry, S., Kokoh, K.B., 2010. Electrochemical characterization of adsorbed bilirubin oxidase on Vulcan XC 72R for the biocathode preparation in a glucose/O_2 biofuel cell. Electrochim. Acta 55, 7701–7705.

Hafid, H.S., Rahman, N.A.A., Shah, U.K.M., Baharuddin, A.S., Ariff, A.B., 2017. Feasibility of using kitchen waste as future substrate for bioethanol production: a review. Renew. Sust. Energ. Rev. 74, 671–686.

Ichi-Ribault, S.E., Zebda, A., Laaroussi, A., Reverdy-Bruas, N., Chaussy, D., Belgacem, M.N., Suherman, A.L., Cinquin, P., Martin, D.K., 2016. Laccase-based biocathodes: comparison of chitosan and Nafion. Anal. Chim. Acta 937, 43–52.

Jain, L., Agrawal, D., 2018. Performance evaluation of fungal cellulases with dilute acid pretreated sugarcane bagasse: a robust bioprospecting strategy for biofuel enzymes. Renew. Energy 115, 978–988.

Jambo, S.A., Abdulla, R., Azhar, S.H.M., Marbawi, H., Gansau, J.A., Ravindra, P., 2016. A review on third generation bioethanol feedstock. Renew. Sust. Energ. Rev. 65, 756–769.

Johnson, D.R., Curtright, A.E., Willis, H.H., 2013. Identifying key drivers of greenhouse gas emissions from biomass feedstocks for energy production. Environ. Sci. Pol. 33, 109–119.

Jung, H.J., Kim, J., Lee, C.S., 2016. ontinuous anaerobic co-digestion of Ulva biomass and cheese whey at varying substrate mixing ratios: different responses in two reactors with different operating regimes. Bioresour. Technol. 221, 366–374.

Kalyani, D.C., Zamanzadeh, M., Müller, G., Horn, S.J., 2017. Biofuel production from birch wood by combining high solid loading simultaneous saccharification and fermentation and anaerobic digestion. Appl. Energy 193, 210–219.

Karnicka, K., Miecznikowski, K., Kowalewska, B., Skunik, M., Opallo, M., Rogalski, J., Schuhmann, W., Kulesza, P.J., 2008. ABTS-modified multiwalled carbon nanotubes as an effective mediating system for bioelectrocatalytic reduction of oxygen. Anal. Chem. 80, 7643–7648.

Karray, R., Hamza, M., Sayadi, S., 2015. Evaluation of ultrasonic, acid, thermo-alkaline and enzymatic pre-treatments on anaerobic digestion of *Ulva rigida* for biogas production. Bioresour. Technol. 187, 205–213.

Kihara, T., Liu, X.Y., Nakamura, C., Park, K.M., Han, S.W., Qian, D.J., Kawasaki, K., Zorin, N.A., Yasuda, S., Hata, K., Wakayama, T., Miyake, J., 2011. Direct electron transfer to hydrogenase for catalytic hydrogen production using a single-walled carbon nanotube forest. Int. J. Hydrog. Energy 36, 7523–7529.

Kim, J., Jia, H., Wang, P., 2006. Challenges in biocatalysis for enzyme-based biofuel cells. Biotechnol. Adv. 24, 296–308.

Könst, P., Mireles, I.H., van der Stel, R., van Os, P., Goetheer, E., 2017. Integrated system for capturing CO_2 as feedstock for algae production. Energy Procedia 114, 7126–7132.

Kontani, R., Tsujimura, S., Kano, K., 2009. Air diffusion biocathode with CueO as electrocatalyst adsorbed on carbon particle-modified electrodes. Bioelectrochemistry 76, 10–13.

Krzyżaniak, M., Stolarski, M.J., Szczukowski, S., Tworkowski, J., Bieniek, A., Mleczek, M., 2015. Willow biomass obtained from different soils as a feedstock for energy. Ind. Crop. Prod. 75, Part B. 114–121.

Kumar, V., Nanda, M., Joshi, H.C., Singh, A., Sharma, S., Verma, M., 2018. Production of biodiesel and bioethanol using algal biomass harvested from fresh water river. Renew. Energy 116 (Part A), 606–612.

Lee, M.J., Cho, S.H., Kim, J.Y., 2017. A comprehensive model for design and analysis of bioethanol production and supply strategies from lignocellulosic biomass. Renew. Energy 112, 247–259.

Lesniewski, A., Paszewski, M., Opallo, M., 2010. Gold-carbon three dimensional film electrode prepared from oppositely charged conductive nanoparticles by layer-by-layer approach. Electrochem. Commun. 12, 435–437.

Liew, K.B., Daud, W.R.W., Ghasemi, M., Leong, J.X., Lim, S.S., Ismail, M., 2014. Non-Pt catalyst as oxygen reduction reaction in microbial fuel cells: a review. Int. J. Hydrog. Energy 39 (10), 4870–4883.

Lillebø, A.I., Pita, C., Rodrigues, J.G., Ramos, S., Villasante, S., 2017. How can marine ecosystem services support the Blue Growth agenda? Mar. Policy 81, 132–142.

Lim, S.S., Yu, E.H., Daud, W.R.W., Kim, B.H., Scott, K., 2017. Bioanode as a limiting factor to biocathode performance in microbial electrolysis cells. Bioresour. Technol. 238, 313–324.

Limayem, A., Ricke, S.C., 2012. Lignocellulosic biomass for bioethanol production: current perspectives, potential issues and future prospects. Prog. Energy Combust. Sci. 38 (4), 449–467.

Liu, Y., Wang, M., Zhao, F., Liu, B., Dong, S., 2005. A low-cost biofuel cell with pH-dependent power output based on porous carbon as matrix. Chem. Eur. J. 11, 4970–4974.

Loman, A.A., Ju, L.K., 2016. Soybean carbohydrate as fermentation feedstock for production of biofuels and value-added chemicals. Process Biochem. 51 (8), 1046–1057.

Lu, X., Zhang, Q., Zhang, L., Li, J., 2006. Direct electron transfer of horseradish peroxidase and its biosensor based on chitosan and room temperature ionic liquid. Electrochem. Commun. 8, 874–878.

Macazo, F.C., Minteer, S.D., 2017. Enzyme cascades in biofuel cells. Curr. Opin. Electrochem. 5 (1), 114–120.

Majidian, P., Tabatabaei, M., Zeinolabedini, M., Naghshbandi, M.P., Chisti, Y., 2018. Metabolic engineering of microorganisms for biofuel production. Renew. Sust. Energ. Rev. 82 (Part 3), 3863–3885.

Martín, M., Grossmann, I.E., 2014. Design of an optimal process for enhanced production of bioethanol and biodiesel from algae oil via glycerol fermentation. Appl. Energy 135, 108–114.

Md Isa, K., Abdullah, T.A.T., Md Ali, U.F., 2018. Hydrogen donor solvents in liquefaction of biomass: a review. Renew. Sust. Energ. Rev. 81 (Part 1), 1259–1268.

Menon, V., Rao, M., 2012. Trends in bioconversion of lignocellulose: biofuels, platform chemicals & biorefinery concept. Prog. Energy Combust. Sci. 38 (4), 522–550.

Milner, E.M., Popescu, D., Curtis, T., Head, I.M., Scott, K., Yu, E.H., 2016. Microbial fuel cells with highly active aerobic biocathodes. J. Power Sources 324, 8–16.

Moehlenbrock, M.J., Minteer, S.D., 2008. Extended lifetime biofuel cells. Chem. Soc. Rev. 37, 1188–1196.

Mohapatra, S., Mishra, C., Behera, S.S., Thatoi, H., 2017. Application of pretreatment, fermentation and molecular techniques for enhancing bioethanol production from grass biomass—a review. Renew. Sust. Energ. Rev. 78, 1007–1032.

Nazaruk, E., Karaskiewicz, M., Żelechowska, K., Biernat, J.F., Rogalski, J., Bilewicz, R., 2012. Powerful connection of laccase and carbon nanotubes. Electrochem. Commun. 14, 67–70.

Nazaruk, E., Sadowska, K., Madrak, K., Biernat, J.F., Rogalski, J., Bilewicz, R., 2009. Composite bioelectrodes based on lipidic cubic phase with carbon nanotube network. Electroanalysis 21, 507–511.

Nelson, D.L., Cox, M.M., Lehninger, A.L., 2008. Principles of biochemistry. fourth ed Freeman, New York, NY.

Neto, S.A., Forti, J.C., Andrade, A.R., 2010. An overview of enzymatic biofuel cells. Electrocatalysis 1, 87–94.

Nguyen, Q.A., Yang, J.M., Bae, H.J., 2017. Bioethanol production from individual and mixed agricultural biomass residues. Ind. Crop. Prod. 95, 718–725.

Nigam, P.S., Singh, A., 2010. Production of liquid biofuels from renewable resources. Prog. Energy Combust. Sci. 37, 52–68.

Palmore, G.T.R., Bertschy, H., Bergens, S.H., Whitesides, G.M., 1998. A methanol/dioxygen biofuel cell that uses NAD^+-dependent dehydrogenases as catalysts: application of an electro-enzymatic method to regenerate nicotinamide adenine dinucleotide at low overpotentials. J. Electroanal. Chem. 443, 155–161.

Paramjeet, S., Manasa, P., Korrapati, N., 2018. Biofuels: production of fungal-mediated ligninolytic enzymes and the modes of bioprocesses utilizing agro-based residues. Biocatal. Agric. Biotechnol. 14, 57–71.

Pereira, A.R., de Souza, J.C.P., Iost, R.M., Sales, F.C.P.F., Crespilho, F.N., 2016. Application of carbon fibers to flexible enzyme electrodes. J. Electroanal. Chem. 780, 396–406.

Plotkin, E.V., Higgins, I.J., Hill, H.A.O., 1981. Methanol dehydrogenase bioelectrochemical cell and alcohol detector. Biotechnol. Lett. 3, 187–192.

Prajapati, S.K., Kaushik, P., Malik, A., Vijay, V.K., 2013. Phycoremediation coupled production of algal biomass, harvesting and anaerobic digestion: possibilities and challenges. Biotechnol. Adv. 31 (8), 1408–1425.

Raj, K., Krishnan, C., 2018. High sugar yields from sugarcane (*Saccharum officinarum*) bagasse using low-temperature aqueous ammonia pretreatment and laccase-mediator assisted enzymatic hydrolysis. Ind. Crop. Prod. 111, 673–683.

Ramasamy, R.P., Luckarift, H.R., Ivnitski, D.M., Atanassov, P.B., Johnson, G.R., 2010. High electrocatalytic activity of tethered multicopper oxidase—carbon nanotube conjugates. Chem. Commun. 46, 6045–6047.

Rempel, A., Machado, T., Treichel, H., Colla, E., Margarites, A.C., Colla, L.M., 2018. Saccharification of *Spirulina platensis* biomass using free and immobilized amylolytic enzymes. Bioresour. Technol. 263, 163–171.

Rimboud, M., Bergel, A., Erable, B., 2016. Multiple electron transfer systems in oxygen reducing biocathodes revealed by different conditions of aeration/agitation. Bioelectrochemistry 110, 46–51.

Rios, M.L.V., González, A.M., Lora, E.E.S., del Olmo, O.A.A., 2018. Reduction of tar generated during biomass gasification: a review. Biomass Bioenergy 108, 345–370.

Rizza, L.S., Smachetti, M.E.S., Do Nascimento, M., Salerno, G.L., Curatti, L., 2017. Bioprospecting for native microalgae as an alternative source of sugars for the production of bioethanol. Algal Res. 22, 140–147.

Rodionova, M.V., Poudyal, R.S., Tiwari, I., Voloshin, R.A., Zharmukhamedov, S.K., Nam, H.G., Zayadan, B.K., Bruce, B.D., Hou, H.J.M., Allakhverdiev, S.I., 2017. Biofuel production: challenges and opportunities. Int. J. Hydrog. Energy 42 (12), 8450–8461.

Rodrigues, B., Lima-Costa, M.E., Constantino, A., Raposo, S., Felizardo, C., Gonçalves, D., Fernandes, T., Dionísio, L., Peinado, J.M., 2016. Growth kinetics and physiological behavior of co-cultures of *Saccharomyces cerevisiae* and *Kluyveromyces lactis*, fermenting carob sugars extracted with whey. Enzym. Microb. Technol. 92, 41–48.

Sadowska, K., Stolarczyk, K., Biernat, J.F., Roberts, K.P., Rogalski, J., Bilewicz, R., 2010. Derivatization of single-walled carbon nanotubes with redox mediator for biocatalytic oxygen electrodes. Bioelectrochemistry 80, 73–80.

Salinas, A., French, C.E., 2017. The enzymatic ulvan depolymerisation system from the alga-associated marine flavobacterium *Formosa agariphila*. Algal Res. 27, 335–344.

Santoro, C., Arbizzani, C., Erable, B., Ieropoulos, I., 2017. Microbial fuel cells: from fundamentals to applications. A review. J. Power Sources 356, 225–244.

Secanell, M., Wishart, J., Dobson, P., 2011. Computational design and optimization of fuel cells and fuel cell systems: a review. J. Power Sources 196 (8), 3690–3704.

Seghetta, M., Østergård, H., Bastianoni, S., 2014. Energy analysis of using macroalgae from eutrophic waters as a bioethanol feedstock. Ecol. Model. 288, 25–37.

Shemer, H., Semiat, R., 2017. Sustainable RO desalination—energy demand and environmental impact. Desalination 424, 10–16.

Shleev, S., Shumakovich, G., Morozova, O., Yaropolov, A., 2010. Stable 'floating' air diffusion biocathode based on direct electron transfer reactions between carbon particles and high redox potential laccase. Fuel Cells 10, 726–733.

Shleev, S., Tkac, J., Christenson, A., Ruzgas, T., Yaropolov, A.I., Whittaker, J.W., Gorton, L., 2005. Direct electron transfer between copper-containing proteins and electrodes. Biosens. Bioelectron. 20, 2517–2554.

Shuba, E.S., Kifle, D., 2018. Microalgae to biofuels: 'promising' alternative and renewable energy, review. Renew. Sustain. Energy Rev. 81 (Part 1), 743–755.

Sirajunnisa, A.R., Surendhiran, D., 2016. Algae—a quintessential and positive resource of bioethanol production: a comprehensive review. Renew. Sust. Energ. Rev. 66, 248–267.

Sivaramakrishnan, R., Incharoensakdi, A., 2018. Utilization of microalgae feedstock for concomitant production of bioethanol and biodiesel. Fuel 217, 458–466.

Skolrud, T.D., Galinato, G.I., Galinato, S.P., Shumway, C.R., Yoder, J.K., 2016. The role of federal renewable fuel standards and market structure on the growth of the cellulosic biofuel sector. Energy Econ. 58, 141–151.

Smolander, M., Boer, H., Valkiainen, M., Roozeman, R., Bergelin, M., Eriksson, J.E., Zhang, X.C., Koivula, A., Viikari, L., 2008a. Development of a printable laccase-based biocathode for fuel cell applications. Enzym. Microb. Technol. 43 (2), 93–102.

Smolander, M., Boer, H., Valkiainen, M., Roozeman, R., Bergelin, M., Eriksson, J.E., Zhang, X.C., Koivula, A., Viikari, L., 2008b. Development of a printable laccase-based biocathode for fuel cell applications. Enzym. Microb. Technol. 43, 93–102.

Sóti, V., Lenaerts, S., Cornet, I., 2018. Of enzyme use in cost-effective high solid simultaneous saccharification and fermentation processes. J. Biotechnol. 270, 70–76.

Stephen, J.L., Periyasamy, B., 2018. Innovative developments in biofuels production from organic waste materials: a review. Fuel 214, 623–633.

Sudhakar, K., Mamat, R., Samykano, M., Azmi, W.H., Ishak, W.F.W., Yusaf, T., 2018. An overview of marine macroalgae as bioresource. Renew. Sust. Energ. Rev. 91, 165–179.

Suganya, T., Varman, M., Masjuki, H.H., Renganathan, S., 2016. Macroalgae and microalgae as a potential source for commercial applications along with biofuels production: a biorefinery approach. Renew. Sust. Energ. Rev. 55, 909–941.

Szot, K., Nogala, W., Niedziolka-Jönsson, J., Jönsson-Niedziolka, M., Marken, F., Rogalski, J., Kirchner, C.N., Wittstock, G., Opallo, M., 2009. Hydrophilic carbon nanoparticle-*laccase* thin film electrode for mediatorless dioxygen reduction. Electrochim. Acta 54, 4620–4625.

Taner, T., 2018. Energy and exergy analyze of PEM fuel cell: a case study of modeling and simulations. Energy 143, 284–294.

Tgarguifa, A., Abderafi, S., Bounahmidi, T., 2017. Modeling and optimization of distillation to produce bioethanol. Energy Procedia 139, 43–48.

Thomas, C.S., Glassman, M.J., Olsen, B.D., 2011. Solid-state nanostructured materials from self-assembly of a globular protein-polymer diblock copolymer. ACS Nano 5, 5697–5707.

Urbaniec, K., Bakker, R.R., 2015. Biomass residues as raw material for dark hydrogen fermentation—a review. Int. J. Hydrog. Energy 40 (9), 3648–3658.

Vassilev, S.V., Vassileva, C.G., 2016. Composition, properties and challenges of algae biomass for biofuel application: an overview. Fuel 181, 1–33.

Vaz-Dominguez, C., Campuzano, S., Rüdiger, O., Pita, M., Gorbacheva, M., Shleev, S., Fernandez, V.M., De Lacey, A.L., 2008. Laccase electrode for direct electrocatalytic reduction of O_2 to H_2O with high-operational stability and resistance to chloride inhibition. Biosens. Bioelectron. 24, 531–537.

Veeranan, T., Kasirajaan, R., Gurunathan, B., Sahadevan, R., 2018. A novel approach for extraction of algal oil from marine macroalgae *Ulva fasciata*. Renew. Energy 127, 64–73.

Voloshin, R.A., Rodionova, M.V., Zharmukhamedov, S.K., Veziroglu, T.N., Allakhverdiev, S.I., 2016. Review: biofuel production from plant and algal biomass. Int. J. Hydrog. Energy 41 (39), 17257–17273.

Wang, X., Sjöberg-Eerola, P., Immonen, K., Bobacka, J., Bergelin, M., 2011. Immobilization of Trametes hirsuta laccase into poly(3,4-ethylenedioxythiophene) and polyaniline polymer-matrices. J. Power Sources 196, 4957–4964.

Widjaya, E.R., Chen, G., Bowtell, L., Hills, C., 2018. Gasification of non-woody biomass: a literature review. Renew. Sust. Energ. Rev. 89, 184–193.

Wu, J.F., Xu, M.Q., Zhao, G.C., 2010. Graphene-based modified electrode for the direct electron transfer of Cytochrome c and biosensing. Electrochem. Commun. 12, 175–177.

Yahiro, A.T., Lee, S.M., Kimble, D.O., 1964. Enzyme utilizing bio-fuel cell studies. Biochim. Biophys. Acta (BBA) 88, 375–383.

Yahmed, N.B., Carrere, H., Marzouki, M.N., Smaali, I., 2017. Enhancement of biogas production from *Ulva* sp. by using solid-state fermentation as biological pretreatment. Algal Res. 27, 206–214.

Zabed, H., Sahu, J.N., Boyce, A.N., Faruq, G., 2016. Fuel ethanol production from lignocellulosic biomass: an overview on feedstocks and technological approaches. Renew. Sust. Energ. Rev. 66, 751–774.

Zollmann, M., Traugott, H., Chemodanov, A., Liberzon, A., Golberg, A., 2018. Exergy efficiency of solar energy conversion to biomass of green macroalgae *Ulva* (Chlorophyta) in the photobioreactor. Energy Convers. Manag. 167, 125–133.

CHAPTER

16

Algal Cellulases

*Priyanka Arora**, Vipin Kumar Shukla*[†], *Archana Tiwari*[‡]

*School of Sciences, Noida International University, Greater Noida, India [†]Department of Biotechnology, C.C.S University, Meerut, India [‡]Amity Institute of Biotechnology, Amity University, Noida, India

16.1 INTRODUCTION

Cellulases are the enzymes that catalyze the breakdown of cellulose. The major sources of this enzyme are protozoans, bacteria, fungi, etc. Cellulases play their role via disintegration of a few polysaccharide units of cellulose to simpler units. The composite of these types of enzymes shows an interactive and steadier ability to degrade the material, which is mainly cellulosic. The crack-up of the cellulose molecules to simple sugars, i.e., monosaccharides, takes place with the help of cellulase enzymes.

The degradation of cellulose offers a huge commercial interest because it is a major element of plants, which present a barrier for intake and are also used in most of the fine chemical responses. The particular reaction that takes place is hydrolysis of a 1,4-β-D-glycosidic linkage present inside the cellulosic material such as lichenin, hemicelluloses, and cereal β-D-glucans (Barkalow and Whistler, 2008).

β-Glucose is the product formed by the breakdown of cellulose in the presence of an enzyme cellulase, commonly produced by bacteria, which occur symbiotically in most herbivores, but animals as well as humans do not form cellulase so they are unable to utilize the catalytic activity in respective materials, i.e., plants.

Cellulase belongs to a group of enzymes whose main purpose is the hydrolysis of cellulose. These cellulase enzymes are scattered end to end in ecosystems and are frequently represented in most of the microbes. The cellulase enzyme is often produced by bacteria, which are symbiotic. Also, some termites are known to produce cellulase enzymes (Watanabe et al., 1998).

There are different types of cellulases available. They differ in structure and mechanism of action. Some of the cellulase enzymes are specific, called cellulase, which include endo-1,4-glucanase, 1,4-beta-D-glucan 4-glucanohydrolase, carboxymethyl cellulase (CMCase), celludextrinase, Avicelase, etc.. Lignin is cleaved by enzymes that were previously referred to as cellulases, but nowadays they are known as lignin-modifying enzymes. One of the most renewable biological sources of energy is cellulose (Lynd et al., 2008). Relevant benefits can be brought about by the fabrication of biobased products and bioenergy through the use of renewable materials like lignocelluloses, which in turn can benefit national energy security, local economies, and the environment (Zhang, 2008).

Biomass biorefineries are restricted in terms of commercialization because of high costs of cellulases. Biomass saccharification requires a large amount of cellulases, for example, for the production of 1 gal of cellulosic ethanol, ~100 g of enzymes are required (Zhang et al., 2006; Zhu et al., 2009). Currently, consolidated bioprocessing has been projected to be suitable for decreasing the use of cellulases and increasing productivity volumetrically. It also helps in reducing capital investment. This technology aims to integrate production and hydrolysis of cellulose along with ethanol fermentation of algae in a single stride (Lynd et al., 2002, 2008).

β-1,4-Linkages in cellulosic chains are hydrolyzed by cellulase enzymes. Protozoans, animals, fungi, bacteria, and plants produce this enzyme. Based on amino acid sequences and their corresponding crystal structures, classification

From Cellulose to Cellulase: Strategies to Improve Biofuel Production
https://doi.org/10.1016/B978-0-444-64223-3.00016-3

of cellulase modules into several families has been carried out (Henrissat, 1991). Numerous carbohydrate-binding modules (CBMs) having noncatalytic activity along with several other functionally known or unknown modules are present in cellulase enzymes. These modules might be situated at the amino (N-) or carboxy (C-) terminus of the module having catalytic activity.

The major types of cellulases that participate during cellulose hydrolysis are:

1. Endoglucanases (EC 3.2.1.4)
2. Exoglucanases (EC 3.2.1.91)
3. β-Glucosidase (EC 3.2.1.21)

Almost all the cellulases except β-glucosidase are secreted by microorganisms. These cellulases may be free or cell surface bound and aim at hydrolyzing and metabolizing insoluble cellulose. Cellulase found its major application in industries, namely the paper and pulp industry, textile industry, and food industry. Cellulases are also used in detergents as an additive. Digestibility of animal feeds can also be improved by using such enzymes. That is why industrial cellulases hold a considerable share of the global industrial enzyme market. Considering the present scenario, which indicates depletion of crude oil and greenhouse gas emissions, which are major sources of pollution in the world, the production of bioethanol through hydrolysis of lignocelluloses by cellulase enzymes has emerged as a prominent technology for the production of renewable sources from biomass (Bayer et al., 2007; Himmel et al., 1999; Zaldivar et al., 2001).

16.1.1 Mechanism of Action of Cellulases

Cellulases have been divided into five general types based on the type of reaction they catalyze and their mechanisms of action:

1. **Endocellulases** (EC 3.2.1.4): The crystalline structure of cellulose is broken down by these cellulases. They do so by breaking up internal bonds and as a result individual polysaccharide chains of cellulose are exposed.
2. **Exocellulases** (EC 3.2.1.91): The exposed polysaccharide chains of cellulose (obtained by endocellulase) are cleaved 2–4 units from the end units resulting in the formation of tetrasaccharides or disaccharides like cellobiose. Exocellulases are later categorized into two types known as cellobiohydrolases: type I works consistently from the reducing end of the chain of cellulose and in contrast type II works actively on the nonreducing end of cellulose.
3. **Cellobioses** (EC 3.2.1.21): These are also known as β-glucosidases, which hydrolyze the product of exocellulase individually into simple sugars.
4. **Oxidative cellulases**: These catalyze radical reactions, which depolymerize cellulose, for example, cellobiose dehydrogenase.
5. **Cellulose phosphorylases**: Cellulases are depolymerized most commonly by using phosphates rather than water.

Avicel, a prominent microcrystalline substrate, has the most exclusive exocellulase activity. Cellulases may also be classified as progressive and nonprogressive types. Those cellulases that consistently interact with single polysaccharide strands are known as progressive cellulases, whereas those that interact with different polysaccharide strands one after another are known as nonprogressive cellulases. Animals can break down cellulose only partially through fermentation because animals are not capable of producing cellulase in their bodies. This limits the ability of animals to utilize the maximum energy available in fibrous plant material.

16.1.2 Cellulase Enzyme Systems for Cellulose Hydrolysis

For complete utilization of cellulases, a microorganism producing the concerned enzyme uses two strategies. One is discrete noncomplexed cellulases and the other is complexed cellulases (Lynd et al., 2002). A set of individual cellulases is secreted by most aerobic cellulolytic microorganisms, which helps in the degradation of cellulose.

Generally, two two-domain structures are present in fungal cellulases, one being the catalytic domain and the other the cellulose-binding domain. Both these domains are connected by a linker peptide, which is flexible. The standard structure of the cellulases produced from multiple types of bacteria depicts the CBMs present either on the C-terminus or N-terminus of the module, which is catalytic in nature but its function is not concerned with its location (Wilson, 2008). In contrast, anaerobic microorganisms produce a complex of multienzymes known as cellulosomes (Bayer et al., 2004). CMBs are present in some of the enzymes of cellulosomes. Others are found connected to scaffoldin protein having CBMs. Cellulosomes and cellulases can both be produced by certain anaerobic bacteria (Gilad et al., 2003; Doi and Kosugi, 2004; Berger et al., 2007). Many reviewers have also shed some light on the structural design and purpose of cellulosomes (Bayer et al., 1998a,b, 2004; Doi and Tamaru, 2001; Doi, 2008; Doi and Kosugi, 2004; Doi et al., 1998, 2003).

16.1.3 Classification of Cellulases

Cellulolysis is the process of cellulose hydrolysis, which is often carried out by cellulase enzymes. Cellulase is a complex enzyme system that consists of exo-1,4-β-D-glucanases, endo-1,4-β-D-glucanases, as well as cellobiose (β-D-glucoside glucanohydrolase).

16.1.3.1 Types of Cellulases

It has been demonstrated that based on fractionation studies on culture filtrate, there are basically three major types of enzymes that are involved in native cellulose hydrolysis to glucose, while some are produced by some plants and animals.

16.1.3.1.1 Endoglucanase

Endoglucanase is mostly active in response to cellulose that is acid swollen, derived soluble cellulose such as carboxymethyl cellulose (CMC), and cellooligosaccharides (Wood, 1989). Endoglucanase (also known as endo-β-1,4-glucanase/endo-β-1,4-D-glucan-4-glucanohydrolase, EC 3.2.1.4) is often known as CMCase; they hydrolyze CMC due to which chain length decreases rapidly with reducing groups increasing slowly. Cellodextrins, which are the mediatory products of the hydrolysis of cellulose, also act as endoglucanase. They can convert cellulose to cellobiose and glucose. They can cut cellulose polysaccharides and oligosaccharides at interior sites into various lengths at consequential new chain ends. They are inactive against CMC and cellooligosaccharides (cellobiose) but active against a pellucid substrate (Avicel).

16.1.3.1.2 Exoglucanases

Exoglucanases (exo-β-1,4-D-glucan-4-cellobiohydrolase, EC 3.2.1.91) split off cellobiose units from the nonreducing end of the chain. Cotton is not degraded by cellobiohydrolase rapidly but saccharification of microcrystalline celluloses can be considerably affected.

16.1.3.1.3 β-Glucosidase/Cellobiose

β-Glucosidase cleaves cellobiose units and removes glucose units from the oligosaccharides, in particular nonreducing ends, thereby completing the process of hydrolysis of cellulose so an interactive action of all the components is required for overall degradation from cellulose to glucose. Various phosphorolytic activities are catalyzed for the cleaving of cellobiose. It was first discovered in *Ruminococcus flavefaciens* cells by Ayers (1959).

16.1.3.1.4 Cellodextrin Phosphorylase or 1,4-D-Oligoglucan Orthophosphate α-D-Glucosyl Transferase (EC 2.3.1.49)

It is present in *Clostridium thermocellum* cells (Sheth and Alexander, 1969). It does not proceed on cellobiose, but catalyzes the varying phosphorolytic cleaving of cellodextrin ranking to cellohexose from cellotriose.

16.1.4 Microbial Strains Producing Cellulases

Bacteria: *Acinetobacter junii*, *Bacillus flexus*, *Acinetobacter amitratus*, *Anoxybacillus* sp., *Bacillus pumilus*, *Bacillus amyloliquefaciens*, *Bacillus circulans*, *Acinetobacter junii*, *Acidothermus cellulolyticus*, *Anoxybacillus* sp., *Paenibacillus curdlanolyticus*, *Geobacillus* sp., *Bacillus subtilis*, *Bacillus licheniformis*, *Fibrobacter succinogens*, *Acetivibrio cellulolyticus*, *Clostridium thermocellum*, *Butyrivibrio fibrisolvens*, *Clostridium acetobutylium*, *Clostridium papyrosolvens*, *Ruminococcus albus*, *Clostridium cellulolyticum*, etc.
Fungi: *H. grisea*, *Penicillium occitanis*, *Melanocarpus albomyces*, *Penicillium decumbans*, *Trichoderma reesei*, *Trichoderma harzianum*, *Penicillium brasilianum*, *Chaetomium cellulyticum*, *Chaetomium thermophilum*, *Neurospora crassa*, *Trichoderma attroviride*, *Penicillium fumigosum*, *Penicillium janthinellum*, *Paecilomyces inflatus*, *Penicillium echinulatum*, *Thermoascus aurantiacus*, *Acetivibrio cellulolyticus*, *Clostridium acetobutylium*, *Butyrivibrio fibrisolvens*, etc.
Algae: *Chlamydomonas*, *Scenedesmus*, *Dunaliella*, etc.

16.1.5 Improvement of Strains Producing Cellulases via Chemical and Physical Mutagens

1. The chemical analogs that can be used may be 5-chlorouracil, 5-bromouracil, 2-aminopurines deamination agent, and hydroxylamine.
2. The physical mutagens for improvement of strains may be X-rays, γ-rays, and UV rays, which are responsible for breakage of single or double strands and dimerization of DNA pyrimidine.
3. Nitrous acids in the form of alkylating agents may be *N*-nitrosoguanidine; ethyl methane sulfonate can also be used.
4. Acridine dyes and ethidium bromide intercalate between two base pairs serving the same purpose.

Apart from these mutagens, strains can be improved by employing techniques like gene cloning, genome shuffling, mutagenesis, and selection.

16.1.6 Sources of Cellulases

Types of cellulases are based on their sources. They can be plant-, animal-, or microbe-based cellulases.

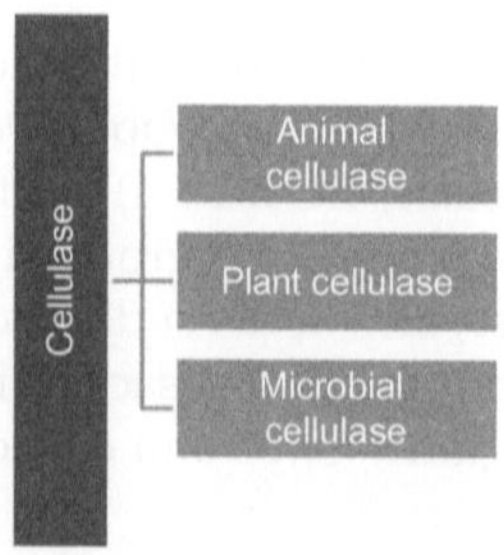

16.1.6.1 Animal Cellulases

Some cellulolytic insects are known to produce their own cellulases. Cellulolytic microorganisms grind the biomass material into smaller pieces leading to degradation of the cell wall of plant cells. This makes hydrolysis of simple polymers easy for enzymes involved in cell wall degradation, which in turn makes it more accessible for symbiotic microorganisms. Metagenome sequencing was performed from termite's hindgut, which depicted that many different enzymes helpful in the degradation of the cell wall of plant cells were present. Also, cellulolytic fungus *Termitomyces*, which is an aerobic symbiotic fungus, can be utilized by cellulolytic termites Macrotermitinae for plant material breakdown in their habitat, after which the termites eat the residual plant material as well as fungi used for this purpose. Many other types of mollusks, specifically slugs, snails, and shipworms, are also known to produce cellulases.

16.1.6.2 Plant Cellulases

Plants produce many different types of cellulases and these cellulases perform diverse roles in plant physiology. Apart from their direct function of degrading cellulases, cellulases may be involved in the abscission of leaves or ripening of fruits. They may also be involved in the growth and remodeling of the cell wall of plants.

16.1.6.3 Algal Cellulases

Algae contribute to a group of microorganisms that are adapted to wide habitats, which include terrestrial, aquatic, and extreme environments. They have the ability to perform oxygenic photosynthesis and they are morphologically diverse. Some of them may be unicellular, colonial, or filamentous forms. Various photosynthetic pigments like carotenoids, chlorophyll, and phycobiliproteins are present in algae. They contain variable amounts of metabolites and N-assimilatory enzymes.

Algae are adaptable to temperature extremes, illumination, desiccation, salinity, toxicants, pH, and availability of nutrients. An alga plays a functional role in:

1. *Bioindustry*: Algae can be used as a source of pharmaceuticals, drugs, and biofuels.
2. *Environmental management*: Algae are ameliorants of degraded wasteland and water-polluted bodies, biofertilizers, soil conditioners, and scavengers of metals, which are heavy.

16.1.6.3.1 Biofuel Production From Microalgae

In microalgae, biofuel production is light driven and therefore photosynthesis plays a very important role. Fermentation is the procedure utilized for bioethanol extraction from biomass. Apart from this, a hydrogenase system is utilized for the production of biohydrogen. Biodiesel from lipids can also be extracted from lipids using this system.

16.1.6.3.2 Advantages of Microalgae

- Photosynthetic efficiency is higher
- Growth is faster and a large biomass is obtained
- Mass culture is easy
- Environmental conditions are easily controllable
- Production is year-round

These are environmentally friendly and nonpolluting and do not contribute to atmospheric CO_2 levels (Fig. 16.1).

16.1.6.3.3 Selection of Algal Culture

Three types of selection are most commonly used:

- Mutation selection
- Natural selection
- Recombination selection

16.1.6.3.4 Mutation Selection

This can be achieved based on mutation in algae.

Natural selection: This can be achieved via inherent variability, which can be genetic and metabolic.

Recombination selection: Parameters for types of selection can be:

- Stability of culture
- Nutrient for growth requirements
- Condition of culture

16.1.6.3.5 Isolation and Purification of Algal Cellulases

Water samples collection: A loopful of algal growth is suspended in sterilized distilled water and homogenized. This culture is surface plated on suitable agar media for growth and incubated. The isolated colonies are observed through a microscope. These colonies are picked and transferred to agar slants.

Soil samples collection: Soil is collected after removing the upper layer (approximately 1 cm) of earth crust. These samples are dried and preserved for isolation purposes.

Culture media: Culture media consisting of nitrogen are utilized for isolation and maintenance of algal culture.

16.1.6.3.6 Preservation of Algal Cultures

The algal population can be preserved in a viable state. Algal growth slows down during preservation and these preserved cultures can be reactivated using suitable growth conditions. There are several technologies that can be utilized for preservation:

- Cryopreservation
- Lyophilization
- Immobilization

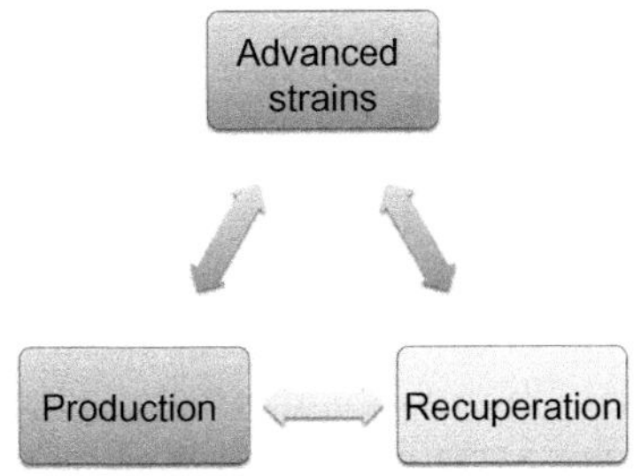

FIG. 16.1 Challenges for microalgal biofuel production.

CCUBGA, IARI, has been recognized as a large national repository in India containing freshwater cyanobacteria. It houses more than 550 strains. They mainly focus on isolating, identifying, and strengthening germplasms of cyanobacteria. They have characterized various cyanobacterial isolates as well.

Subculturing of large numbers of heterocystous and nonheterocystous strains is done in individual forms and they are maintained. This culture serves as a repository for the majority of freshwater blue-green algae (heterocystous and nonheterocystous).

16.1.6.3.7 Purification

Commonly there are three methods for the purification of algal culture:

- Pour plate
- Streak plate
- Spread plate

These techniques result in the separation of cyanobacterium individually from the mixture, and they can be picked as they form individual colonies and utilized for making pure cultures.

16.1.6.3.8 Purification Techniques

As mentioned earlier there are three techniques for the purification of algal cell cultures:

- Pour plate
- Streak plate
- Spread plate

All the techniques are aimed at reducing the number of cells, which makes it feasible to isolate full cultures. In the pour plate method, the blue-green algae are diluted and suspended in the pour agar, which eventually results in the separation of individual colonies. The individual cells are trapped in place on addition of solidifying agent over liquid medium, which contains blue-green algae. In this method, fixed colonies of cells or filaments are produced, which grow as separate colonies.

In the streak plate method, a loopful cyanobacterial culture is taken and streaked onto an agar plate. The number is reduced toward the end of the streak, thereby separating the colonies. In the spread plate method, the cynobacterial culture is spread over the solidified agar medium leading to separation of colonies.

Most of the bioethanol-producing industries are based on hydrolysis of cellulose and lignocelluloses. However, these industries face major constraints due to the high cost of production of enzyme cellulases required for hydrolyzing these celluloses and hemicelluloses. The production of cellulase enzyme from algae has the ability to lower the cost of bioethanol production by 1000–5000 times as compared to other sources (Subhadra and George, 2010) (Fig. 16.2).

16.1.6.3.9 Depiction of Algae

Algae are eukaryotic chlorophyll-containing photosynthetic organisms. The chlorophyll provides them with a green color. They are veiled by a range of photosynthetic pigments, which give them characteristic colors, giving them key distribution (Menetrez, 2012). Based on their morphology and size, there are two major classes of algae, namely macroalgae and microalgae (Fig. 16.3).

Macroalgae: also called seaweed. These are multicellular, and in turn organize themselves in the form of structures, which resemble stems, leaves, and roots of higher plants. Some species are provided with buoyancy because they have gas-filled structures (Chen et al., 2009).

On the basis of their pigmentation, algae are classified broadly into three categories namely:

1. Phaeophyceae (brown seaweed): including large kelps
2. Chlorophyceae (green seaweed): e.g., sea lettuce
3. Rhodophyceae (red seaweed): a widely diverse group (Singh et al., 2011)

Microalgae: These are unicellular, photosynthetic organisms, microscopic in nature (Chen et al., 2009). In contrast to macroalgae, stems, roots, and leaves are not present in microalgae. By photosynthesis, approximately half of the atmospheric oxygen is produced by these algae and the carbon dioxide produced simultaneously is utilized by them for their phototropic growth. In terms of their abundance, microalgae have been classified into three major classes, namely golden algae (Chrysophyceae), green algae (Chlorophyceae), and diatoms (Bacillariophyceae) (Carlsson et al., 2007).

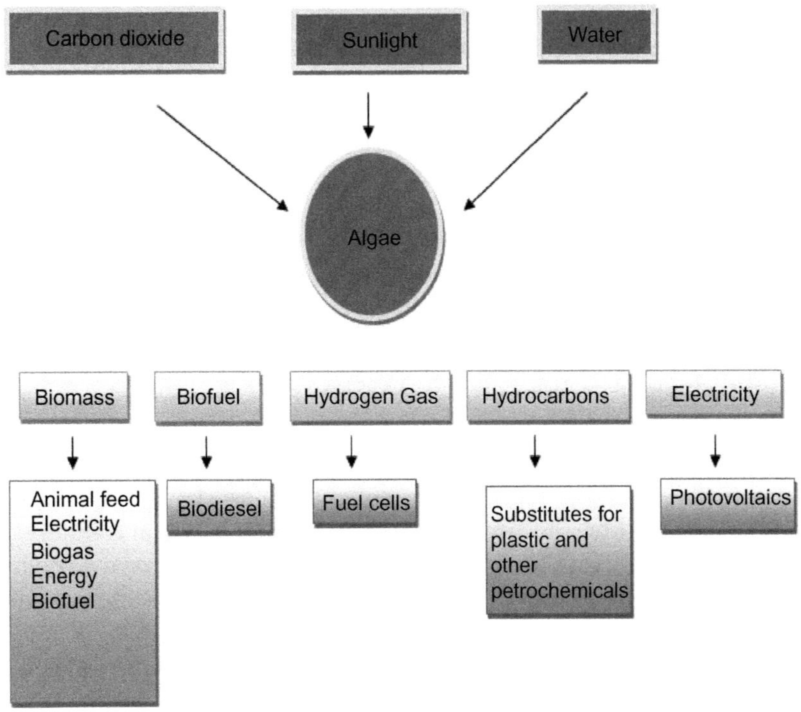

FIG. 16.2 Algae and their bioenergy products.

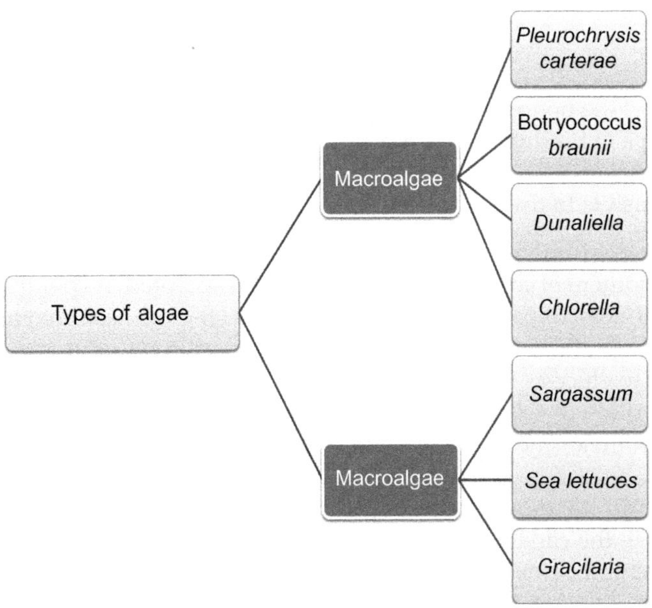

FIG. 16.3 Algal classification and biofuel production.

16.1.6.3.10 Bioethanol Production From Algae

Algae bioethanol production offers several advantages over other sources of bioethanol production, which includes:

Carbon dioxide from power plants and the atmosphere could be taken up by algal cells and hence could be utilized for the production of algae bioethanols yielding greenhouse gas reductions, which are comparative to other biobased fuels and fossil fuels (Vasudevan et al., 2012; Singh et al., 2011).

Also, very low levels of hemicelluloses are present and moreover no lignin production has been found in these cells. This increases the efficiency of its hydrolysis and yields of fermentation (Eshaq et al., 2011; Choi et al., 2012) leading to the production of low-cost bioethanol.

Algae cells are rich in carbohydrates (specifically starch). So, this characteristic of algal cells could be utilized for bioethanol production by performing fermentation of these biomolecules.

Freshwater as well as marine algae could be used for algae bioethanol production, therefore no compromise on food production is imposed because there is no specific requirement in terms of land or water (Singh et al., 2011).

Stable expression of cellulases/hemicellulases from genetically modified strains of algae (for instance, *Dunaliella* and *Chlamydomonas*) could prove to be a boon to cellulose-based industries because it could reduce the cost of bioethanol production commercially (Subhadra and George, 2010).

Therefore, the current chapter is focused on the potential of algae being utilized as substrate for the production of cellulase enzyme, which in turn can be used for various commercial purposes specifically in the production of low-cost biofuels.

16.1.7 Algal Cellulase Production

The bioethanol industries based on cellulose and lignocelluloses are currently facing several issues relating to technologies as well as infrastructure (Rejinders, 2010). Cellulosic material can be hydrolyzed by enzyme cellulases, which can lead to the production of carbon, which is a renewable source. Furthermore, products such as hexoses, pentoses, etc., which are low molecular weight, are also obtained by hydrolysis of cellulose (Spano, 1978). The most expensive step during the course of ethanol production is the production of enzyme "cellulases," therefore to enhance the viability of biofuel production commercially, low-cost cellulase production is needed.

However, the accessibility of cellulases/hemicellulases from inexpensive sources can hydrolyze cellulose to simple sugars at much lower cost leading to success in industry. As discussed earlier also, generally, three different types of cellulases are required for cellulose bioconversion, namely:

1. Exo-β-glucanase (EC 3.2.1.4)
2. Endo-β-glucanase (EC 3.21.91)
3. β-Glucosidase (EC 3.2.1.21)

Their modes of action are:

1. Exo-β-glucanase: In this case, the nonreducing end of cellulase is attacked exogenously.
2. Endo-β-glucanase: This helps in random scission of the cellulase chain, which in turn enhances production of glucose and cellooligosaccharides.
3. β-Glucosidase: This performs via hydrolysis of cellobiose to glucose.

Exo-β-glucanase leads to the disruption of H-bonding in cellulose. Afterward, endo-β-glucanase performs hydrolysis. The proposal for the development of cellulase enzymes has been accepted. As per this proposition, in a sequence of interactive events, the cellulosic chain is randomly attacked by endo-β-glucanase enzymes, whereas the exposed ends of the chain are attacked by exo-β-glucanase resulting in cellobiose splitting or glucose, which in turn is hydrolyzed subsequently by β-glucosidase to glucose. However, Reese et al. (1950) proposed the opposite of this hypothesis, which indicates that for the decomposition of cellulosic biomass, three enzymes are required in contrast to two enzymes.

Cellulases are being used in major industries for the recycling of paper, processing of cotton, animal feed additives, detergent enzymes, and extraction of juices. This has upgraded cellulases as the third largest enzyme for industrial applications in the world. Moreover, cellulases can become ubiquitous industrial enzymes if their potential is fully utilized. Cellulases have the capability to produce ethanol, butanol, etc. from biomasses, which can also be used as fuel. In the current scenario, *Humicola insolens* or *Hypocrea jecorina*, aerobic cellulolytic fungi, is utilized for industrial cellulase production. Their strains are engineered genetically in such a manner that they produce enormous quantities of crude cellulases, which have high specific activity. These strains are altered specifically so that they provide specified uses for industries.

The cost of production in cellulose-based industries can be reduced by 1000–5000 times if extraction of cellulases is performed using algal biomass. It has been reported in recent research that in plant expression systems, protein (enzyme) production cost is estimated as $0.05 g^{-1}, whereas fortunately the production of enzymes from algae is stated to be $0.002 g^{-1} (Franklin and Mayfield, 2004). A proposal for the production of biodiesel, recombinant proteins, and unsaturated fatty acids by algal biomass processing has already been processed (Subhadra, 2010).

Algal strains of *Chlamydomonas* and *Dunaliella*, which can be genetically modified in such a manner that they express cellulase enzymes stably, unlock several ways for enzymes to be produced as a coproduct during biofuel production from algae. This coproduct can then be utilized for the processing of cellulose-based feedstocks through hydrolysis (Fig. 16.4).

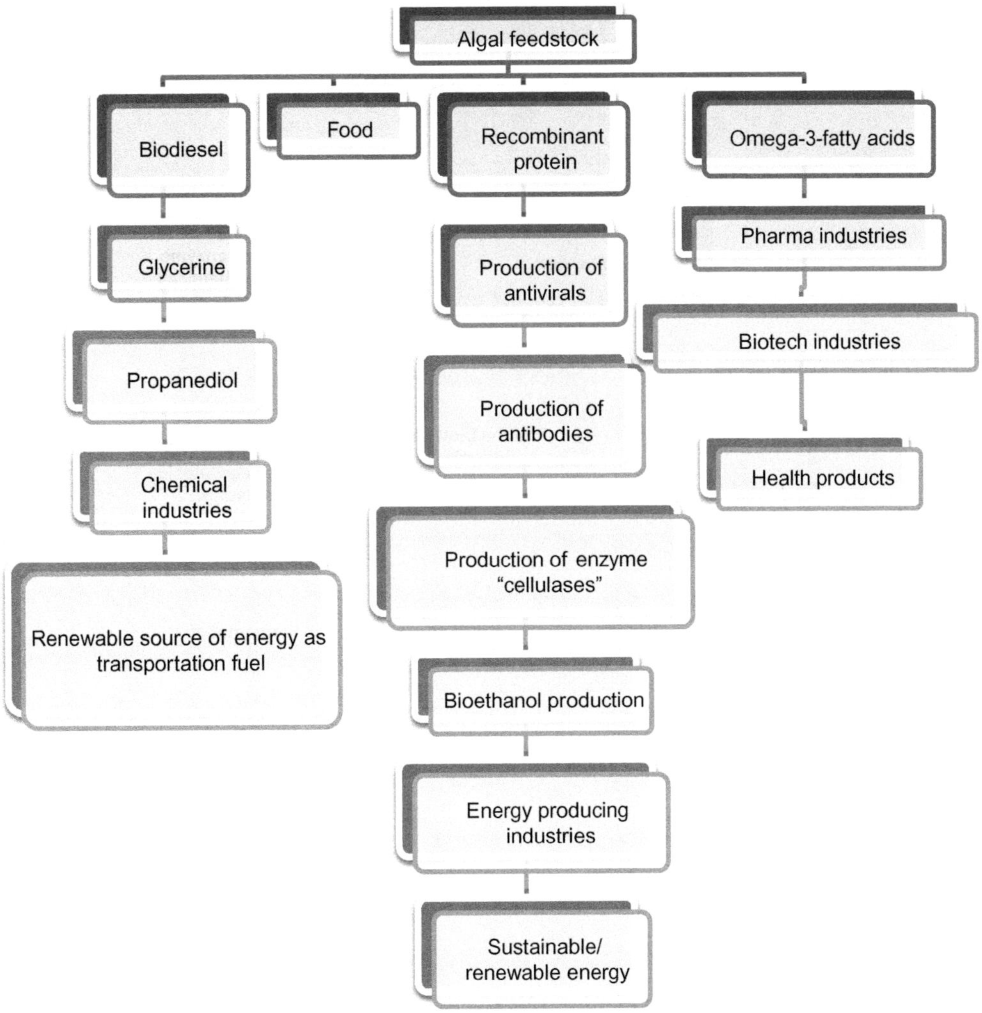

FIG. 16.4 Algal feedstock could be used for the production of high-value coproducts, food, and fuel sustainably.

The storage of purified enzymes produced from algal biomass is also possible. Whenever required they can be propelled into cellulosic feedstocks and hydrolysis can be carried out (Fig. 16.5). This is how the algal sector is contributing to commercial viability.

The production of ethanol, Single cell proteins (SCPs) (which are produced from unicellular micro-organisms) and a range of other chemicals from cellulosic materials economically requires cellulase enzyme production. As per recent investigations, cellulase production has been carried out through submerged culture processes, whereas the high cost of producing enzymes has put a restriction on cellulose bioconversion, thereby hindering its industrial application (Pandey et al., 2000).

Moreover, solid-state fermentation (SSF) has emerged as an eminent procedure for the production of enzyme cellulase because it offers low capital investment. Also, SSF offers lower operating expenses (Cen and Xia, 1999). In SSF, microbes are cultivated on solid media (Pandey, 2003). Recent studies show that in comparison to conventional submerged fermentation, SSF is much more advantageous (Hölker and Lenz, 2005). However, due to the extent of technological difficulties, there are not many opportunities in production industries for these processes. They face difficulties in optimization and operation of SSF bioreactors on a large scale (Hölker and Lenz, 2005).

Researchers are now focusing on the production of enzymes through microbes by utilizing agro-industrial-based residues (Ramachandran et al., 2004; Botella et al., 2007; Shankar and Mulimani, 2007; Sun et al., 2007). One of the most important microbes, namely *Cellulomonas uda*, increases the production of cellulase (Poulsen et al., 2016). Microbes have the capability to utilize algal biomass as a substrate and produce enzyme cellulase.

In recent research, the National Chemical Laboratory, Pune, a division of the National Centre for Industrial Microorganisms, provided *C. uda* NCIM 2353 and its inocula were prepared in the laboratory.

Algae used for substrate was dried and then ground, and salt solution was added. Sterilization was performed using an autoclave. The algal substrate was then inoculated with *C. uda* inocula for 2 days at 37°C. After this, enzyme

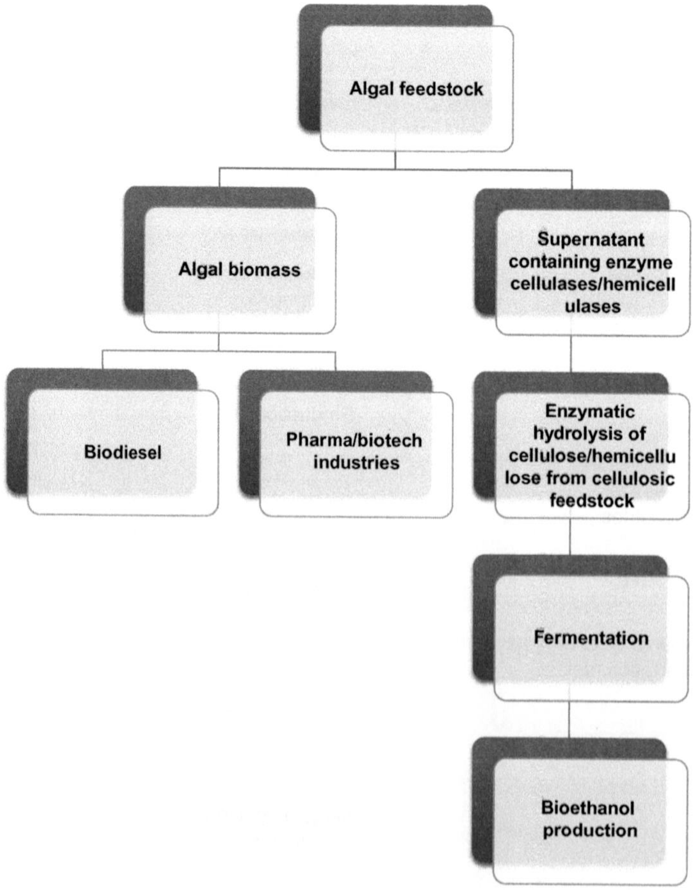

FIG. 16.5 Flow chart representing the processing of cellulosic feedstocks utilizing cellulases/hemicellulases produced from algal feedstock for bioethanol production.

extraction was performed. The fermented algal substrates were filtered and centrifuged. Crude enzyme collected as substrate was then utilized for cellulase activity determination (Shafique et al., 2004).

Enzyme activity assay was carried out using the method of Ray et al. in 1993. Cellulose can now be degraded using this cellulase, which produces simple sugars (glucose). This was assayed using 3,5-dinitrosalicylic acid (Bernfeld, 1955). Absorbance was measured at 575 nm. The result of this study performed by Pramod et al. (2014) indicated that *C. uda* (NCIM 2353) has immense potential for cellulase production to utilize algal substrate using SSF.

Cellulase production can be carried out by solid or submerged fermentations. However, production through submerged fed-batch fermentation is preferred by most industries due to its low cost because more than 100 g of crude cellulase by weight per liter of broth are produced by this method. It has been estimated that on the basis of dry protein weight, the cost of production may be as low as $10–40 per kg of dry protein weight.

Trichoderma and *Aspergillus* or their derivative strains, except Dyadic's *Chrysosporium lucknowense*, are utilized by industries like Novozymes, Genencor, Iogen, etc. to produce cellulases commercially. Reduction of cellulase production by 20- to 30-fold per gallon of cellulosic ethanol has been claimed by these industries (Himmel et al., 2007).

16.1.8 Applications

The bioethanol industries based on cellulose and lignocelluloses have been facing many constraints relating to technology and infrastructure (Rejinders, 2010). Hydrolysis of cellulose and hemicelluloses by inexpensive sources is the major goal of cellulose-based industries. Major types of cellulases needed for cellulose conversion to glucose are glucosidase, endoglucanase, and exoglucanase or cellobiohydrolase. Cellulases are now being used in the processing of cotton, recycling of paper, extraction of juices, and animal feed additives, therefore it has developed into the third largest industrial enzyme worldwide. Bioethanol is produced by fermentation of sugar with the help of enzyme cellulases. So, cellulases will become the largest industrial enzyme if the bioethanol produced with it can be used as a major transportation fuel.

Cellulases have been found to be produced from fungi such as *Humicola insolens* or *Hypocrea jecorina*, which are aerobic cellulolytics. An extremely large amount of cellulases have been found to be produced by using genetically engineered strains of these microorganisms. The crude cellulases thus produced have relatively high specific activity. Moreover, these microorganisms can be genetically modified, which results in the production of a set of enzymes that can be used for specific purposes. The cost of production of bioethanol can be lowered by 1000–5000 times by using algae in comparison to further sources. In the existing scenario, the cost of bioethanol manufacturing from other sources has been found to be $0.05 g^{-1}, whereas from algal proteins the cost is $0.002 g^{-1} (Franklin and Mayfield, 2004). The production of multiple products such as, for example, unsaturated fatty acids, biodiesel, and recombinant proteins through integrated processing of algae biomass sequentially has been proposed (Subhadra, 2010). A stable expression of cellulase/hemicellulase enzymes has been seen from the genetically modified strains of algae such as *Dunaliella* and *Chlamydomonas*.

These cellulases can be utilized for hydrolysis of cellulose in cellulose-based industries. They help in the processing of cellulosic-based feedstock. Use of biomass of algae can bring down the cost of production of cellulase much more than any other platforms meant for the production of the same enzyme (Subhadra, 2010). Also, these algal cellulases can be stored and used whenever required for hydrolysis of cellulosic feedstock. Commercial viability can hence be enhanced using lignocellulosic and algal sectors.

16.1.8.1 Industrial Integration for Food and Fuel Production

Many environmental benefits are associated with algal biofuel production, which can provide a net gain in energy. The major advantage of algal cellulases is that they can be produced in huge quantities offering no hindrance to food supplies. So, use of refineries based on algal biofuel can prove to be of large benefit in this sector. They can also offer:

Minimization and effective utilization of resources: In comparison to other platforms, biofuel production from algae reduces the usage of energy, water, and land. The wastewater from dairy-based industries, municipal waste, and various other agricultural practices is loaded in nutrients and can be used for algal biomass production. This wastewater otherwise could eventually lead to polluting various useful resources. Apart from pollution control benefits, the use of this wastewater in algal production will prevent the use of land sources, which could be utilized for the production of food. Moreover, they will offer no harmful discharge, no chemicals, and no antibiotics would be required. Because of this their location can be close to market, which will even reduce the cost of transportation (Subhadra and George, 2010).

Integration of sectors: The algal sector could be integrated with cellulosic/lignocellulosic industries. They can serve as a substrate for cellulase/hemicellulase production for the hydrolysis of cellulose/hemicelluloses, thereby producing low-cost bioethanol. This will increase the viability of both sectors commercially (Subhadra and George, 2010).

Carbon efficacy: Due to the consistent increase in the global population as well as fossil fuel production, CO_2 emissions are steadily increasing in the atmosphere. Algae have the capacity to capture approximately 2 pounds of carbon dioxide per pound of algae production. They can effectively utilize that CO_2 providing viable products as well as services (Subhadra and George, 2010).

16.2 CONCLUSION

Greenhouse gas emissions could be reduced by reducing the intensity of carbon released in several industrial sectors from various energy sources. Food and fuel production are strongly unified. The production of one leads to compromising the production of others. So, a planned and integrated approach is thus needed. Algal cellulase production and simultaneous low-cost biofuel production could provide a better platform for both food and fuel production. As discussed in this chapter, use of algal cellulases could help in integrating biofuel and food production sectors.

In future, a wide green industrial consortium will be developed by government and nongovernment sectors, which will sustainably solve many key issues being faced by humans today.

References

Ayers, W.A., 1959. Phosphorolysis and synthesis of cellobiose by cell extracts from *Ruminococcus flavefaciens*. Biochem. Biophys. Res. Commun. 140, 219–229.

Barkalow, D.G., Whistler, R.L., 2008. Cellulose. Access Science, McGraw-Hill.

Bayer, E.A., Chanzy, H., Lamed, R., Shoham, Y., 1998a. Cellulose, cellulases and cellulosomes. Curr. Opin. Struct. Biol. 8, 548–557.

Bayer, E.A., Shimon, L.J., Shoham, Y., Lamed, R., 1998b. Cellulosomes-structure and ultrastructure. J. Struct. Biol. 124, 221–234.

Bayer, E.A., Belaich, J.P., Shoham, Y., Lamed, R., 2004. The cellulosomes: multienzyme machines for degradation of plant cell wall polysaccharides. Annu. Rev. Microbiol. 58, 521–554.

Bayer, E.A., Lamed, R., Himmel, M.E., 2007. The potential of cellulases and cellulosomes for cellulosic waste management. Curr. Opin. Biotechnol. 18, 237–245.

Berger, E., Zhang, D., Zverlov, V.V., Schwarz, W.H., 2007. Two noncellulosomal cellulases of *Clostridium thermocellum*, Cel9I and Cel48Y, hydrolyse crystalline cellulose synergistically. FEMS Microbiol. Lett. 268, 194–201.

Bernfeld, P., 1955. Amylases, alpha and beta. In: Colowick, S.P., Kalpan, N.O. (Eds.), Methods in Enzymology. vol. 1, pp. 149–151.

Botella, C., Diaz, A., de Ory, I., Webb, C., Blandino, A., 2007. Xylanase and pectinase production by *Aspergillus awamori* on grape pomace in solid state fermentation. Process Biochem. 42, 98–101.

Carlsson, A.S., Van Beilen, J.B., Möller, R., Clayton, D., 2007. In: Bowles, D. (Ed.), Micro- and Macro-Algae: Utility for Industrial Applications. CPL, Newbury, UK.

Cen, P., Xia, L., 1999. Production of cellulase by solid-state fermentation. Adv. Biochem. Eng. Biotechnol. 65, 68–92.

Chen, P., Min, M., Chen, Y., Wang, L., Li, Y., Chen, Q., Wang, C., Wan, Y., Wang, X., Cheng, Y., Deng, S., Hennessy, K., Lin, X., Liu, Y., Wang, Y., Martinez, B., Ruan, R., 2009. Review of the biological and engineering aspects of algae to fuels approach. Int. J. Agric. Biol. Eng. 2 (4), 1–30. https://doi.org/10.3965/j.issn.1934-6344.2009.04.001-030.

Choi, W.Y., Han, J.G., Lee, C.G., Song, C.H., Kim, J.S., Seo, Y.C., Lee, S.E., Jung, K.H., Kang, D.H., Heo, S.J., Cho, J.S., Lee, H.Y., 2012. Bioethanol production from Ulva pertusa Kjellman by high-temperature liquefaction. Chem. Biochem. Eng. Q. 26 (1), 15–21.

Doi, R.H., 2008. Cellulases of mesophilic microorganisms: cellulosome and noncellulosome producers. Ann. N. Y. Acad. Sci. 1125, 267–279.

Doi, R.H., Kosugi, A., 2004. Cellulosomes: plant-cell-wall degrading enzyme complexes. Nat. Rev. Microbiol. 2, 541–551.

Doi, R.H., Tamaru, Y., 2001. The *Clostridium cellulovorans* cellulosome: an enzyme complex with plant cell wall degrading activity. Chem. Rec. 1, 24–32.

Doi, R.H., Park, J.S., Liu, C.C., Malburg, L.M., Tamaru, Y., Ichiishi, A., Ibrahim, A., 1998. Cellulosome and noncellulosomal cellulases of *Clostridium cellulovorans*. Extremophiles 2, 53–60.

Doi, R.H., Kosugi, A., Murashima, K., Tamaru, Y., Han, S.O., 2003. Cellulosomes from mesophilic bacteria. J. Bacteriol. 185, 5907–5914.

Eshaq, F.S., Ali, M.N., Mohd, M.K., 2011. Production of bioethanol from next generation feed-stock alga Spirogyra species. Int. J. Eng. Sci. Technol. 3 (2), 1749–1755.

Franklin, S.E., Mayfield, S.P., 2004. Prospects for molecular farming in the green alga *Chlamydomonas reinhardtii*. Curr. Opin. Plant Biol. 7, 159–165.

Gilad, R., Rabinovich, L., Yaron, S., Bayer, E.A., Lamed, R., Gilbert, H.J., Shoham, Y., 2003. CelI, a noncellulosomal family 9 enzyme from *Clostridium thermocellum*, is a processive endoglucanase that degrades crystalline cellulose. J. Bacteriol. 185, 391–398.

Henrissat, B., 1991. A classification of glycosyl hydrolases based on amino acid sequence similarities. Biochem. J. (2), 309–316.

Himmel, M.E., Ruth, M.F., Wyman, C.E., 1999. Cellulase for commodity products from cellulosic biomass. Curr. Opin. Biotechnol. 10, 358–364.

Himmel, M.E., Ding, S.Y., Johnson, D.K., Adney, W.S., Nimlos, M.R., Brady, J.W., Foust, T.D., 2007. Biomass recalcitrance: engineering plants and enzymes for biofuels production. Science 315 (5813), 804–807.

Hölker, U., Lenz, J., 2005. Solid state fermentation is there any biotechnological advantages? Curr. Opin. Microbiol. 8 (3), 301–306.

Lynd, L.R., Weimer, P.J., van Zyl, W.H., Pretorius, I.S., 2002. Microbial cellulose utilization: fundamentals and biotechnology. Microbiol. Mol. Biol. Rev. 66, 506–577.

Lynd, L.R., Laser, M.S., Bransby, D., Dale, B.E., Davison, B., Hamilton, R., Himmel, M., Keller, M., McMillan, J.D., Sheehan, J., Wyman, C.E., 2008. How biotech can transform biofuels. Nat. Biotechnol. 26, 169–172.

Menetrez, M.Y., 2012. An overview of algae biofuel production and potential environmental impact. Environ. Sci. Technol. 46 (13), 7073–7085. https://doi.org/10.1021/Es300917r.

Pandey, A., 2003. Solid state fermentation. Biochem. Eng. J. 13 (2–3), 81–84.

Pandey, A., Soccol, C.R., Mitchell, D., 2000. New developments in solid state fermentation: I—bioprocesses and products. Process Biochem. 35, 1153–1169.

Poulsen, H.V., Willink, F.W., Ingvorsen, K., 2016. Aerobic and anaerobic cellulase production by *Cellulomonas uda*. Arch. Microbiol. 198 (8), 1–11.

Pramod, B.P., Payal, A.P., Rutuja, R.D., Sharanappa, A., Patil, I.D., 2014. Bioprocessing of algal waste for cellulase production by *Cellulomonas uda* (NCIM 2353). Int. J. Adv. Biotechnol. Res. 5 (3), 547–551.

Ramachandran, S., Patel, A.K., Nampoothiri, K.M., Francis, F., Nagy, V., Szakacs, G., Pandey, A., 2004. Coconut oil cake a potential raw material for the production of amylase. Bioresour. Technol. 93, 169–174.

Reese, E.T., Siu, R.G.H., Levinson, H.S., 1950. The biological degradation of soluble cellulose derivatives and its relationship to the mechanism of cellulose hydrolysis. J. Bacteriol. 59 (4), 485–497.

Rejinders, L., 2010. Transport biofuel yields from food and lignocellulosic C4 crops. Biomass Bioenergy 34, 152–155.

Shafique, S., Asgher, M., Sheikh, M.A., Asad, M., 2004. Solid state fermentation of banana stalk for exoglucanase production. Int. J. Agric. Biol. 6 (3), 488–491.

Shankar, S.K., Mulimani, V.H., 2007. Galactosidase production by *Aspergillus oryzae* in solid-state fermentation. Bioresour. Technol. 98, 958–961.

Sheth, K., Alexander, J.K., 1969. Purification and properties of ß-1, 4-oligoglucan: orthophosphate glucosyl transferase from *Clostridium thermocellum*. J. Biol. Chem. 244, 457–464.

Singh, A., Olsen, S.I., Nigam, P.S., 2011. A viable technology to generate third-generation biofuel. J. Chem. Technol. Biotechnol. 86 (11), 1349–1353. https://doi.org/10.1002/Jctb.2666.

Spano, L., 1978. Enzymatic hydrolysis of cellulose to fermentable sugar for production of ethanol. In: Shuster, W.W. (Ed.), Proceeding of 2nd Annual Symposium on Fuels from Biomass. John Willey and Sons, New York, pp. 671–684.

Subhadra, B., 2010. Sustainability of algal biofuel production using integrated renewable energy park (IREP) and algal biorefinery approach. Energy Policy 38, 5892–5901.

Subhadra, B., George, G., 2010. Algal biorefinery-based industry: an approach to address fuel and food insecurity for a carbon-smart world. J. Sci. Food Agric. https://doi.org/10.1002/jsfa.4207.

Sun, H., Ge, X., Zhang, W., 2007. Production of a novel raw starch digesting glucoamylase by *Penicillium* sp. X-1 under solid state fermentation and its use in direct hydrolysis of raw starch. World J. Microbiol. Biotechnol. 23, 603–613.

Vasudevan, V., Stratton, R.W., Pearlson, M.N., Jersey, G.R., Beyene, A.G., Weissman, J.C., Rubino, M., Hileman, J.I., 2012. Environmental performance of algal biofuel technology options. Environ. Sci. Technol. 46 (4), 2451–2459. https://doi.org/10.1021/es2026399.

Watanabe, H., Noda, H., Tokuda, G., Lo, N., 1998. A cellulase gene of termite origin. Nature 394 (6691), 330–331.

Wilson, D.B., 2008. Three microbial strategies for plant cell wall degradation. Ann. N. Y. Acad. Sci. 1125, 289–297.

Wood, T.M., 1989. Mechanisms of cellulose degradation by enzymes from aerobic and anaerobic fungi. In: Coughlan, M.P. (Ed.), Enzyme Systems for Lignocellulose Degradation. Elsevier Applied Science, London, pp. 17–35.

Zaldivar, J., Nielsen, J., Olsson, L., 2001. Fuel ethanol production from lignocellulose: a challenge for metabolic engineering and process integration. Appl. Microbiol. Biotechnol. 56, 17–34.

Zhang, H.P., 2008. Reviving the carbohydrate economy via multi-product biorefineries. J. Ind. Microbiol. Biotechnol. 35, 367–375.

Zhang, H.P., Himmel, M., Mielenz, J.R., 2006. Outlook for cellulase improvement: screening and selection strategies. Biotechnol. Adv. 24, 452–481.

Zhu, Z., Sathitsuksanoh, N., Zhang, H.P., 2009. Direct quantitative determination of adsorbed cellulase on lignocellulosic biomass with its application to study cellulase desorption for potential recycling. Analyst 134, 2267–2272.

Index

Note: Page numbers followed by *f* indicate figures and *t* indicate tables.

C

D

E

9780444642233